Workbook
Electro-Hydraulic Components and Systems

Dr. Medhat Kamel Bahr Khalil, Ph.D, CFPHS, CFPAI.
Director of Professional Education and Research Development,
Applied Technology Center, Milwaukee School of Engineering,
Milwaukee, WI, USA.

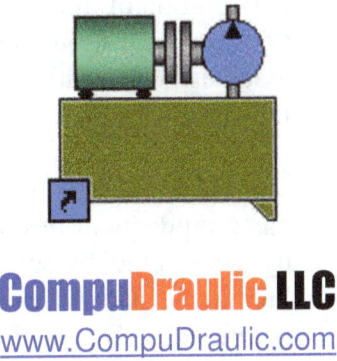

www.CompuDraulic.com

CompuDraulic LLC

Workbook
Electro-Hydraulic Components and Systems

ISBN: 978-0-9977816-2-5

Copyright © 2017 by CompuDraulic LLC.
3850 Scenic Way, Franksville, WI, 53126 USA.
www.compudraulic.com

Revised by July 2018

All rights reserved

No part of this book may be reproduced or utilized in any form or by any means, electronic or mechanical, including photocopying and microfilm, without written permission from CompuDraulic LLC at the address above.

Printed in the United States of America

Disclaimer

Any portion of information presented in this book could apply on some applications and not on other ones due to various reasons, and since errors can occur in circuits, tables, and text, the publisher assumes no liability for the safe and/or satisfactory operation of any system designed based on the information in this text.

The publisher does not endorse or recommend any brand name product by including such brand name products in this book. Conversely the publisher does not disapprove any brand name product by not including such brand name in this book. The publisher obtained data from catalogs, literatures, and material from hydraulic components and systems manufacturers based on their permissions. The publisher welcomes additional data from other sources for future editions. s

Workbook
Electro-Hydraulic Components and Systems

PREFACE, 4

Chapter 1: Hydraulic versus Electrical Systems, 5

Chapter 1: Reviews and Assignments, 22

Chapter 2: Hydro-Mechanical versus Electro-Hydraulic Solutions, 27

Chapter 2: Reviews and Assignments, 46

Chapter 3: Switching Valves Construction and Operation, 51

Chapter 3: Reviews and Assignments, 84

Chapter 4: Electrical Circuits for Switching Valves, 89

Chapter 4: Reviews and Assignments, 121

Chapter 5: Proportional Valves, 127

Chapter 5: Reviews and Assignments, 169

Chapter 6: Servo Valves, 179

Chapter 6: Reviews and Assignments, 210

Chapter 7: Electro-Hydraulic Valve Selection Criteria, 217

Chapter 7: Reviews and Assignments, 267

Chapter 8: Open-Loop versus Closed-Loop EH Applications, 271

Chapter 8: Reviews and Assignments, 295

Chapter 9: Control Electronics for Electro-Hydraulic Valves, 301

Chapter 9: Reviews and Assignments, 352

Chapter 10: Electro-Hydraulic Valves Commissioning and Maintenance, 357

Chapter 10: Reviews and Assignments, 365

Answers to Chapters Reviews, 369

PREFACE

This Workbook is a complementary part to the textbook of the same title. This book is used as a workbook for students to take notes during the course delivery. It contains colored printout of the PowerPoint slides that are designed to present the course. Each chapter is followed by a number of review questions and assignments for homework.

Dr. Medhat Kamel Bahr Khalil

Chapter 1
Hydraulic versus Electrical Systems

Objectives:

This chapter presents structural and operational comparisons between hydraulic and electrical power transmission and control systems. Similarities and differences between the inductive, capacitive, and resistive elements are discussed.

Brief Contents:

1.1- Structural Analogy between Hydraulic and Electrical Systems.

1.2- Operational Analogy between Hydraulic and Electrical Systems.

Workbook: Electro-Hydraulic Components and Systems
Chapter 1: Hydraulic versus Electrical Systems

1.1- Structural Analogy between Hydraulic and Electrical Systems

Hydraulic and electrical power transmission and control systems, among other systems, are

the two most comparable systems

because both:

- Have various controlling and actuating elements.
- Can transmit the same magnitude of power.
- Can be used in Industrial and Mobile applications.

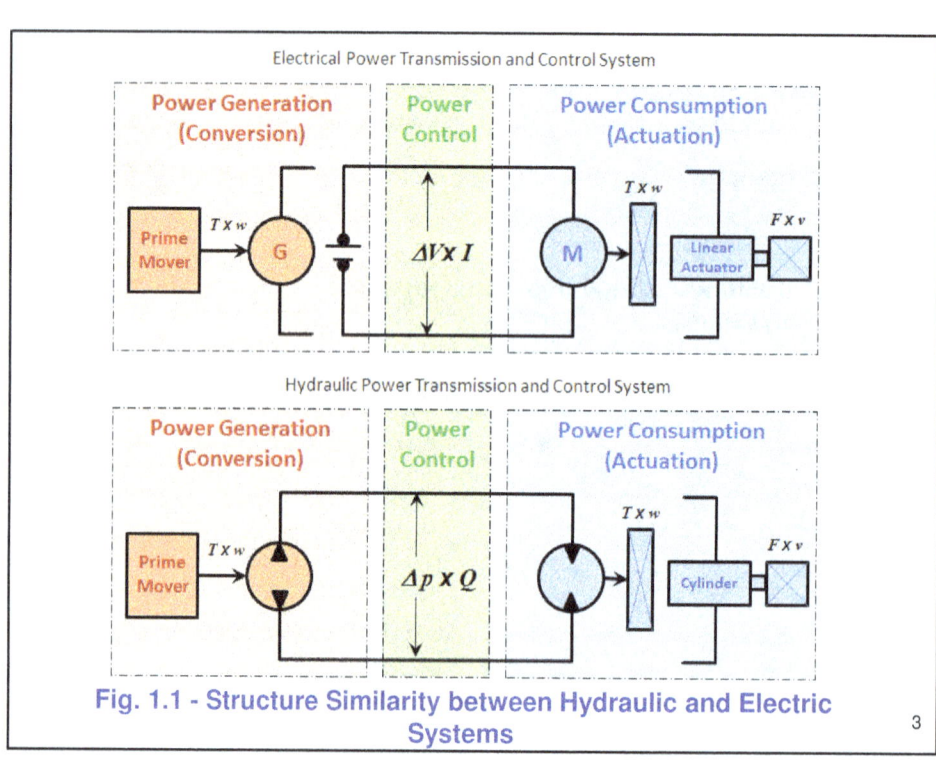

Fig. 1.1 - Structure Similarity between Hydraulic and Electric Systems

Workbook: Electro-Hydraulic Components and Systems
Chapter 1: Hydraulic versus Electrical Systems

1.1.1- Power Generation (Conversion)
1.1.1.1- Electrical Power Generators

1. Wall outlet.
2. Small electrical generator.
3. Medium size mobile electric generator.
4. Large side stationary electric generator.

Fig. 1.2 - Electrical Power Generators

input mechanical power (Torque T x Angular Speed ω)
→ output electrical power (Voltage Difference ΔV x Electrical Current I).

1.1.1.2- Hydraulic Power Generators

Fig. 1.3 - Hydraulic Power Generators (Pumps)

input mechanical power (Torque T x Angular Speed ω)
→ output hydraulic power (Differential Pressure Δp x flow Q).

1.1.2- Power Control

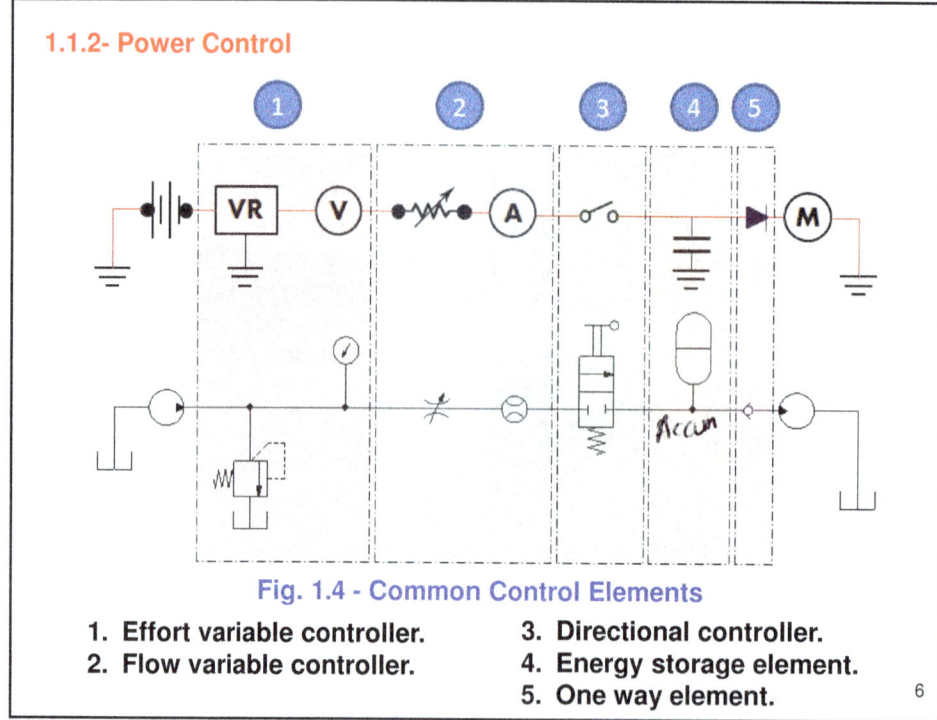

Fig. 1.4 - Common Control Elements

1. Effort variable controller.
2. Flow variable controller.
3. Directional controller.
4. Energy storage element.
5. One way element.

1.1.2.1- Power Quantification

Effort × Flow

$$Electrical\ Power:\ P_E\ (kW) = \frac{\Delta V\ (V) \times I(A)}{1000} \qquad 1.1$$

$$Hydraulic\ Power:\ P_H\ (kW) = \frac{\Delta p\ (bar) \times Q(\frac{lit}{min})}{600} \qquad 1.2$$

1.1.2.2- Effort Variable Controllers (Overload Protection)

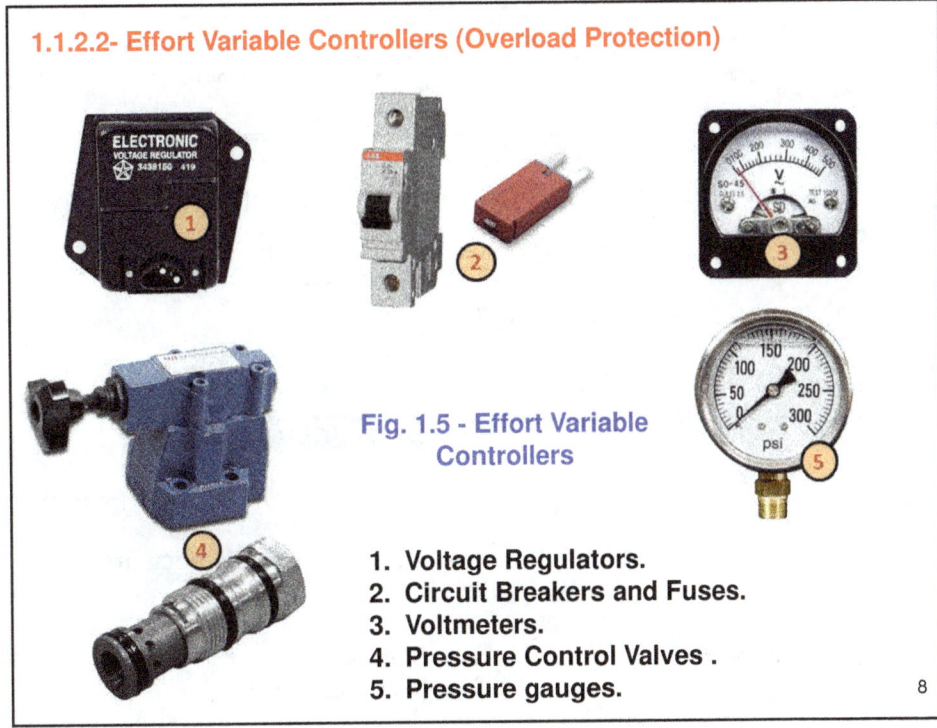

Fig. 1.5 - Effort Variable Controllers

1. Voltage Regulators.
2. Circuit Breakers and Fuses.
3. Voltmeters.
4. Pressure Control Valves.
5. Pressure gauges.

1.1.2.3- Flow Variable Controllers (Resistive Elements)

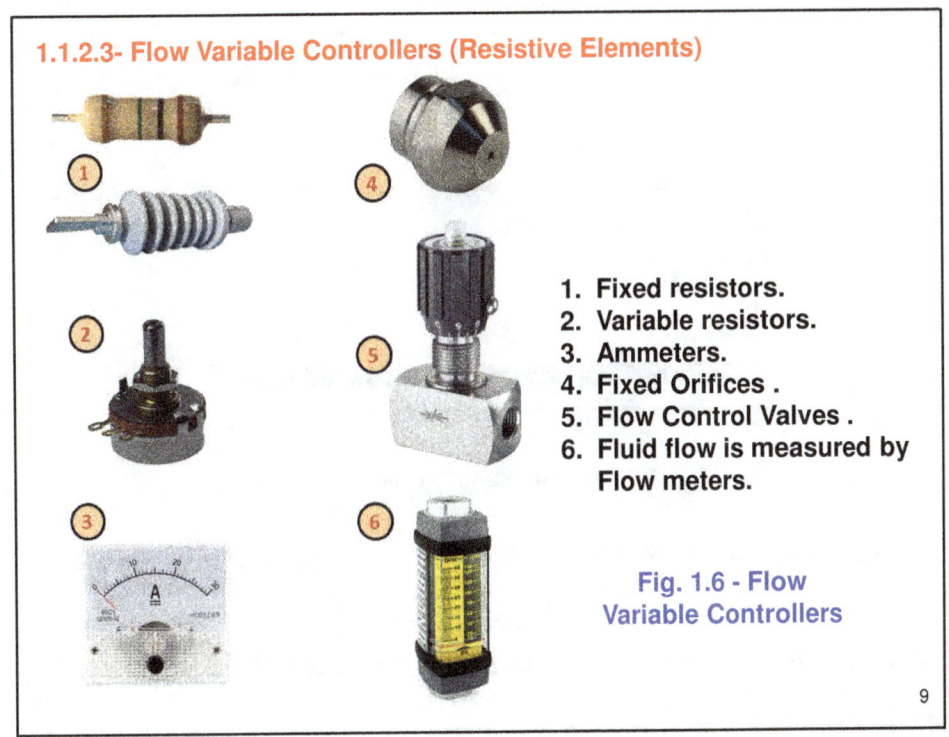

1. Fixed resistors.
2. Variable resistors.
3. Ammeters.
4. Fixed Orifices.
5. Flow Control Valves.
6. Fluid flow is measured by Flow meters.

Fig. 1.6 - Flow Variable Controllers

Workbook: Electro-Hydraulic Components and Systems
Chapter 1: Hydraulic versus Electrical Systems

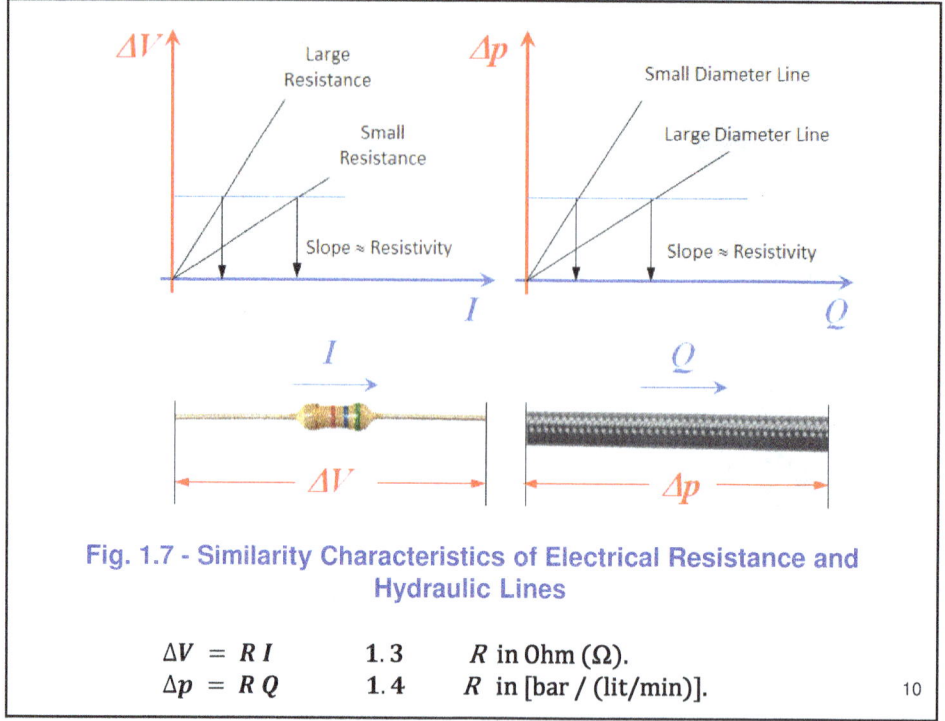

Fig. 1.7 - Similarity Characteristics of Electrical Resistance and Hydraulic Lines

$$\Delta V = R\, I \quad\quad 1.3 \quad\quad R \text{ in Ohm } (\Omega).$$
$$\Delta p = R\, Q \quad\quad 1.4 \quad\quad R \text{ in [bar / (lit/min)]}.$$

Linear Resistances in Series:

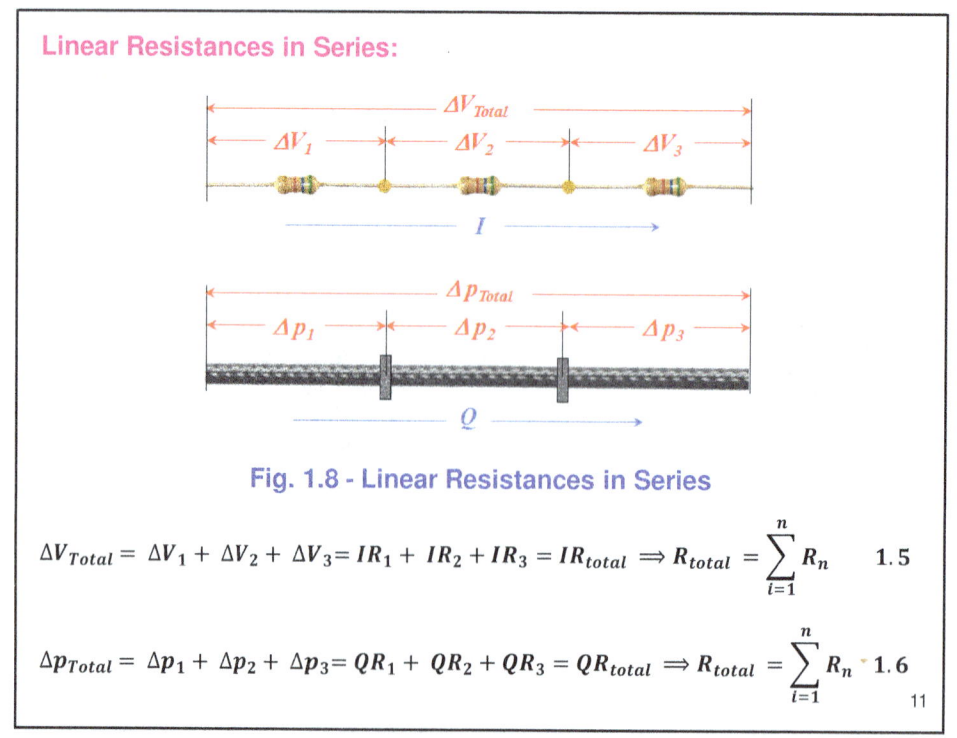

Fig. 1.8 - Linear Resistances in Series

$$\Delta V_{Total} = \Delta V_1 + \Delta V_2 + \Delta V_3 = IR_1 + IR_2 + IR_3 = IR_{total} \Rightarrow R_{total} = \sum_{i=1}^{n} R_n \quad 1.5$$

$$\Delta p_{Total} = \Delta p_1 + \Delta p_2 + \Delta p_3 = QR_1 + QR_2 + QR_3 = QR_{total} \Rightarrow R_{total} = \sum_{i=1}^{n} R_n \quad 1.6$$

Workbook: Electro-Hydraulic Components and Systems
Chapter 1: Hydraulic versus Electrical Systems

Linear Resistances in Parallel:

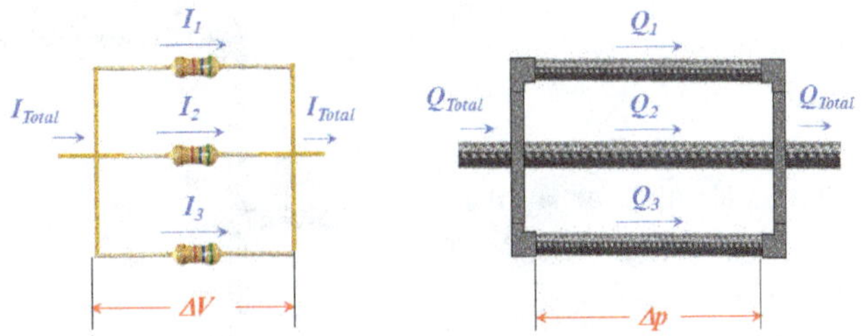

Fig. 1.9 - Linear Resistances in Parallel

$$I_{Total} = I_1 + I_2 + I_3 = \frac{\Delta V}{R_1} + \frac{\Delta V}{R_2} + \frac{\Delta V}{R_3} = \frac{\Delta V}{R_{Total}} \Rightarrow \frac{1}{R_{Total}} = \sum_{i=1}^{n} \frac{1}{R_n} \quad 1.7$$

$$Q_{Total} = Q_1 + Q_2 + Q_3 = \frac{\Delta p}{R_1} + \frac{\Delta p}{R_2} + \frac{\Delta p}{R_3} = \frac{\Delta p}{R_{Total}} \Rightarrow \frac{1}{R_{Total}} = \sum_{i=1}^{n} \frac{1}{R_n} \quad 1.8$$

Nonlinear Hydraulic Resistances:

$$Q = C_v \sqrt{\frac{\Delta p}{SG}} \quad 1.9$$

Size of ↙ ↑ *Oil or water*

Hydraulic Nonlinear Resistances in Series:

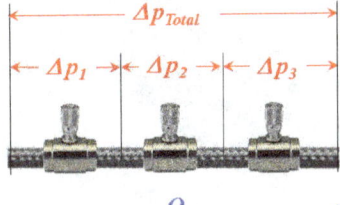

Fig. 1.10 - Nonlinear Hydraulic Resistances in Series

$$\Delta p_{Total} = \Delta p_1 + \Delta p_2 + \Delta p_3 = Q^2 \cdot SG \left(\frac{1}{C_{v1}^2} + \frac{1}{C_{v2}^2} + \frac{1}{C_{v3}^2} \right) \quad 1.10A$$

$$Q = \sqrt{\frac{\Delta p_{Total}}{\left(\frac{1}{C_{v1}^2} + \frac{1}{C_{v2}^2} + \frac{1}{C_{v3}^2} \right) \times SG}} = C_{ves} \sqrt{\frac{\Delta p_{Total}}{SG}} \quad 1.10B$$

$$C_{ves} = \sqrt{\frac{1}{\sum_{1}^{n} \frac{1}{C_{vn}^2}}} \quad 1.10C$$

Hydraulic Nonlinear Resistances in Parallel:

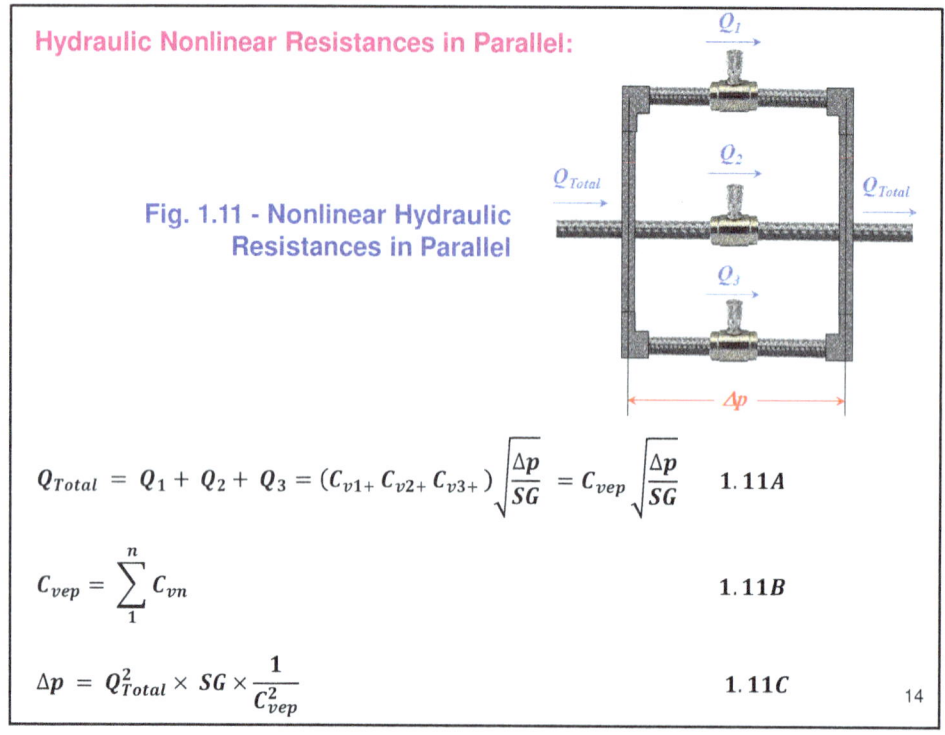

Fig. 1.11 - Nonlinear Hydraulic Resistances in Parallel

$$Q_{Total} = Q_1 + Q_2 + Q_3 = (C_{v1} + C_{v2} + C_{v3})\sqrt{\frac{\Delta p}{SG}} = C_{vep}\sqrt{\frac{\Delta p}{SG}} \quad 1.11A$$

$$C_{vep} = \sum_{1}^{n} C_{vn} \quad 1.11B$$

$$\Delta p = Q_{Total}^2 \times SG \times \frac{1}{C_{vep}^2} \quad 1.11C$$

1.1.2.4- Directional Controllers

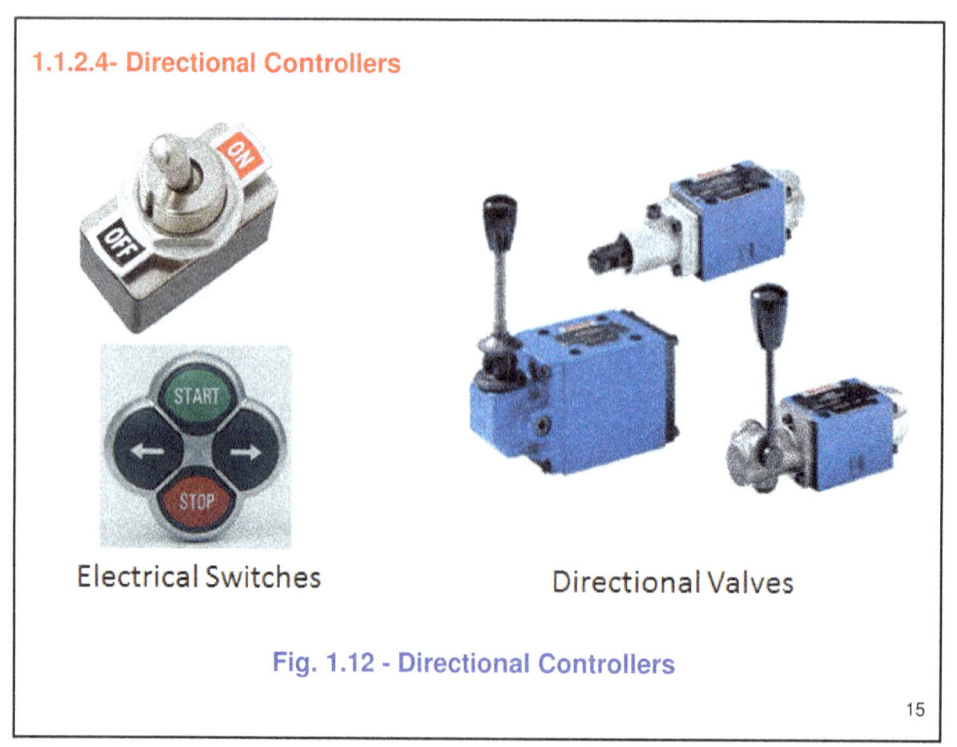

Electrical Switches   Directional Valves

Fig. 1.12 - Directional Controllers

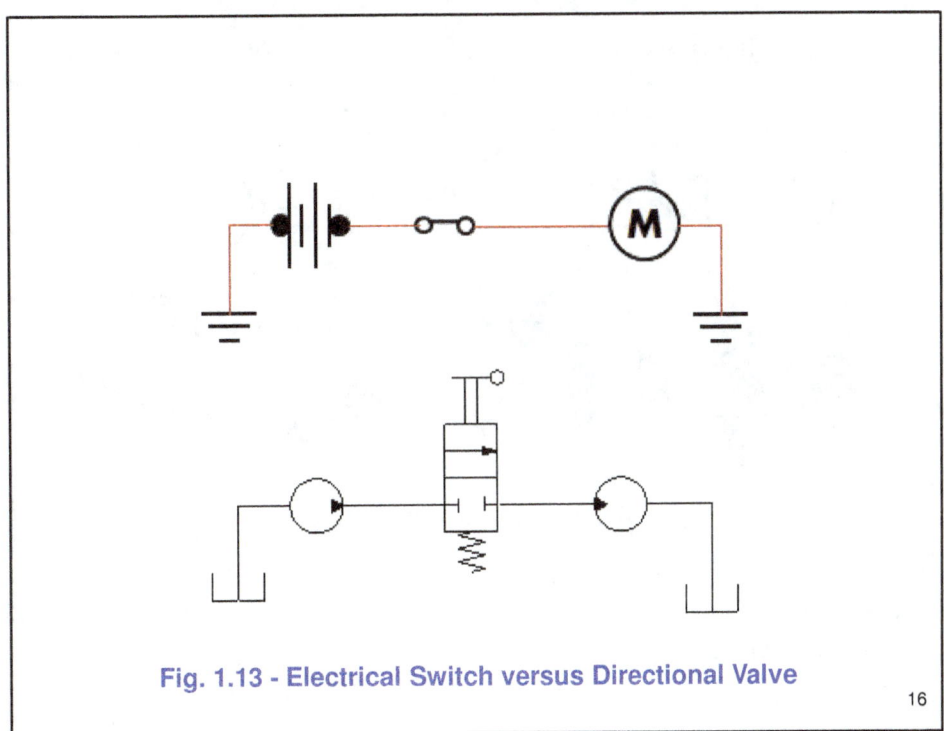

Fig. 1.13 - Electrical Switch versus Directional Valve

1.1.2.5- Energy Storage Elements (Capacitive Elements)

Energy Density?

1. Capacitors.
2. Batteries.
3. Accumulators.

Fig. 1.14 - Energy Storage Elements

1.1.2.6- Non-Return Elements

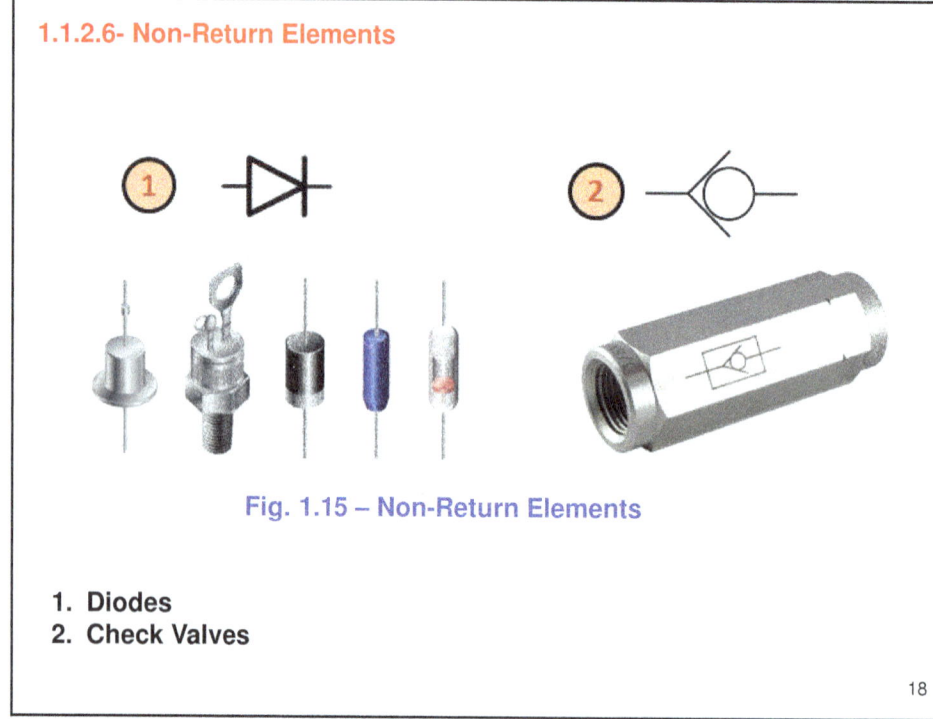

Fig. 1.15 – Non-Return Elements

1. Diodes
2. Check Valves

1.1.2.7- Electrical versus Hydraulic Power Control Elements

❏ **Controllability:**
- Electrical (Digital, smarter, reprogrammable control).

❏ **Closed Loop Control:**
- Hydro-Mechanical (perform limited closed control with mechanical feedback such as in pressure compensated, power limited and load sense pumps).

❏ **Size:**
- Electrical (smaller in size for the same power).

❏ **Material:**
- Electrical (produced from less amount of material).

❏ **Production Process:**
- Electrical (require less physical production process).
- Hydraulic (require casting, machining, forging, heat treatment, etc).

1.1.3- Power Consumption (Actuation)
1.1.3.1- Rotational Actuators

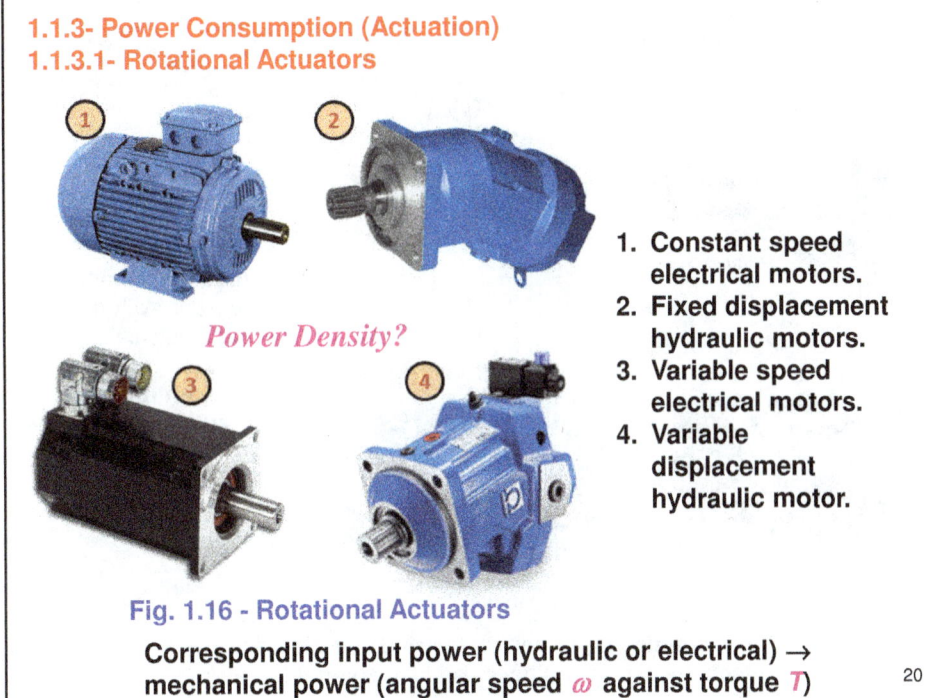

Power Density?

1. Constant speed electrical motors.
2. Fixed displacement hydraulic motors.
3. Variable speed electrical motors.
4. Variable displacement hydraulic motor.

Fig. 1.16 - Rotational Actuators

Corresponding input power (hydraulic or electrical) → mechanical power (angular speed ω against torque T)

Similarities:
- Both systems have motors of different sizes.
- Both systems have constant and variable speed motors.

Differences:
- **Types:** Electrical motors could be of AC or DC type.
- **Nmax:** Electric Motors > hydraulic motors (fluid friction)
- **Nmin:** Electric Motors < hydraulic motors (erratic & inefficient)
- **Power Density:** Hydraulic motors > electric motors.
- **Saturation:**
- Hydraulic motors are not saturated as electric motors.
- A Hydraulic motor torque = f (Δp and size).
- Limitations are (maximum pressure and maximum torque).
- Hydraulic motors have better starting torque and better dynamic response as compared to the electrical motors.
- **Cost:** Electric motors are commonly less expensive for the same power.

1.1.3.2 - Linear Actuators

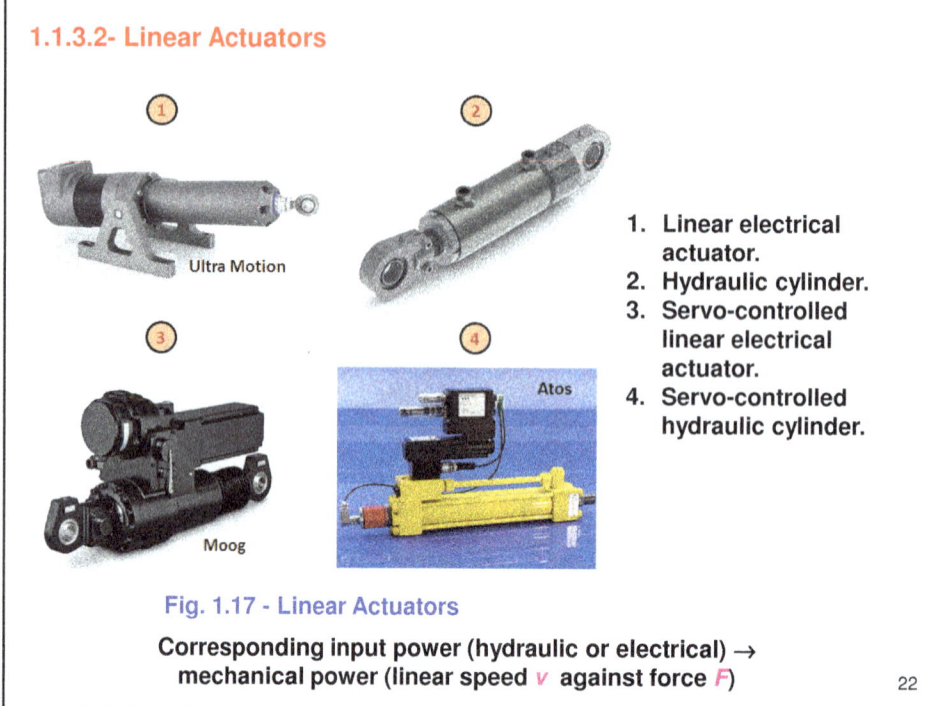

1. Linear electrical actuator.
2. Hydraulic cylinder.
3. Servo-controlled linear electrical actuator.
4. Servo-controlled hydraulic cylinder.

Fig. 1.17 - Linear Actuators

Corresponding input power (hydraulic or electrical) → mechanical power (linear speed v against force F)

Similarities:

- Both systems have linear actuators of different sizes. They are called "*Cylinders*" in hydraulic systems.
- Both systems have servo-controlled linear actuators.

Differences:

- Mechanisms: A Hydraulic cylinder has no intermediate mechanism to produce linear motion.
- Strokes: Hydraulic cylinders offer longer strokes than electrical linear actuators.
- Power Density: For the same physical size, hydraulic cylinders carry larger loads.
- Cost: Hydraulic cylinders are commonly less expensive for the same size.

Workbook: Electro-Hydraulic Components and Systems
Chapter 1: Hydraulic versus Electrical Systems

1.2- Operational Analogy between Hydraulic and Electrical System

1.2.1- Energy Transmission in a Circuit

- **Flow Variable:**

- **Effort Variable:**

- **Power of Each Element:** gained, consumed, or wasted.

- **Circuit Power:**
- Power gained by the generator = Sum of the power consumed or wasted by the rest of elements in the circuit + the line losses.

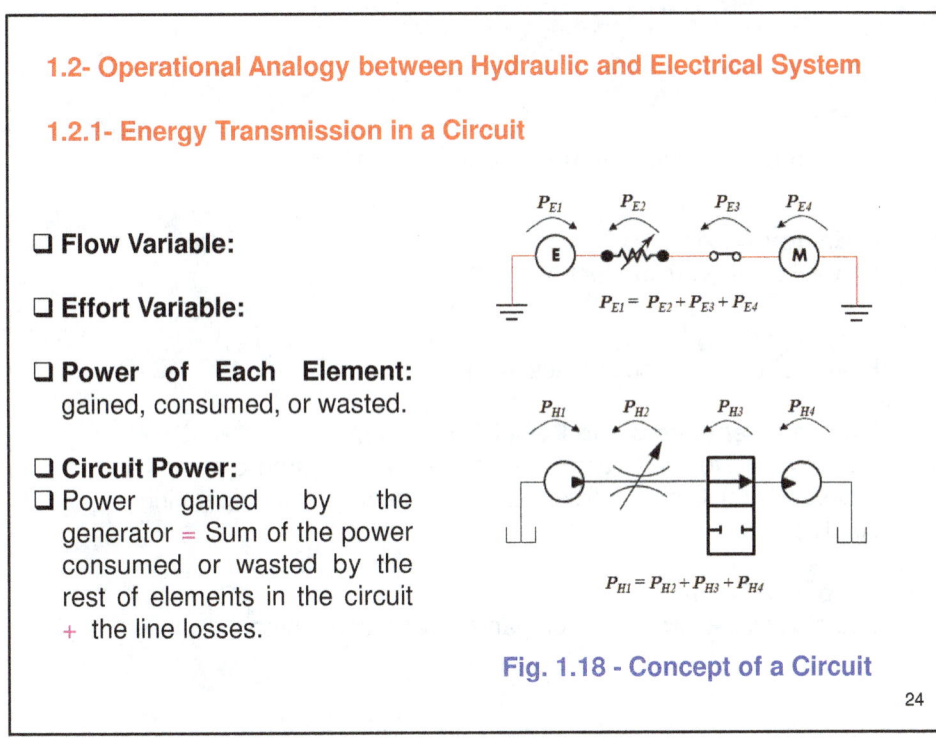

Fig. 1.18 - Concept of a Circuit

Fluid to fluid friction

1.2.2- Energy Transmission Efficiency

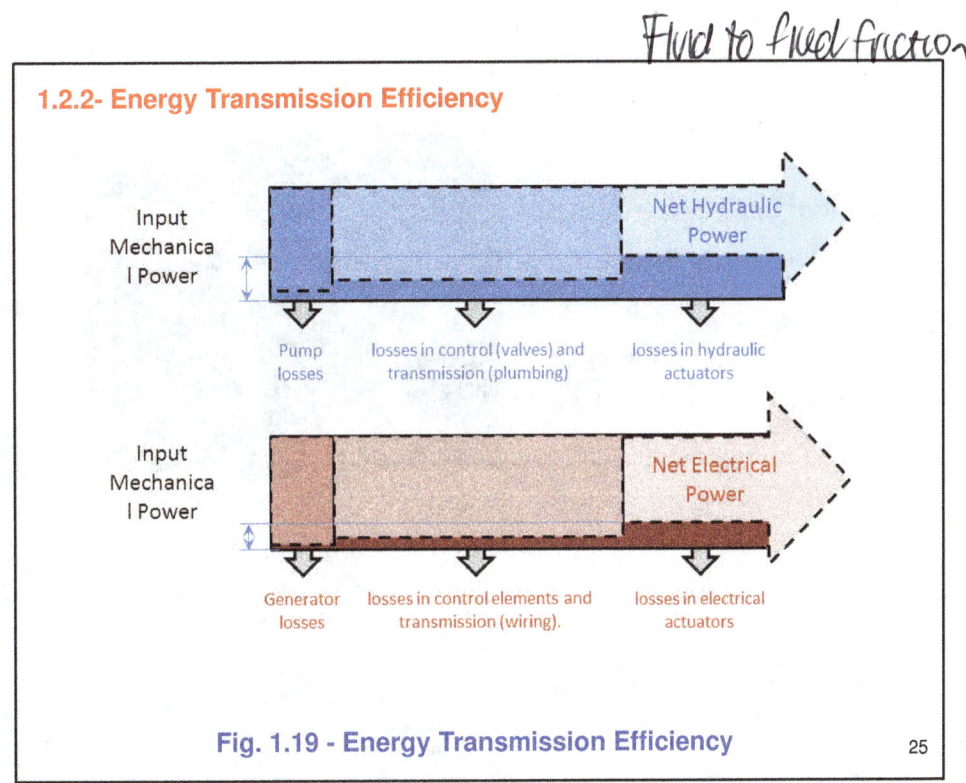

Fig. 1.19 - Energy Transmission Efficiency

much less power lost in an electrical

An electrical system generally wastes less energy because:

❑ **Energy Carrier:**
Hydraulic fluid versus Electrons.

❑ **Transmitting Lines:**
Pipes, tubes and hoses versus Wires.

❑ **Power Generators and Power Consumers:**
As compared to equivalent electrical elements, hydraulic pumps and motors are relatively inefficient & highly affected by the operating conditions.

❑ **Control Elements:**
Use of valves is always accompanied by pressure drop.

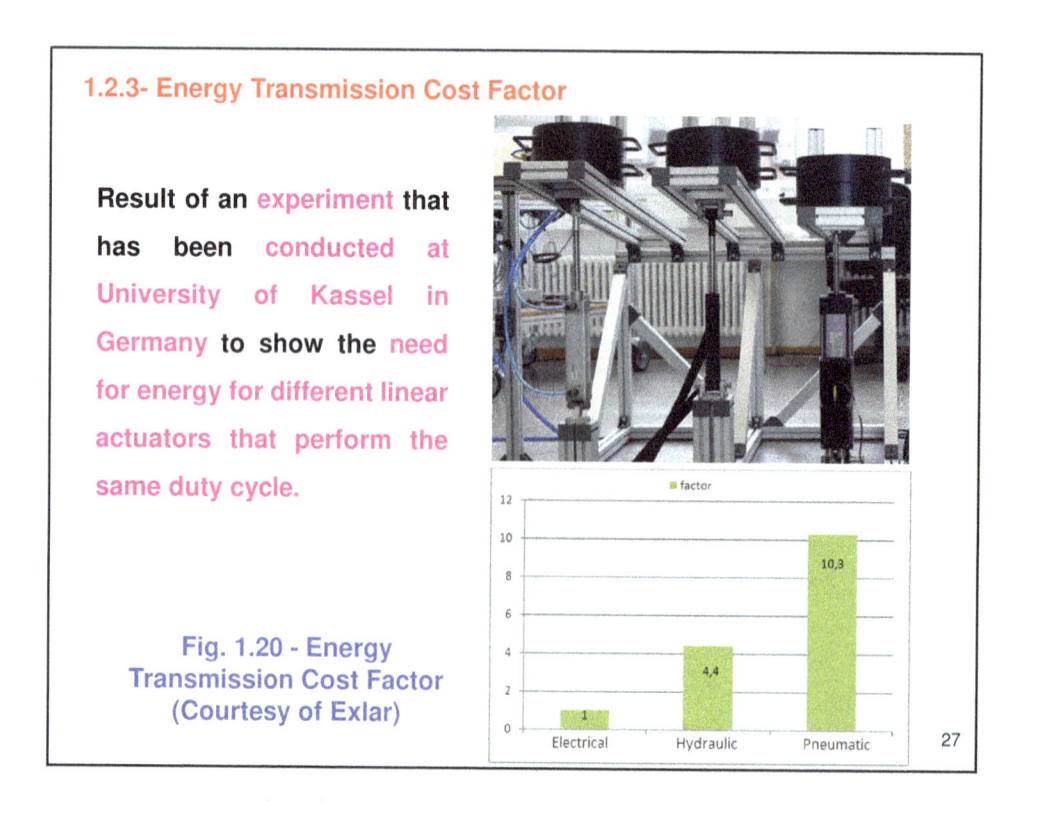

1.2.3- Energy Transmission Cost Factor

Result of an experiment that has been conducted at University of Kassel in Germany to show the need for energy for different linear actuators that perform the same duty cycle.

Fig. 1.20 - Energy Transmission Cost Factor (Courtesy of Exlar)

System	Factor
Electrical	1
Hydraulic	4,4
Pneumatic	10,3

1.2.4- Energy Transmission Distance

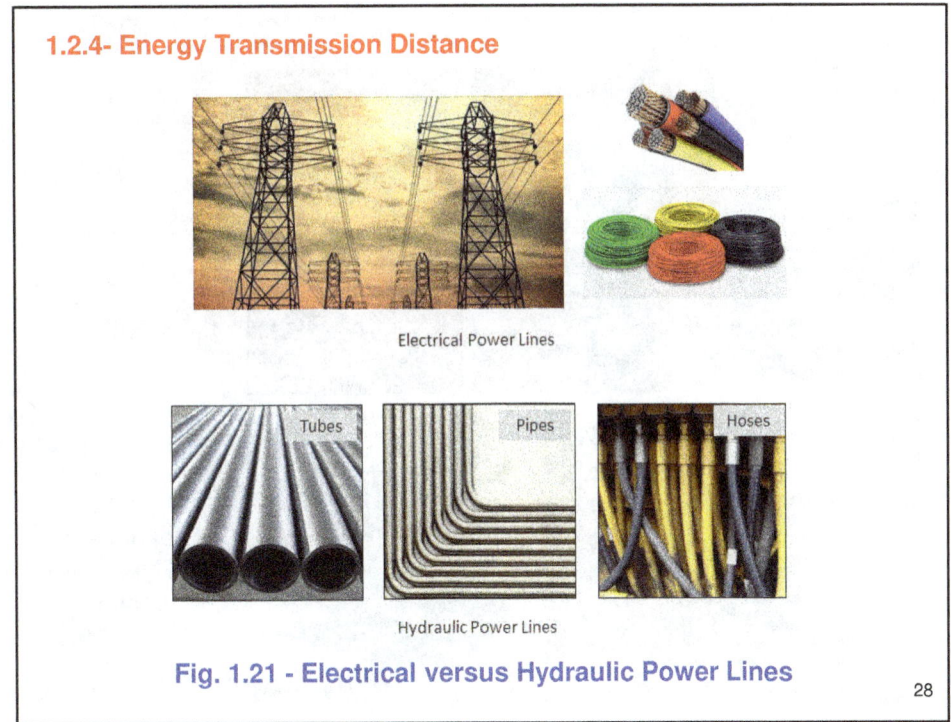

Fig. 1.21 - Electrical versus Hydraulic Power Lines

1.2.5- Readiness of Operation

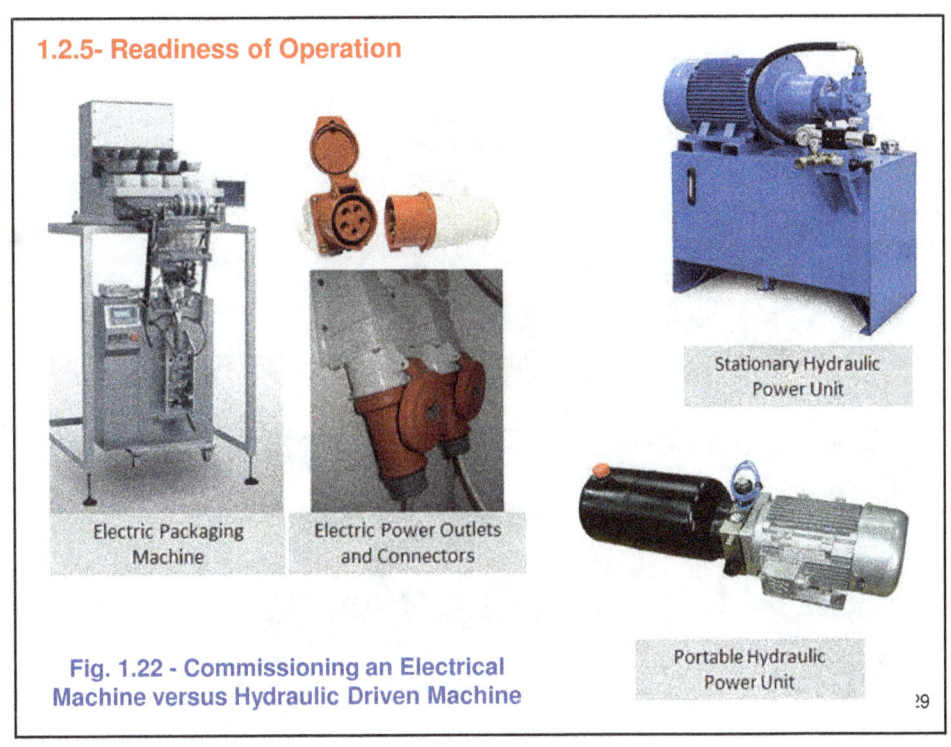

Fig. 1.22 - Commissioning an Electrical Machine versus Hydraulic Driven Machine

Fig. 1.23 - Cleanliness Of Hydraulic-Driven Machine

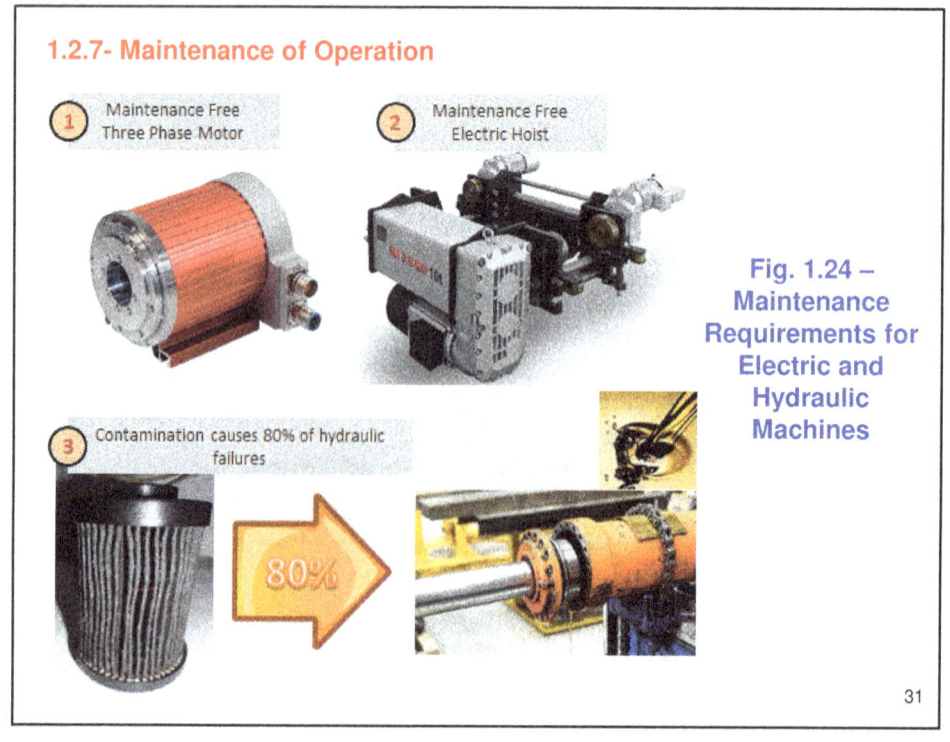

Fig. 1.24 – Maintenance Requirements for Electric and Hydraulic Machines

1.2.8- Safety of Operation under High Temperature

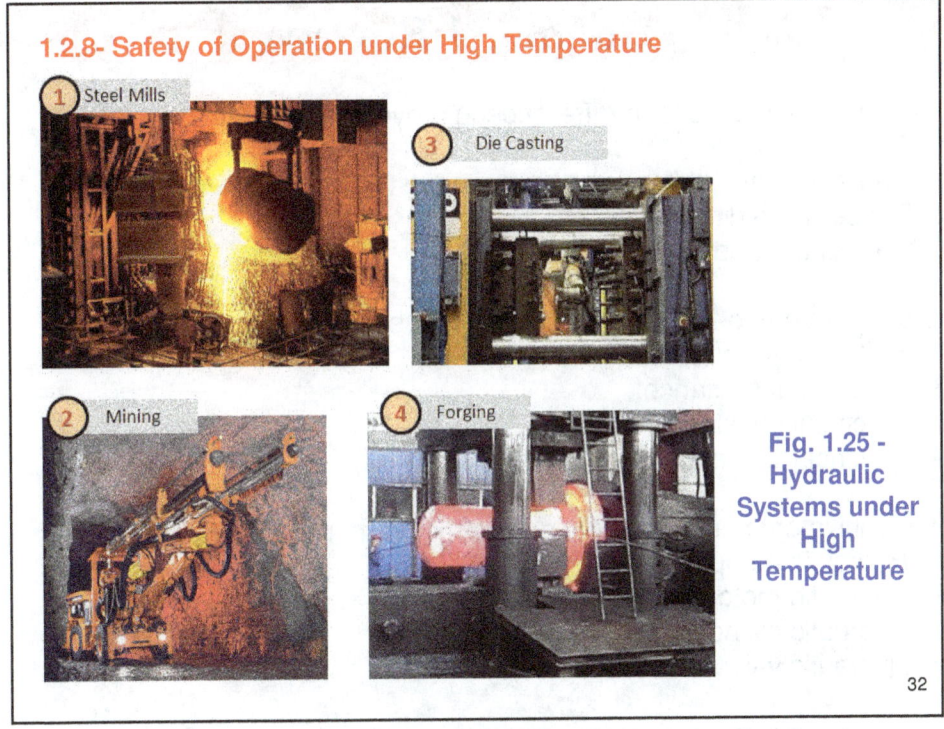

Fig. 1.25 - Hydraulic Systems under High Temperature

1.2.9- Noise of Operation

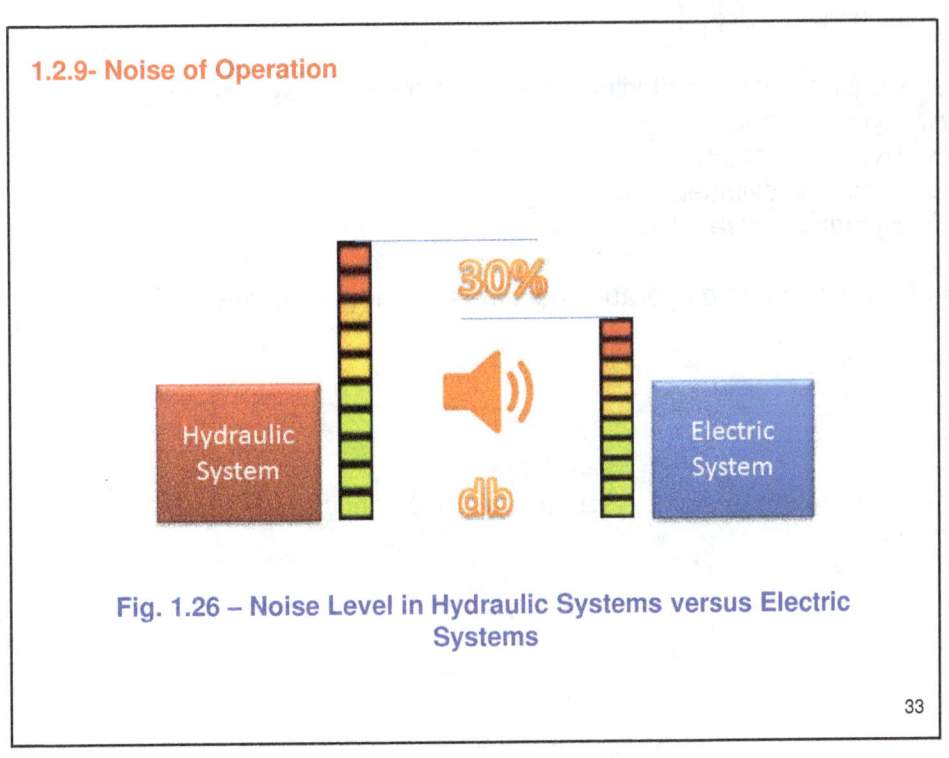

Fig. 1.26 – Noise Level in Hydraulic Systems versus Electric Systems

Workbook: Electro-Hydraulic Components and Systems
Chapter 1: Reviews and Assignments

Chapter 1 Reviews

1. What is equivalent to voltage difference in a hydraulic system?
    A. Flow rate.
    B. Differential pressure across components.
    C. Temperature difference.
    D. None of the above.

2. What is equivalent to flow rate in an electrical system?
    A. Voltage difference.
    B. Electrical inductance.
    C. Electrical resistance
    D. Electrical Current.

3. What is equivalent to an electrical generator in a hydraulic system?
    A. Hydraulic pump.
    B. Hydraulic motor.
    C. Hydraulic cylinder.
    D. Hydraulic valve.

4. What is equivalent to an electrical motor in a hydraulic system?
    A. Hydraulic pump.
    B. Hydraulic motor.
    C. Hydraulic cylinder.
    D. Hydraulic valve.

5. What is equivalent to an electrical relay in a hydraulic system?
    A. Hydraulic pump.
    B. Hydraulic motor.
    C. Hydraulic cylinder.
    D. Hydraulic valve.

6. What is equivalent to a hydraulic accumulator in an electrical system?
    A. Fuse.
    B. Diode.
    C. Electrical resistance.
    D. Capacitor.

7. What is equivalent to a check valve in an electrical system?
    A. Fuse.
    B. Diode.
    C. Electrical resistance.
    D. Capacitor.

8. What is equivalent to pressure relief valve in an electrical system?
    A. Circuit breaker.
    B. Toggle switch.
    C. Electrical resistance.
    D. Capacitor.

9. Which of the following elements has the higher energy density?
    A. Battery.
    B. Accumulator.
    C. Capacitor.
    D. Spring.

10. Which of the following elements has the higher power density?
    A. Electrical motor.
    B. Pneumatic motor.
    C. Hydraulic motor.
    D. They are all equal.

Workbook: Electro-Hydraulic Components and Systems
Chapter 1: Reviews and Assignments

Chapter 1 Assignment

Student Name: -- Student ID: -------------------

Date: -- Score: ------------------------

Find the differential pressure, Δp, in the system shown below where:

$Q_{Total} = 5\ GPM,\ SG = 1,\ C_{v1} = 1,\ C_{v2} = 0.8,\ and\ C_{v3} = 1.1.$

Units of valve coefficient is $GPM/(PSI)^{-0.5}$

Find the differential pressure if the same valves are placed in series.

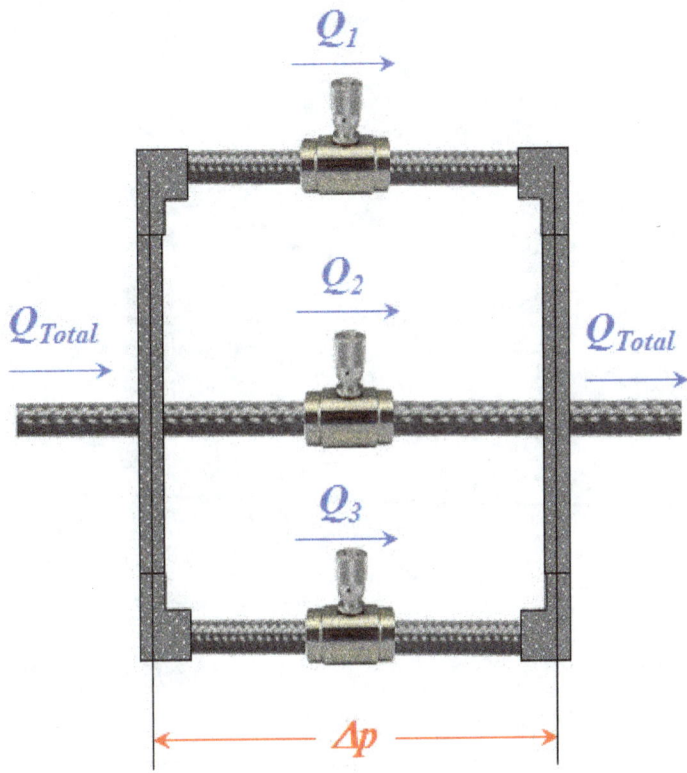

Chapter 2
Hydro-Mechanical versus Electro-Hydraulic Solutions

Objectives:

This chapter explores the features and challenges of electro-hydraulic systems. Additionally the chapter introduces the benefits of converting the classical hydro-mechanical systems into electro-hydraulic ones. Pressure control, flow control, power control, and sequence control are among the systems discussed.

Brief Contents:

2.1- Features and Challenges of Electro-Hydraulic Systems
2.2- Pressure Control Solutions
2.3- Flow Control Solutions
2.4- Power Control Solutions
2.5- Sequence Control Solutions
2.6- Hydraulic Deceleration System
2.7- Cylinder Speed Synchronization
2.8- Accumulator Charging
2.9- Control of Overrunning Loads
2.10- Control Block Actuation in Mobile Applications

2.1 - Features and Challenges of Electro-Hydraulic Systems

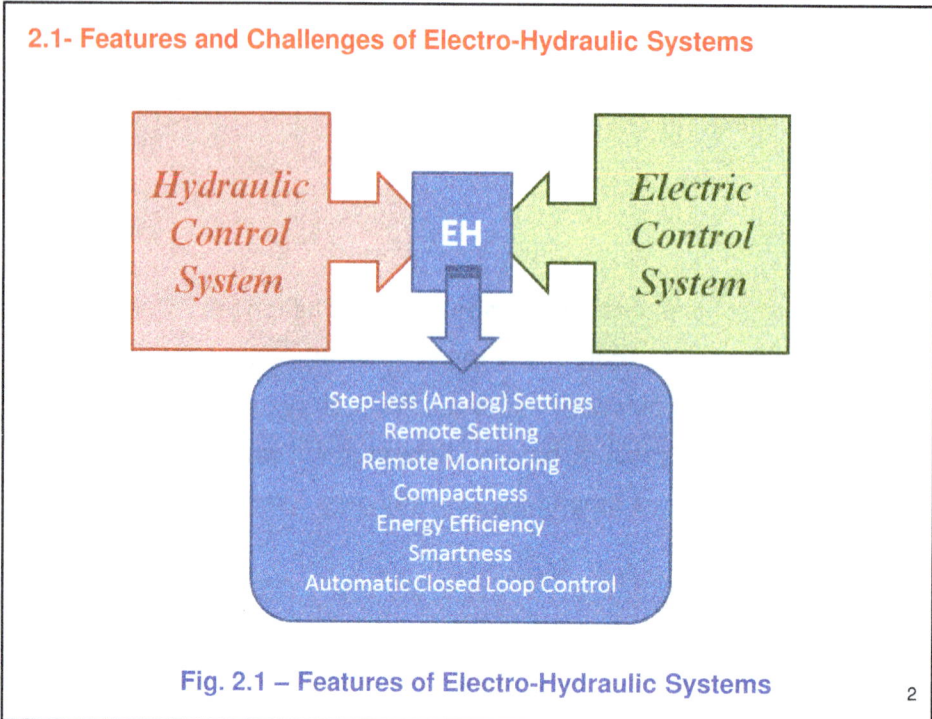

Fig. 2.1 – Features of Electro-Hydraulic Systems

Challenges:

- **Reliability:** EH systems are highly sensitive to contamination.

- **Know-How:** programming, diagnostics, troubleshooting, and know how it works.

- **System Tuning:** requires certain minimum level of experience and education.

- **Single Point Failure:** an EH system commonly consists of small number of components. As a result, if any of these devices fail, the rest of the system fails.

- **Cost:** EH systems are pricey.

2.2- Pressure Control Solutions
2.2.1- Pressure Control using Hydro-Mechanical Pressure Relief Valve

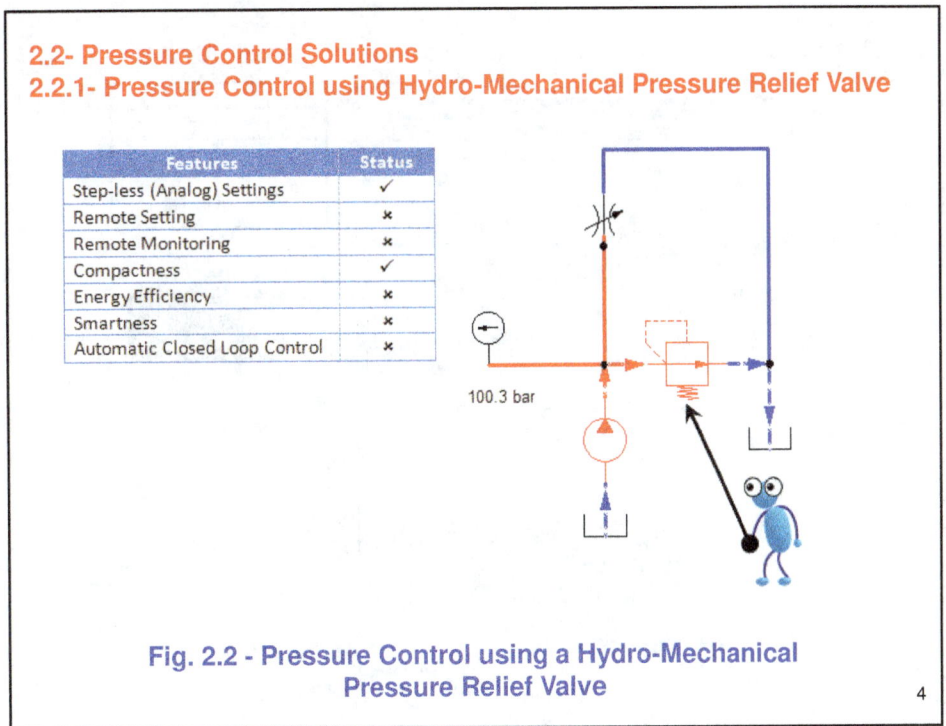

Fig. 2.2 - Pressure Control using a Hydro-Mechanical Pressure Relief Valve

2.2.2- Pressure Control using Hydro-Mechanical Pressure Compensated Pump

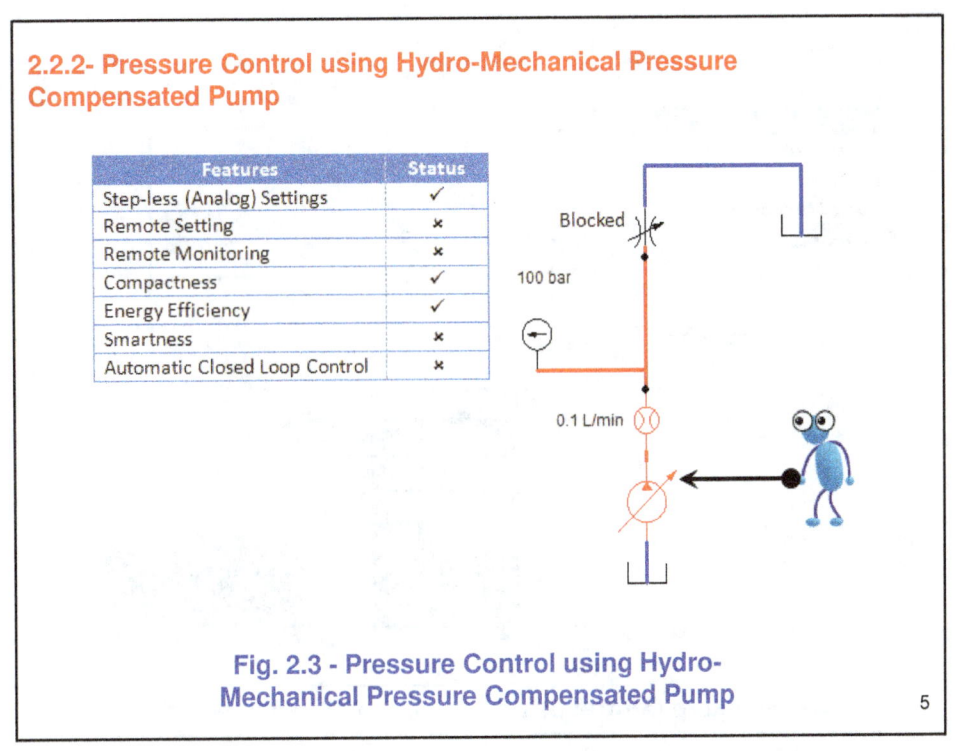

Fig. 2.3 - Pressure Control using Hydro-Mechanical Pressure Compensated Pump

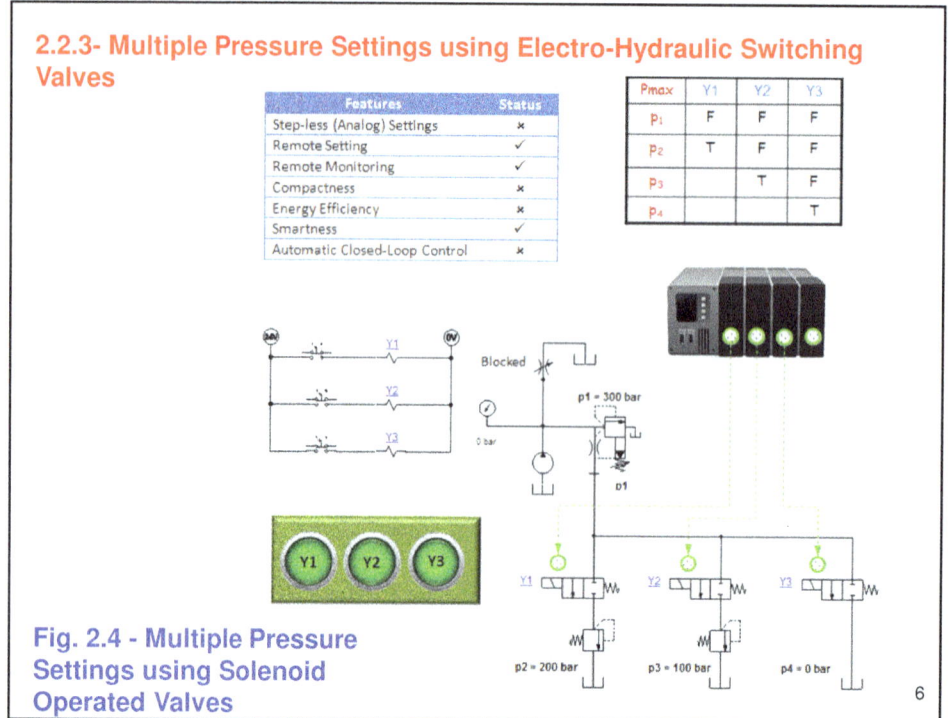

Fig. 2.4 - Multiple Pressure Settings using Solenoid Operated Valves

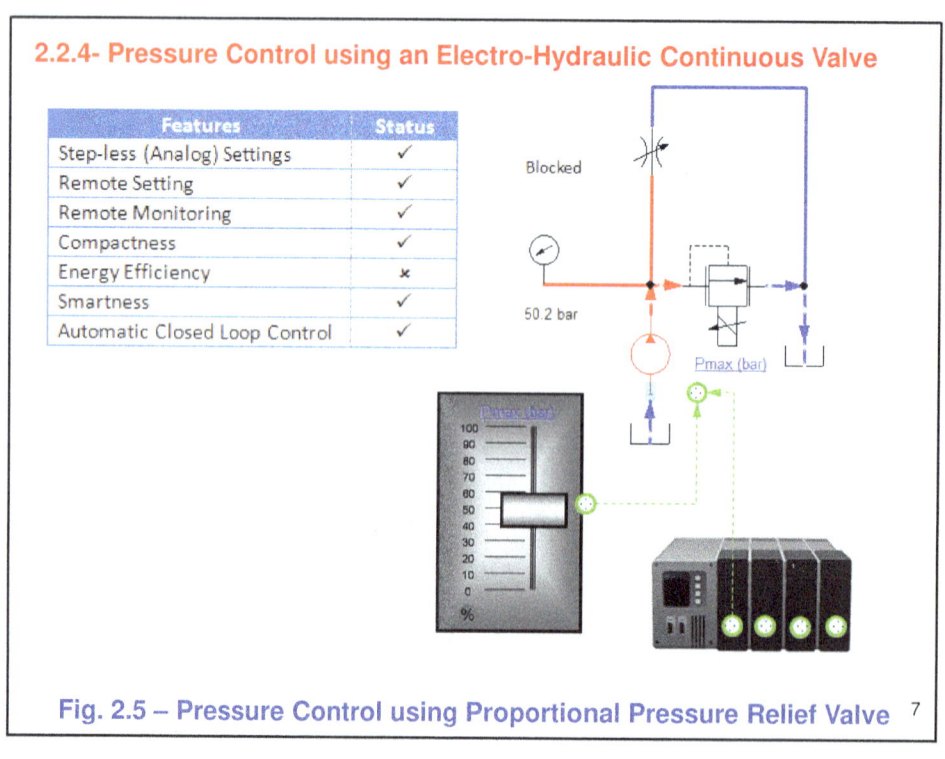

Fig. 2.5 – Pressure Control using Proportional Pressure Relief Valve

2.2.5- Pressure Control using an Electro-Hydraulic Pressure Compensated Pump

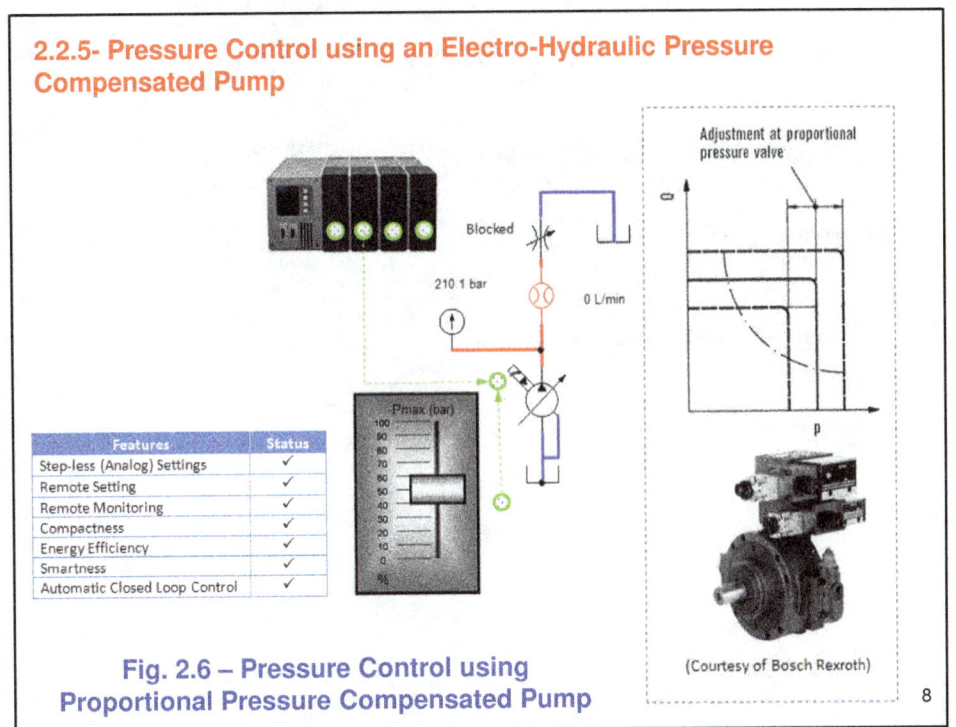

Fig. 2.6 – Pressure Control using Proportional Pressure Compensated Pump

2.3- Flow Control Solutions
2.3.1- Flow Control using Hydro-Mechanical Flow Control Valve

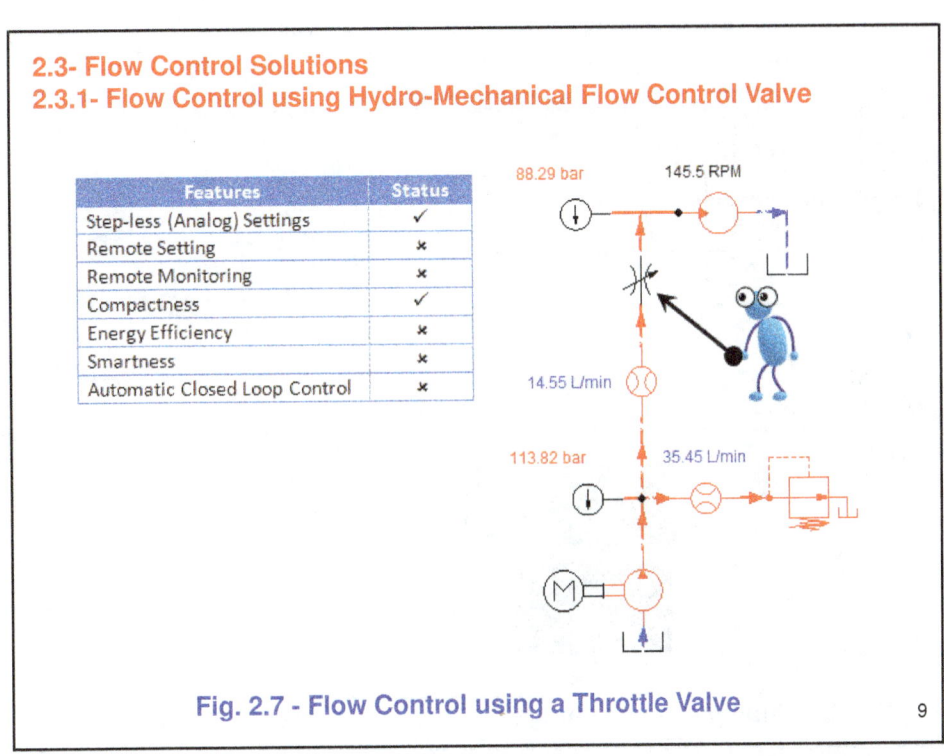

Fig. 2.7 - Flow Control using a Throttle Valve

2.3.2- Flow Control using Hydro-Mechanical Displacement Controlled Pump

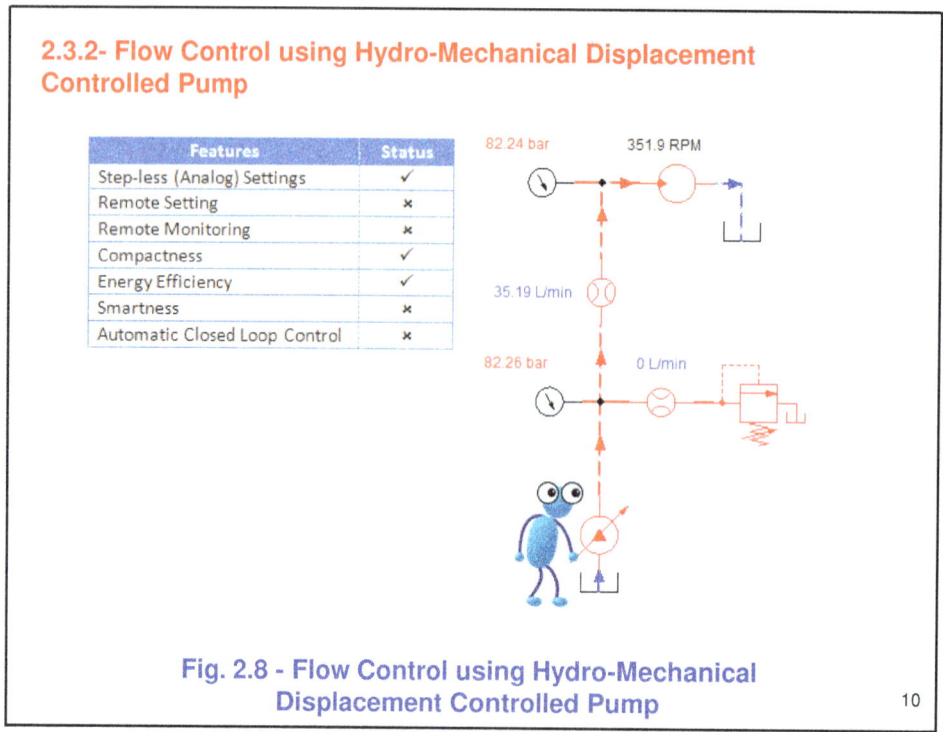

Fig. 2.8 - Flow Control using Hydro-Mechanical Displacement Controlled Pump

2.3.3- Multiple Flow Settings using Electro-Hydraulic Switching Valves

$$z = 2^n - 1 \quad 2.1$$

↓ Number of lines e.g. 3

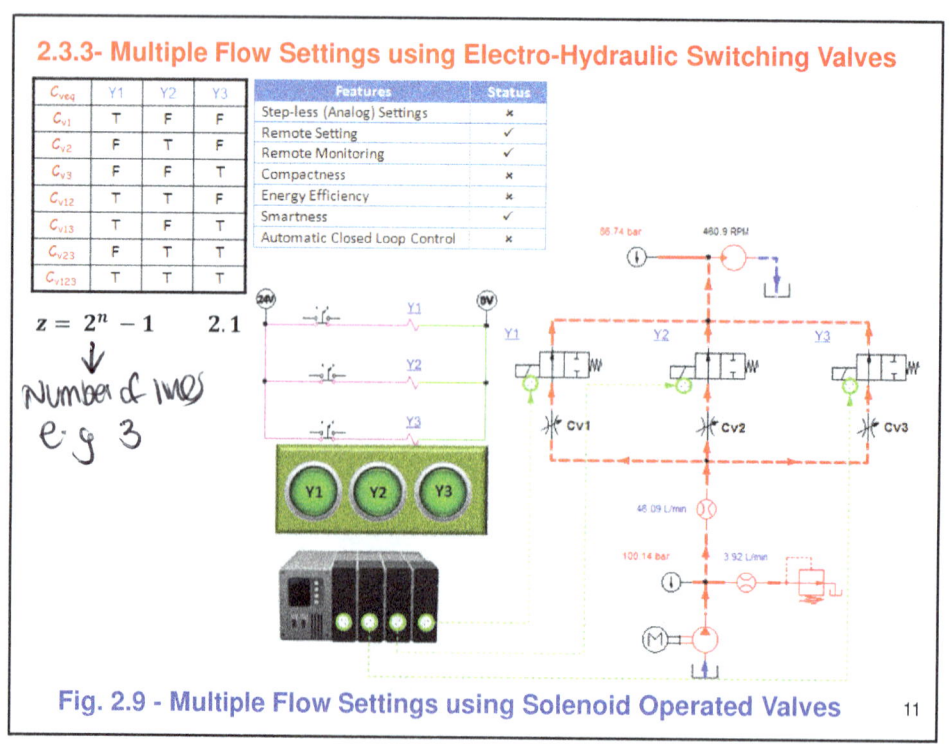

Fig. 2.9 - Multiple Flow Settings using Solenoid Operated Valves

Workbook: Electro-Hydraulic Components and Systems
Chapter 2: Hydro-Mechanical versus Electro-Hydraulic Solutions

2.3.4- Flow Control using an Electro-Hydraulic Continuous Valve

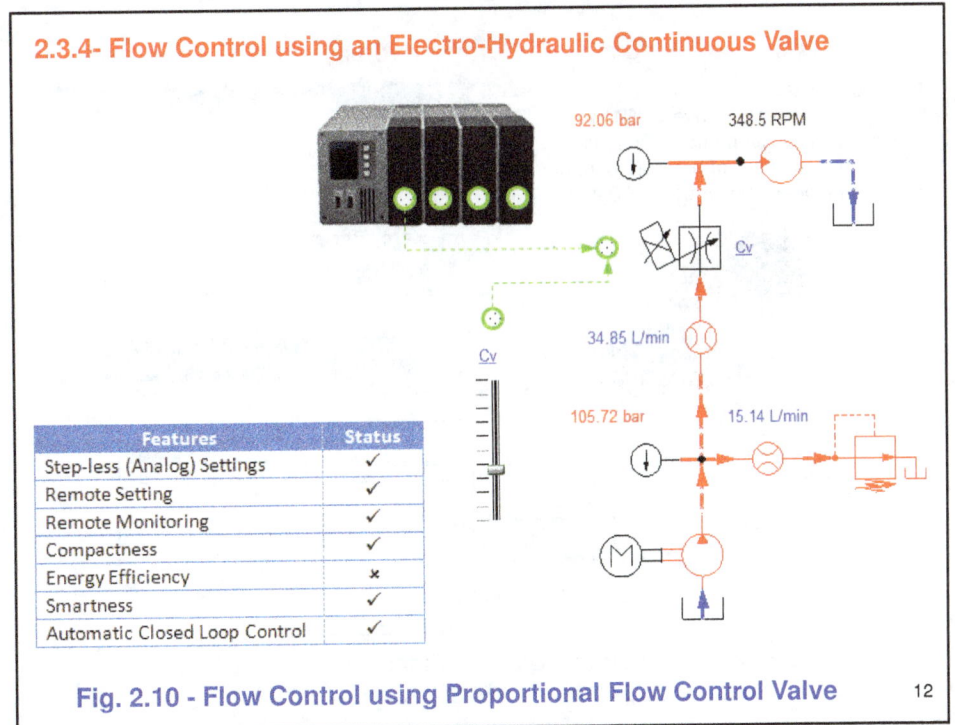

Fig. 2.10 - Flow Control using Proportional Flow Control Valve

2.3.5- Flow Control using an Electro-Hydraulic Displacement Controlled Pump

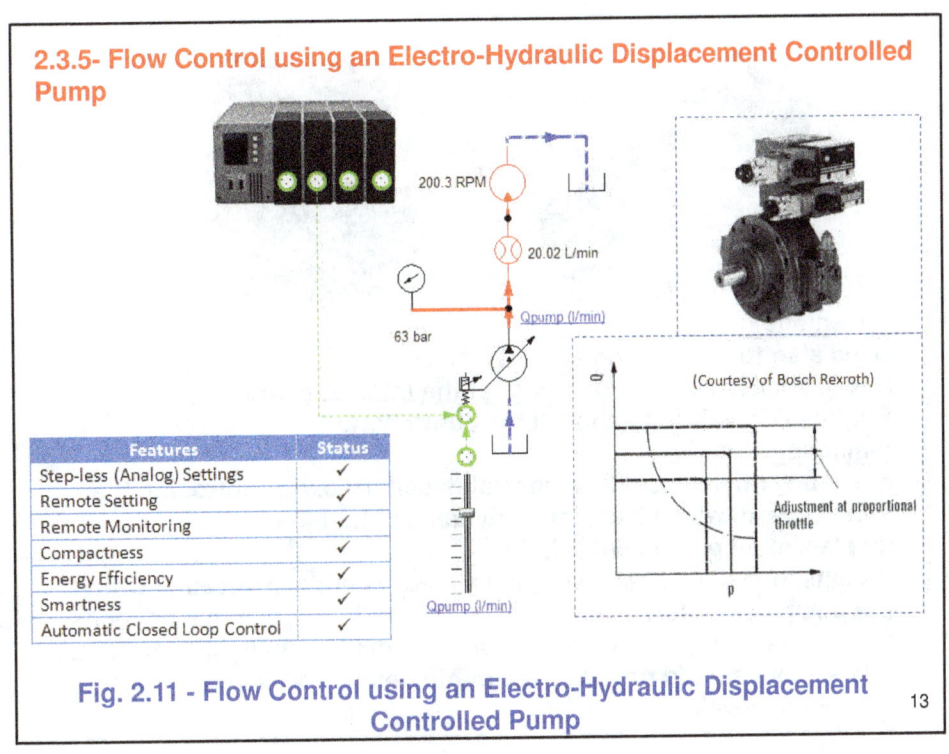

Fig. 2.11 - Flow Control using an Electro-Hydraulic Displacement Controlled Pump

2.3.6- Flow Control using a Fixed Pump Driven by Variable Speed Motor

Technical Data:
- Maximum power per drive = 250 kW
- Maximum flow per pump = 625 l/min
- Maximum accuracy of pressure control = ± 1 bar
- Maximum operating pressure = 345 bar

Features	Status
Step-less (Analog) Settings	✓
Remote Setting	✓
Remote Monitoring	✓
Compactness	✓
Energy Efficiency	✓
Smartness	✓
Automatic Closed Loop Control	✓

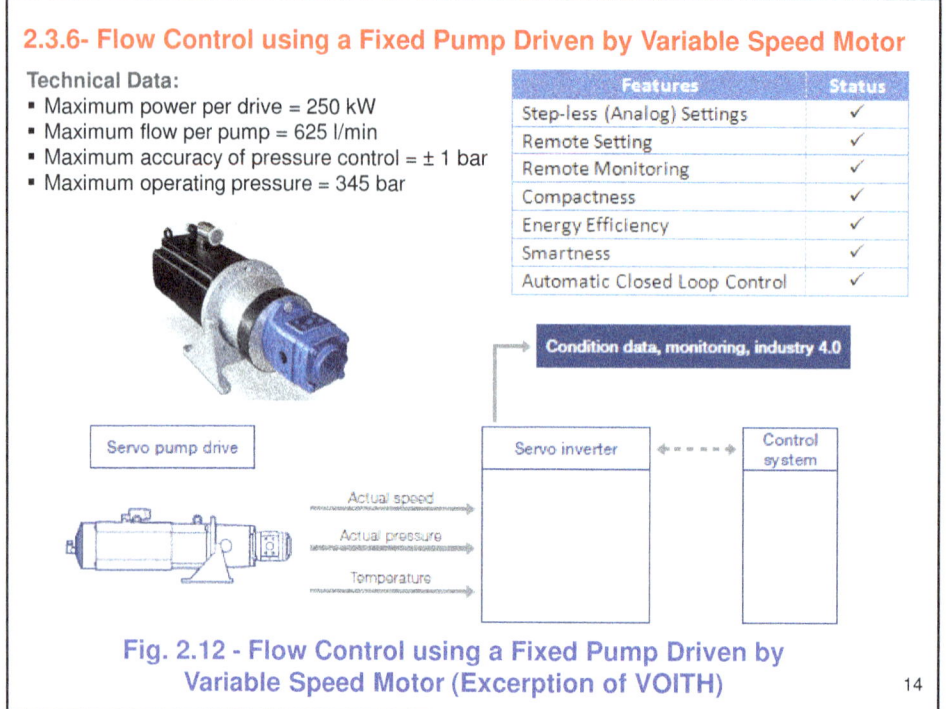

Fig. 2.12 - Flow Control using a Fixed Pump Driven by Variable Speed Motor (Excerption of VOITH)

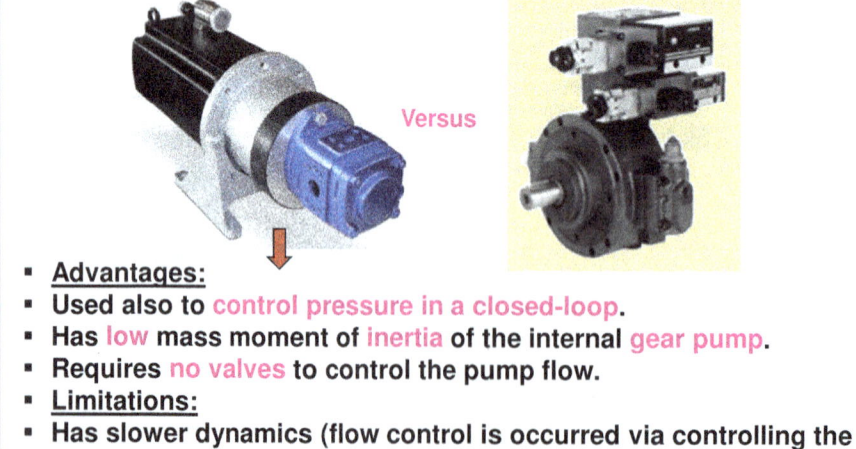

- **Advantages:**
- Used also to control pressure in a closed-loop.
- Has low mass moment of inertia of the internal gear pump.
- Requires no valves to control the pump flow.
- **Limitations:**
- Has slower dynamics (flow control is occurred via controlling the motor speed, which has slower dynamics than the variable displacement pump controller).
- Results in inefficient pump at low speed. There is a speed at which the pump outlet flow is equal to its internal leakage. If this happens at high pressure, the pump will overheat in few minutes. Therefore, this system should not drive the pump below a certain specified minimum speed.

2.4- Power Control Solutions

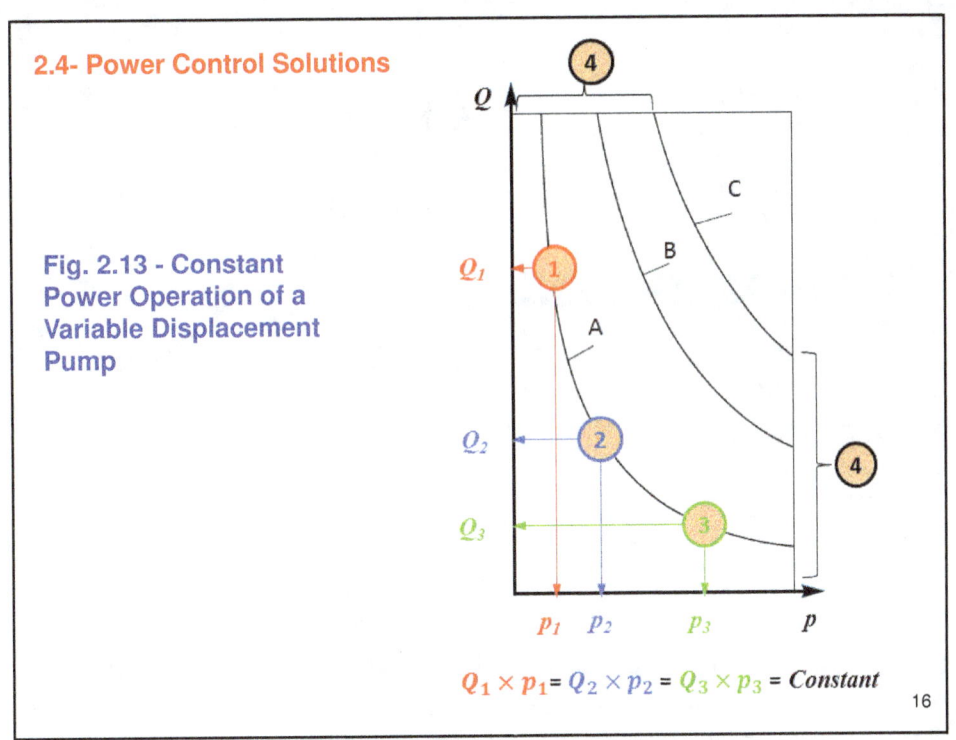

Fig. 2.13 - Constant Power Operation of a Variable Displacement Pump

$$Q_1 \times p_1 = Q_2 \times p_2 = Q_3 \times p_3 = \text{Constant}$$

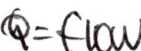
Q = flow

2.4.1- Hydro-Mechanical Power Controlled Pumps

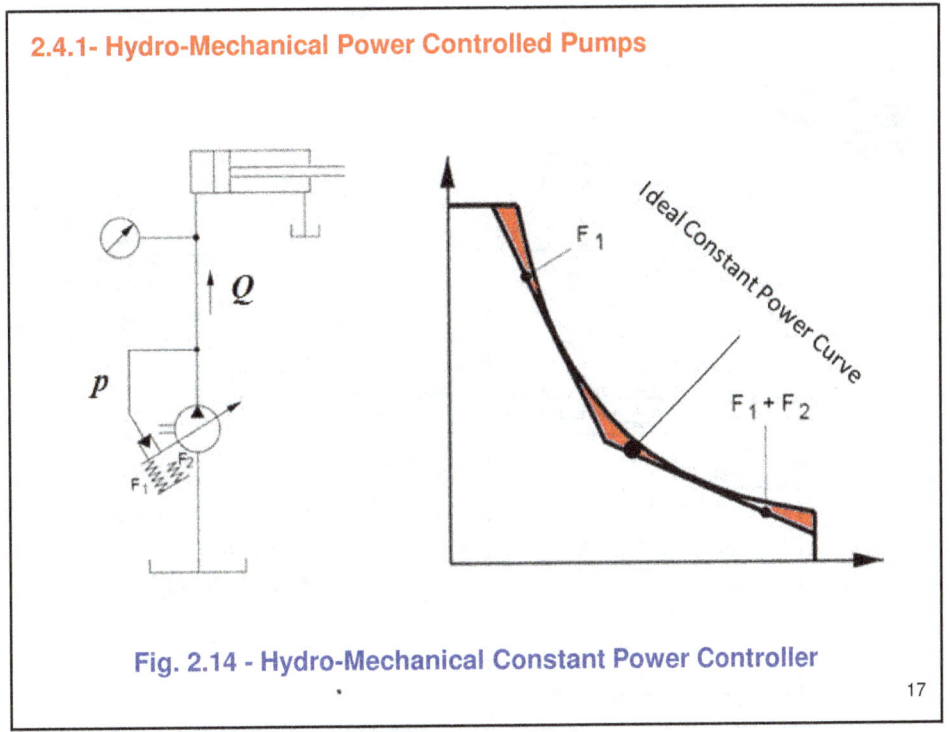

Fig. 2.14 - Hydro-Mechanical Constant Power Controller

- Springs are a linear element

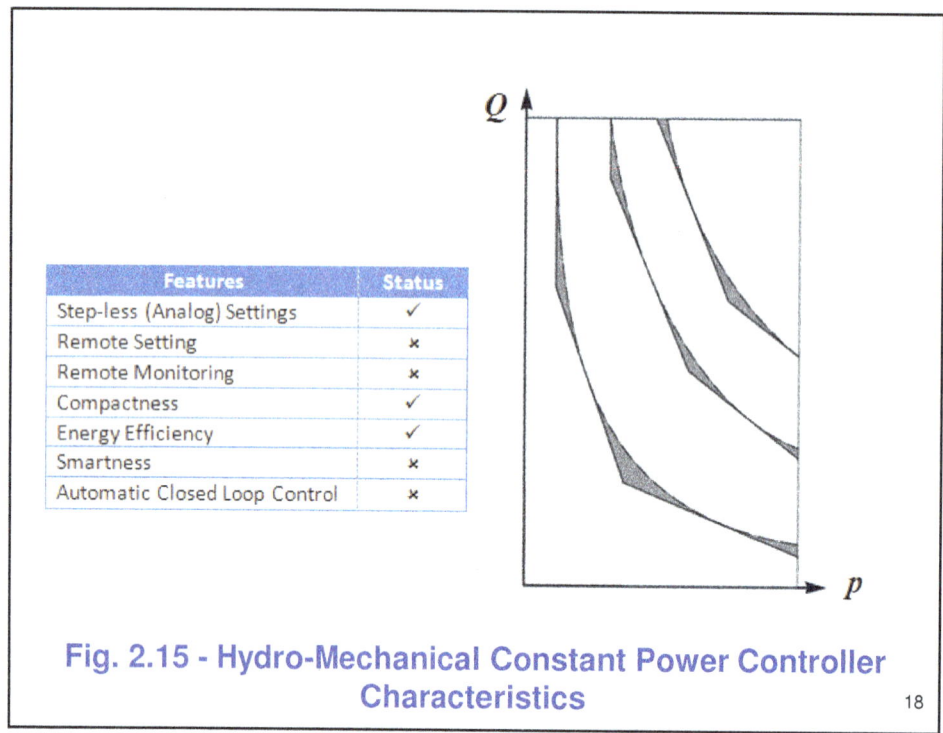

Fig. 2.15 - Hydro-Mechanical Constant Power Controller Characteristics

2.4.2 - Electro-Hydraulic Power Controlled Pumps

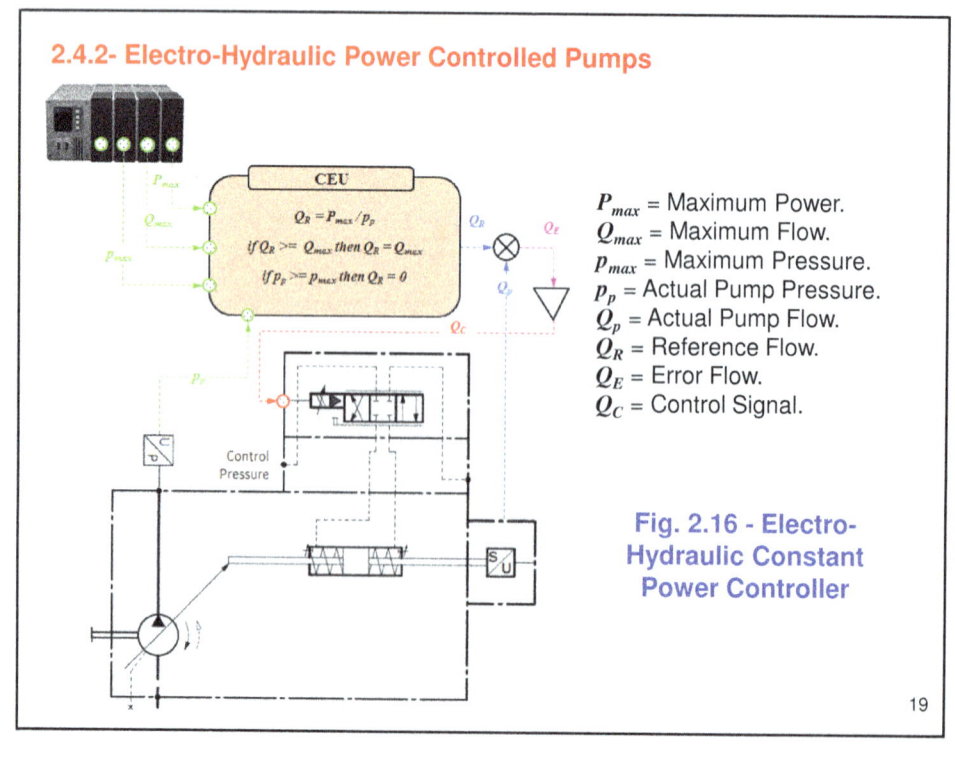

P_{max} = Maximum Power.
Q_{max} = Maximum Flow.
p_{max} = Maximum Pressure.
p_p = Actual Pump Pressure.
Q_p = Actual Pump Flow.
Q_R = Reference Flow.
Q_E = Error Flow.
Q_C = Control Signal.

Fig. 2.16 - Electro-Hydraulic Constant Power Controller

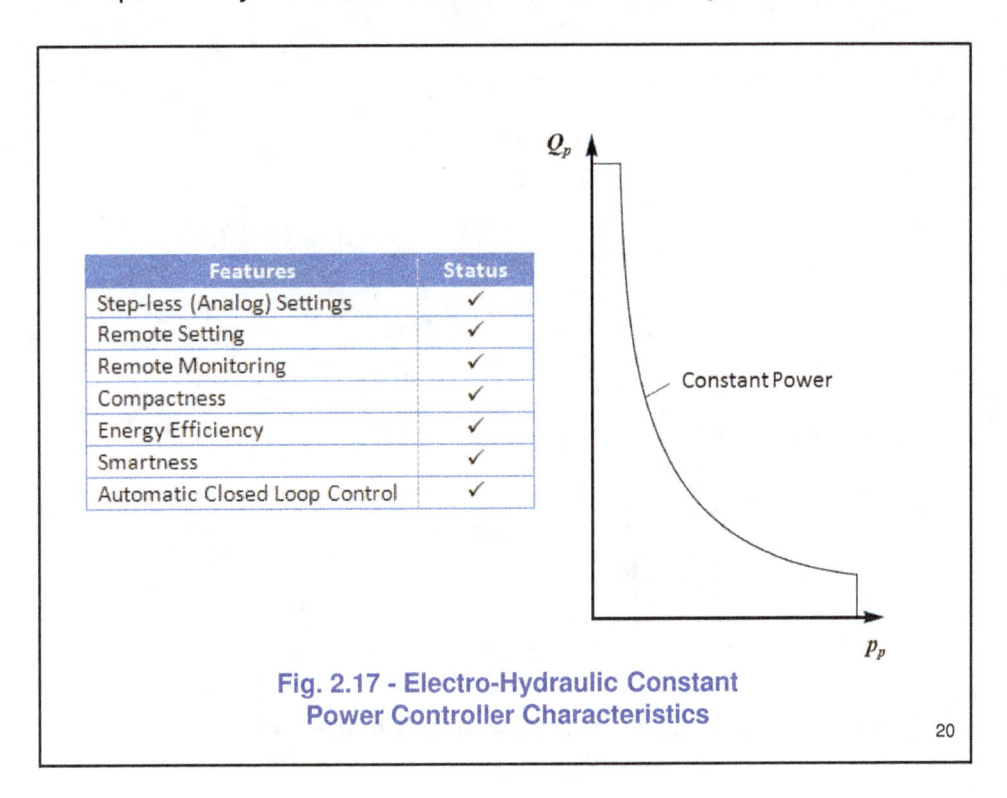

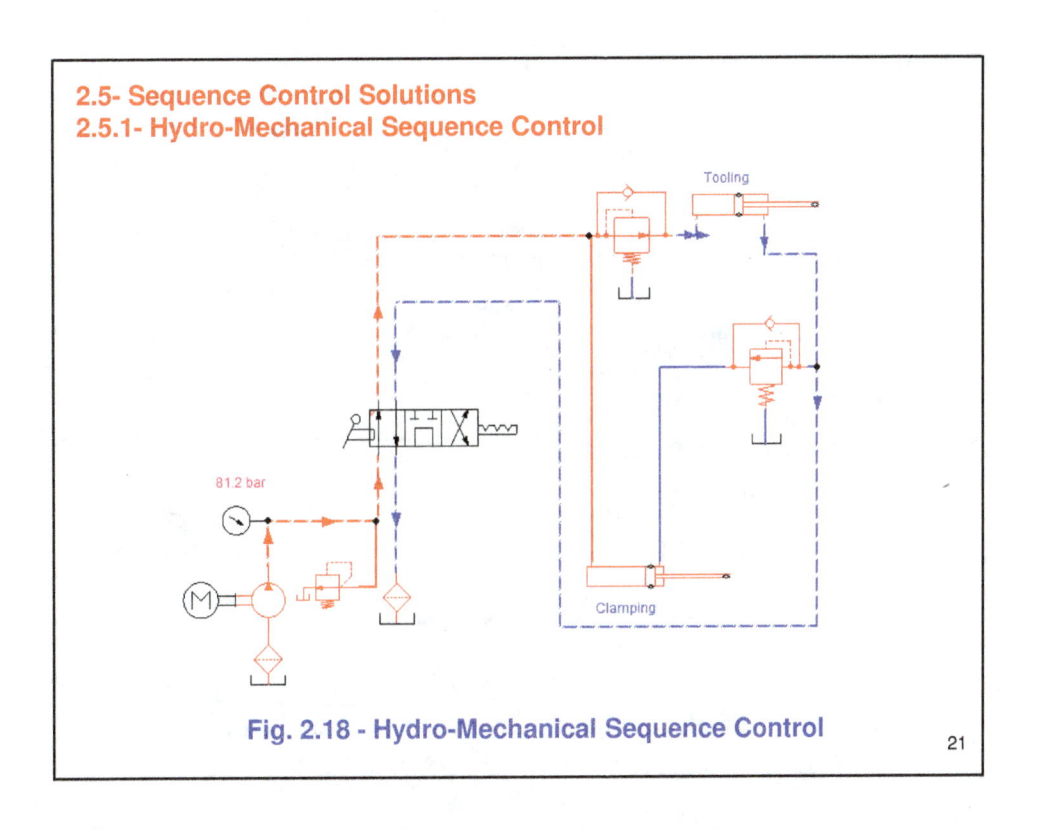

Features	Status
Step-less (Analog) Settings	✓
Remote Setting	✓
Remote Monitoring	✓
Compactness	✓
Energy Efficiency	✓
Smartness	✓
Automatic Closed Loop Control	✓

Fig. 2.17 - Electro-Hydraulic Constant Power Controller Characteristics

2.5- Sequence Control Solutions
2.5.1- Hydro-Mechanical Sequence Control

Fig. 2.18 - Hydro-Mechanical Sequence Control

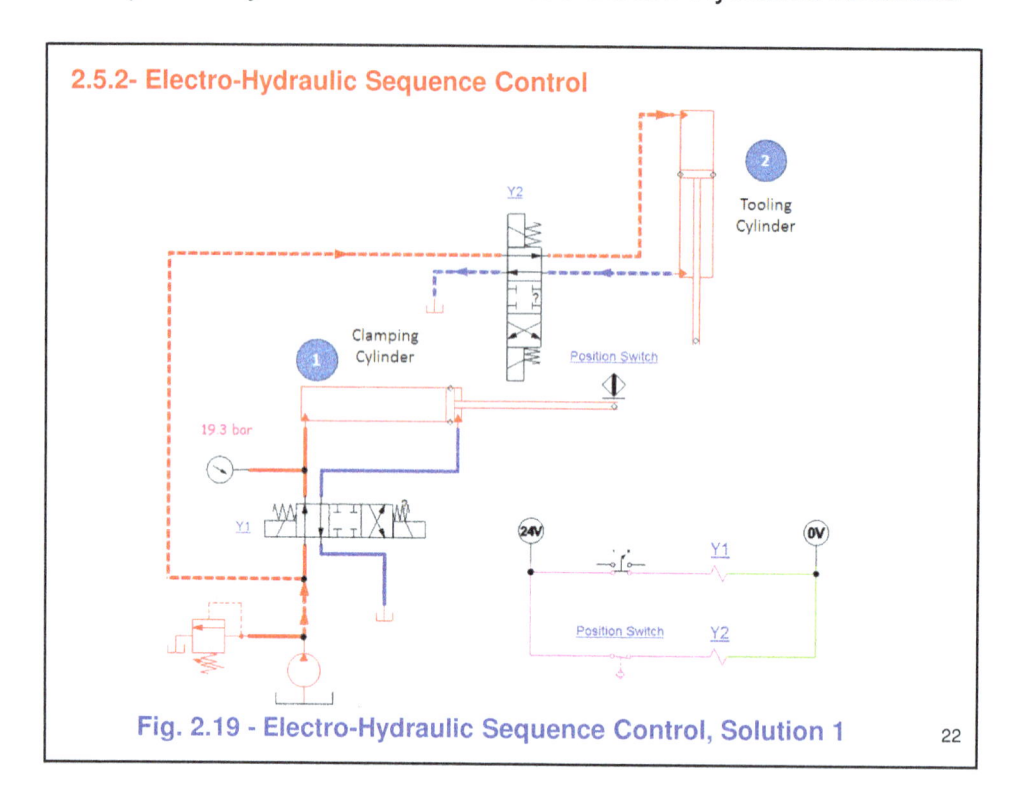

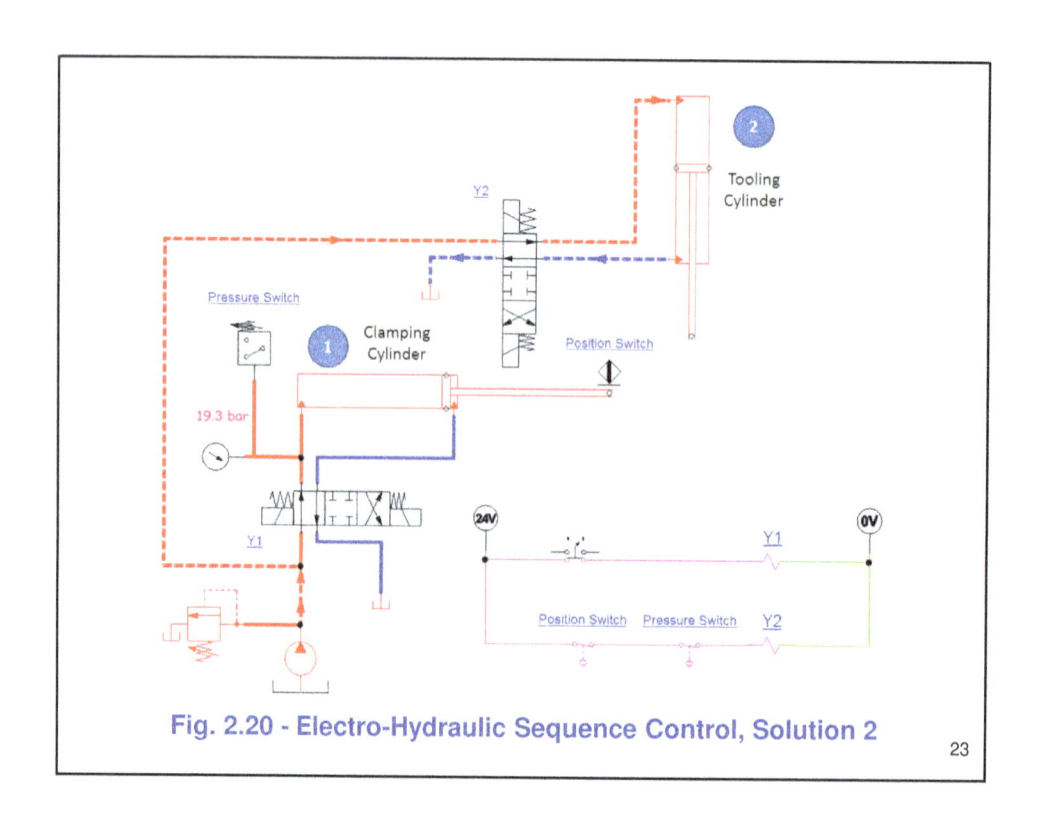

2.5.2 - Electro-Hydraulic Sequence Control

Fig. 2.19 - Electro-Hydraulic Sequence Control, Solution 1

Fig. 2.20 - Electro-Hydraulic Sequence Control, Solution 2

2.6- Hydraulic Deceleration System
2.61- Hydro-Mechanical Deceleration System

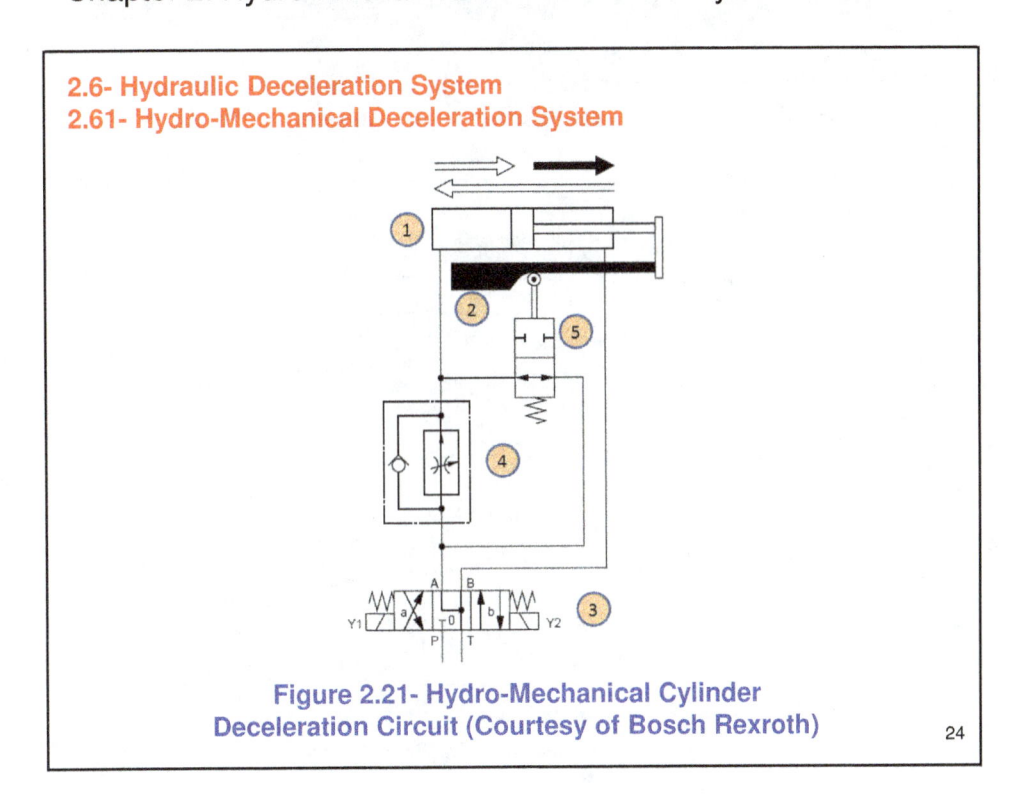

Figure 2.21- Hydro-Mechanical Cylinder Deceleration Circuit (Courtesy of Bosch Rexroth)

2.6.2- Electro-Hydraulic Deceleration System

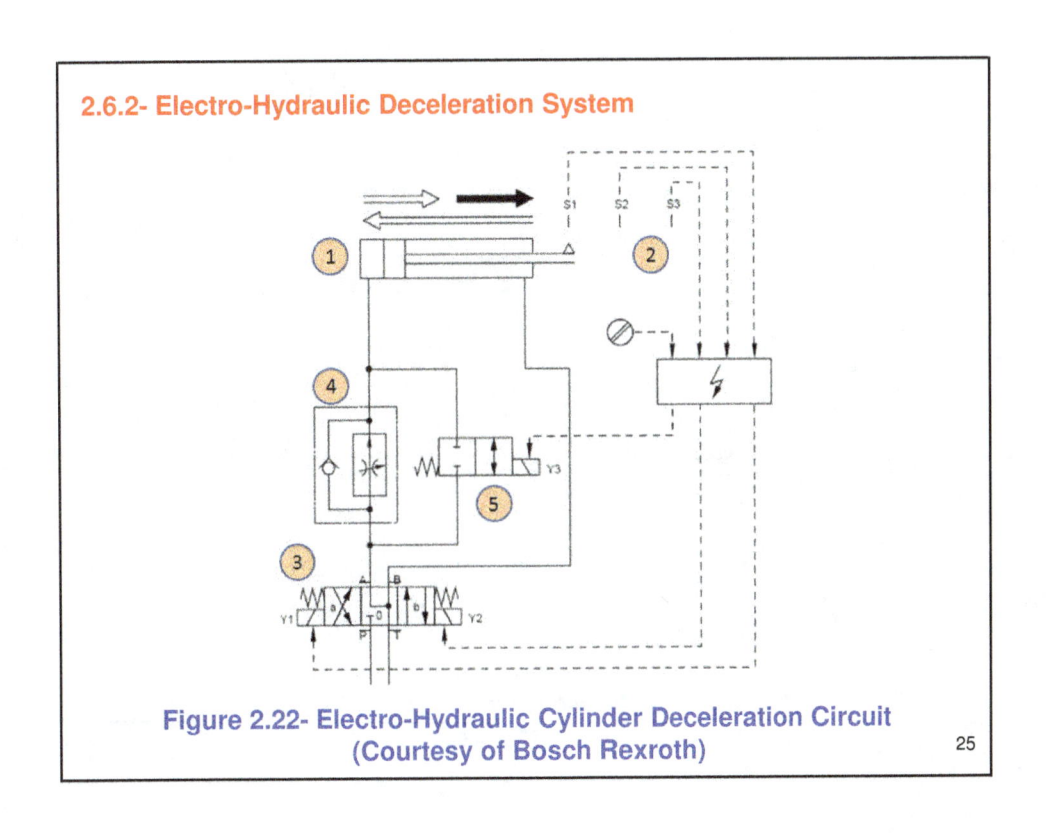

Figure 2.22- Electro-Hydraulic Cylinder Deceleration Circuit (Courtesy of Bosch Rexroth)

Workbook: Electro-Hydraulic Components and Systems
Chapter 2: Hydro-Mechanical versus Electro-Hydraulic Solutions

2.7- Cylinder Speed Synchronization
2.7.1- Hydro-Mechanical Cylinder Speed Synchronization

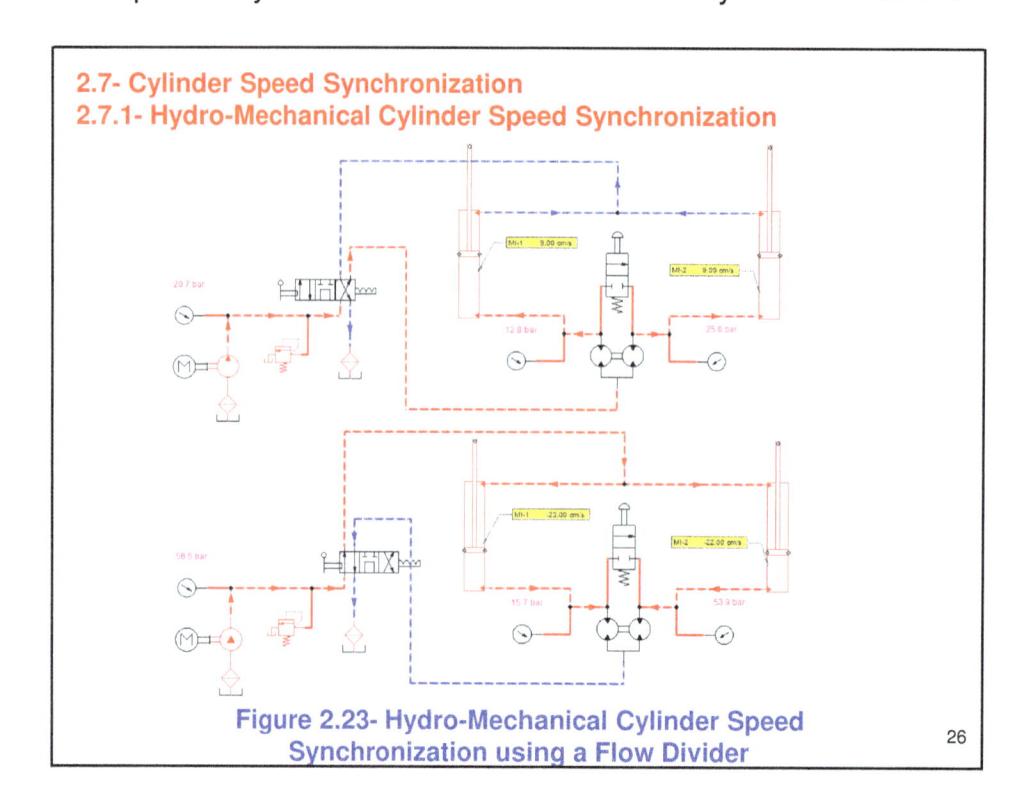

Figure 2.23- Hydro-Mechanical Cylinder Speed Synchronization using a Flow Divider

2.7.2- Electro-Hydraulic Cylinder Speed Synchronization

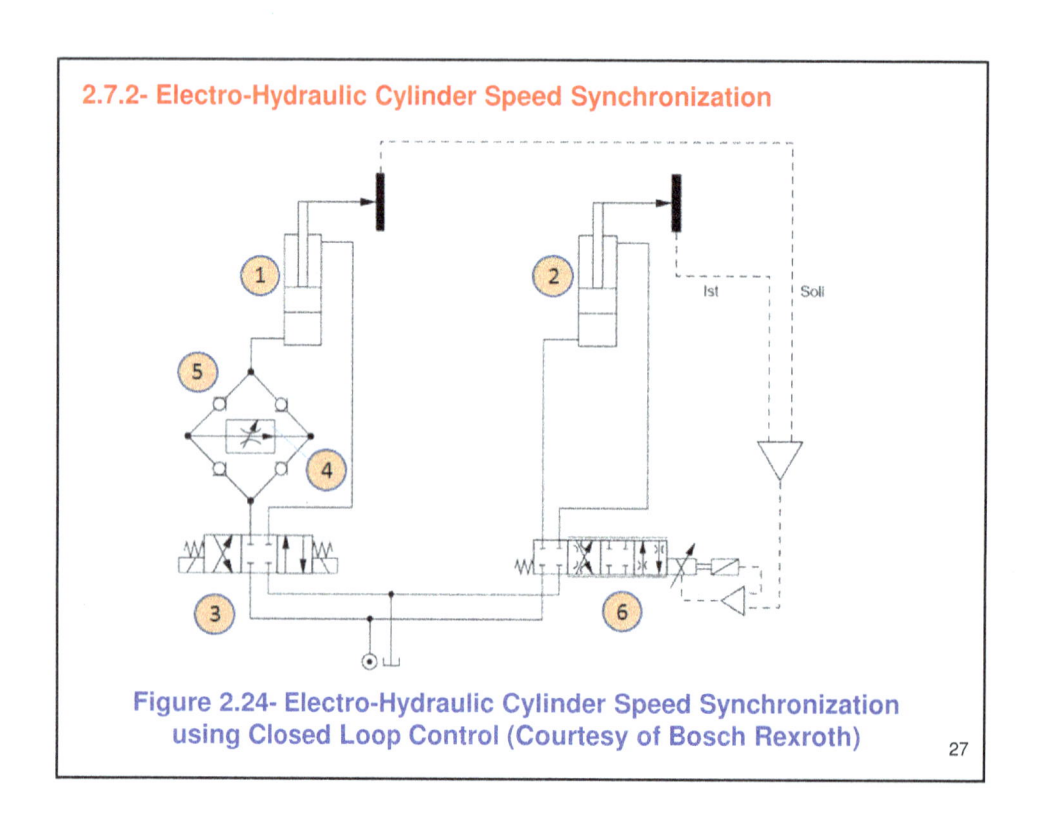

Figure 2.24- Electro-Hydraulic Cylinder Speed Synchronization using Closed Loop Control (Courtesy of Bosch Rexroth)

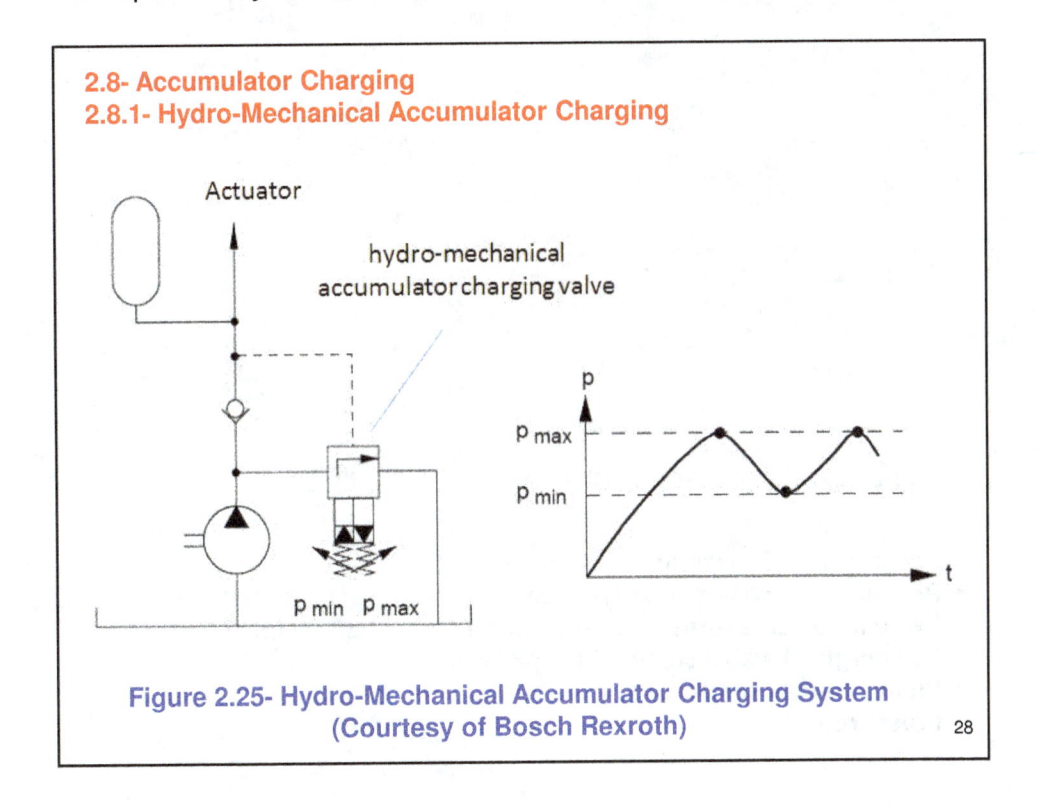

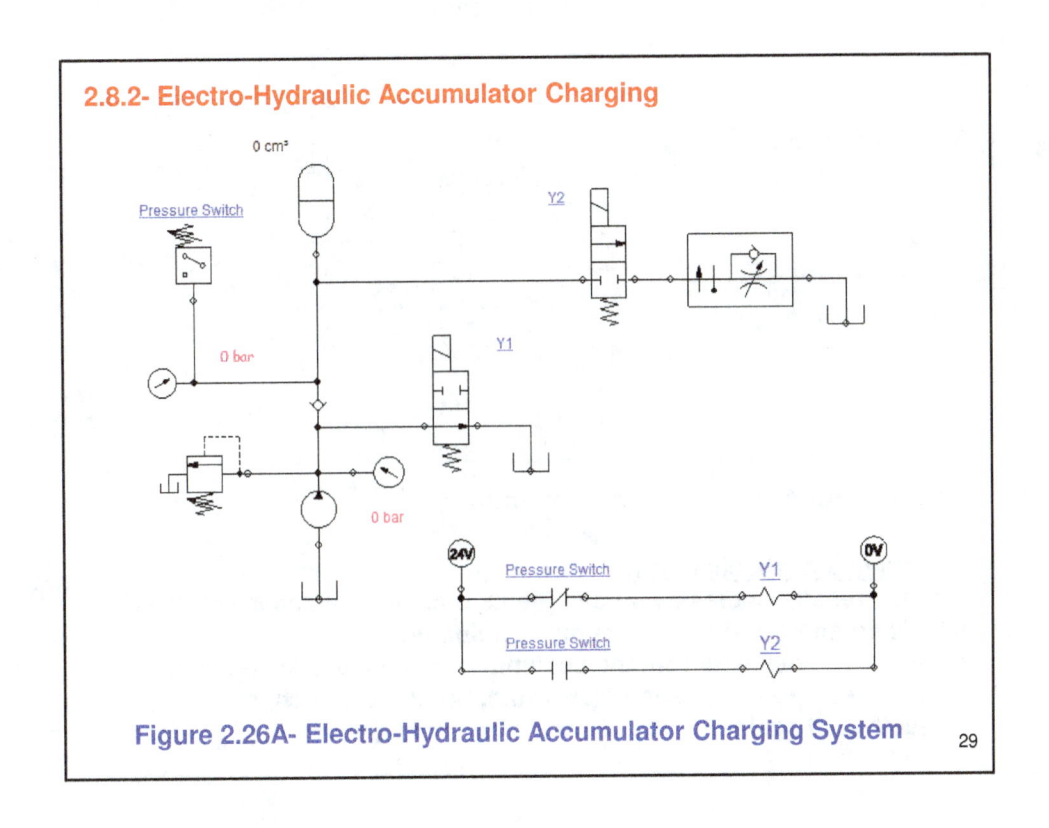

Workbook: Electro-Hydraulic Components and Systems
Chapter 2: Hydro-Mechanical versus Electro-Hydraulic Solutions

2.8- Accumulator Charging
2.8.1- Hydro-Mechanical Accumulator Charging

Figure 2.25- Hydro-Mechanical Accumulator Charging System (Courtesy of Bosch Rexroth)

2.8.2- Electro-Hydraulic Accumulator Charging

Figure 2.26A- Electro-Hydraulic Accumulator Charging System

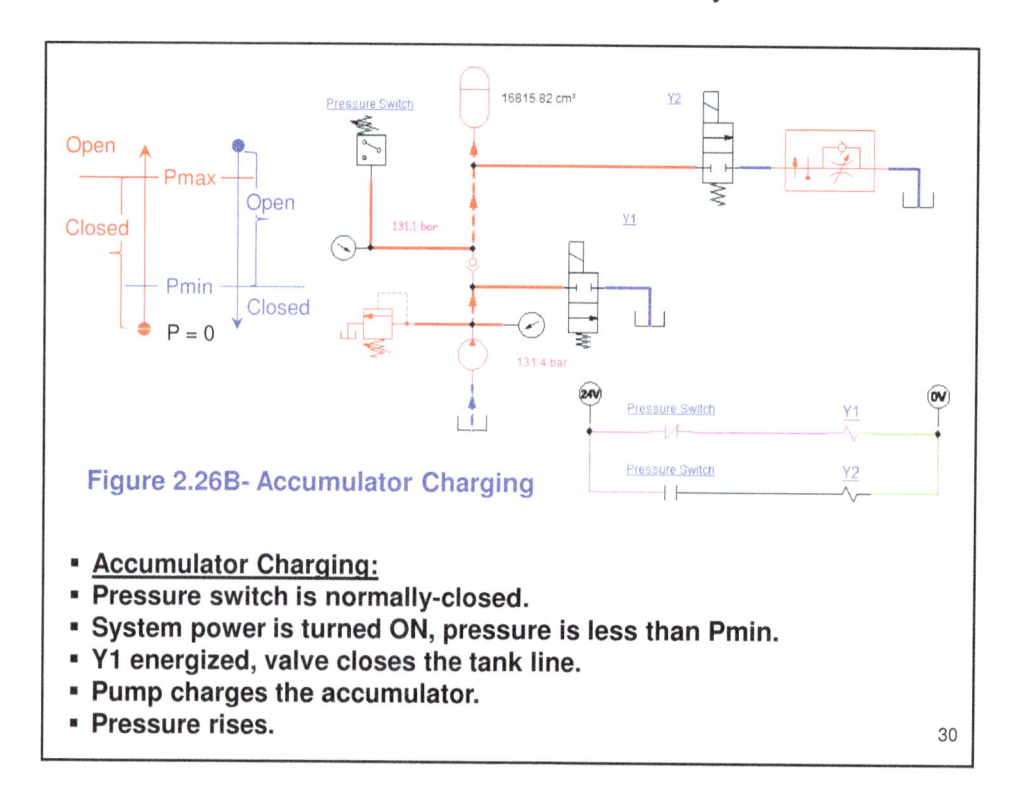

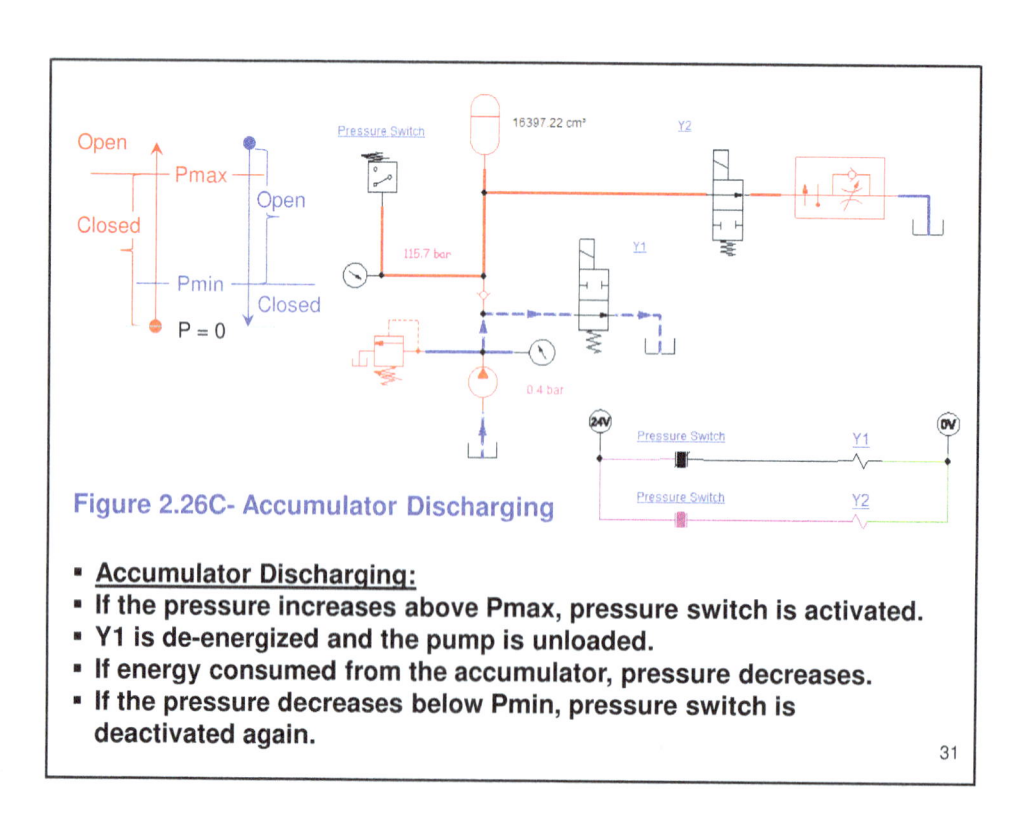

Workbook: Electro-Hydraulic Components and Systems
Chapter 2: Hydro-Mechanical versus Electro-Hydraulic Solutions

Figure 2.26B- Accumulator Charging

- **Accumulator Charging:**
- Pressure switch is normally-closed.
- System power is turned ON, pressure is less than Pmin.
- Y1 energized, valve closes the tank line.
- Pump charges the accumulator.
- Pressure rises.

Figure 2.26C- Accumulator Discharging

- **Accumulator Discharging:**
- If the pressure increases above Pmax, pressure switch is activated.
- Y1 is de-energized and the pump is unloaded.
- If energy consumed from the accumulator, pressure decreases.
- If the pressure decreases below Pmin, pressure switch is deactivated again.

2.9- Control of Overrunning Loads
2.9.1- Hydro-Mechanical Overrunning Load Control

- Prevents overrunning the load.
- Gives a chance for the pump to make up the oil in the piston side of the cylinder.
- Avoids creating cavitation in the piston side of the cylinder.
- Controls the lowering speed of the load.
- If the pump fails, the load will be suspended with creep-free conditions (pilot operated check valve only)

Fig. 2.27 - Hydro-Mechanical Overrunning Load Control

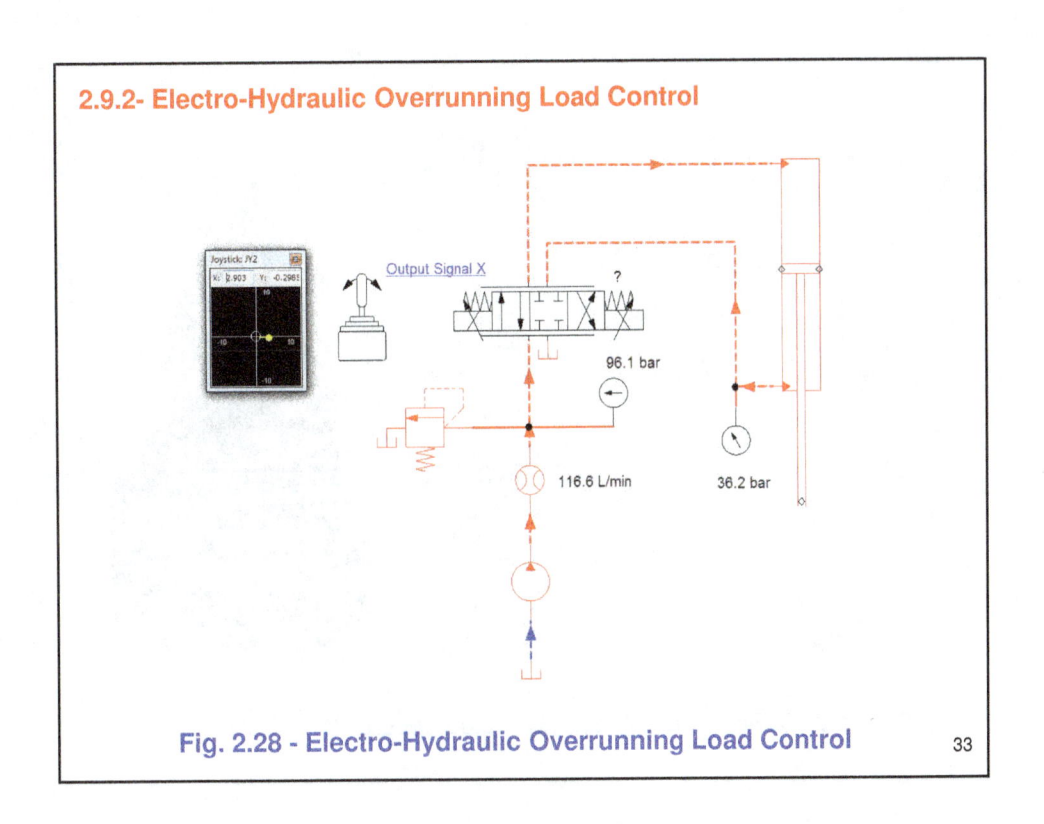

2.9.2- Electro-Hydraulic Overrunning Load Control

Fig. 2.28 - Electro-Hydraulic Overrunning Load Control

Workbook: Electro-Hydraulic Components and Systems
Chapter 2: Hydro-Mechanical versus Electro-Hydraulic Solutions

2.10- Control Block Actuation in Mobile Applications
2.10.1- Hydro-Mechanical Control Block Actuation

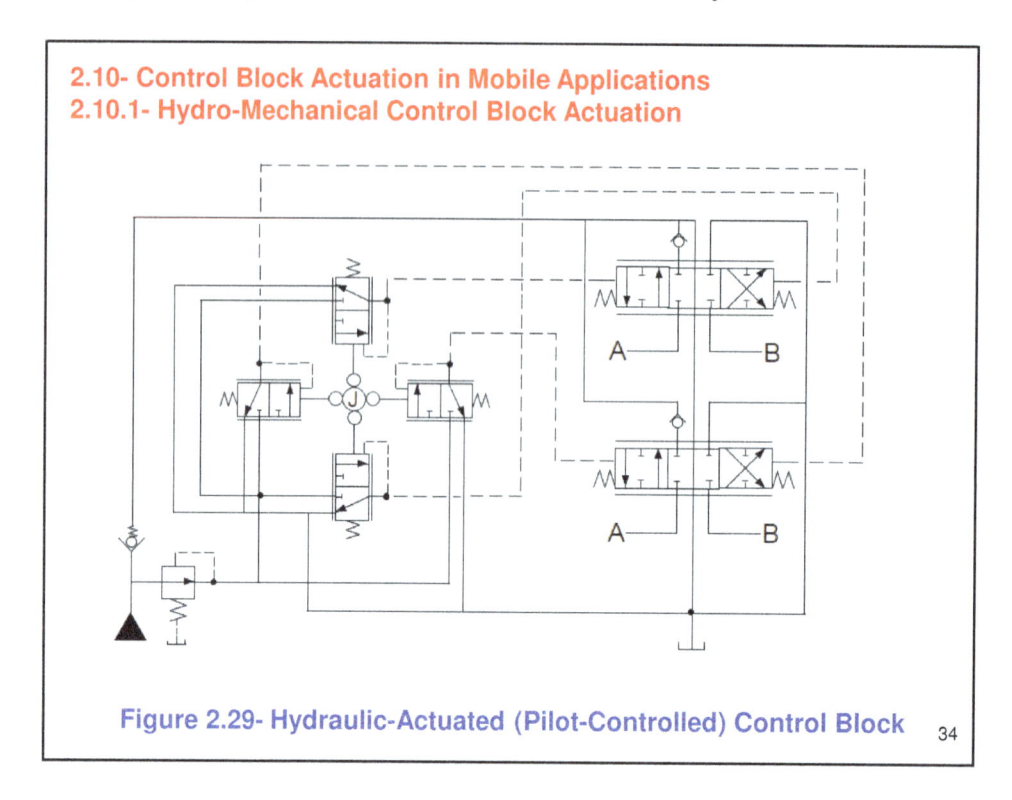

Figure 2.29- Hydraulic-Actuated (Pilot-Controlled) Control Block

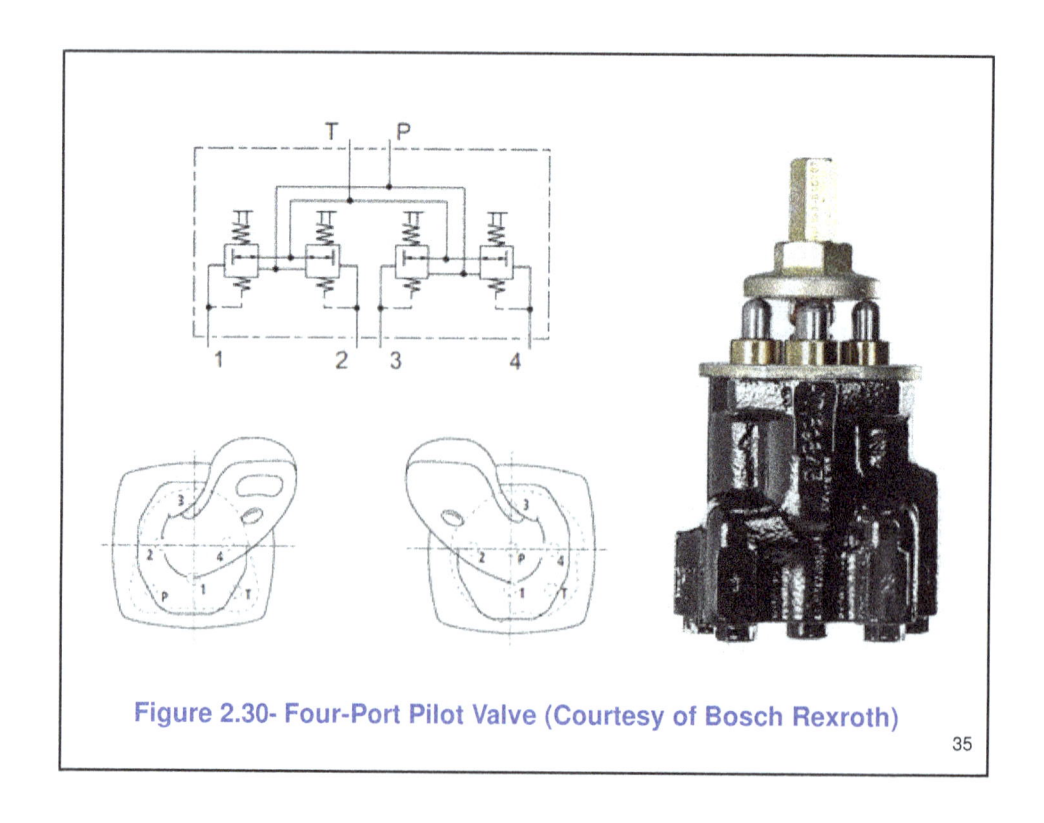

Figure 2.30- Four-Port Pilot Valve (Courtesy of Bosch Rexroth)

2.10.2- Electro-Hydraulic Control Block Actuation

- The most **space saving** and flexibility in laying out the system.
- **Less energy consumption** as compared to hydraulic-operated systems.
- Requires **less labor in assembling** the system.
- Upgradable to features of **artificial intelligence**.
- **High maintenance cost** due to the lack of know-how.

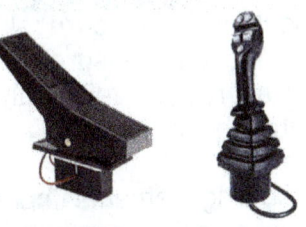

Figure 2.31- Electric-Actuated Control Block (Courtesy of Bosch Rexroth)

Chapter 2 Reviews

1. Converting a hydro-mechanical system to an electro-hydraulic system will result in**?**
    A. Ability to control the system remotely.
    B. Reduce the number of components in the system.
    C. Ability to control the system automatically and digitally.
    D. All of the above.

2. Which of the following elements can be used to do remote control of the maximum system pressure?
    A. Hydro-mechanical pressure relief valve.
    B. Hydro-mechanical pressure compensated pump.
    C. Proportional pressure relief valve.
    D. None of the above.

3. Which of the following elements improves the system efficiency?
    A. Hydro-mechanical pressure relief valve.
    B. Hydro-mechanical pressure compensated pump.
    C. Proportional pressure relief valve.
    D. None of the above.

4. An electro-hydraulic pressure compensated pump has the following features?
    A. Can be used to set the system maximum pressure remotely and step-less.
    B. Help build compact and energy efficiency system.
    C. Can be programmed and be used in a closed loop control system.
    D. All of the above.

5. Which of the following elements can be used to do remote control of flow?
    A. Hydro-mechanical flow control valve.
    B. Hydro-mechanical displacement controlled pump.
    C. Proportional flow control valve.
    D. None of the above.

6. Which of the following elements improves the system efficiency?
    A. Hydro-mechanical flow control valve.
    B. Hydro-mechanical displacement controlled pump.
    C. Proportional flow control valve.
    D. None of the above.

7. An electro-hydraulic displacement controlled pump has the following features?
    A. Can be used to set the system flow remotely and step-less.
    B. Help build compact and energy efficiency system.
    C. Can be programmed and be used in a closed loop control system.
    D. All of the above.

8. The curve shown below is a result of using?

   A. An electro-hydraulic pressure compensated pump.
   B. An electro-hydraulic displacement controlled pump.
   C. An electro-hydraulic constant power controlled pump.
   D. None of the above.

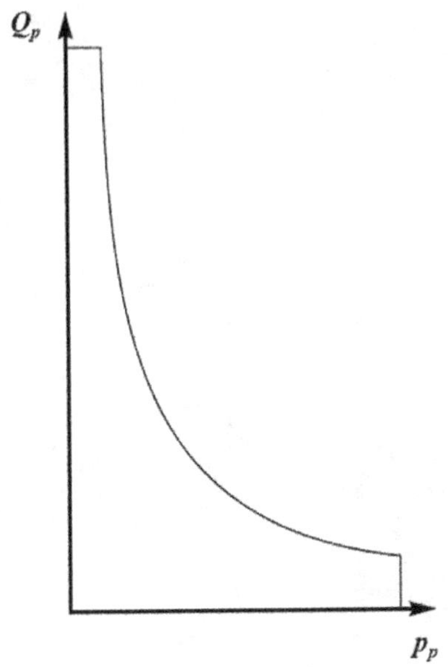

9. Use of an electro-hydraulic controlled pump satisfies many good operational features in a hydraulic system, in your understanding, what are the main limitations on using such pumps?

   A. Higher cost, lack of know-how, and requirements for electronic interface.
   B. The pump is large in size.
   C. The pump is less efficient.
   D. None of the above.

10. How many different speeds does the system shown below have to drive the hydraulic motor?
    A. 3.
    B. 5.
    C. 7.
    D. 9.

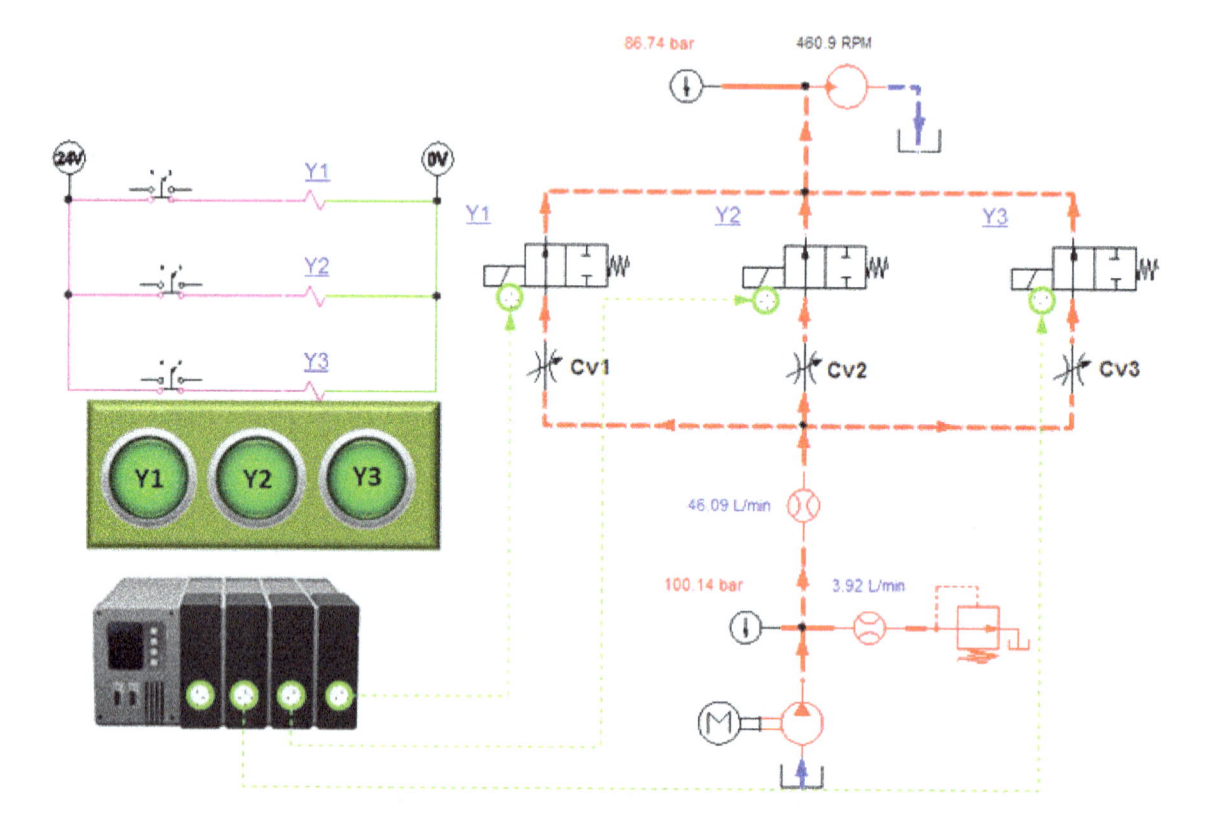

Workbook: Electro-Hydraulic Components and Systems
Chapter 2: Reviews and Assignments

Chapter 2 Assignment

Student Name: --- Student ID: -------------------

Date: --- Score: ------------------------

In the system shown below, propose two alternatives that can improve the energy efficiency of the system.

Features	Status
Step-less (Analog) Settings	✓
Remote Setting	✗
Remote Monitoring	✗
Compactness	✓
Energy Efficiency	✗
Smartness	✗
Automatic Closed Loop Control	✗

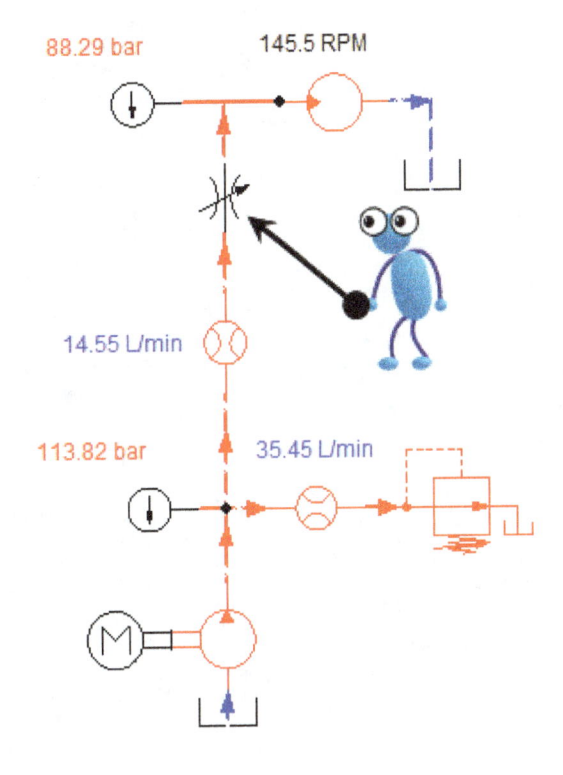

88.29 bar   145.5 RPM

14.55 L/min

113.82 bar   35.45 L/min

Chapter 3
Switching Valves Construction and Operation

Objectives:

This chapter covers the principle of operation and the construction of various types of switching (ON/OFF) solenoids that are used to actuate hydraulic valves. The chapter discusses, qualitatively, the pros and cons of wet type versus dry type switching solenoids and DC type versus AC type switching solenoids. The chapter also discusses the undesirable effects of using AC switching solenoids such as AC hum, eddy current, and inrush current and constructional considerations to minimize their effects. The chapter concludes by presenting examples of using switching solenoids in directional, pressure, and control functions.

Brief Contents:

3.1- Basic Electro-Magnetic Concepts

3.2- Switching Solenoid

3.3- Switching Solenoids for Electro-Hydraulic Valves

3.4- Electro-Hydraulic Switching Directional Control Valves

3.5- Electro-Hydraulic Switching Pressure Control Valves

3.6- Electro-Hydraulic Switching Flow Control Valves

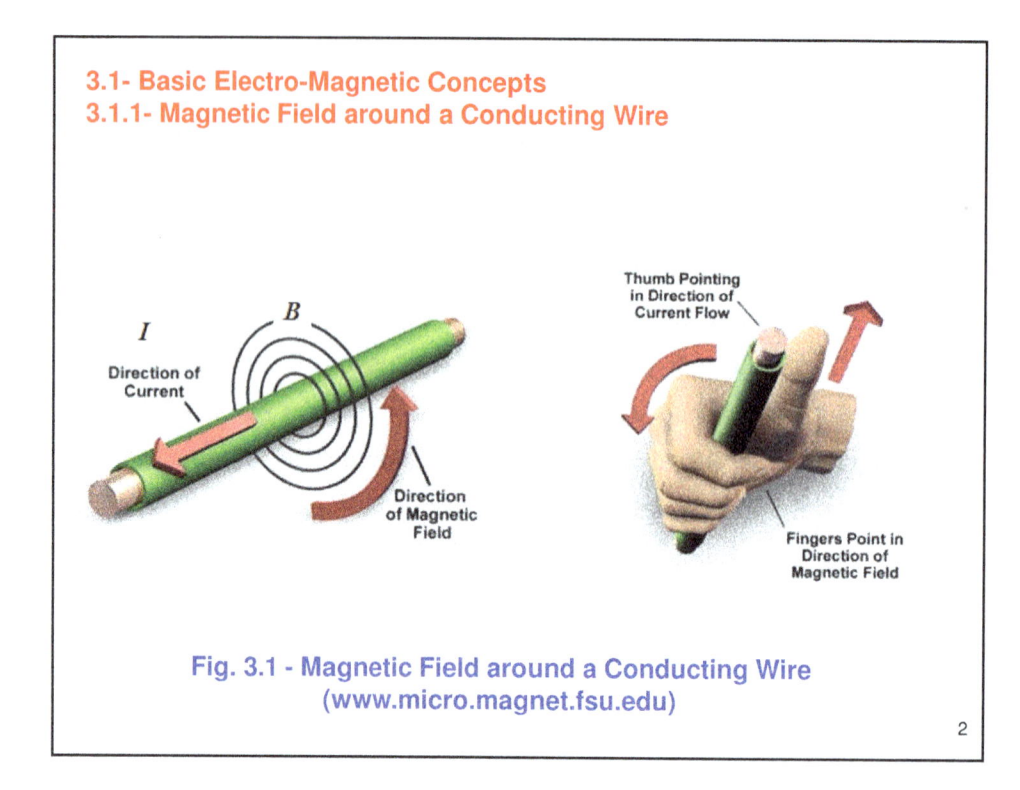

Workbook: Electro-Hydraulic Components and Systems
Chapter 3: Switching Valves Construction and Operation

3.1- Basic Electro-Magnetic Concepts
3.1.1- Magnetic Field around a Conducting Wire

Fig. 3.1 - Magnetic Field around a Conducting Wire
(www.micro.magnet.fsu.edu)

3.1.2- Magnetic Field Measuring Units

$$T\left(\frac{N}{Am}\right) = \frac{Force\ (Newton)}{Current\ (Amplere) \times Length\ of\ Wire\ (Meter)} \qquad 3.1$$

Tesla (T) or **Gauss** (G) where $1T = 10^4\ G$

Magnetic field is typically measured in Gauss because the unit of a Tesla is very large.

To illustrate how large the Tesla is; the magnetic field of the Earth is 0.6 G or 0.00006T.

3.1.3- Magnetic Field Intensity

Fig. 3.2 - Magnetic Field Intensity (Courtesy of Hydraforce)

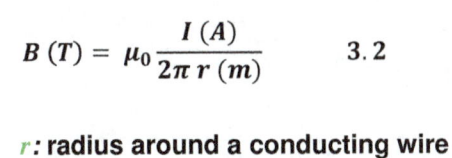

$$B(T) = \mu_0 \frac{I(A)}{2\pi\, r(m)} \qquad 3.2$$

r: radius around a conducting wire
I: current
μ_0: the permeability of the medium that surrounds the wire.

Example:

If μ_0 for air = $4\pi \times 10^{-7}$ N/A², I = 1A, and r = 2.54 cm (1 inch), then:

$B = (4\pi \times 10^{-7})(1) / (2\pi)(0.0254 m) = 0.00000394\, T = 0.0394\, G$

3.1.4- Magnetic Field around a Coil

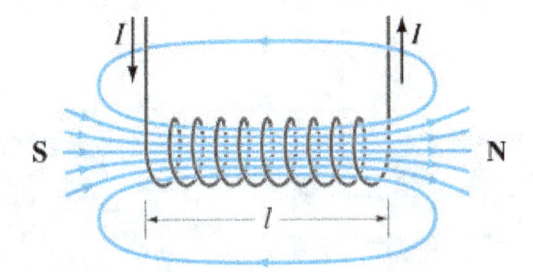

Fig. 3.3 - Magnetic Field around a Coil (www.miniphysics.com)

$$B(T) = \frac{\mu_0(nI)}{l} \qquad 3.3$$

Where: n (number of turns in the coil) and l (length)

Example: If n = 1000, I = 2 A and l = 0.04 m, then

$B = (4\pi \times 10^{-7})(1000)(2) / 0.04 = 0.063\, T = 630\, G$

3.1.5 - Electro-Magnetic Force

- Iron Core ("Armature" or a "Plunger") must be made of a ferromagnetic material.

- Once the current passes through the coil, the armature will become magnetized.

- If the assembly is left unframed, it will be self oriented towards the geographical directions.

- The direction of magnetization depends on the polarity of the current.

- Right hand rule: the thumb points out to the magnetic North Pole that is also the direction of the resulted electromagnetic forces.

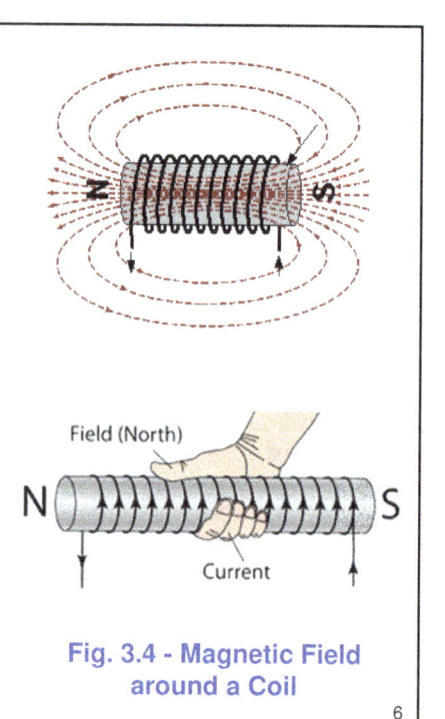

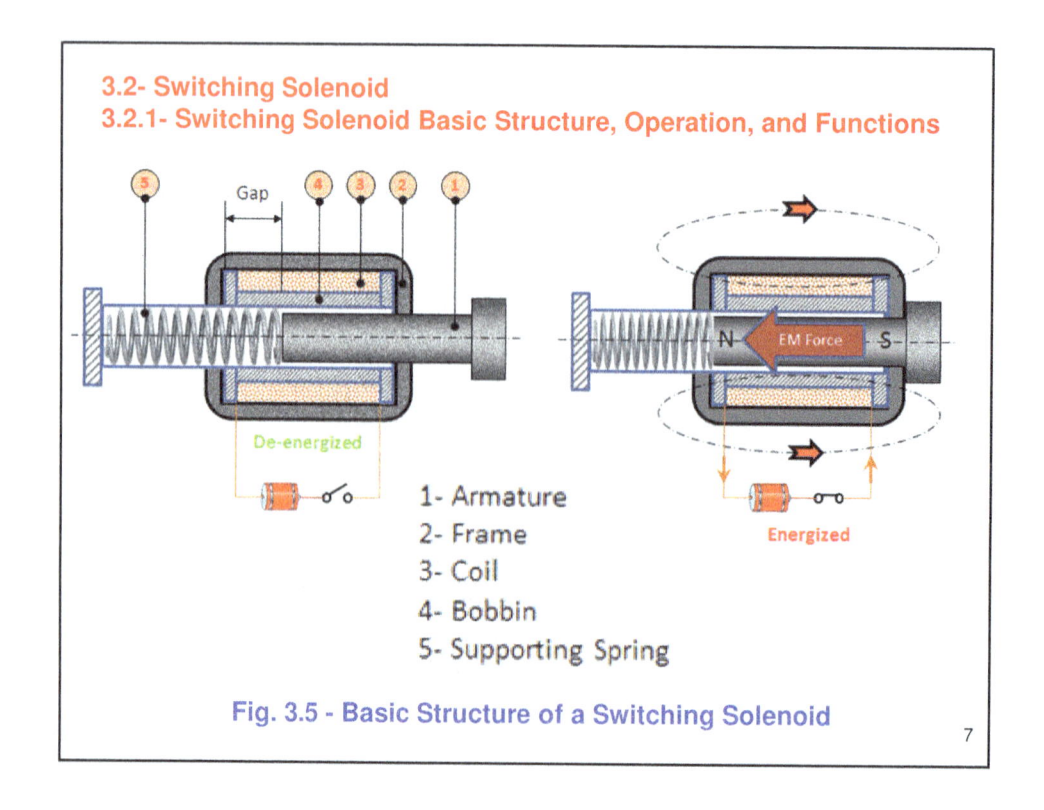

Fig. 3.4 - Magnetic Field around a Coil

3.2 - Switching Solenoid
3.2.1 - Switching Solenoid Basic Structure, Operation, and Functions

1- Armature
2- Frame
3- Coil
4- Bobbin
5- Supporting Spring

Fig. 3.5 - Basic Structure of a Switching Solenoid

Video 082 (0.5 min)

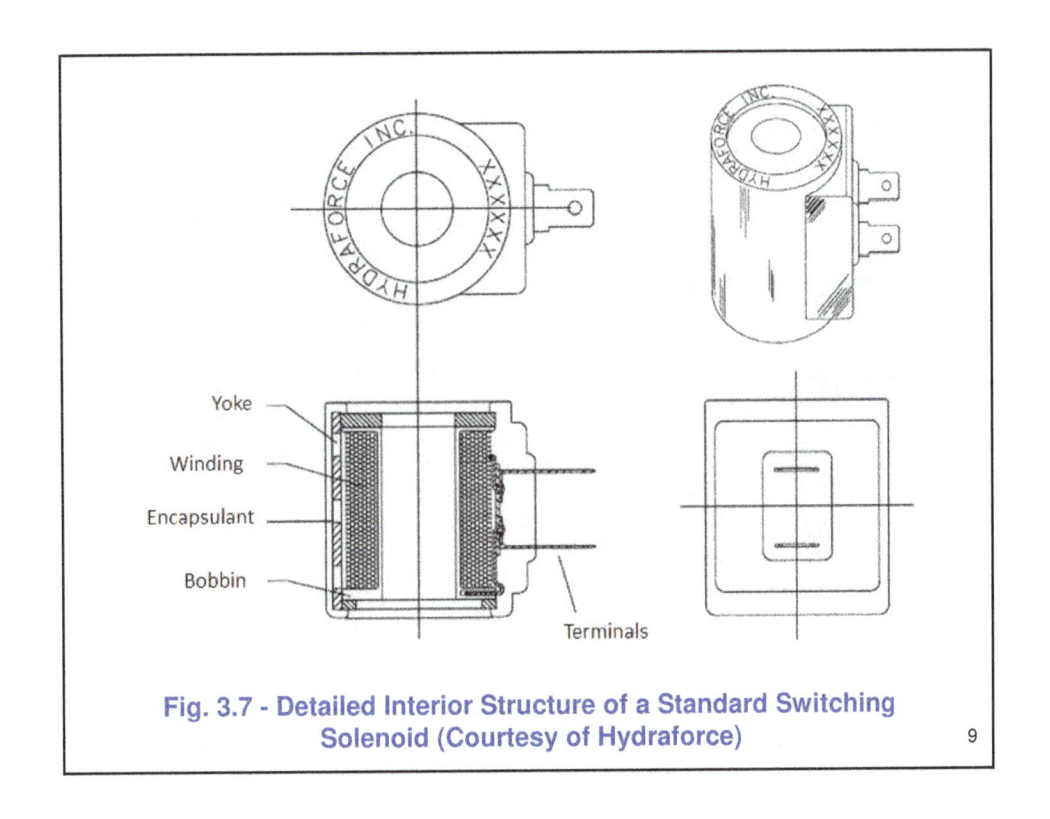

Fig. 3.6 - Various Typical Switching Solenoids

Fig. 3.7 - Detailed Interior Structure of a Standard Switching Solenoid (Courtesy of Hydraforce)

Workbook: Electro-Hydraulic Components and Systems
Chapter 3: Switching Valves Construction and Operation

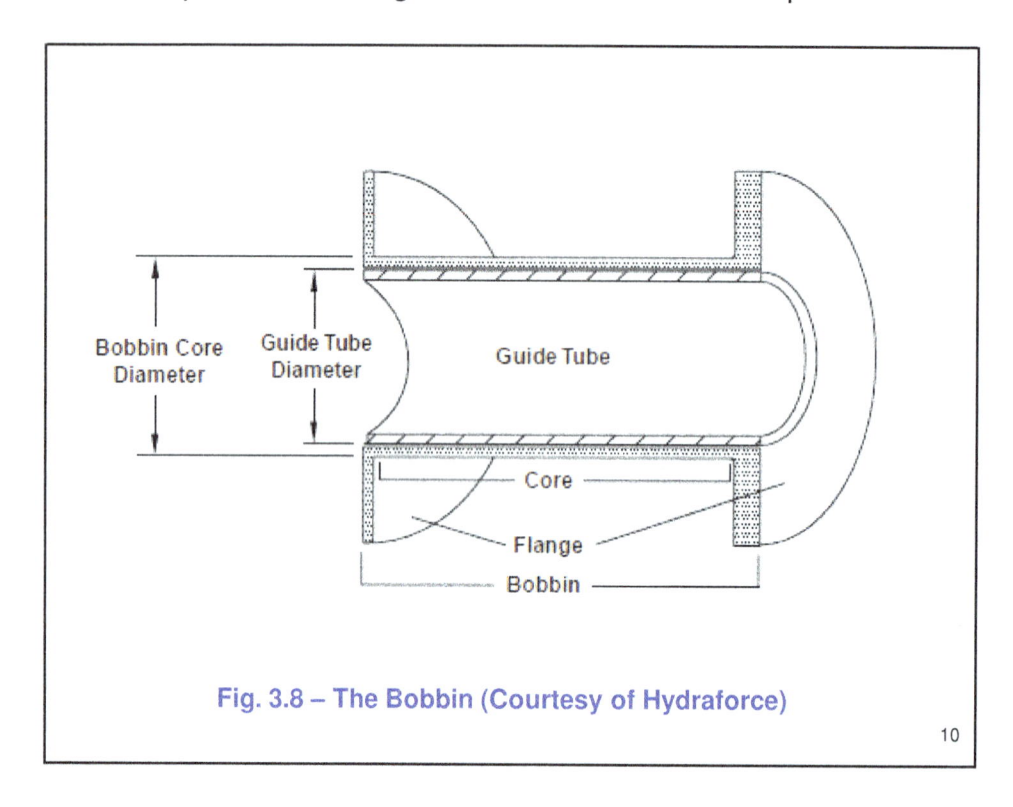

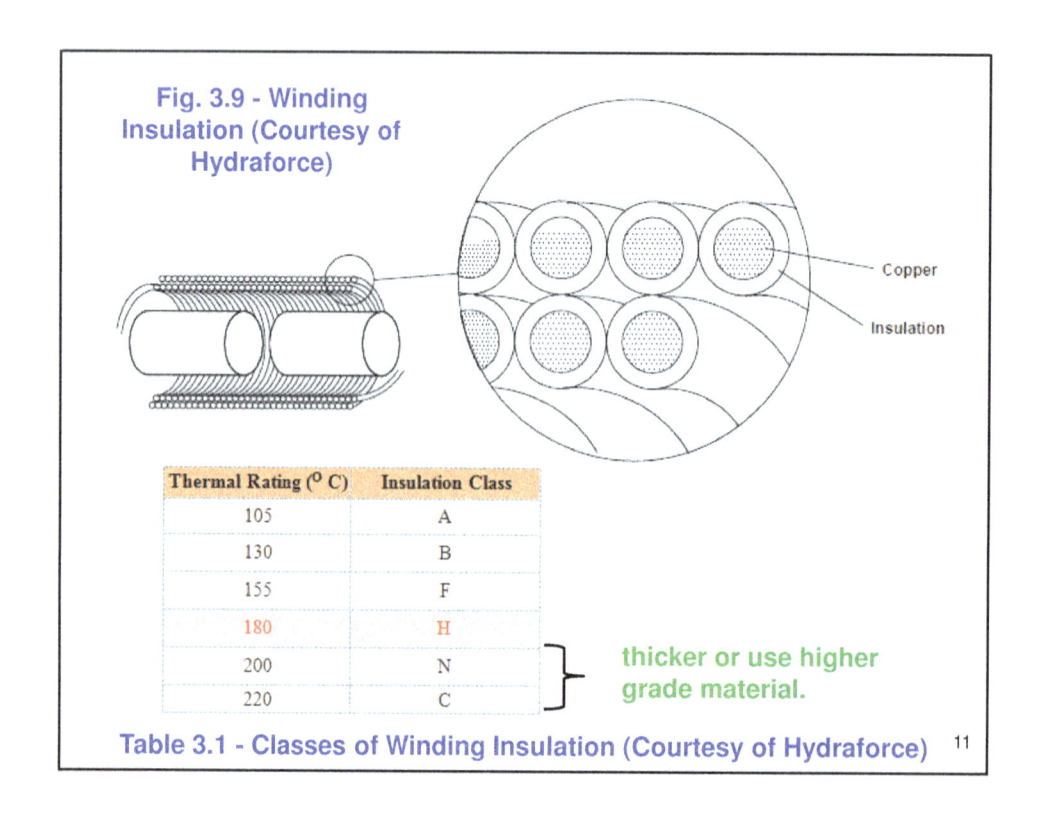

Fig. 3.8 – The Bobbin (Courtesy of Hydraforce)

Fig. 3.9 - Winding Insulation (Courtesy of Hydraforce)

Thermal Rating (° C)	Insulation Class
105	A
130	B
155	F
180	H
200	N
220	C

thicker or use higher grade material.

Table 3.1 - Classes of Winding Insulation (Courtesy of Hydraforce)

3.2.2 - Wiring Methods of a Coil

One Layer of the Solenoid

$$\frac{\text{Core Length}}{\text{Wire Diameter}} = \text{\# of Turns that Fit on a Layer}$$

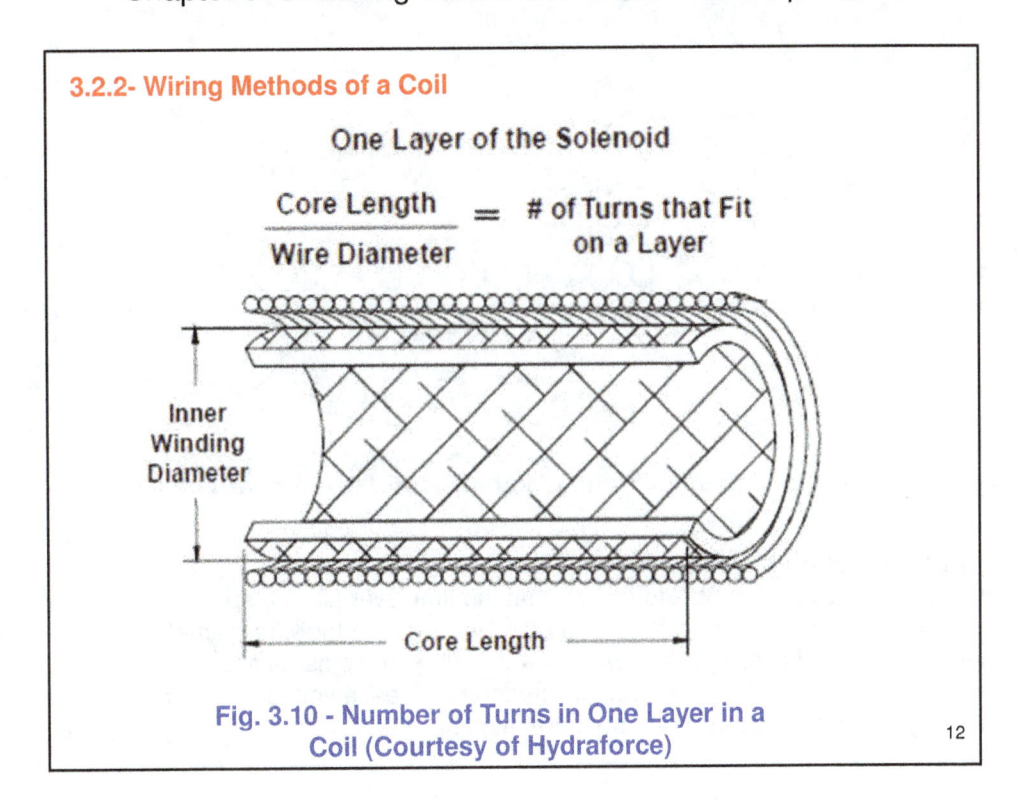

Fig. 3.10 - Number of Turns in One Layer in a Coil (Courtesy of Hydraforce)

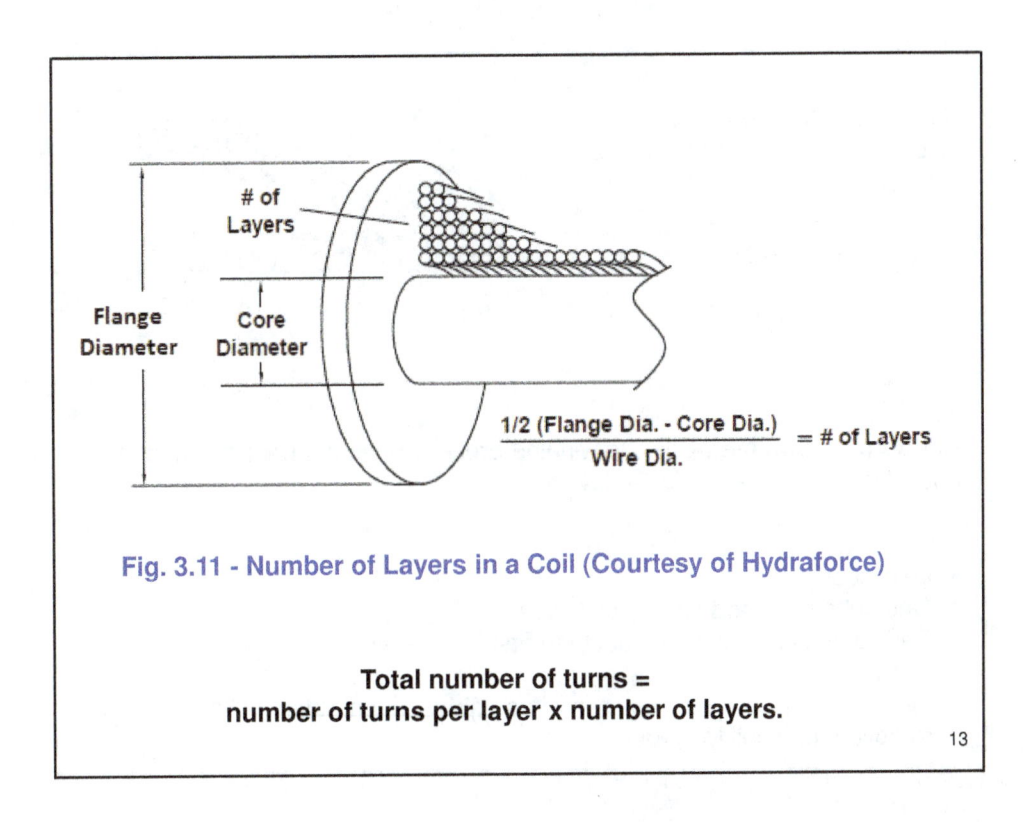

$$\frac{1/2 \, (\text{Flange Dia.} - \text{Core Dia.})}{\text{Wire Dia.}} = \text{\# of Layers}$$

Fig. 3.11 - Number of Layers in a Coil (Courtesy of Hydraforce)

Total number of turns = number of turns per layer x number of layers.

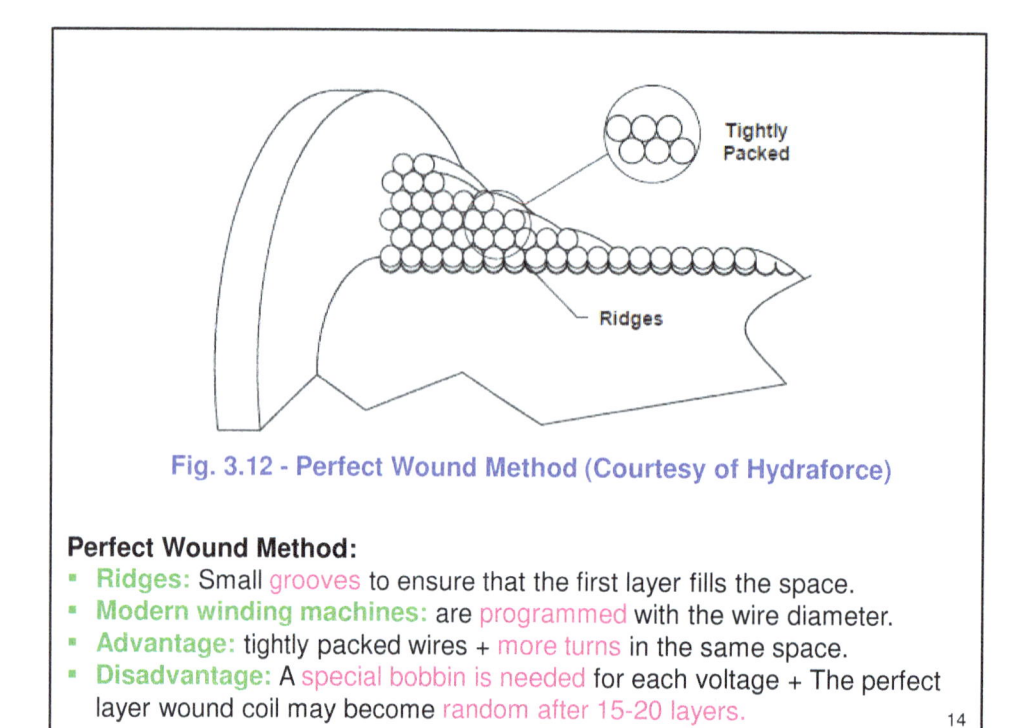

Fig. 3.12 - Perfect Wound Method (Courtesy of Hydraforce)

Perfect Wound Method:
- **Ridges:** Small grooves to ensure that the first layer fills the space.
- **Modern winding machines:** are programmed with the wire diameter.
- **Advantage:** tightly packed wires + more turns in the same space.
- **Disadvantage:** A special bobbin is needed for each voltage + The perfect layer wound coil may become random after 15-20 layers.

Fig. 3.13 – Random Wound Method (Courtesy of Hydraforce)

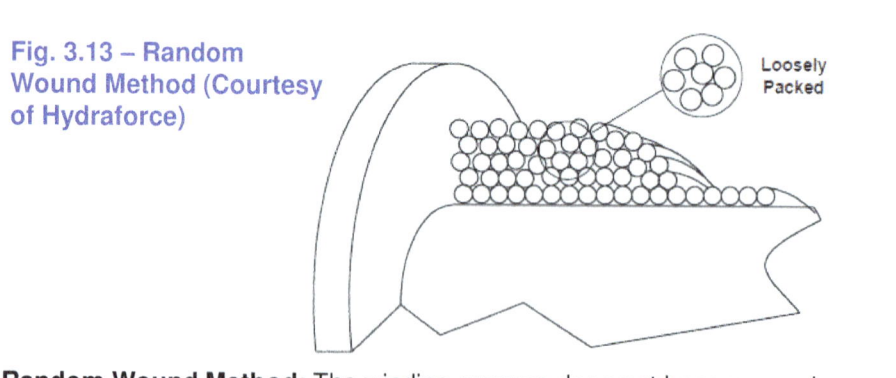

Random Wound Method: The winding process does not become random until after the first seven to ten layers.

Advantage:
- Less cost.
- One bobbin is used for any size wire.
- Perfect wound coils for at least the first 7-10 layers.

Disadvantage:
- Vibrations cause rubbing off the wires against each other, isolation removed, and coil to short.

3.2.3- Switching Solenoid Force-Stroke Relationship

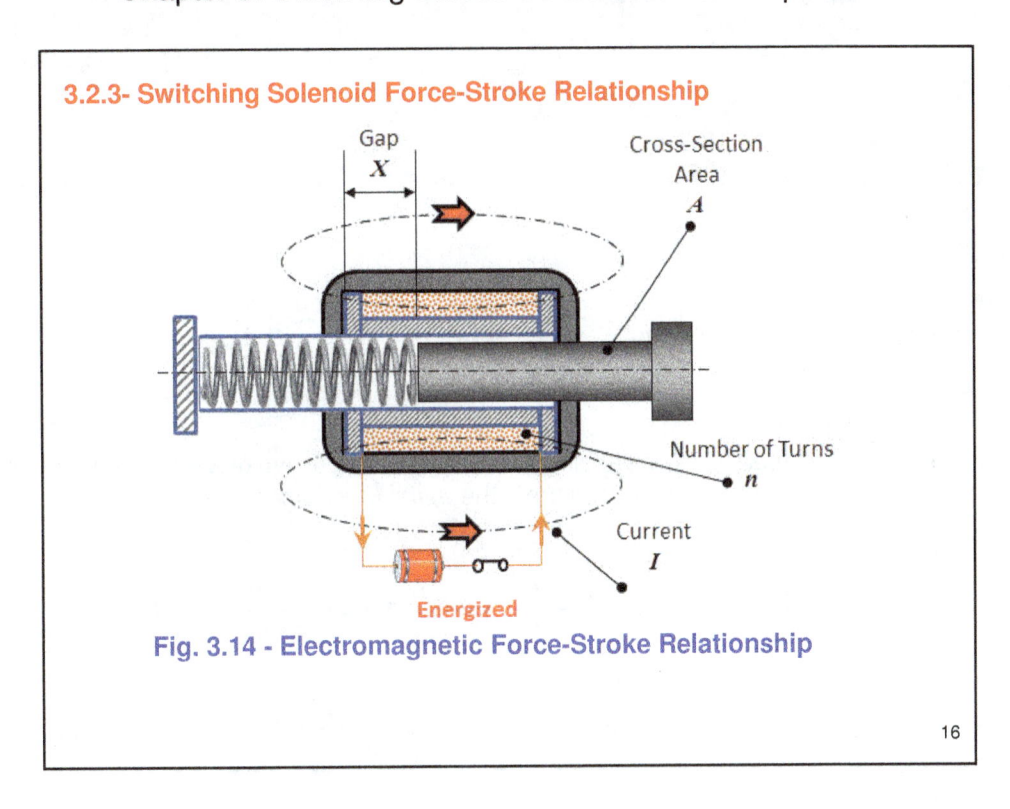

Fig. 3.14 - Electromagnetic Force-Stroke Relationship

Electromagnetic Force:

- $F \propto 1/gap$ (inverse and nonlinear relation).

- **Initial force** increases rapidly with moving the armature under the coil.

- **Maximum (Holding) force** at the rated current and the smallest gap.

- The available commercial switching solenoids produce forces within 60-70 N (13-16) lb.

- **Rated current** produces electromagnetic force can overcome all forces through the full stroke.

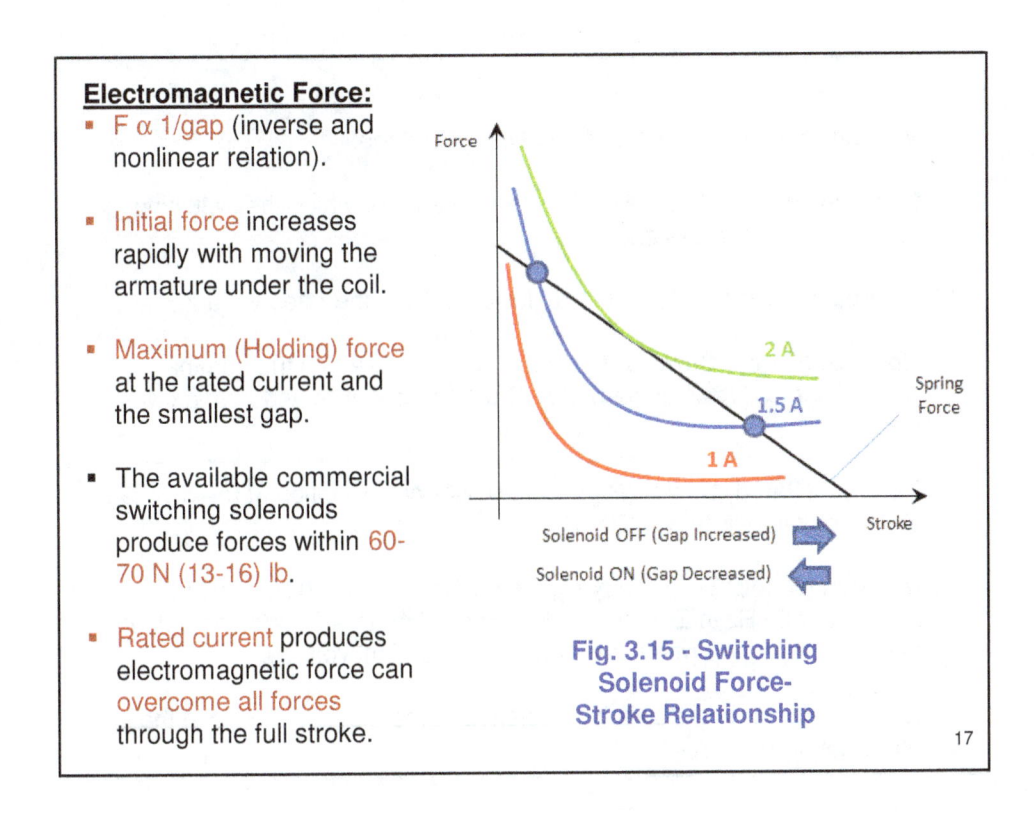

Fig. 3.15 - Switching Solenoid Force-Stroke Relationship

$$F = \frac{1}{2}(nI)^2 \times \frac{\mu_0 A}{X^2} \qquad 3.4$$

Example:
For a coil of 1000 turns, initial air gab of 1 cm, initial current of 2 Amp, and a plunger of 4 cm² cross-section area, the *Initial Electromagnetic Force* is:

$$F = \frac{1}{2} \times \frac{(1000 \; x \; 2)^2 \times (4\pi \times 10^{-7})\, 4}{1} = 10.048 \, N = \frac{10.048}{4.45} = 2.26 \, lb$$

3.2.4- Switching Solenoid Performance Terminologies

- **Voltage Type:** it means DC or AC type of voltage.

- **Nominal (Rated) Voltage (U_N):** is the voltage for which the switching solenoid has been designed.

- **Voltage Tolerance (%):** is the % deviation from the rated voltage.

- **Nominal (Rated) Current (I_N):** is the current based on the nominal voltage and coil resistance, at 20 °C coil temperature and nominal frequency.

- **Nominal (Rated) Power (P_N):** is the power consumption at rated voltage and a coil temperature of 20 °C.

- **Nominal (Rated) Frequency (f_N):** is the frequency (60 Hz in North America or 50 Hz in Europe and ME) for which an AC switching solenoid has been designed when supplied by the nominal voltage.

- **Holding Force:** is the switching solenoid force that is effective at the end of its stroke.

Workbook: Electro-Hydraulic Components and Systems
Chapter 3: Switching Valves Construction and Operation

- ❏ **Residual Force:** is the force that still applies due to any remaining magnetic field after turning the switching solenoid OFF.

- ❏ **Response (Delay) Time:** is the time the coil takes to form the magnetic field and electromagnetic force high enough to move the armature from its initial position. So it is the time between the current application and the initial movement of the armature.

- ❏ **Stroke Time:** is the time the armature takes to move from its initial position to the end of its stroke.

- ❏ **Switching Time:**
  - Response time + stroke time.
  - It is also referred as Pull-in Time or Rise Time.

- ❏ **Switching (Cycling) Rate:** cycling rate means how many cycles the switching solenoid can complete per unit time.

- ❏ **Duty Cycle:** it is the ratio between the ON period and the complete cycle time in %.

- ❏ **Continuous Duty:** it means the coil can turned ON continuously without fear of burning the coil under normal climate conditions. It is to be noted that the coils do heat up during the continuous duty operation that results in loss of power after hours of operation.

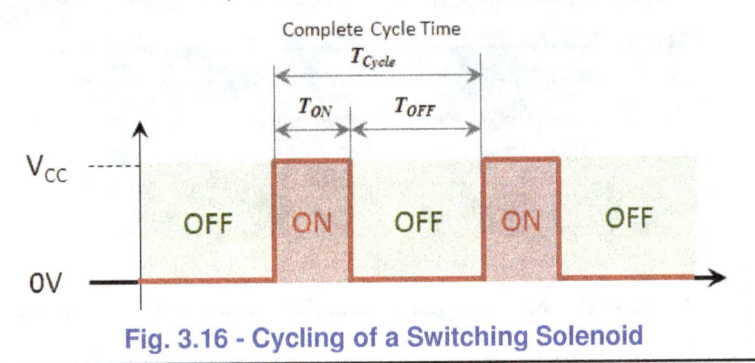

Fig. 3.16 - Cycling of a Switching Solenoid

Rexroth
Bosch Group

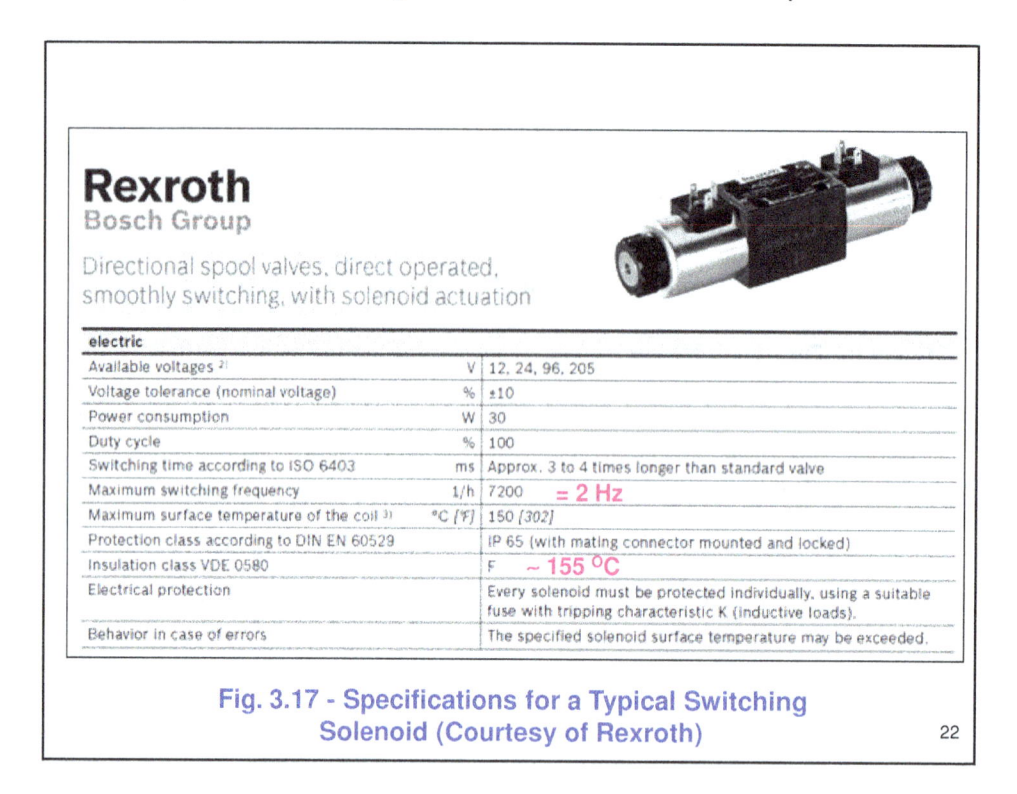

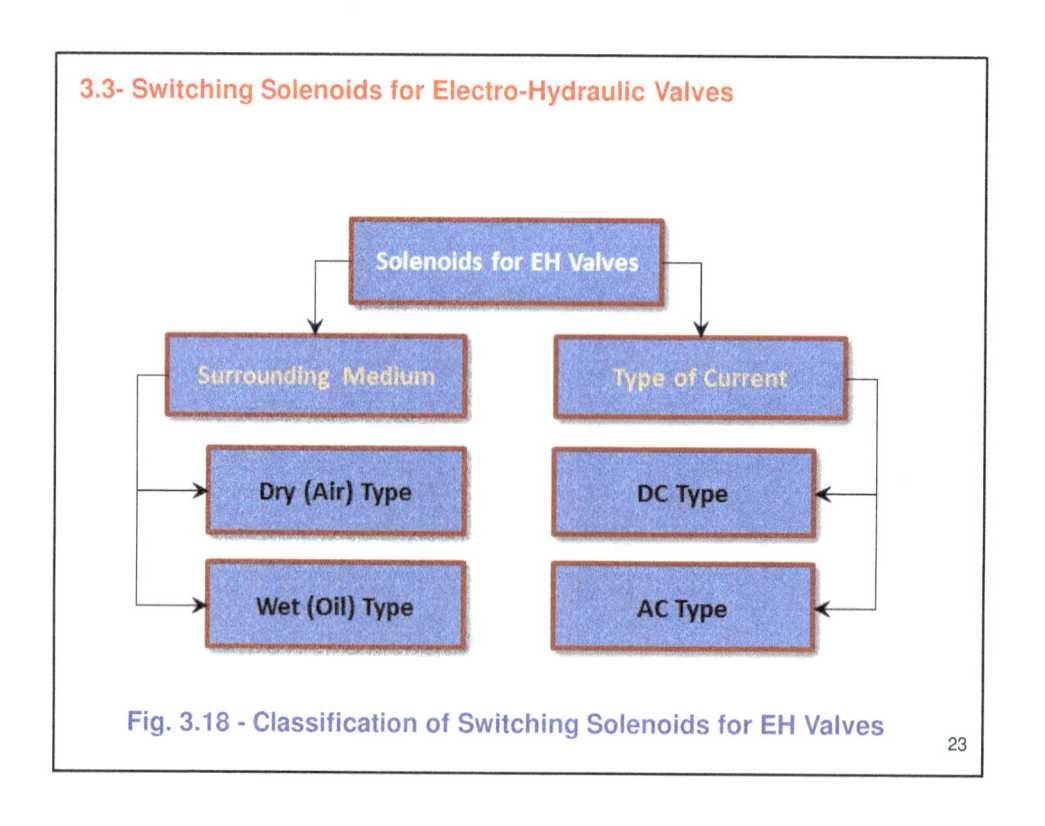

Directional spool valves, direct operated, smoothly switching, with solenoid actuation

electric		
Available voltages [2]	V	12, 24, 96, 205
Voltage tolerance (nominal voltage)	%	±10
Power consumption	W	30
Duty cycle	%	100
Switching time according to ISO 6403	ms	Approx. 3 to 4 times longer than standard valve
Maximum switching frequency	1/h	7200 = 2 Hz
Maximum surface temperature of the coil [3]	°C [°F]	150 [302]
Protection class according to DIN EN 60529		IP 65 (with mating connector mounted and locked)
Insulation class VDE 0580		F ~ 155 °C
Electrical protection		Every solenoid must be protected individually, using a suitable fuse with tripping characteristic K (inductive loads).
Behavior in case of errors		The specified solenoid surface temperature may be exceeded.

Fig. 3.17 - Specifications for a Typical Switching Solenoid (Courtesy of Rexroth)

3.3 - Switching Solenoids for Electro-Hydraulic Valves

Solenoids for EH Valves
- Surrounding Medium
  - Dry (Air) Type
  - Wet (Oil) Type
- Type of Current
  - DC Type
  - AC Type

Fig. 3.18 - Classification of Switching Solenoids for EH Valves

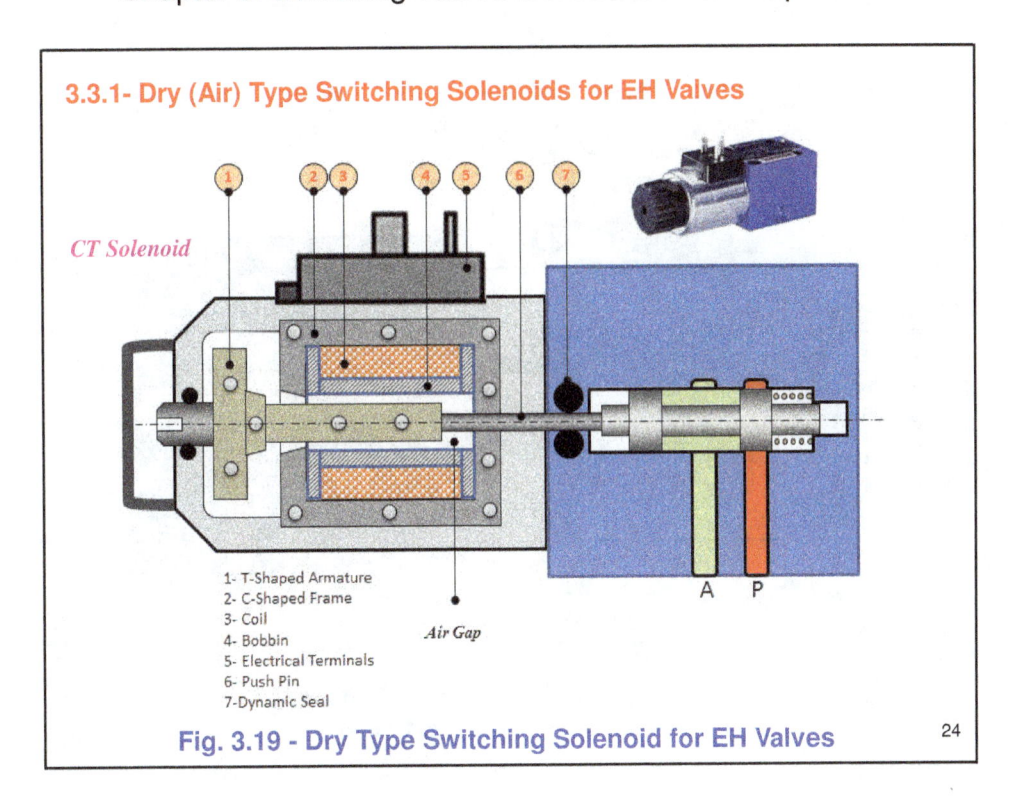

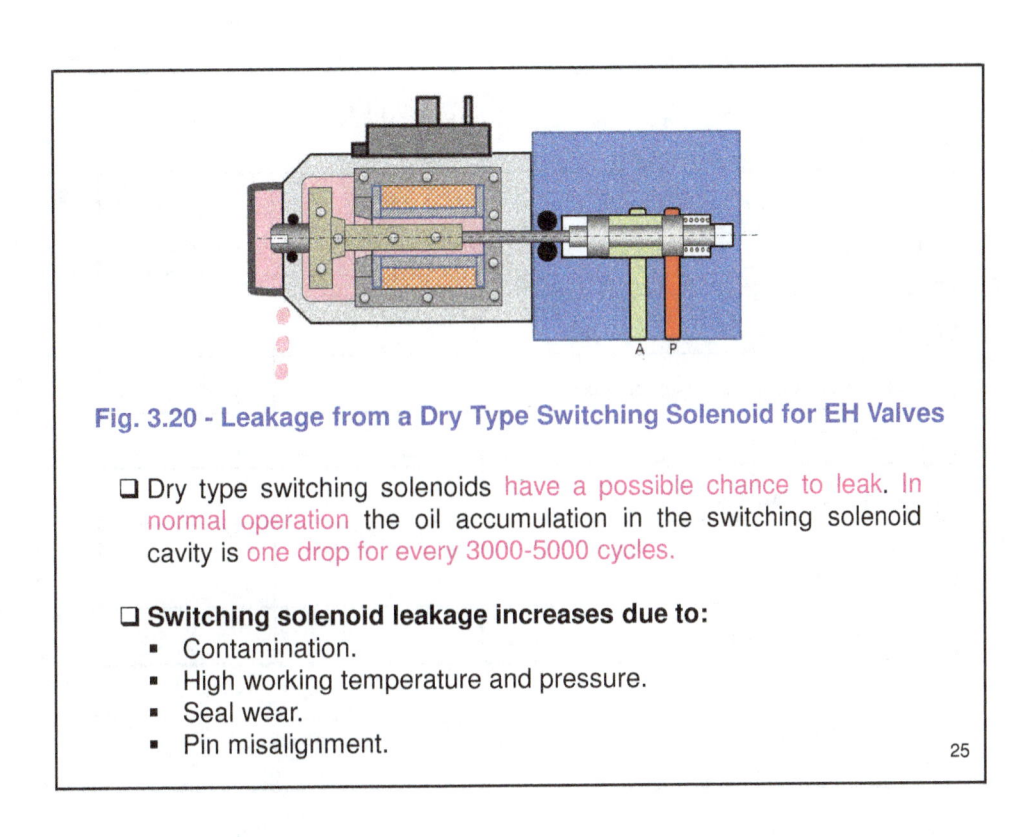

Workbook: Electro-Hydraulic Components and Systems
Chapter 3: Switching Valves Construction and Operation

3.3.1- Dry (Air) Type Switching Solenoids for EH Valves

CT Solenoid

1- T-Shaped Armature
2- C-Shaped Frame
3- Coil
4- Bobbin
5- Electrical Terminals
6- Push Pin
7- Dynamic Seal

Air Gap

Fig. 3.19 - Dry Type Switching Solenoid for EH Valves

Fig. 3.20 - Leakage from a Dry Type Switching Solenoid for EH Valves

❑ Dry type switching solenoids have a possible chance to leak. In normal operation the oil accumulation in the switching solenoid cavity is one drop for every 3000-5000 cycles.

❑ **Switching solenoid leakage increases due to:**
- Contamination.
- High working temperature and pressure.
- Seal wear.
- Pin misalignment.

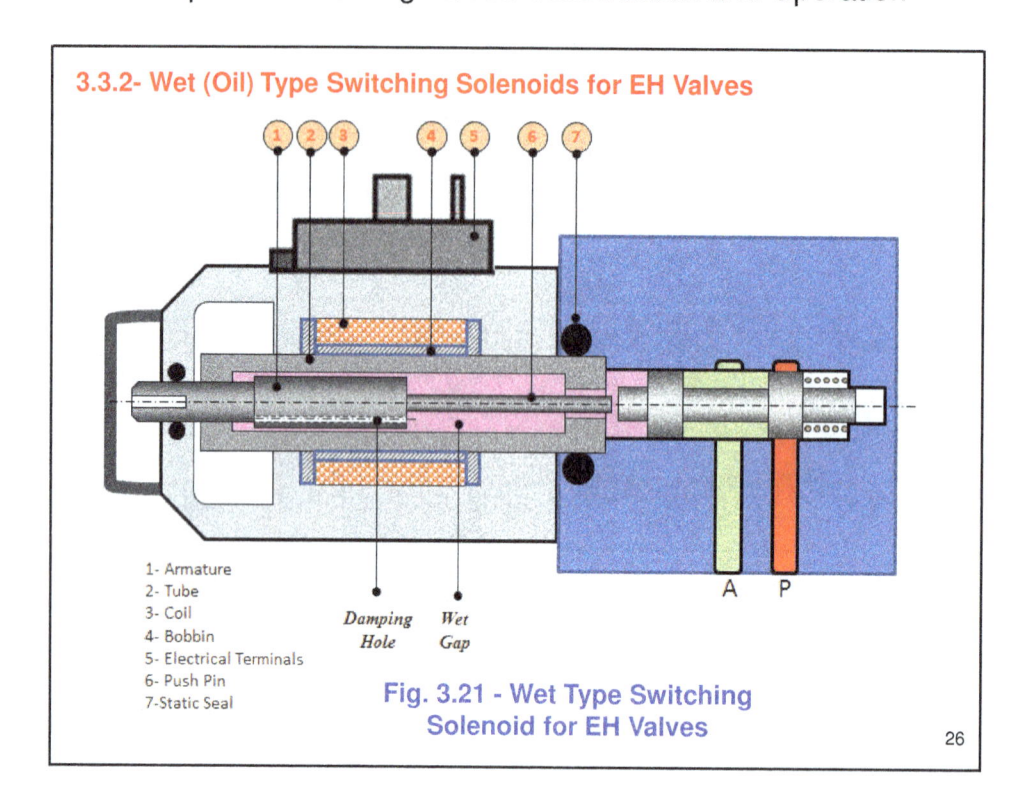

3.3.2- Wet (Oil) Type Switching Solenoids for EH Valves

1- Armature
2- Tube
3- Coil
4- Bobbin
5- Electrical Terminals
6- Push Pin
7- Static Seal

Damping Hole *Wet Gap*

Fig. 3.21 - Wet Type Switching Solenoid for EH Valves

3.3.3- Dry versus Wet Type Switching Solenoids for EH Valves

Item of Comparison	Dry Type Switching solenoid	Wet Type Switching solenoid
Better coil cooling and heat dissipation.	✘	✓
Better valve sealing against air and dust.	✘	✓
Better reliability and longer service life.	✘	✓
Sealing element last longer.	✘	✓
Less noise generation by the switching solenoid.	✘	✓
Lower power consumption.	✓	✘

✓ Advantage   ✘ Disadvantage

Table 3.2 - Dry versus Wet Type Switching Solenoid

Workbook: Electro-Hydraulic Components and Systems
Chapter 3: Switching Valves Construction and Operation

3.3.4- DC Switching Solenoids for EH Valves
3.3.4.1- Construction and Operation of DC Switching Solenoids for EH Valves

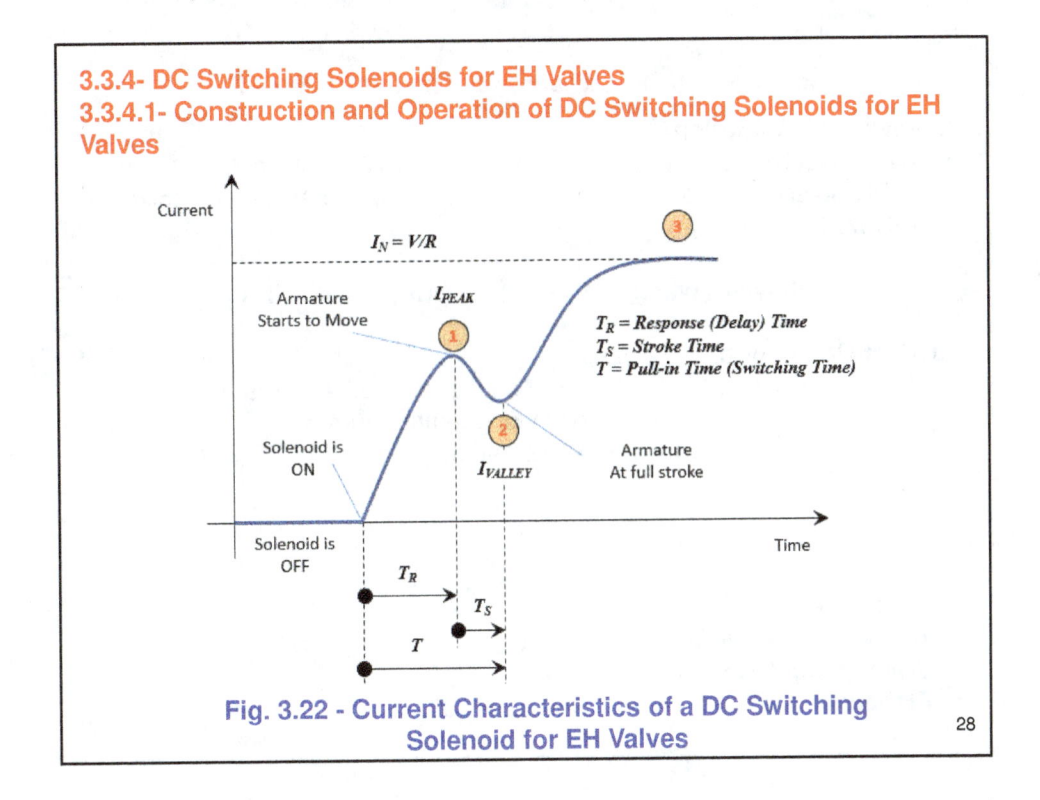

Fig. 3.22 - Current Characteristics of a DC Switching Solenoid for EH Valves

T_R = Response (Delay) Time
T_S = Stroke Time
T = Pull-in Time (Switching Time)

$I_N = V/R$

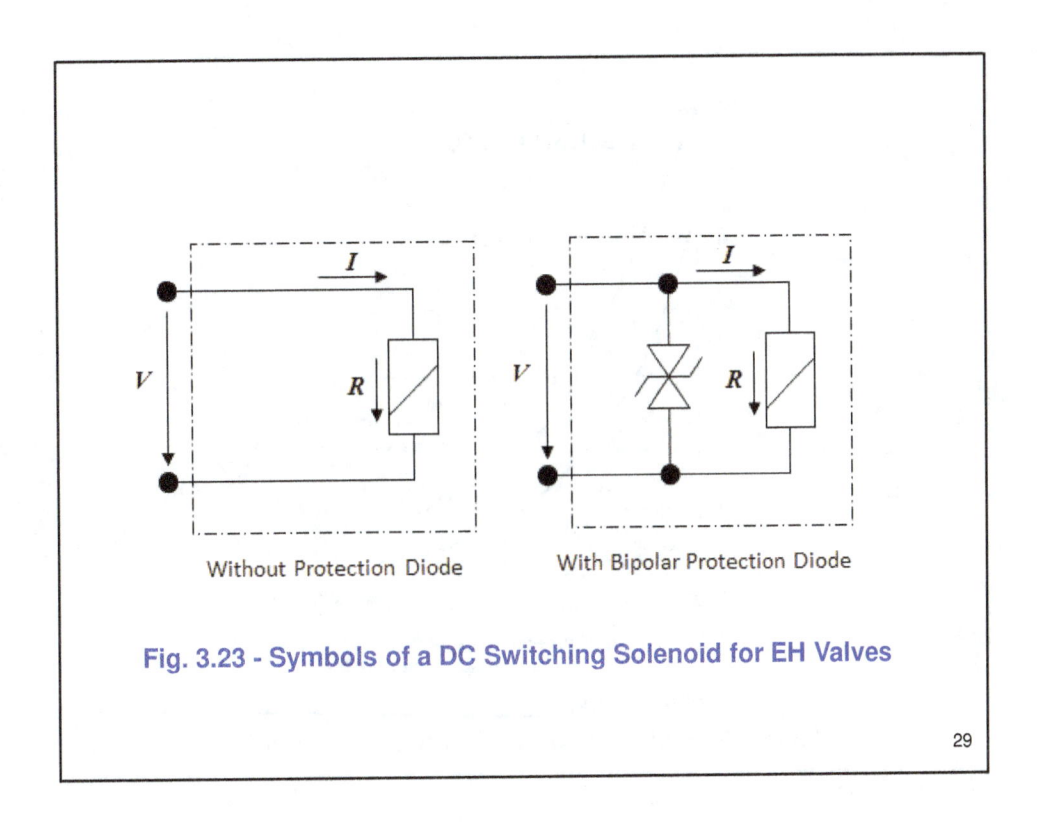

Fig. 3.23 - Symbols of a DC Switching Solenoid for EH Valves

Without Protection Diode | With Bipolar Protection Diode

3.3.4.2- Performance of DC Switching Solenoids for EH Valves

❑ **Power Consumption:**
- A DC solenoid builds up current slower than an AC solenoid.
- A DC solenoid consumes relatively more power than an AC solenoid (AC work on average).

$$\text{Power (Watt)} = \text{V (Volt)} \times \text{I (Amp)} \qquad 3.5$$

❑ **Heat Generation:**
- A DC solenoid dissipates the heat so that the coil temperature does not rise above the maximum rated working temperature.
- If the heat rises above the maximum, current and EM forces are reduced.

Fig. 3.24 - Effect of Temperature on a DC Switching solenoid for EH Valves

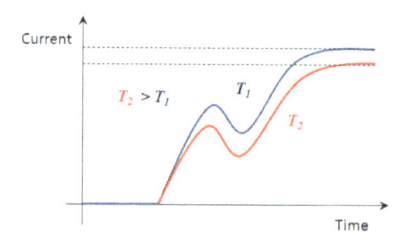

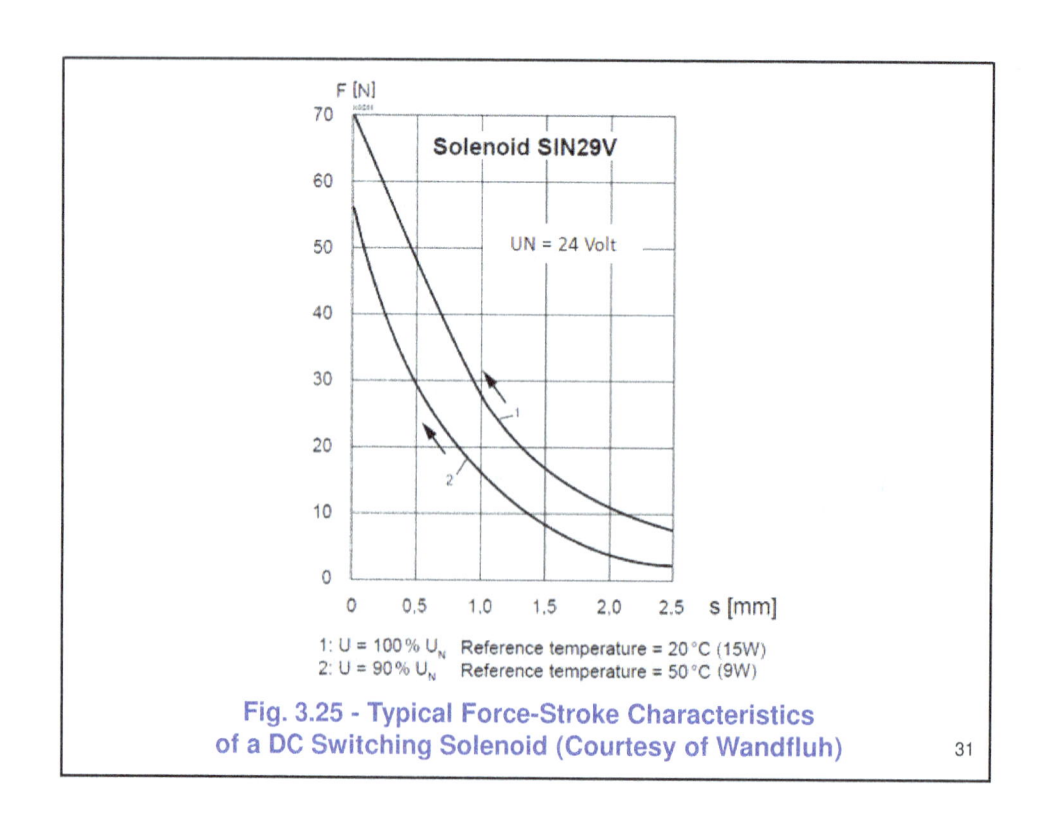

Fig. 3.25 - Typical Force-Stroke Characteristics of a DC Switching Solenoid (Courtesy of Wandfluh)

1: $U = 100\% \ U_N$ Reference temperature = 20 °C (15W)
2: $U = 90\% \ U_N$ Reference temperature = 50 °C (9W)

- ❑ **Switching Time:**
  - The switching time of a DC switching solenoid is relatively longer than in AC switching solenoids.

- ❑ **Switching Rate:**
  - Despite the longer switching time of a DC switching solenoid, it can usually be cycled at a higher rate than an AC switching solenoid.
  - The reason is: no inrush current + better chance to cool faster during the Off time in the cycle.

3.3.4.3- Applications of DC Switching solenoids for EH Valves

- More common in mobile applications since DC power is available.

- Even in industrial applications nowadays, extending DC power lines, hot water, compressed air, and internet communication became a culture in building a factory.

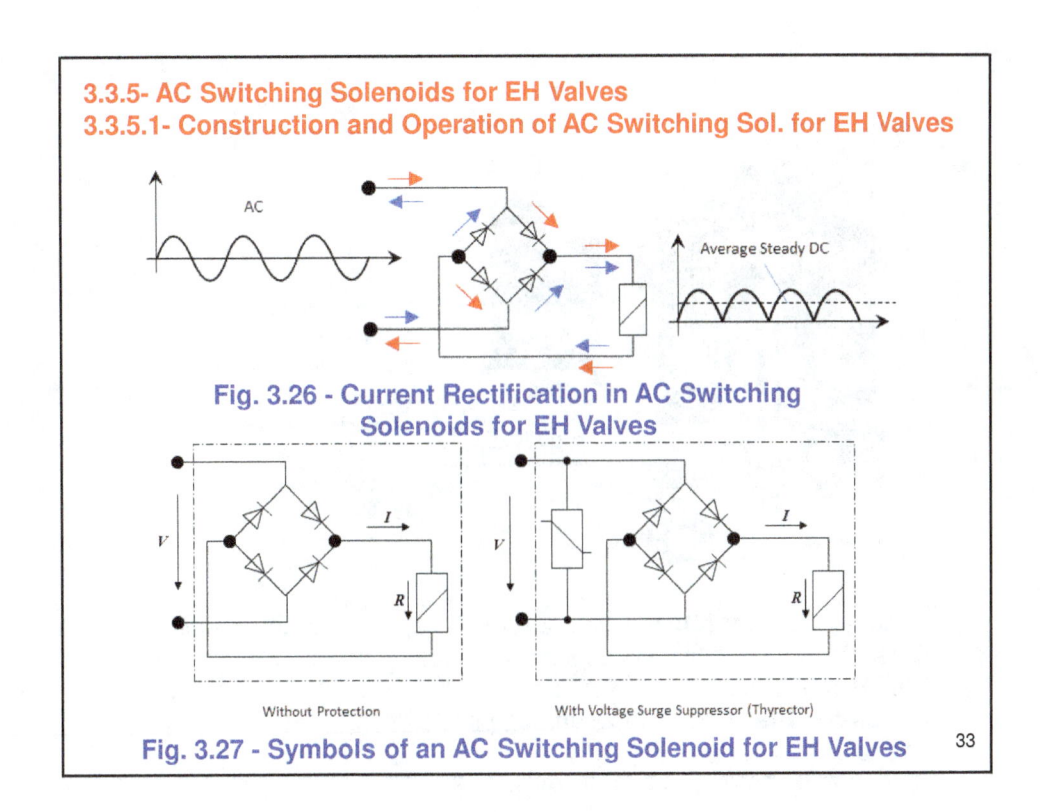

3.3.5- AC Switching Solenoids for EH Valves
3.3.5.1- Construction and Operation of AC Switching Sol. for EH Valves

Fig. 3.26 - Current Rectification in AC Switching Solenoids for EH Valves

Without Protection   With Voltage Surge Suppressor (Thyrector)

Fig. 3.27 - Symbols of an AC Switching Solenoid for EH Valves

3.3.5.2- AC Hum

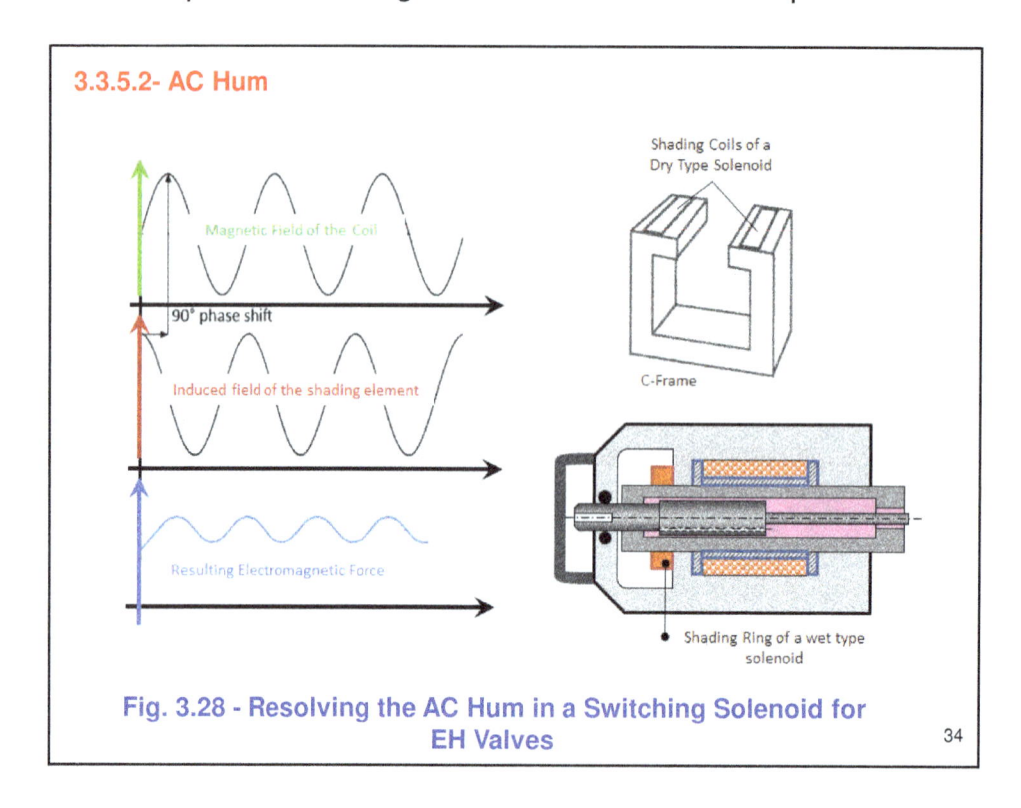

Fig. 3.28 - Resolving the AC Hum in a Switching Solenoid for EH Valves

1. Manual Override
2. Protective Cover
3. Shading Ring
4. Coil
5. Sleeve
6. Coil
7. Hollow Core
8. Solenoid Body
9. Mobile Core
10. Push Rod
11. Disc Holder Spring
12. Perforated Disc
13. Valve Body
14. Earth Pin
15. Electrical Pin (2)
16. Electrical Connector
17. Pole Piece

Fig. 3.29

- **Shading Element (3):** must be accurately aligned with the plunger. Otherwise, if contamination gets between them, the solenoid will vibrate
- **Pole Piece (17):** The core rests on it + it improves the gap closing.
- **Perforated Disc (12):** oil goes through it to lubricate and cool down the solenoid.

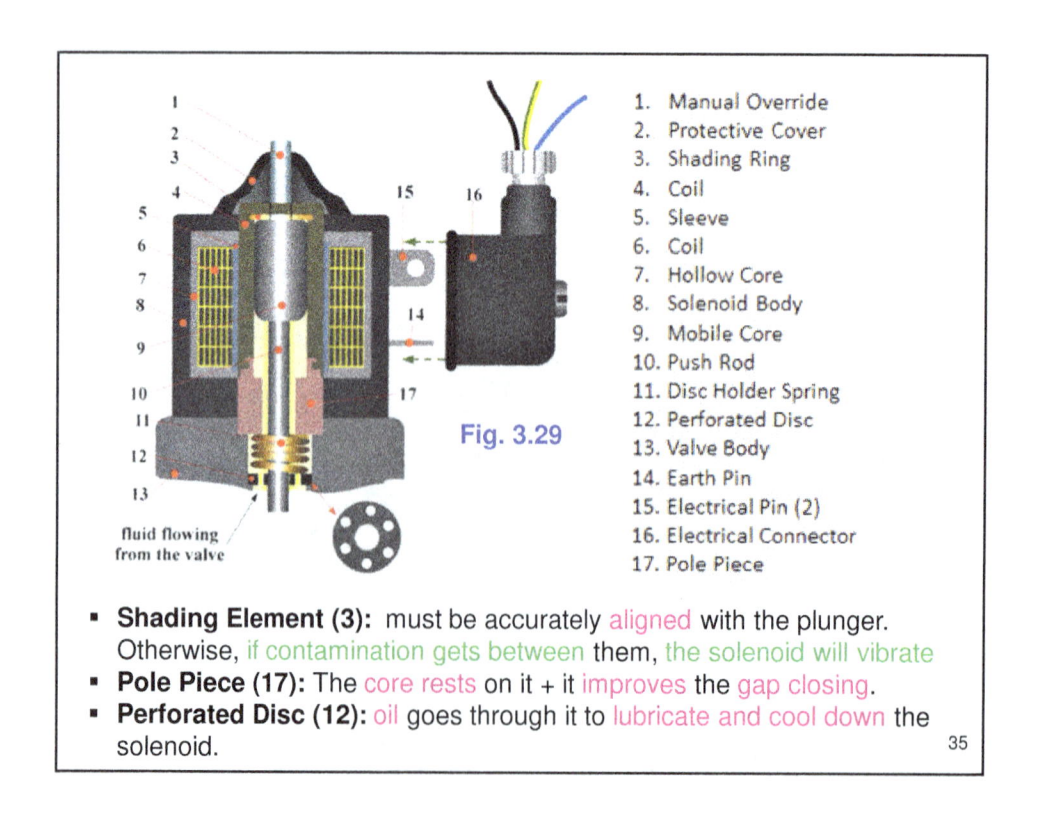

3.3.5.3- Eddy Current

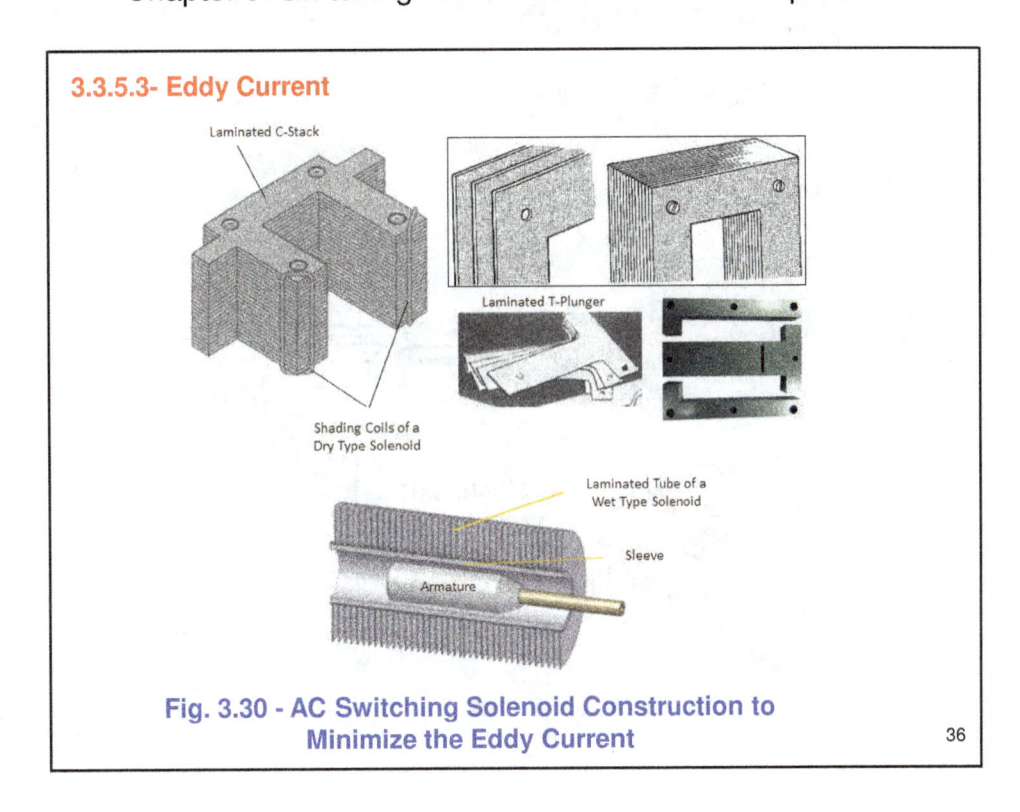

Fig. 3.30 - AC Switching Solenoid Construction to Minimize the Eddy Current

3.3.5.4- Inrush Current

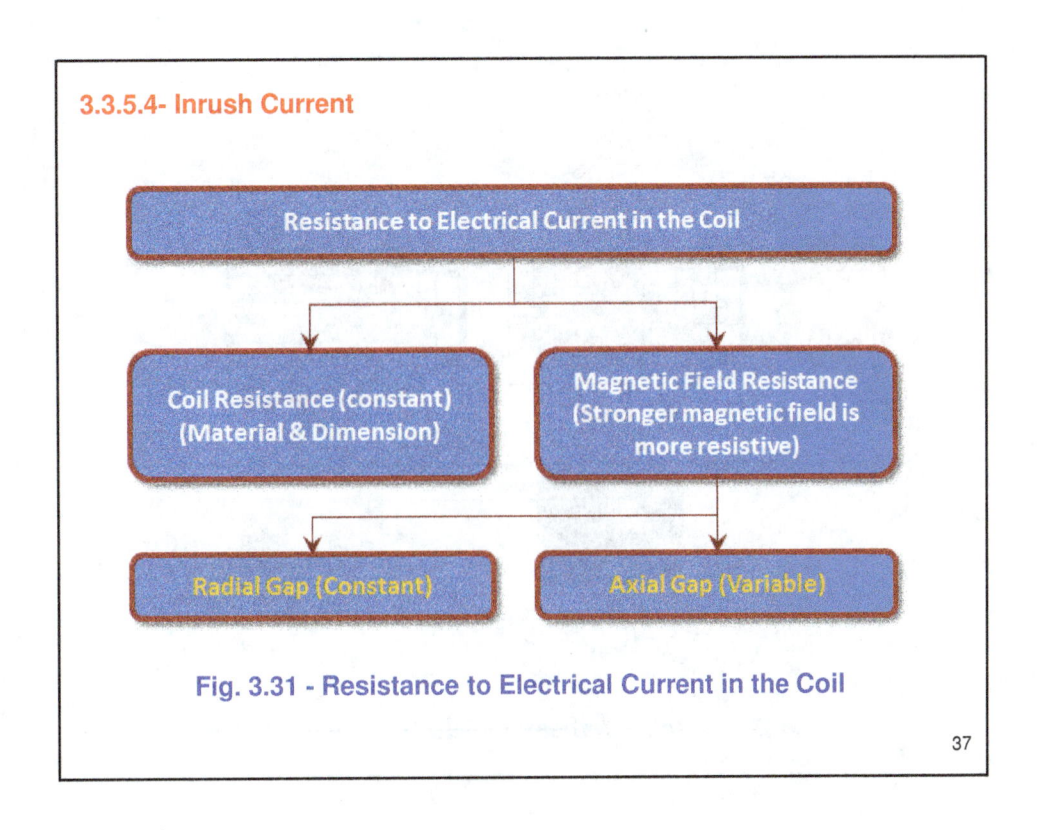

Fig. 3.31 - Resistance to Electrical Current in the Coil

Workbook: Electro-Hydraulic Components and Systems
Chapter 3: Switching Valves Construction and Operation

Inrush Current Development:

inrush current can reach 4-10 ten times the nominal current!!

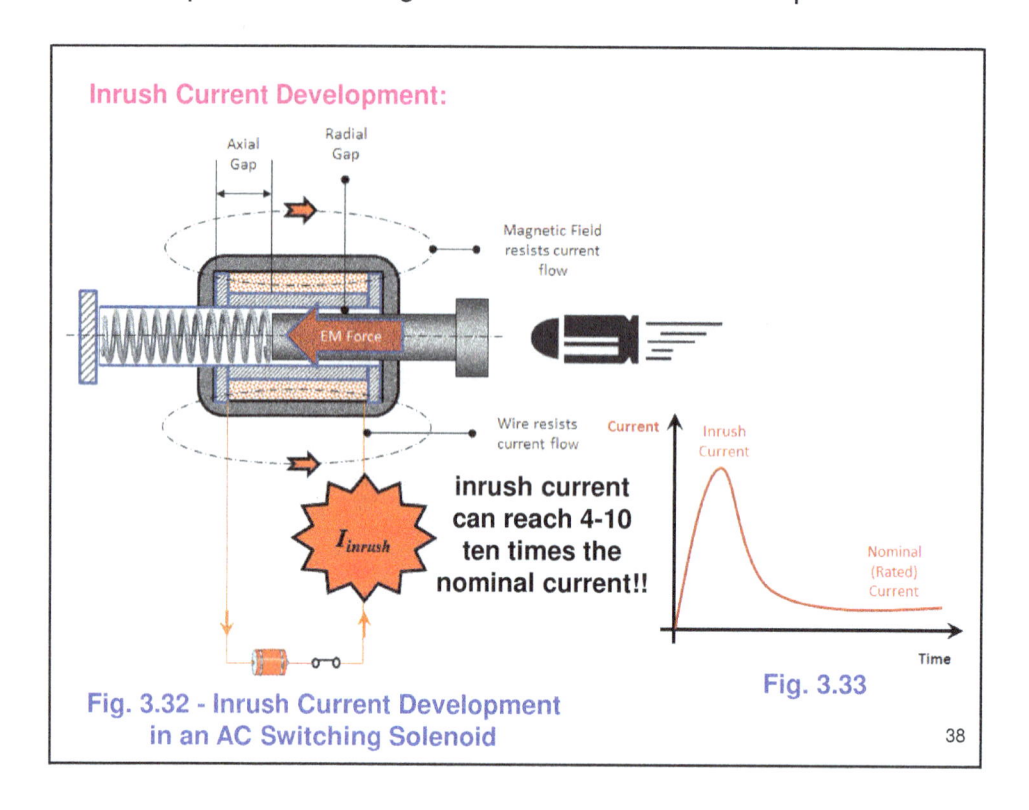

Fig. 3.32 - Inrush Current Development in an AC Switching Solenoid

Fig. 3.33

Inrush Current Consequences:

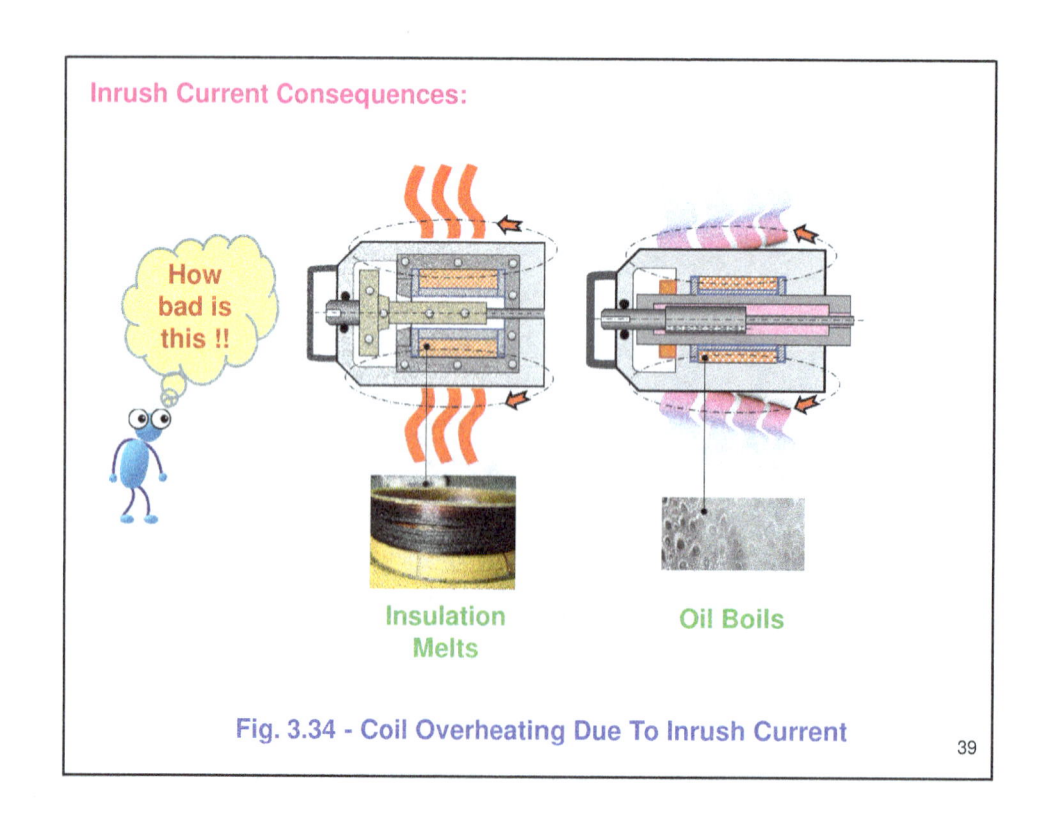

How bad is this !!

Insulation Melts

Oil Boils

Fig. 3.34 - Coil Overheating Due To Inrush Current

How to minimize the effect of the Inrush Current:

Valve Design:
- **Shifting Force:** sufficient to overcome all the opposing forces (return spring + flow force + viscous friction).
- **Shifting Time:** short as possible.
- **Holding Force:** sufficient to hold the core with as little current as possible by designing the coil with sufficient AC resistance which is known as impedance.

Valve Operation:
- **Supply Voltage:** must supply the rated voltage.
- **Maintenance:** periodically to avoid seizing the spool.
- **Cleanliness:** The outside surfaces and the oil.
- **Wiring:** avoid energizing opposing switching solenoids simultaneously.

3.3.5.5- Performance of AC Switching Solenoids for EH Valves

❑ **Power Consumption:**
- An AC solenoid buildup current faster than a DC solenoid.
- An AC solenoid consumes relatively less power than a DC solenoid.

$$Power\ (Watt) = V\ (Volt) \times I\ (Amp) \times \cos\varphi \qquad 3.6$$

φ is the complex power phase angle.

❑ **Heat Generation:**
- An AC switching solenoid generates heat during the inrush current period more than a DC switching solenoid.

❑ **Switching Time:** AC switching solenoid is faster than a DC switching solenoid. Consequently, the shifting time is shorter than in a DC switching solenoid.

Switching Rate:
- Inrush current problem is a limiting factor in the cycling rate of an AC switching solenoid.
- Despite less switching time of an AC solenoid, it can usually be cycled at a lower rate than a DC switching solenoid.
- If the heat generated by cycling the switching solenoid exceeds its ability to dissipate the heat, the switching solenoid will overheat and fail.

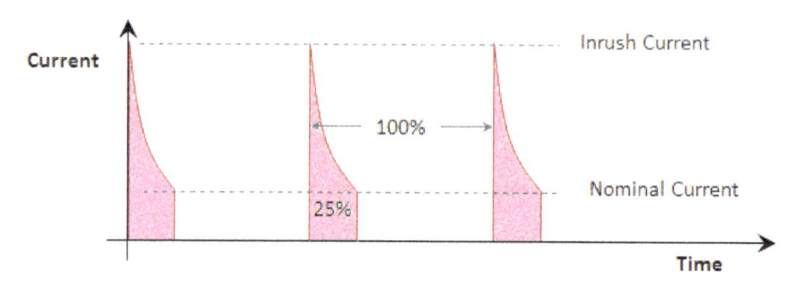

Fig. 3.35 - Switching Rate of an AC Switching solenoid

3.3.5.6- Applications of AC switching Solenoids for EH Valves

AC voltage is very common and readily available to <u>industrial applications</u>. So, AC switching solenoid valves are commonly used in industrial applications.

3.3.6- DC versus AC Switching Solenoid

Item of Comparison	DC Switching solenoid	AC Switching solenoid
Low Power Consumption and higher Holding Force.	✗	✓
Faster Switching Time.	✗	✓
Higher Switching Rate.	✓	✗
Better Reliability and less possible coil overheating.	✓	✗
Less noise generation by the switching solenoid.	✓	✗
Tolerance to contamination.	✓	✗

✓ Advantage   ✗ Disadvantage

Table 3.3 - DC versus AC Switching Solenoid

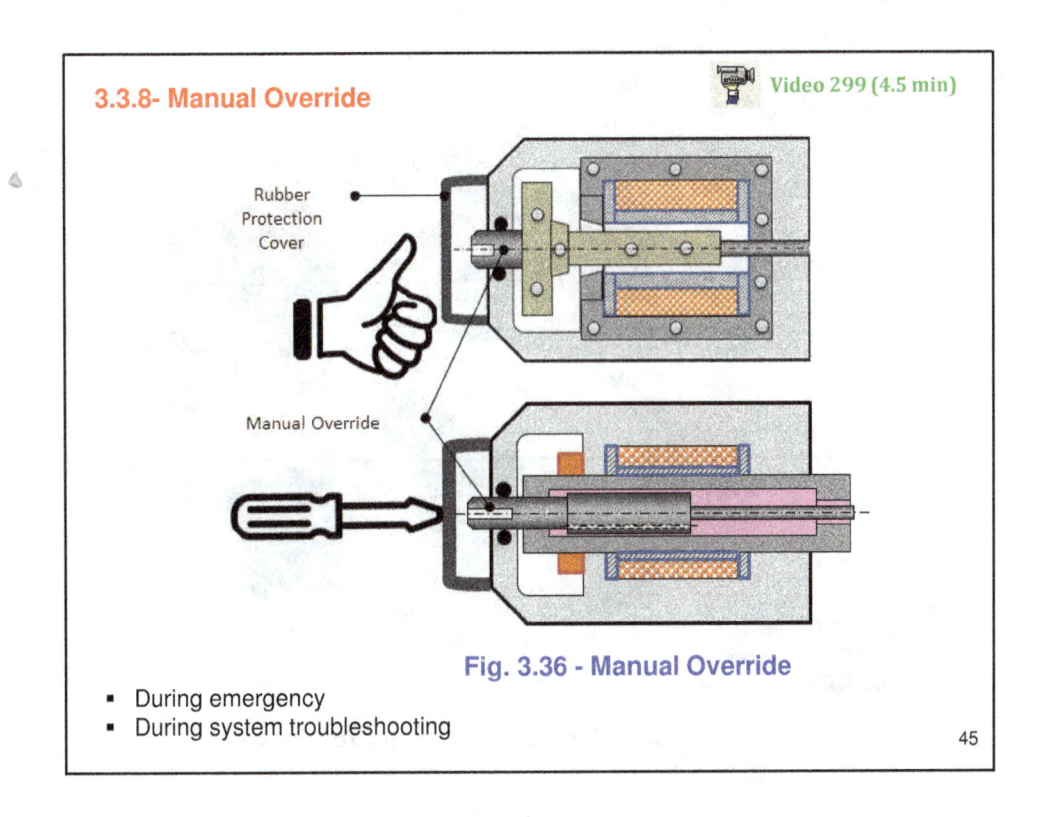

Workbook: Electro-Hydraulic Components and Systems
Chapter 3: Switching Valves Construction and Operation

3.3.7- Electrical Ratings for Switching Solenoids

❑ **Nominal Voltages:**
- DC (l2V, 24V, and 48V) and AC (110V and 220V).
- If a DC switching solenoid is operated on AC power, it could be easily burned out.

❑ **Nominal Currents:**
- (1-3) Amps depends on (nominal "rated" voltage + coil size).

❑ **Nominal Frequencies for AC Switching solenoid:**
- 60 Hz in the USA, Canada, and some South American countries.
- 50 Hz in most other countries.

❑ **Dual-Frequency Switching solenoids:**
- Ideally, a switching solenoid is designed for a specific frequency.
- If a 60 Hz coil works on 50 Hz, it will draw excess current.
- If a 50 Hz coil works on 60 Hz. it will draw less than its rated current, producing a lower electromagnetic force, inrush current period is extended, and the coil becomes less reliable.
- Dual-Frequency solenoids are developed for both 50 and 60 Hz.
- Data sheet for such a dual-frequency solenoids must reflect that.

3.3.8- Manual Override

Video 299 (4.5 min)

Fig. 3.36 - Manual Override

- During emergency
- During system troubleshooting

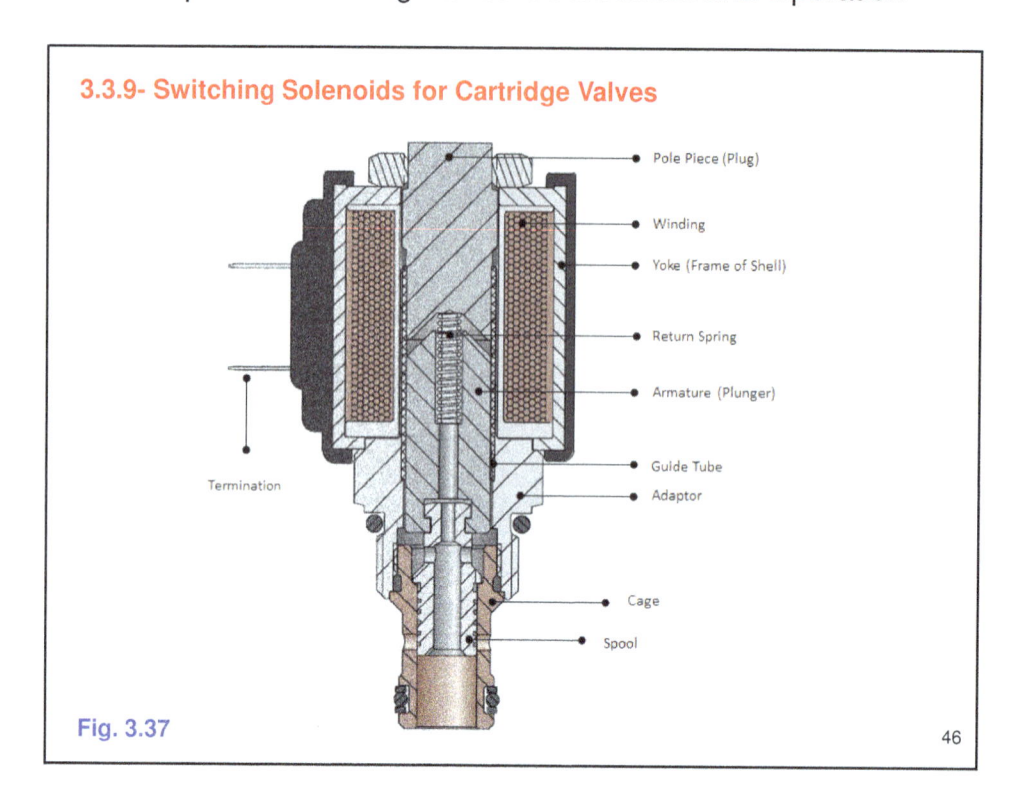

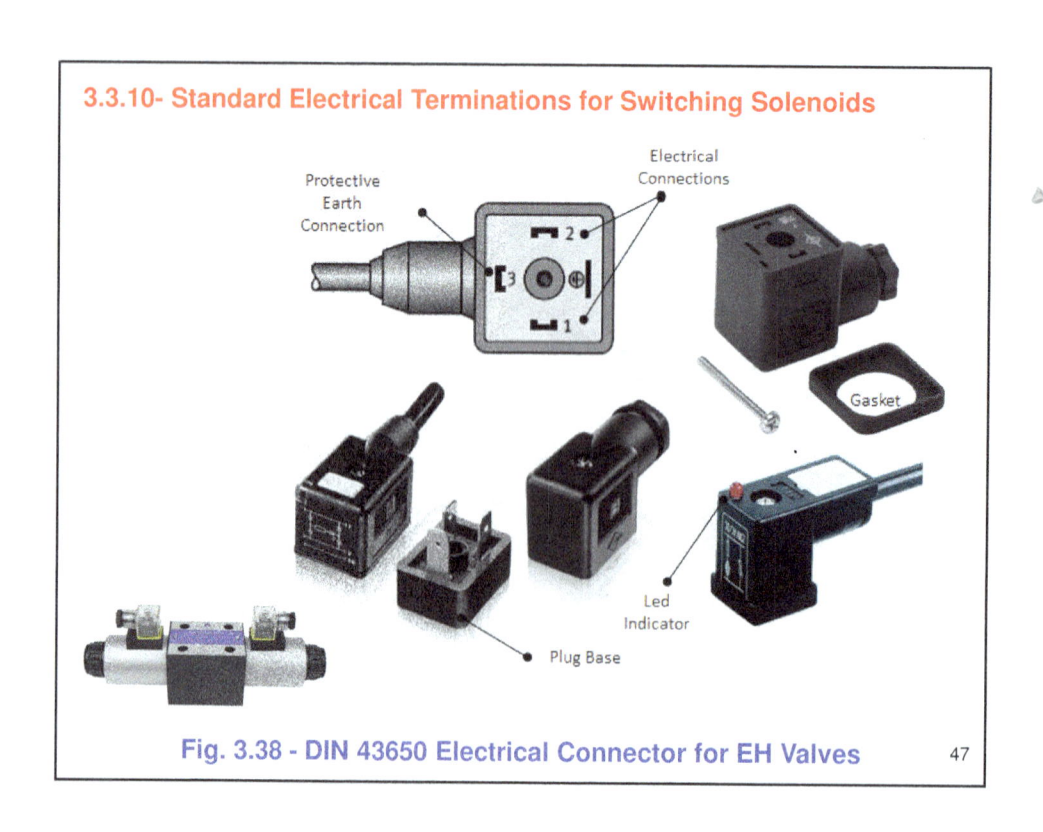

Workbook: Electro-Hydraulic Components and Systems
Chapter 3: Switching Valves Construction and Operation

3.3.9- Switching Solenoids for Cartridge Valves

Fig. 3.37

3.3.10- Standard Electrical Terminations for Switching Solenoids

Fig. 3.38 - DIN 43650 Electrical Connector for EH Valves

Workbook: Electro-Hydraulic Components and Systems
Chapter 3: Switching Valves Construction and Operation

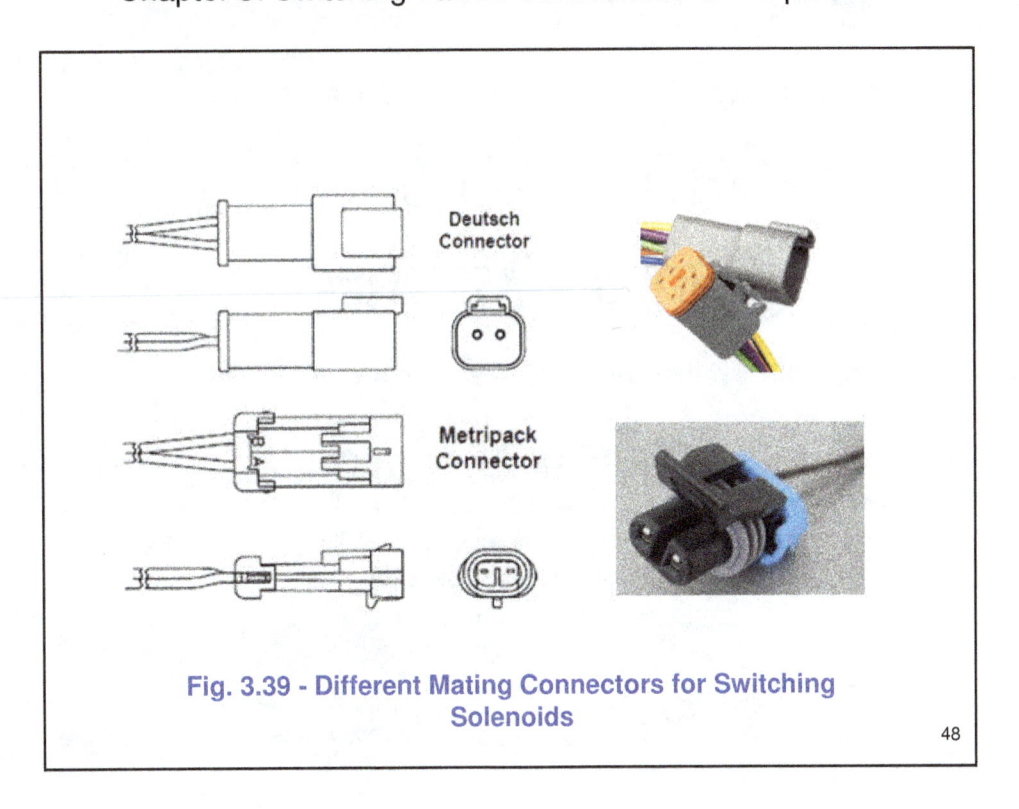

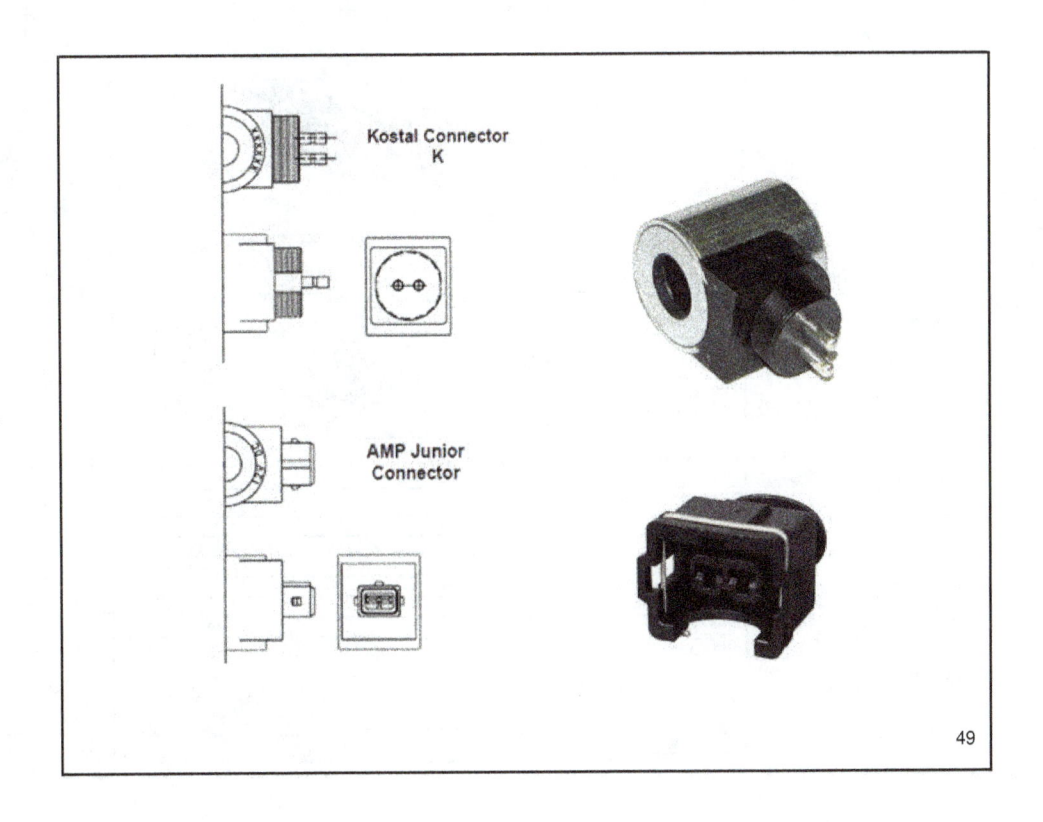

Fig. 3.39 - Different Mating Connectors for Switching Solenoids

Workbook: Electro-Hydraulic Components and Systems
Chapter 3: Switching Valves Construction and Operation

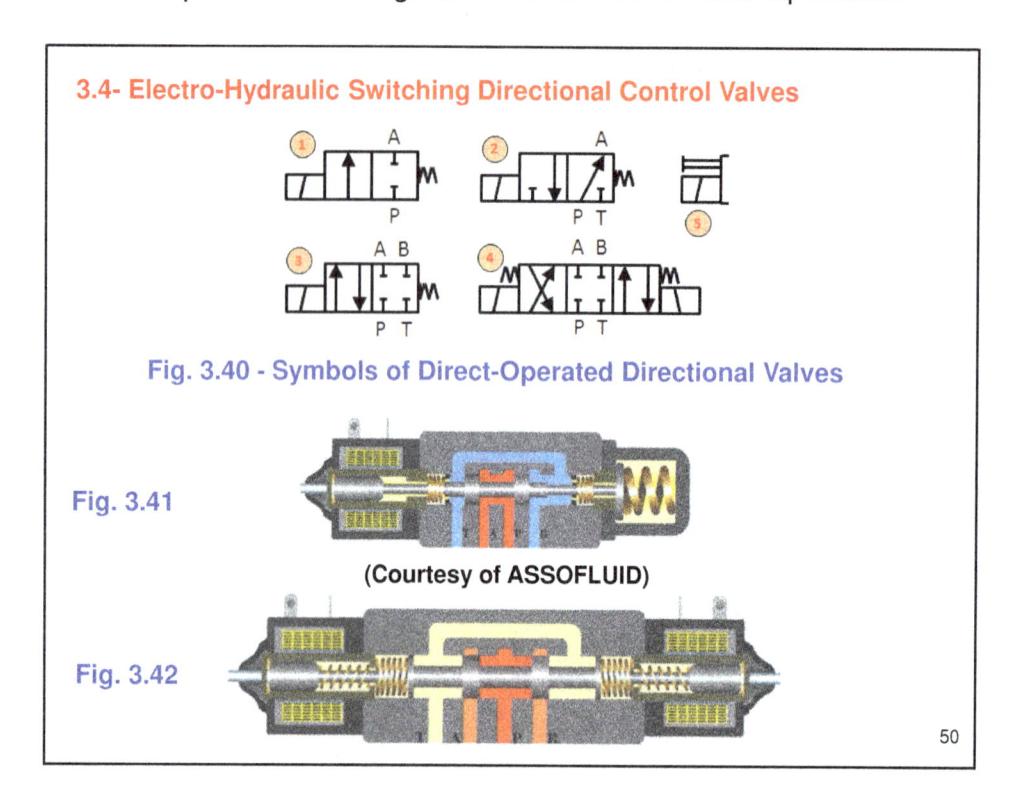

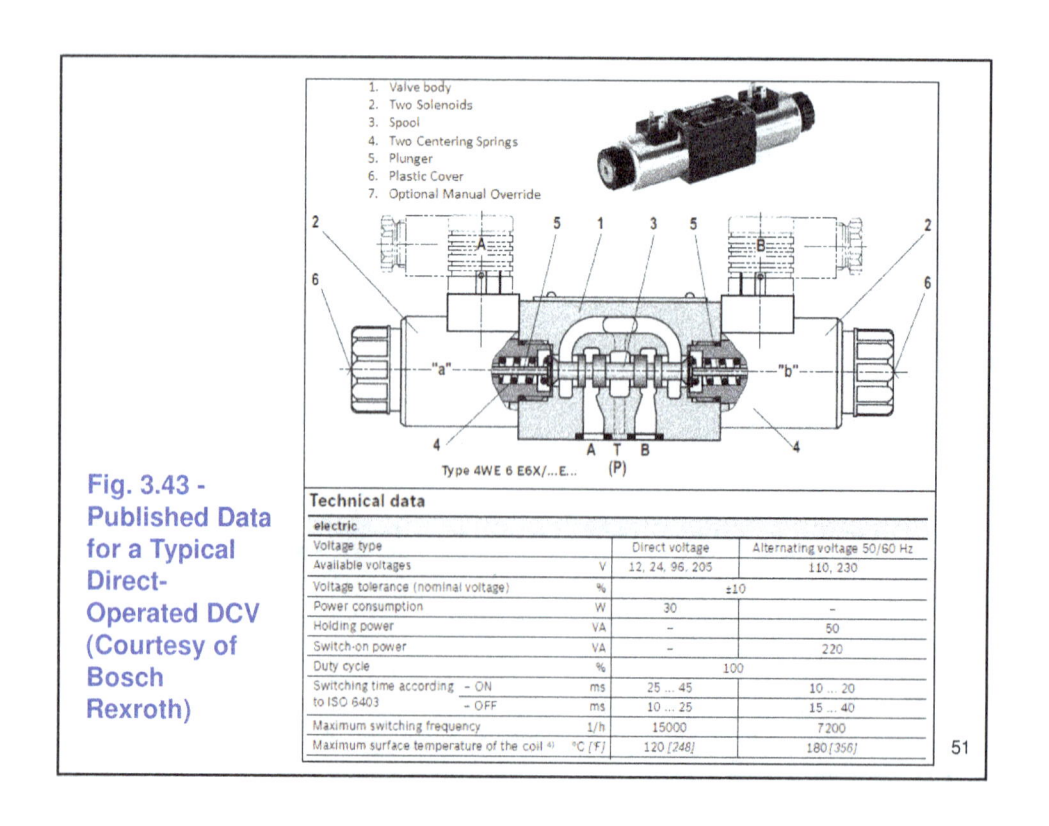

3.4- Electro-Hydraulic Switching Directional Control Valves

Fig. 3.40 - Symbols of Direct-Operated Directional Valves

Fig. 3.41

(Courtesy of ASSOFLUID)

Fig. 3.42

Fig. 3.43 - Published Data for a Typical Direct-Operated DCV (Courtesy of Bosch Rexroth)

1. Valve body
2. Two Solenoids
3. Spool
4. Two Centering Springs
5. Plunger
6. Plastic Cover
7. Optional Manual Override

Type 4WE 6 E6X/...E...

Technical data

electric			
Voltage type		Direct voltage	Alternating voltage 50/60 Hz
Available voltages	V	12, 24, 96, 205	110, 230
Voltage tolerance (nominal voltage)	%	±10	
Power consumption	W	30	–
Holding power	VA	–	50
Switch-on power	VA	–	220
Duty cycle	%	100	
Switching time according to ISO 6403 – ON	ms	25 ... 45	10 ... 20
– OFF	ms	10 ... 25	15 ... 40
Maximum switching frequency	1/h	15000	7200
Maximum surface temperature of the coil [4]	°C [°F]	120 [248]	180 [356]

Workbook: Electro-Hydraulic Components and Systems
Chapter 3: Switching Valves Construction and Operation

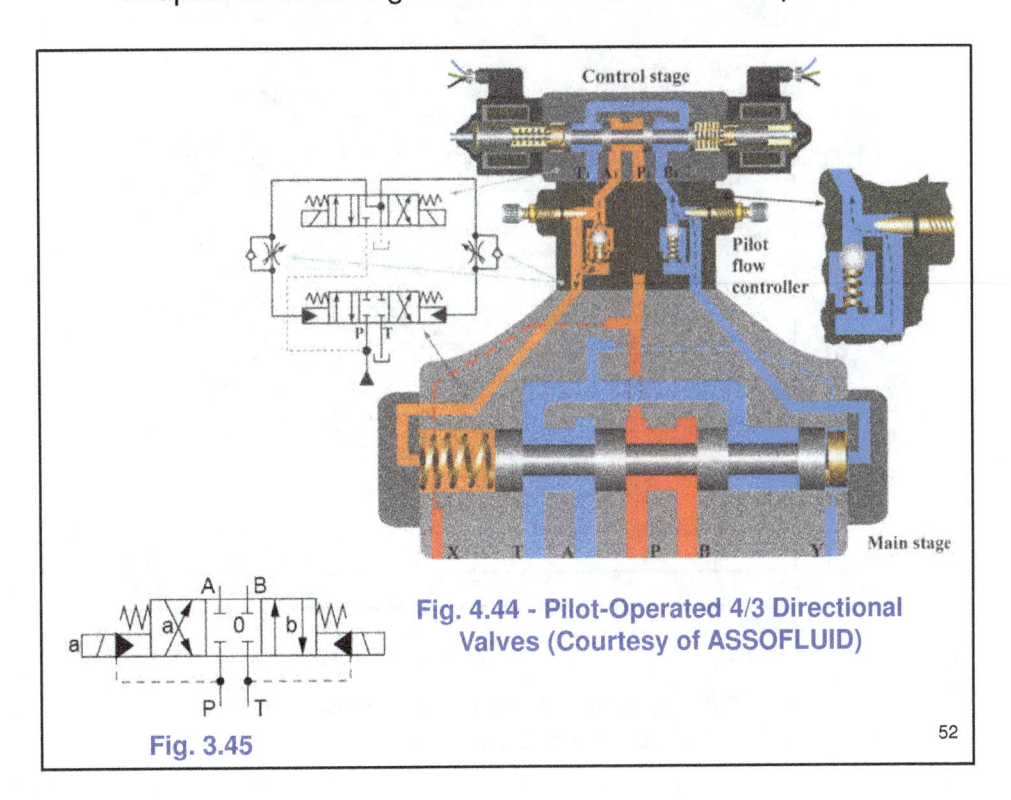

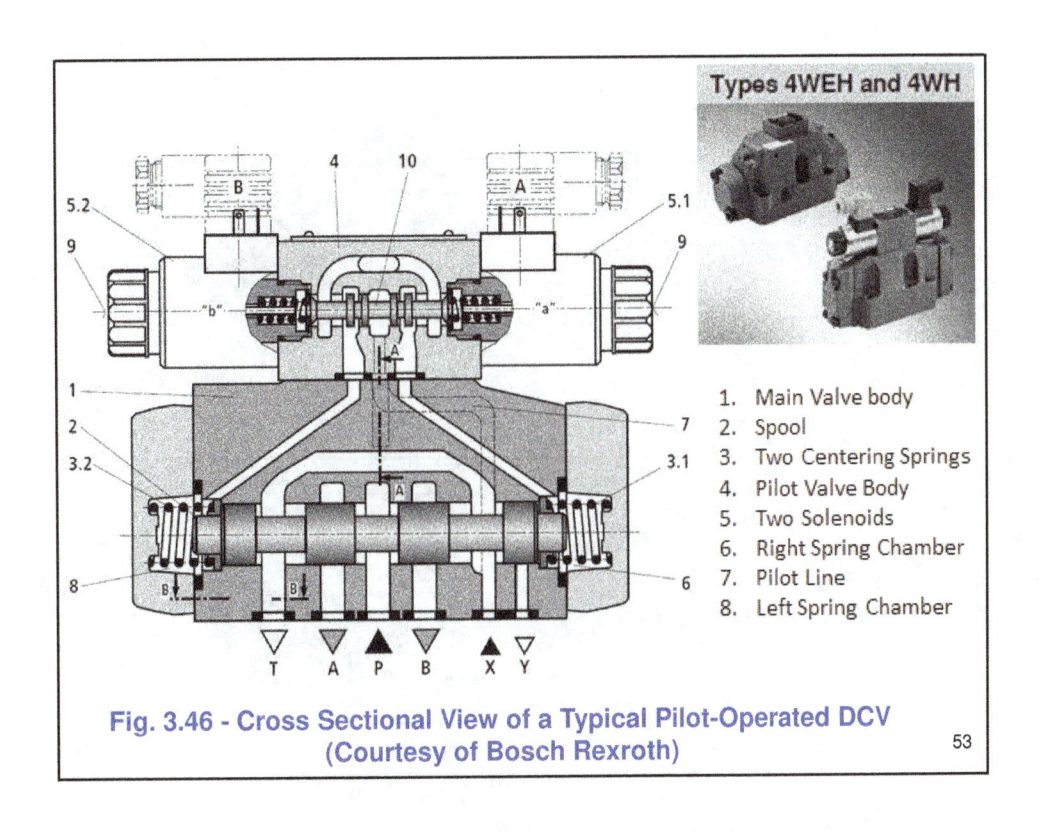

Fig. 4.44 - Pilot-Operated 4/3 Directional Valves (Courtesy of ASSOFLUID)

Fig. 3.45

Types 4WEH and 4WH

1. Main Valve body
2. Spool
3. Two Centering Springs
4. Pilot Valve Body
5. Two Solenoids
6. Right Spring Chamber
7. Pilot Line
8. Left Spring Chamber

Fig. 3.46 - Cross Sectional View of a Typical Pilot-Operated DCV (Courtesy of Bosch Rexroth)

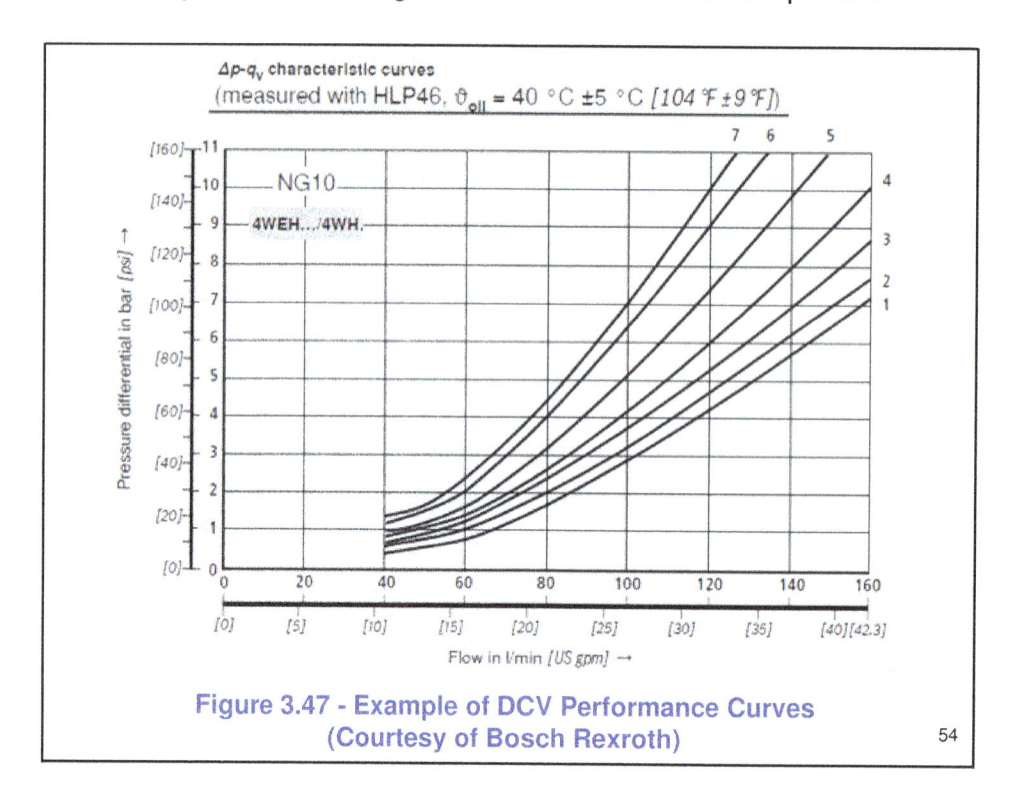

Figure 3.47 - Example of DCV Performance Curves
(Courtesy of Bosch Rexroth)

Fig. 3.48 - Switching Solenoid-Actuated Spool-Type Mobile Control Blocks
(Courtesy of ASSOFLUID)

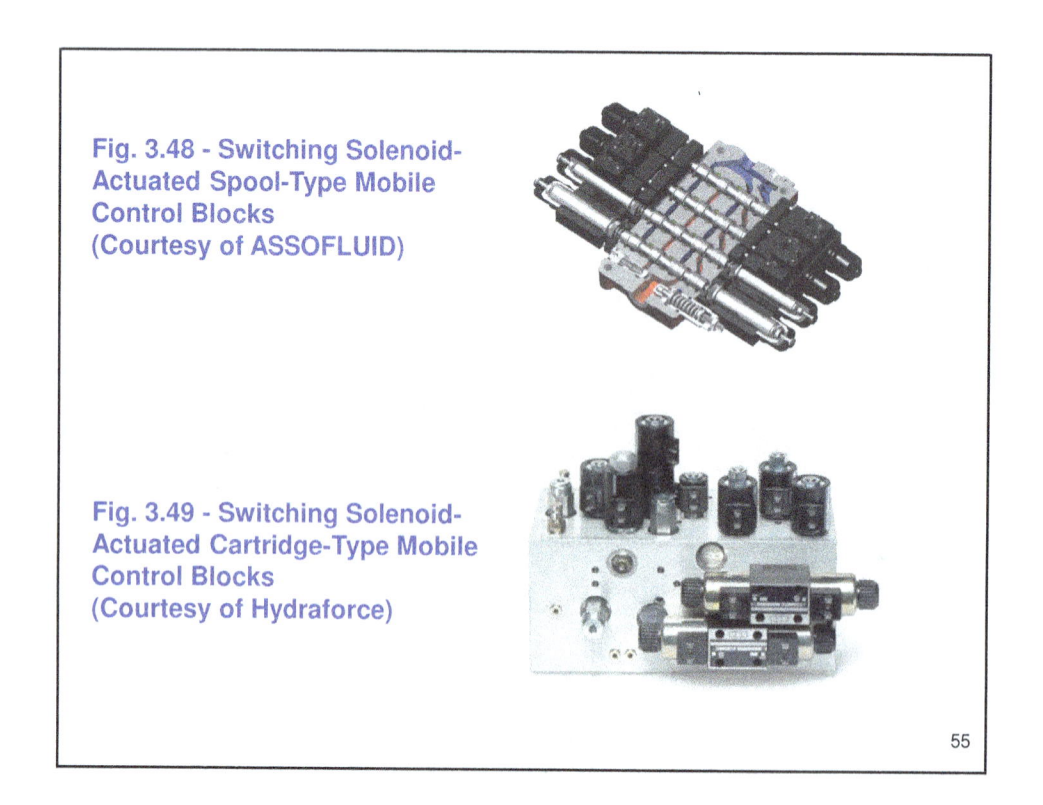

Fig. 3.49 - Switching Solenoid-Actuated Cartridge-Type Mobile Control Blocks
(Courtesy of Hydraforce)

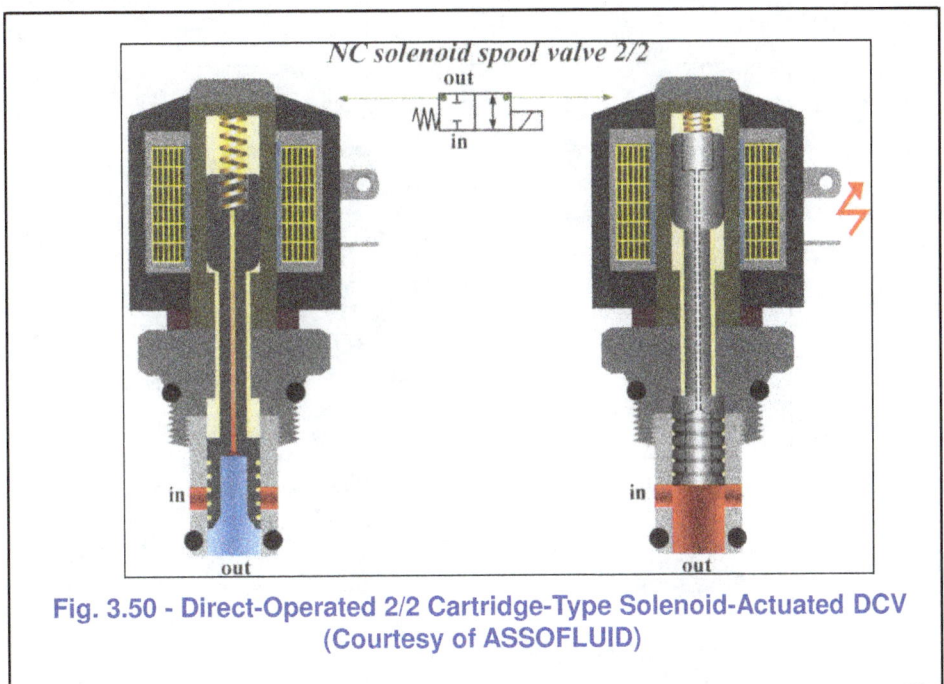

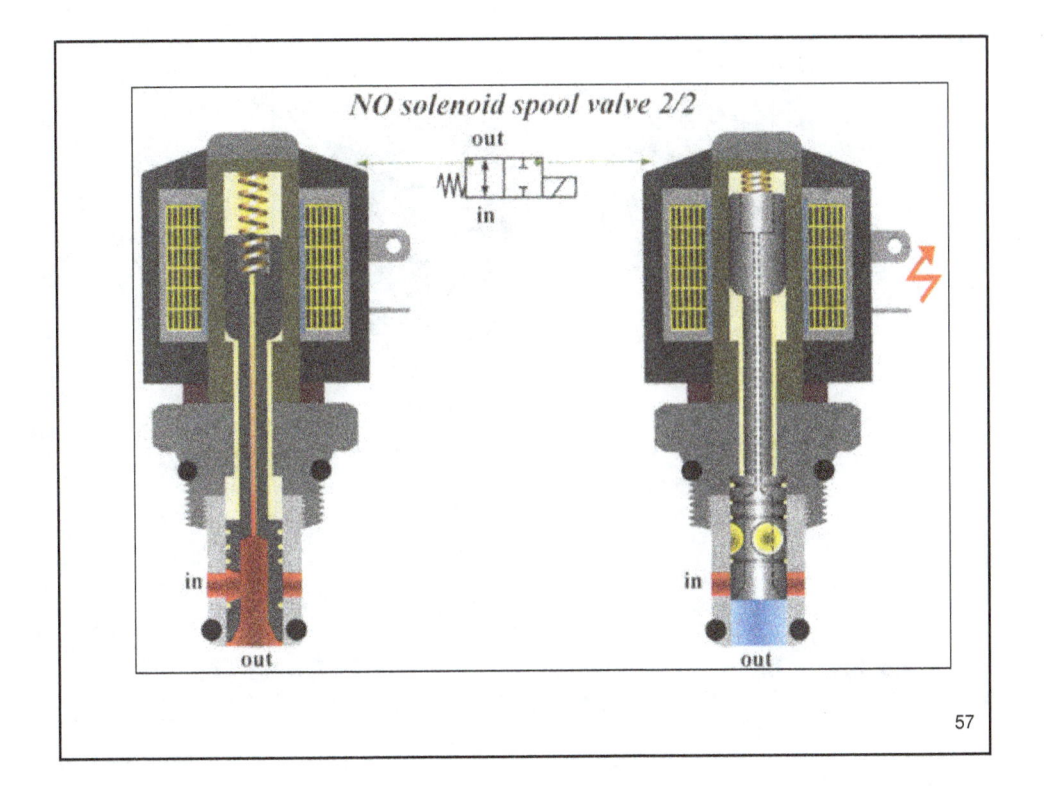

Fig. 3.50 - Direct-Operated 2/2 Cartridge-Type Solenoid-Actuated DCV
(Courtesy of ASSOFLUID)

Workbook: Electro-Hydraulic Components and Systems
Chapter 3: Switching Valves Construction and Operation

Fig. 3.51 - Direct-Operated Normally-Closed 3/2 Cartridge-Type Solenoid-Actuated DCV

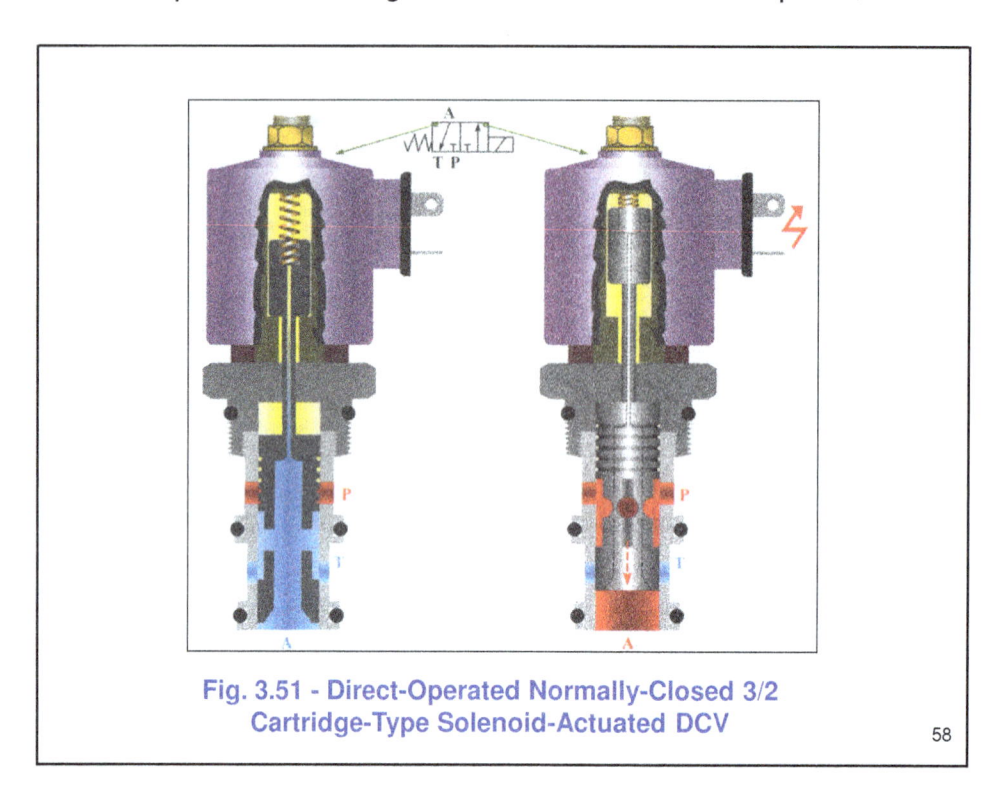

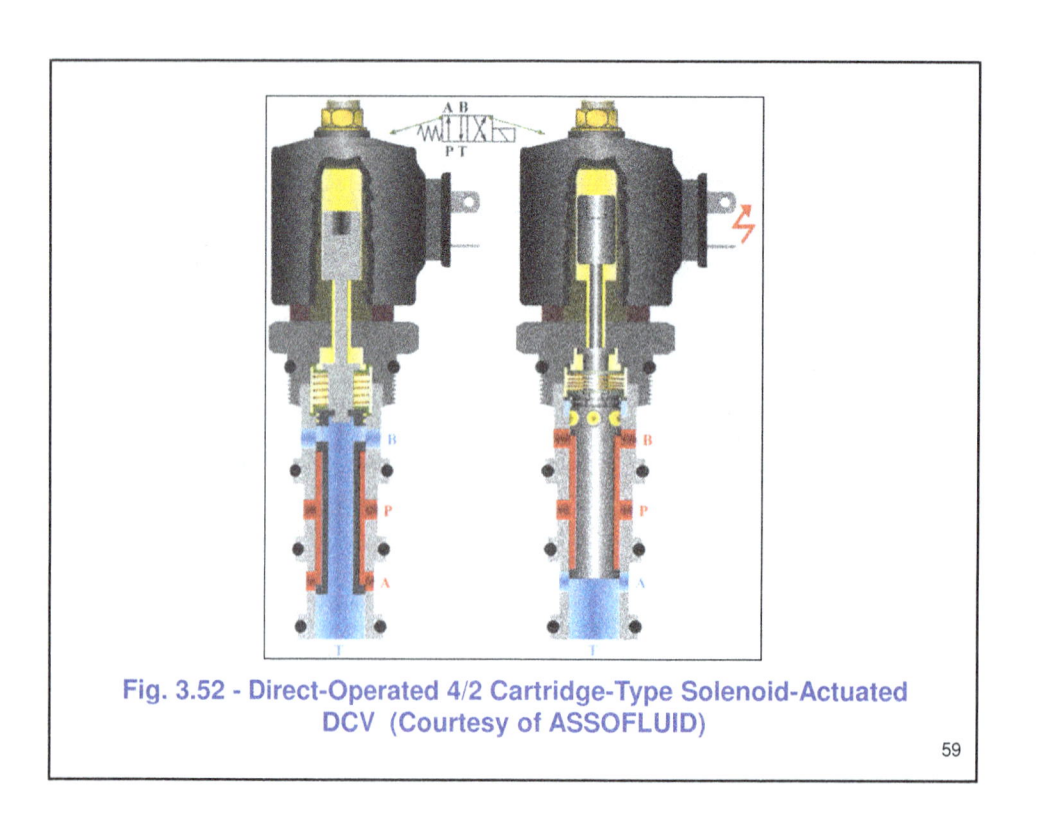

Fig. 3.52 - Direct-Operated 4/2 Cartridge-Type Solenoid-Actuated DCV (Courtesy of ASSOFLUID)

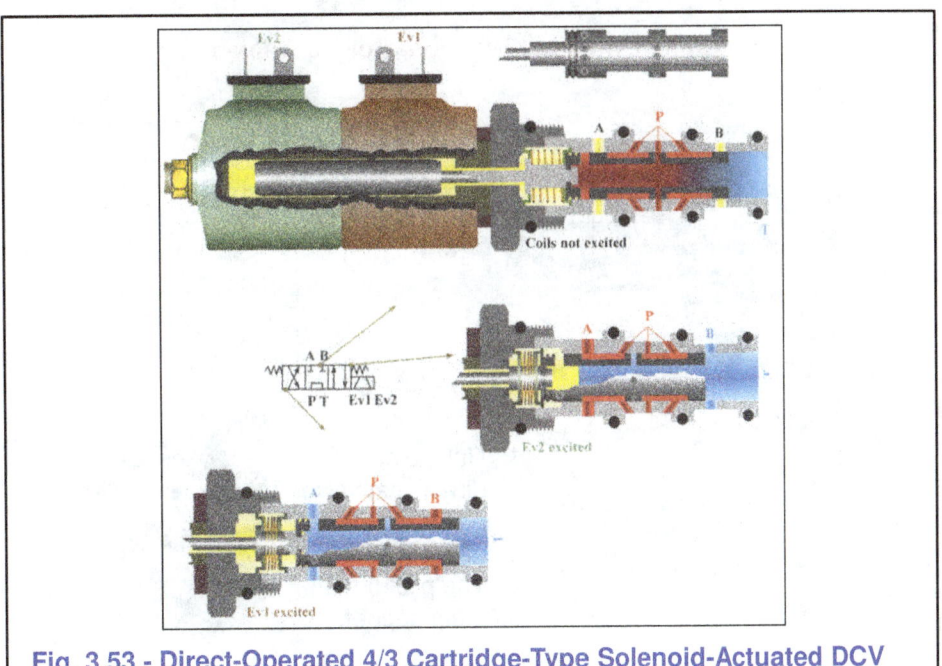

Fig. 3.53 - Direct-Operated 4/3 Cartridge-Type Solenoid-Actuated DCV (Courtesy of ASSOFLUID)

Fig. 3.54 - Solenoid-Actuated Cartridge-Type DCV (Courtesy of Hydraforce)

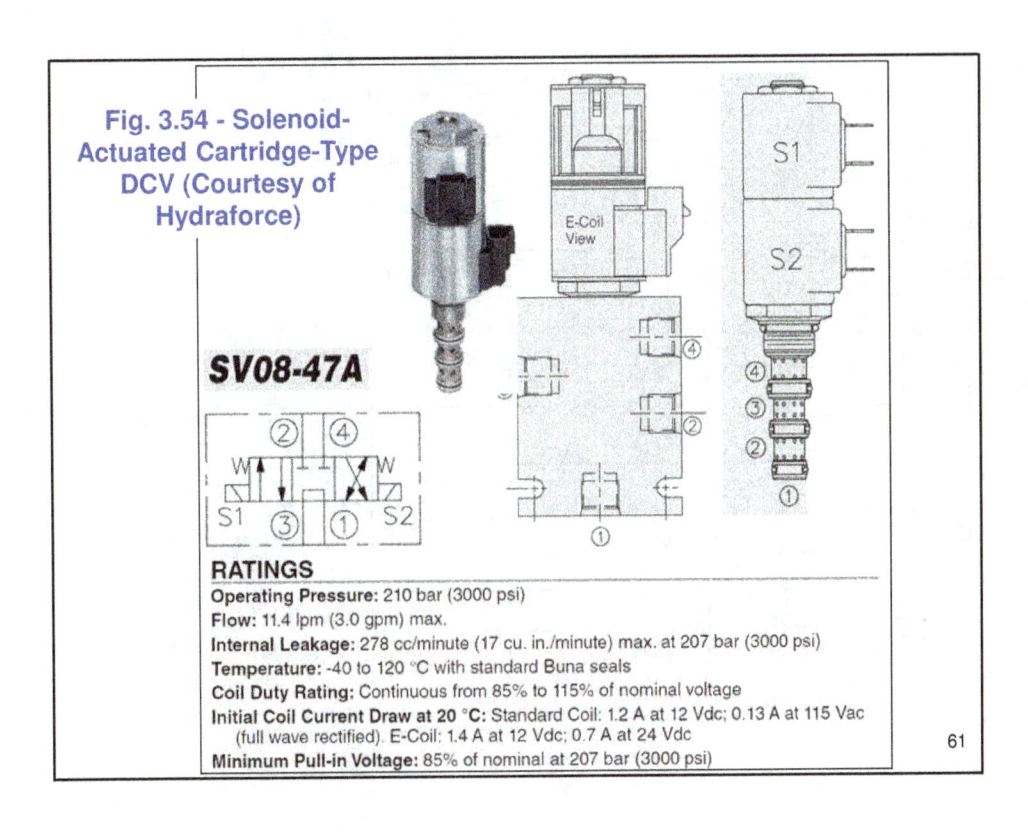

SV08-47A

RATINGS
Operating Pressure: 210 bar (3000 psi)
Flow: 11.4 lpm (3.0 gpm) max.
Internal Leakage: 278 cc/minute (17 cu. in./minute) max. at 207 bar (3000 psi)
Temperature: -40 to 120 °C with standard Buna seals
Coil Duty Rating: Continuous from 85% to 115% of nominal voltage
Initial Coil Current Draw at 20 °C: Standard Coil: 1.2 A at 12 Vdc; 0.13 A at 115 Vac (full wave rectified). E-Coil: 1.4 A at 12 Vdc; 0.7 A at 24 Vdc
Minimum Pull-in Voltage: 85% of nominal at 207 bar (3000 psi)

3.5- Electro-Hydraulic Switching Pressure Control Valves

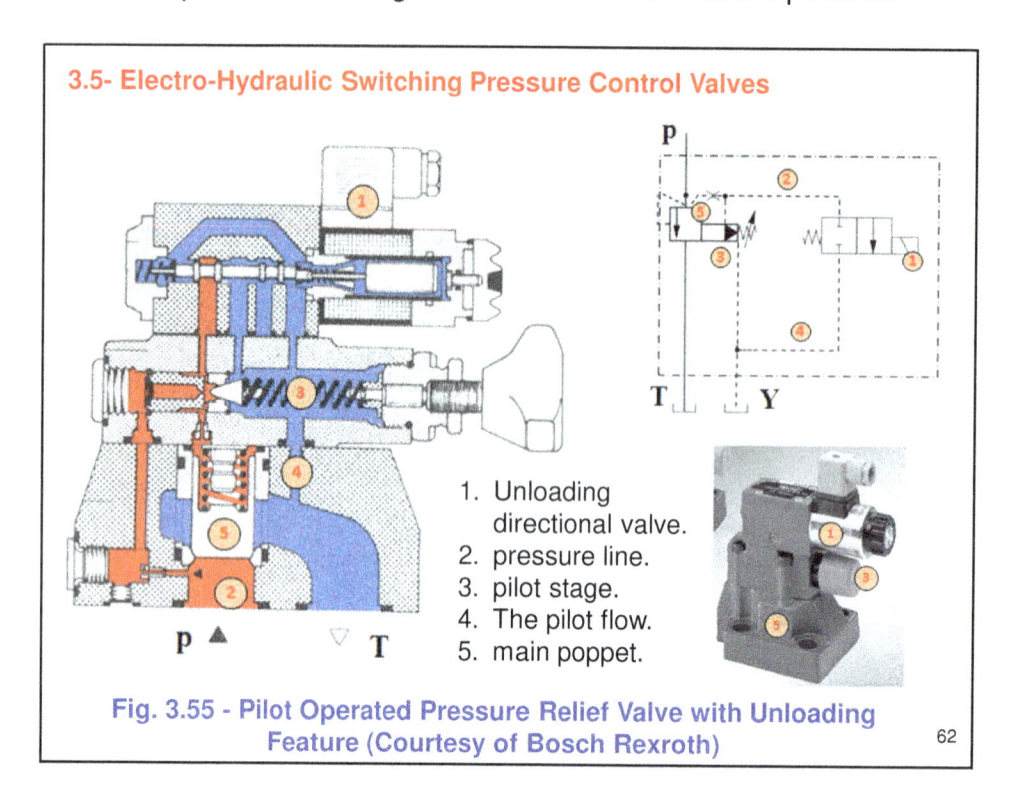

1. Unloading directional valve.
2. pressure line.
3. pilot stage.
4. The pilot flow.
5. main poppet.

Fig. 3.55 - Pilot Operated Pressure Relief Valve with Unloading Feature (Courtesy of Bosch Rexroth)

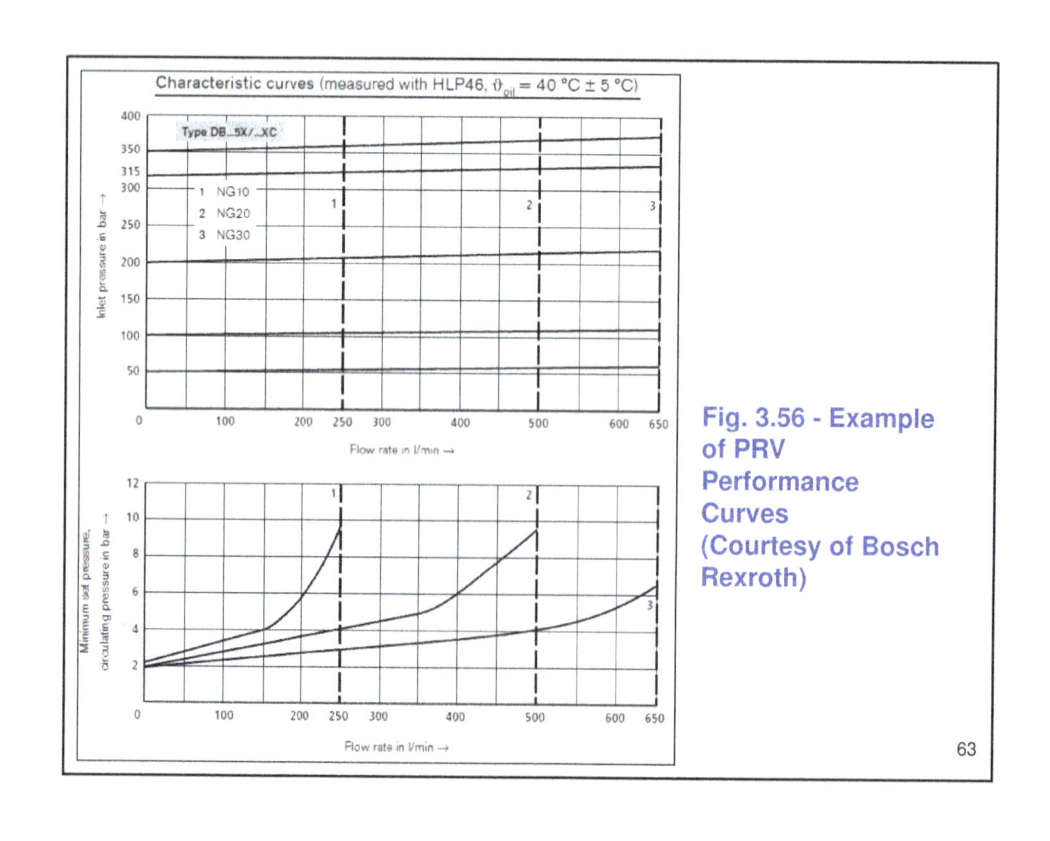

Fig. 3.56 - Example of PRV Performance Curves (Courtesy of Bosch Rexroth)

Workbook: Electro-Hydraulic Components and Systems
Chapter 3: Switching Valves Construction and Operation

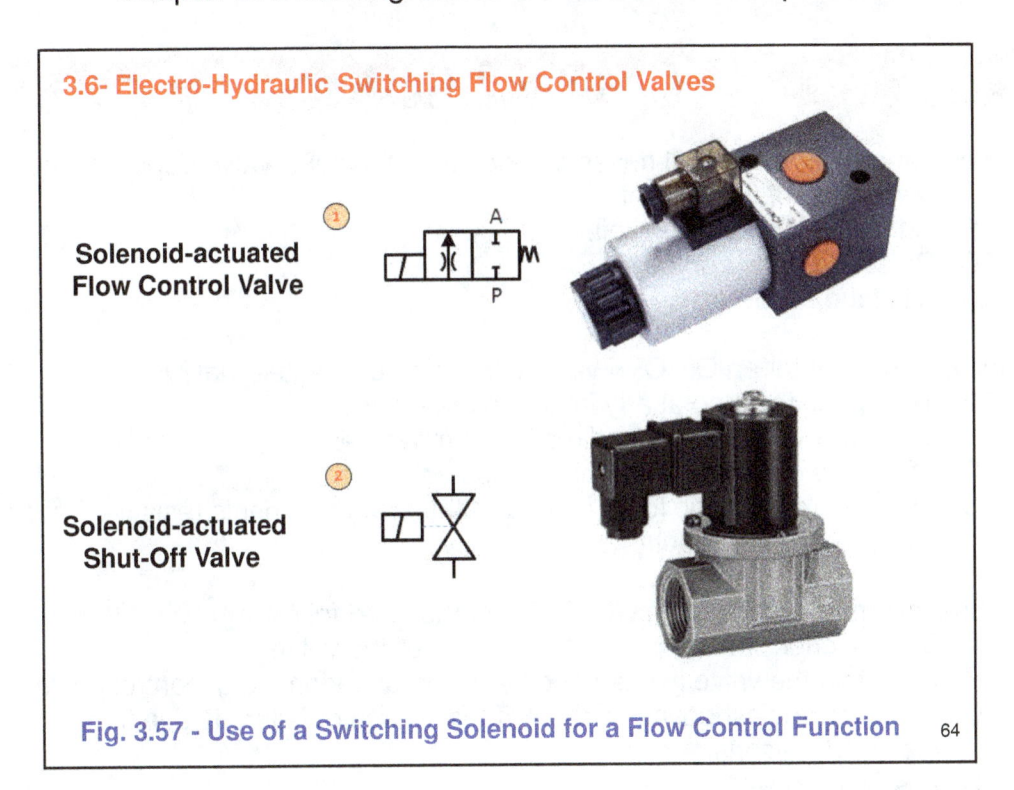

3.6- Electro-Hydraulic Switching Flow Control Valves

① Solenoid-actuated Flow Control Valve

② Solenoid-actuated Shut-Off Valve

Fig. 3.57 - Use of a Switching Solenoid for a Flow Control Function

Chapter 3 Reviews

1. Electromagnetic force that drives a spool of an ON/OFF valve depends on?
    A. Magnitude of the current.
    B. Number of turns of the coil.
    C. Presence of an iron core.
    D. All of the above.

2. Electrical current for an ON/OFF valve is rated to guarantee that?
    A. The spool can stop at any intermediate position.
    B. The spool can change its direction of movement.
    C. The valve can work under high pressure.
    D. The electromagnetic force is greater than return spring's resistive force for the full stroke of the spool.

3. Manual override options in an ON/OFF valve is used for?
    A. Manual checking the proper functioning of the valve.
    B. Actuating the valve in case of emergency situations, e.g. control power outage.
    C. System troubleshooting to check if the source of failure is due to mechanical or electrical reasons.
    D. All of the above.

4. Dry solenoid consumes less power than a wet solenoid because?
    A. A Wet solenoid larger in size than dry solenoid.
    B. A Wet solenoid has larger number of turns than dry solenoid.
    C. A Wet gap has higher resistance to magnetic field than air gap.
    D. A Wet solenoid has better sealing than dry solenoid.

5. Which following statement is True?
    A. Wet solenoids offer better coil cooling as compared to dry solenoids.
    B. Dry solenoids are more reliable than wet solenoid.
    C. Dry solenoids contain static seals only.
    D. None of the above is true.

6. The figure shown is for a?
    A. Wet solenoid.
    B. Dry solenoid.
    C. Proportional solenoid.
    D. Torque motor.

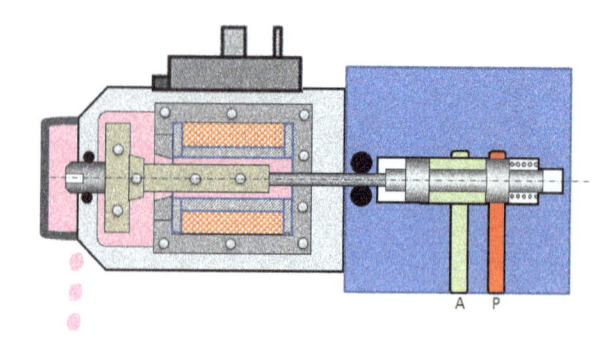

Workbook: Electro-Hydraulic Components and Systems
Chapter 3: Reviews and Assignments

7. The figure shown is for a?
   A. Wet solenoid.
   B. Dry solenoid.
   C. Proportional solenoid.
   D. Torque motor.

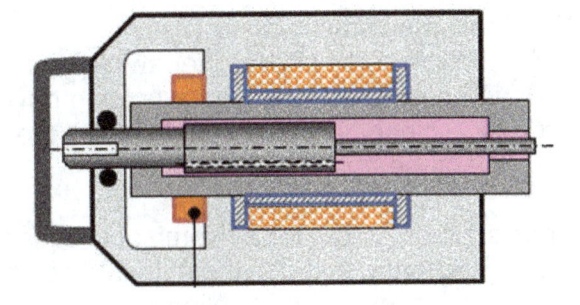

8. AC solenoid operation could be associated with the following phenomena?
   A. Inrush current.
   B. AC Hum.
   C. Eddy current.
   D. All of the above.

9. The following could be a reason for developing inrush current in an AC solenoid operation?
   A. Increased rated current.
   B. The armature is not fully seated because of spool sticking due to oil contamination.
   C. Strong electromagnetic field.
   D. All the above.

10. An AC solenoid can be burned as a consequence of?
    A. Use of incorrect voltage supply.
    B. Eddy current.
    C. Inrush current.
    D. All of the above.

11. Shading coils in dry solenoids and shading rings in wet solenoids are used to?
    A. Prevent developing the inrush current.
    B. Increase the magnetic force.
    C. Eliminate the eddy current.
    D. Minimize the effect of AC Hum.

12. Which of the following constructional arrangements is considered in order to eliminate the eddy current in a dry solenoid?
    A. The CT section is constructed from laminated layers.
    B. Use shading coil.
    C. Use static seals between the valve and the armature.
    D. All of the above.

13. The DC wet solenoid is quieter than the AC dry solenoid because of?
    A. No Inrush current in DC solenoid.
    B. No eddy current in DC solenoid.
    C. No AC Hum phenomena in DC solenoid. Oil in wet solenoid offer better damping for the armature motion.
    D. All of the above.

14. Which following statement is True?
    A. DC solenoids offer higher switching frequency than AC solenoids.
    B. DC solenoids offer higher holding power than DC solenoid.
    C. AC solenoids are more tolerant to the contamination than a DC solenoid.
    D. All of the above.

15. What is meant by the term "Nominal Frequency of an On/OFF solenoid"?
    A. Frequency at which the solenoid can be cycled.
    B. Frequency of the supplied voltage.
    C. The coil can be turned ON continuously without fear of burning the coil under normal climate conditions.
    D. Effective force at the end of the solenoid's stroke.

16. What is meant by the term "Continuous Duty"?
    A. Frequency at which the solenoid can be cycled.
    B. Frequency of the supplied voltage.
    C. The coil can be turned ON continuously without fear of burning the coil under normal climate conditions.
    D. Effective force at the end of the solenoid's stroke.

17. What is meant by the term "Holding Force"?
    A. Frequency at which the solenoid can be cycled.
    B. Frequency of the supplied voltage.
    C. The coil can be turned ON continuously without fear of burning the coil under normal climate conditions.
    D. Effective force at the end of the solenoid's stroke.

18. What is meant by the term "Switching Rate"?
    A. Frequency at which the solenoid can be cycled.
    B. Frequency of the supplied voltage.
    C. The coil can be turned ON continuously without fear of burning the coil under normal climate conditions.
    D. Effective force at the end of the solenoid's stroke.

19. What is meant by "Dual-Frequency Switching Solenoid"?
    A. A solenoid that can move in both directions.
    B. A solenoid that can work on DC or AC current.
    C. A solenoid that can two switching (cycling) frequencies.
    D. A solenoid that is rated for both 50 and 60 Hz systems.

20. What is the measuring unit of magnetic field?
    A. Rad/s.
    B. Hertz.
    C. Tesla T (N/Am).
    D. Ampere.

Workbook: Electro-Hydraulic Components and Systems
Chapter 3: Reviews and Assignments

Chapter 3 Assignment

Student Name: -- Student ID: -------------------

Date: -- Score: ------------------------

Calculate the magnetic field intensity in the cylindrical space around a piece of a wire if:

μ_0 for air $= 4\pi \times 10^{-7}$ N/A^2, $I = 2$A, and $r = 2.54$ cm (1 inch)

Chapter 4
Electrical Circuits for Switching Valves

Objectives:

This chapter covers the basic safety precautions that must be considered when building an electrohydraulic circuit that drives a switching valve. The chapter also presents the electrical devices and their symbols that are most commonly used with switching valves including: switches, buttons, relays, and PLCs. The chapter also covers the rules to read and write various forms of electrical circuit diagrams including: Joint Industrial Council (JIC) schematic diagrams, wiring diagram, and sequence diagram. The chapter concludes with various electrohydraulic circuits that simulate typical applications.

Brief Contents:

4.1- Best Practices for Safe Operation of Electro-Hydraulic Systems

4.2- Basic Electrical Symbols

4.3- Basic Electrical Devices

4.4- Electrical Schematic Diagrams

4.5- Electrical Circuits for Applications of Switching Valves

Workbook: Electro-Hydraulic Components and Systems
Chapter 4: Electrical Circuits for Switching Valves

4.1- Best Practices for Safe Operation of Electro-Hydraulic Systems

Emergency Switch:

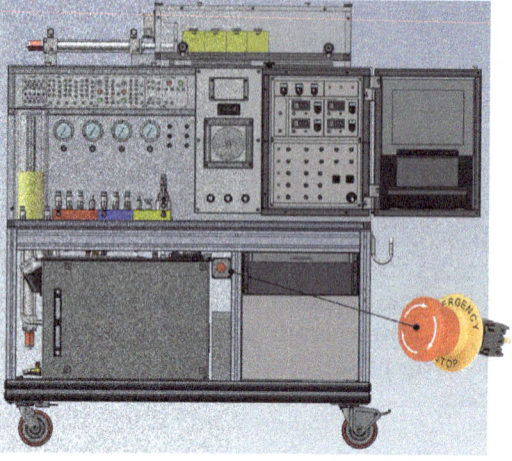

Fig. 4.1 - Use of Emergency Switches for Electro-Hydraulic Systems

Electrical Cable Shielding:

- plastic jacket
- dielectric insulator
- metallic shield
- centre core

Fig. 4.2 - Cable Shielding

Workbook: Electro-Hydraulic Components and Systems
Chapter 4: Electrical Circuits for Switching Valves

Electrical Cable Routing:

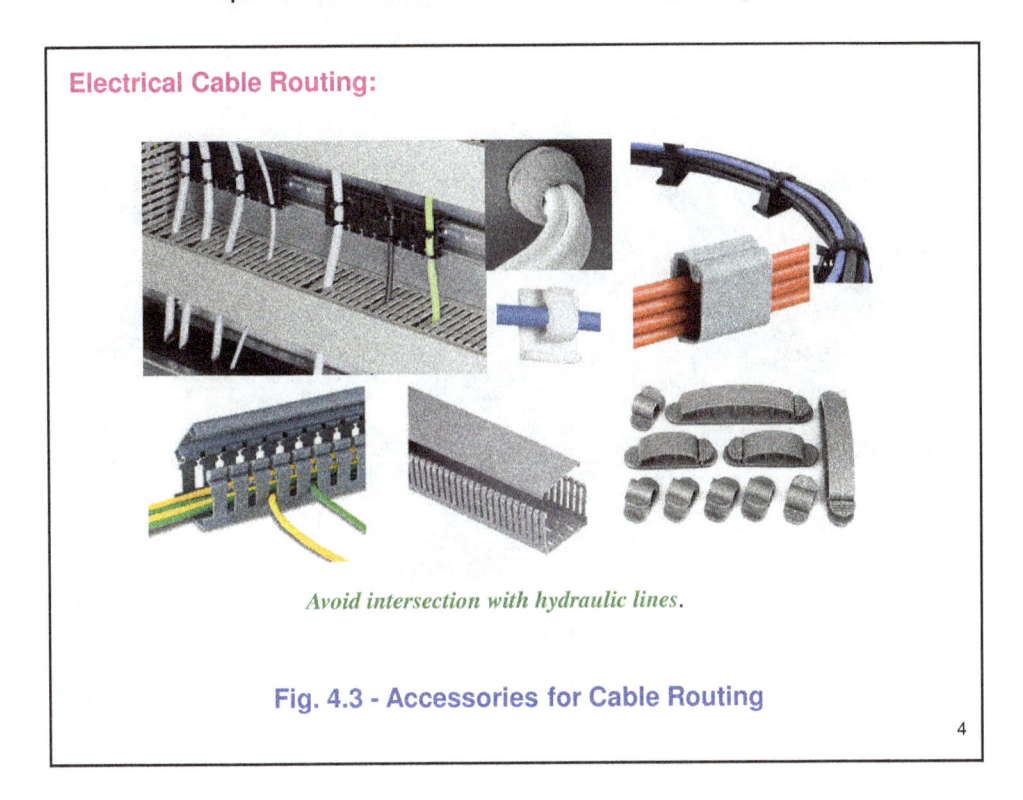

Avoid intersection with hydraulic lines.

Fig. 4.3 - Accessories for Cable Routing

Standard Electrical Diagrams:

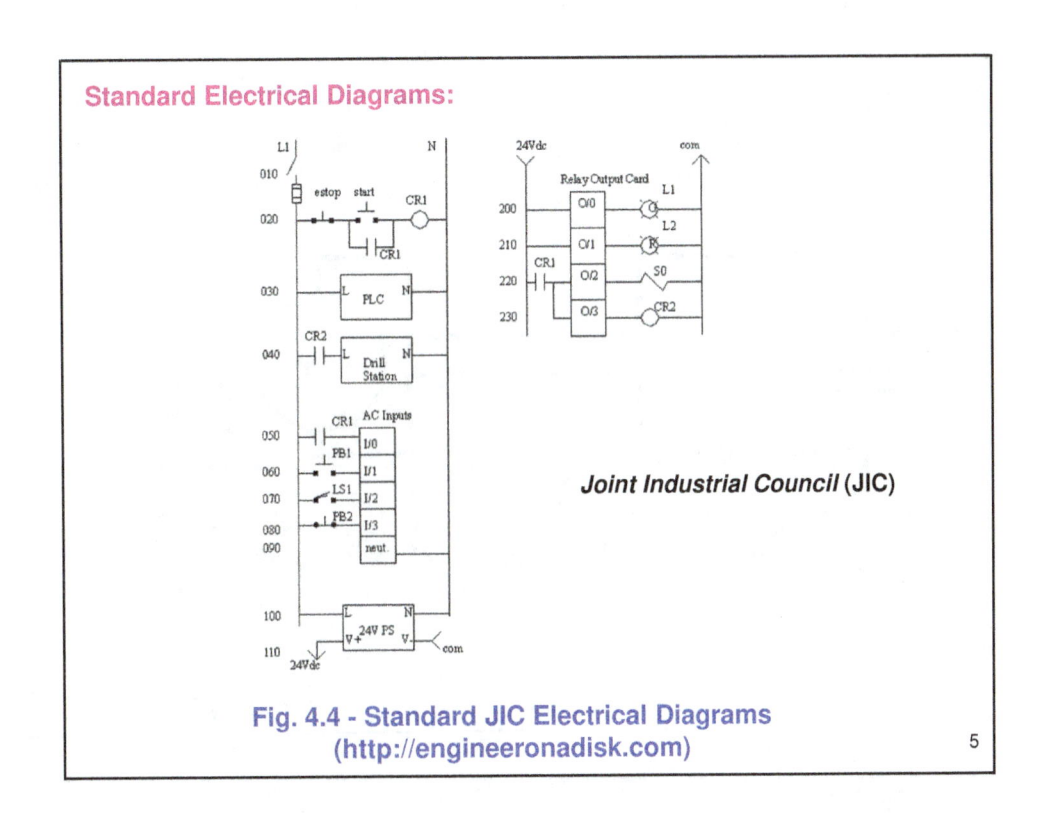

Joint Industrial Council (JIC)

Fig. 4.4 - Standard JIC Electrical Diagrams
(http://engineeronadisk.com)

Workbook: Electro-Hydraulic Components and Systems
Chapter 4: Electrical Circuits for Switching Valves

Avoid Short Circuiting:

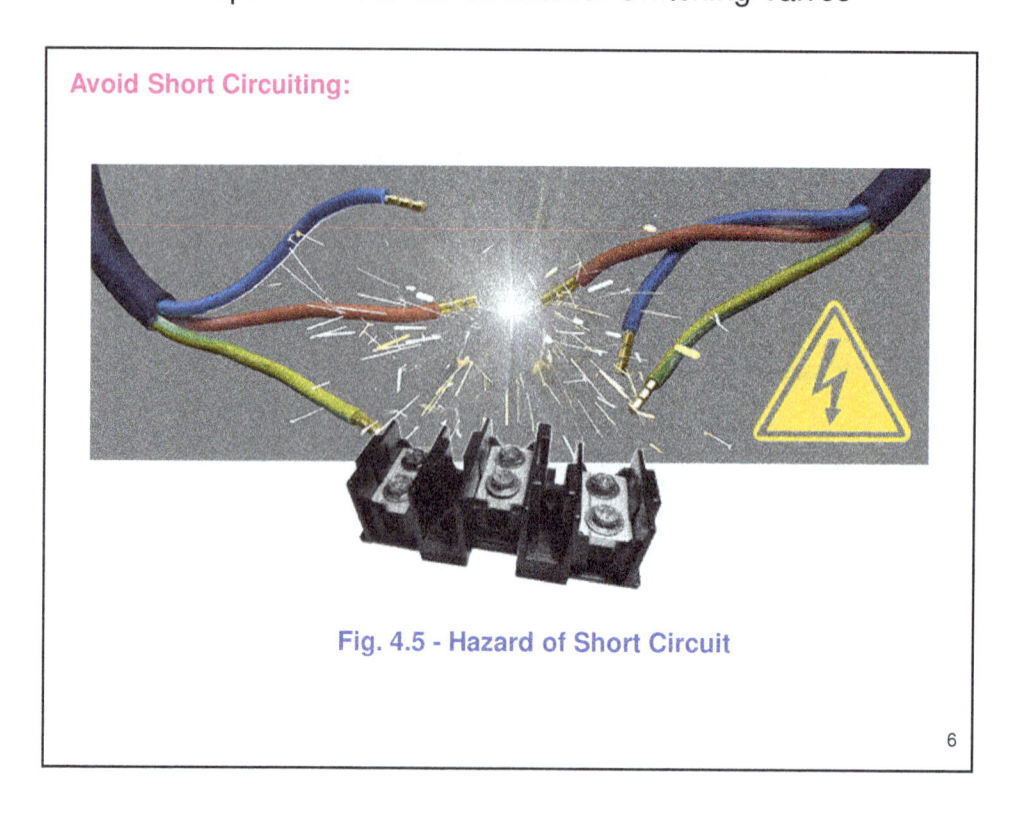

Fig. 4.5 - Hazard of Short Circuit

Proper Grounding:

Ground Point

No Voltage between the case and the ground

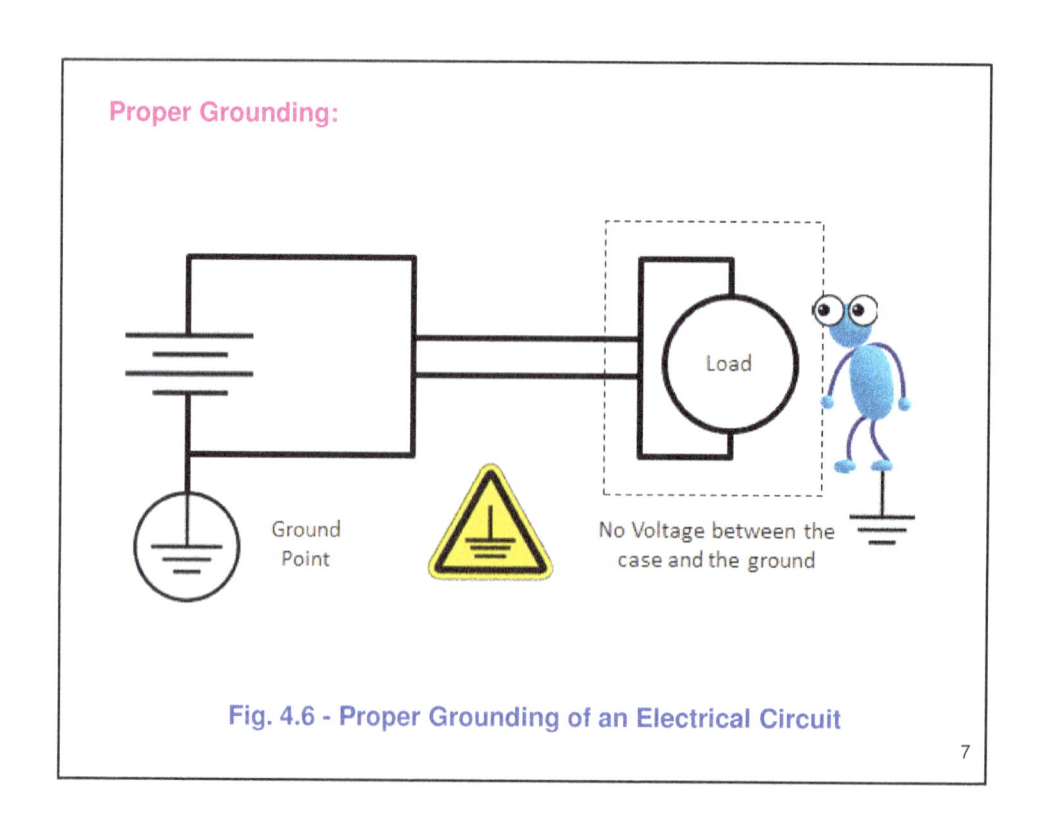

Fig. 4.6 - Proper Grounding of an Electrical Circuit

Workbook: Electro-Hydraulic Components and Systems
Chapter 4: Electrical Circuits for Switching Valves

Proper Powering of Cable Ends :

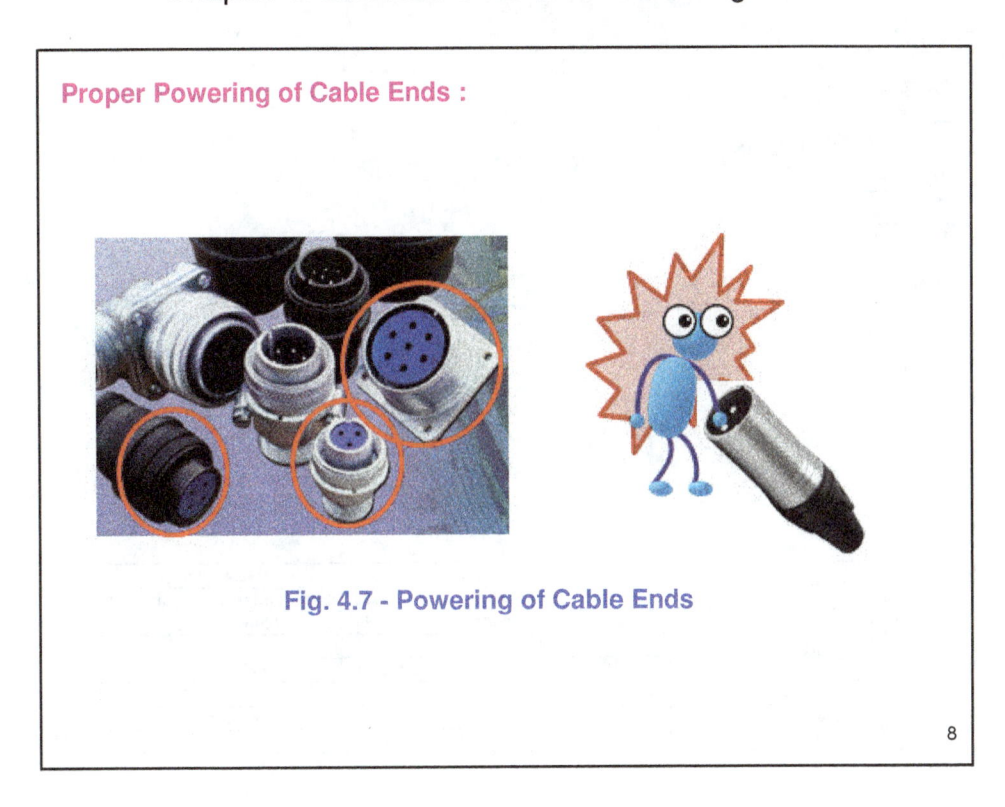

Fig. 4.7 - Powering of Cable Ends

Overload Protection:

Blade Type Fuse Cylindrical Glass Type Fuse Circuit Breaker

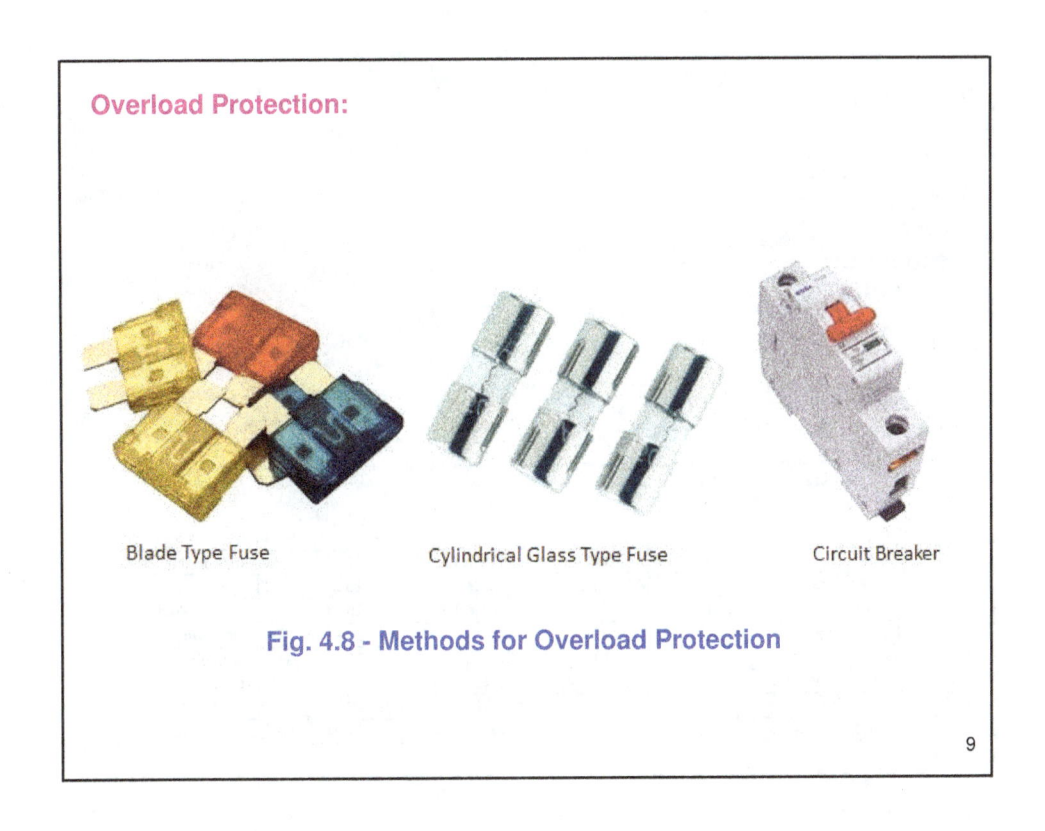

Fig. 4.8 - Methods for Overload Protection

Workbook: Electro-Hydraulic Components and Systems
Chapter 4: Electrical Circuits for Switching Valves

4.2- Basic Electrical Symbols

Fig. 4.9 - Symbols for Electrical Devices Commonly Used for EH Systems

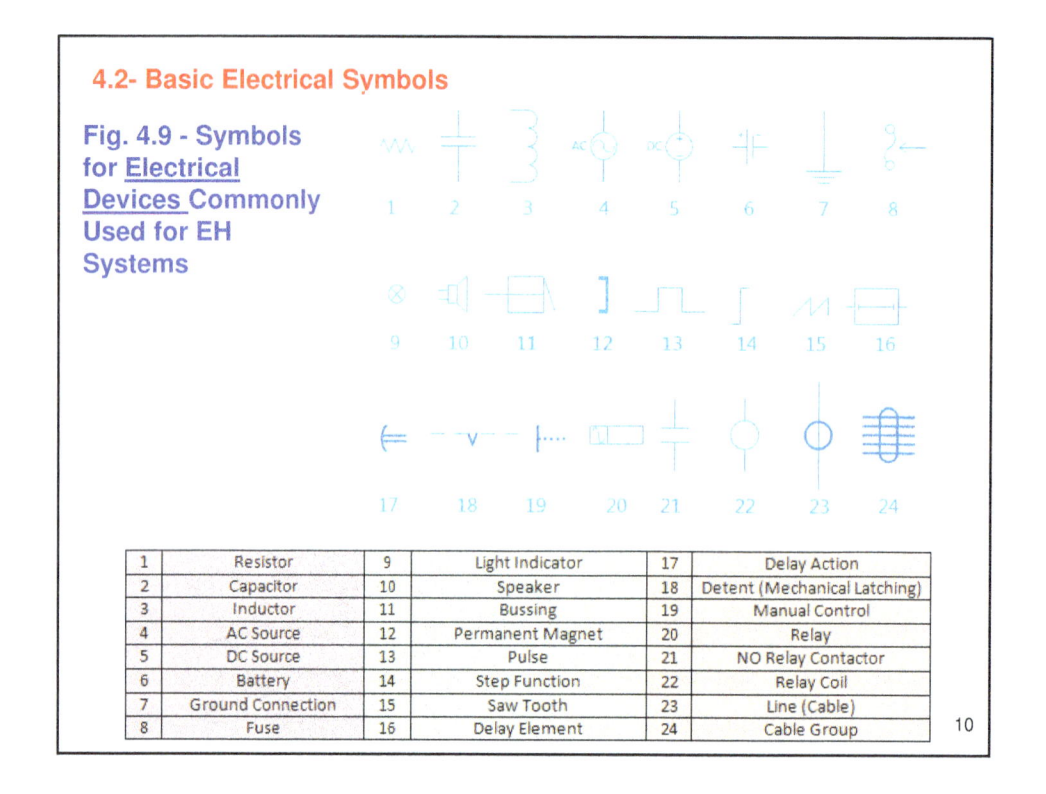

1	Resistor	9	Light Indicator	17	Delay Action
2	Capacitor	10	Speaker	18	Detent (Mechanical Latching)
3	Inductor	11	Bussing	19	Manual Control
4	AC Source	12	Permanent Magnet	20	Relay
5	DC Source	13	Pulse	21	NO Relay Contactor
6	Battery	14	Step Function	22	Relay Coil
7	Ground Connection	15	Saw Tooth	23	Line (Cable)
8	Fuse	16	Delay Element	24	Cable Group

Fig. 4.10 - Symbols for Switches and Relays Commonly Used for EH Systems

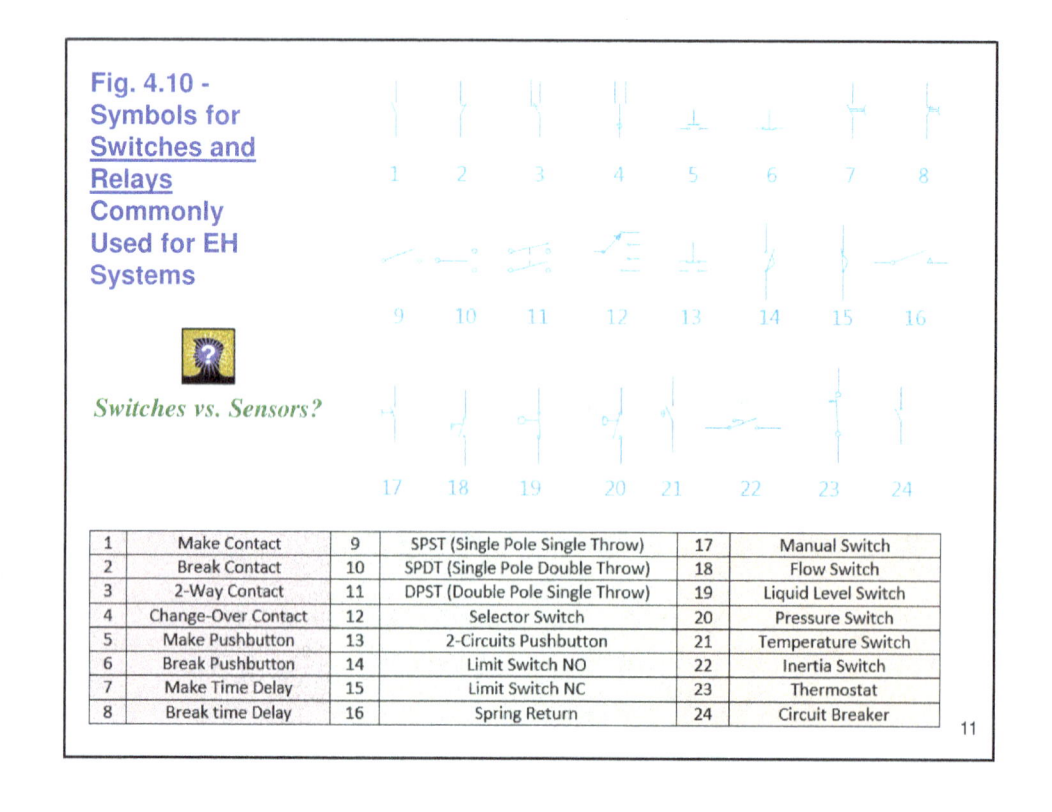

Switches vs. Sensors?

1	Make Contact	9	SPST (Single Pole Single Throw)	17	Manual Switch
2	Break Contact	10	SPDT (Single Pole Double Throw)	18	Flow Switch
3	2-Way Contact	11	DPST (Double Pole Single Throw)	19	Liquid Level Switch
4	Change-Over Contact	12	Selector Switch	20	Pressure Switch
5	Make Pushbutton	13	2-Circuits Pushbutton	21	Temperature Switch
6	Break Pushbutton	14	Limit Switch NO	22	Inertia Switch
7	Make Time Delay	15	Limit Switch NC	23	Thermostat
8	Break time Delay	16	Spring Return	24	Circuit Breaker

Workbook: Electro-Hydraulic Components and Systems
Chapter 4: Electrical Circuits for Switching Valves

4.3- Basic Electrical Devices
4.3.1- Measuring Instruments

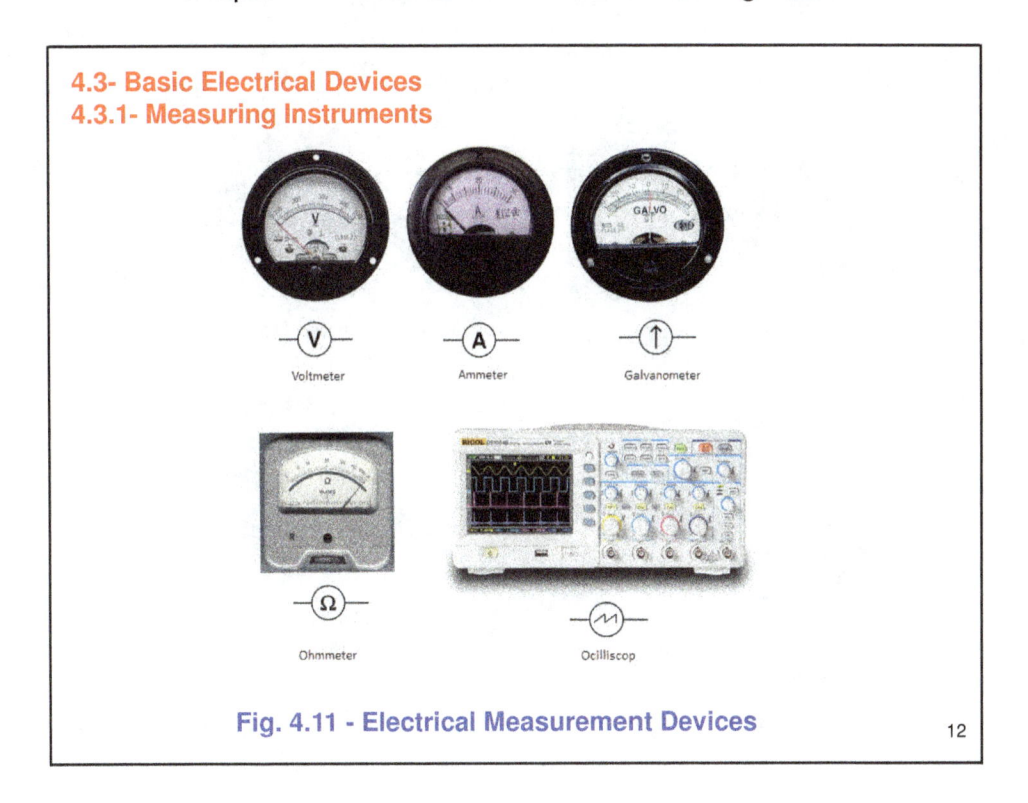

Fig. 4.11 - Electrical Measurement Devices

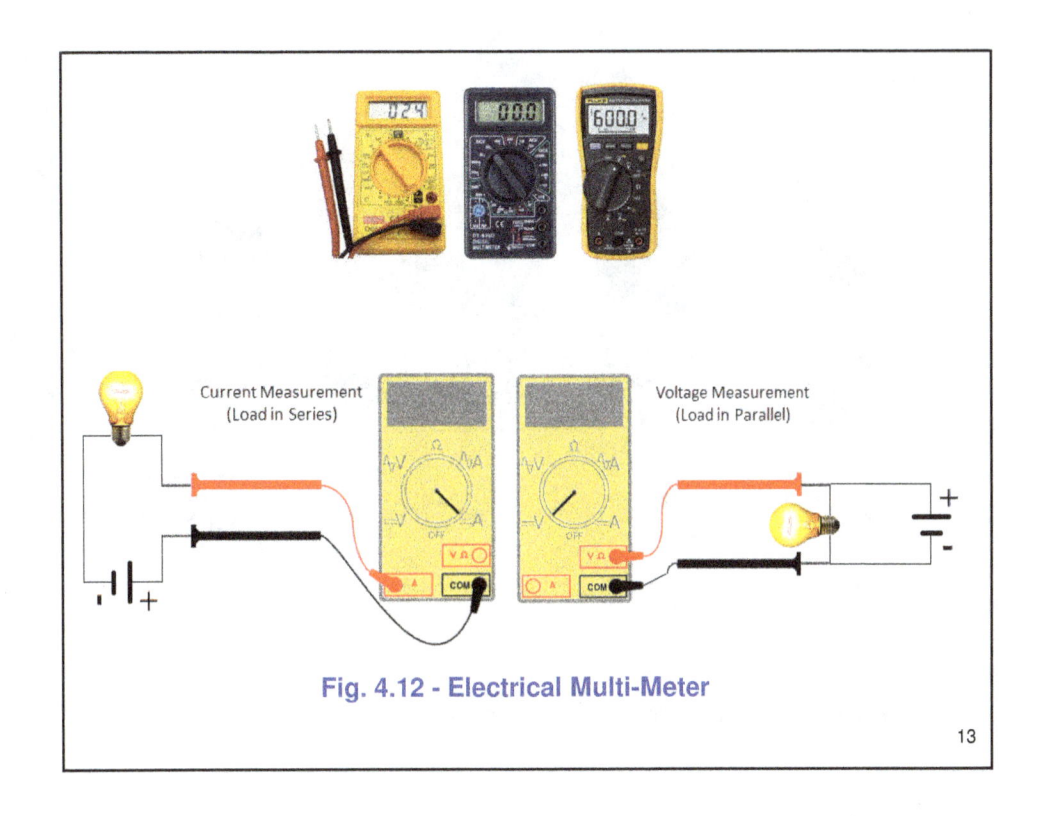

Fig. 4.12 - Electrical Multi-Meter

4.3.2 - Electrical Pushbuttons and Switches

Single Pushbutton   Double Pushbutton   Selector Switch   Combined Function Pushbutton

Key-Locked Pushbutton   Finger Toggle Switch   Push-Pull Toggle Switch

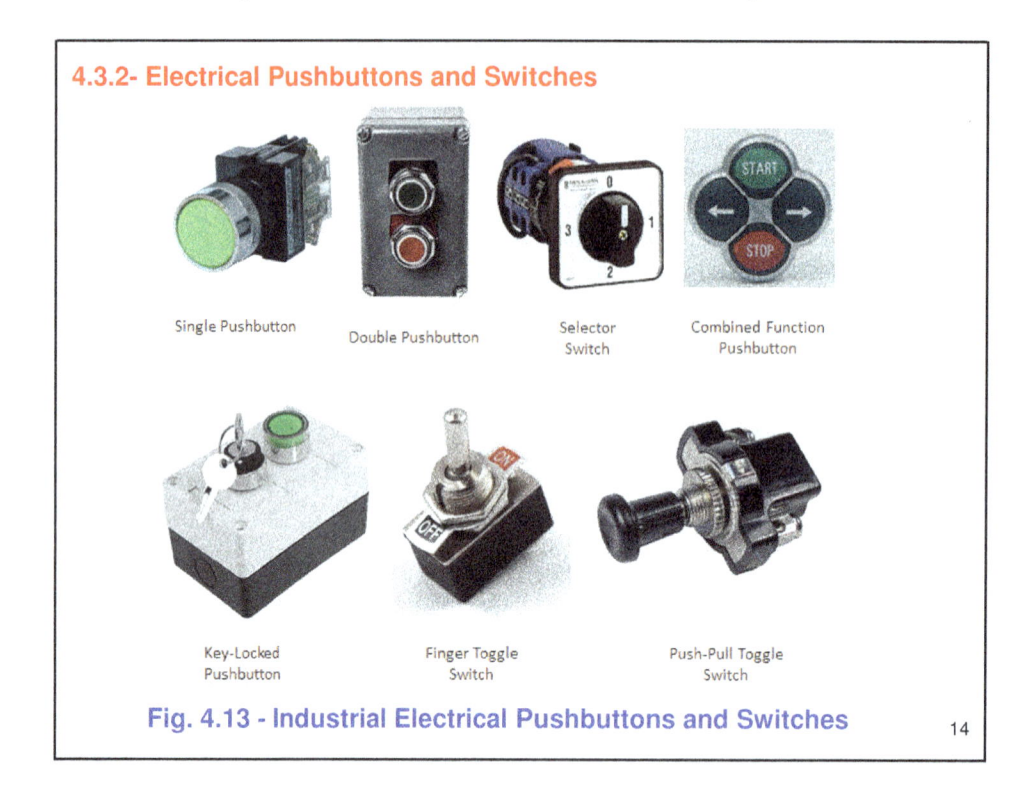

Fig. 4.13 - Industrial Electrical Pushbuttons and Switches

4.3.3 - Limit Switches

Short Roller Lever   Adjustable Roller Lever   Hinge Roller Lever   One-Way Hinge Roller Lever

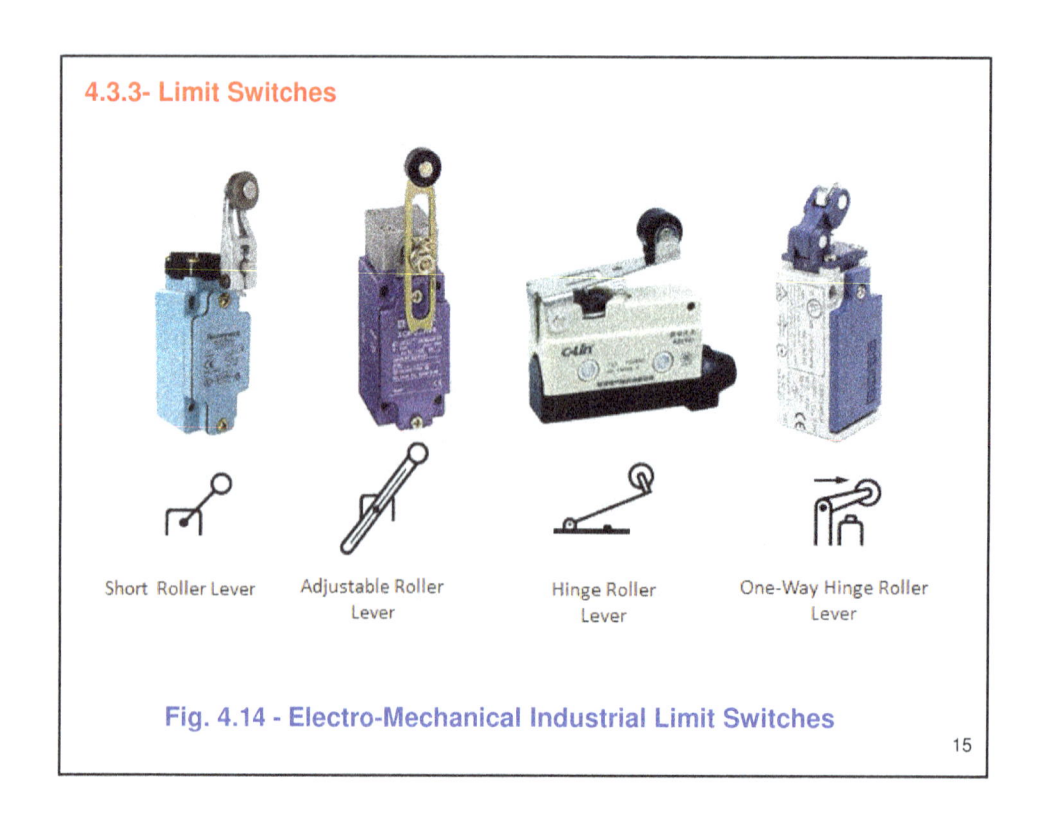

Fig. 4.14 - Electro-Mechanical Industrial Limit Switches

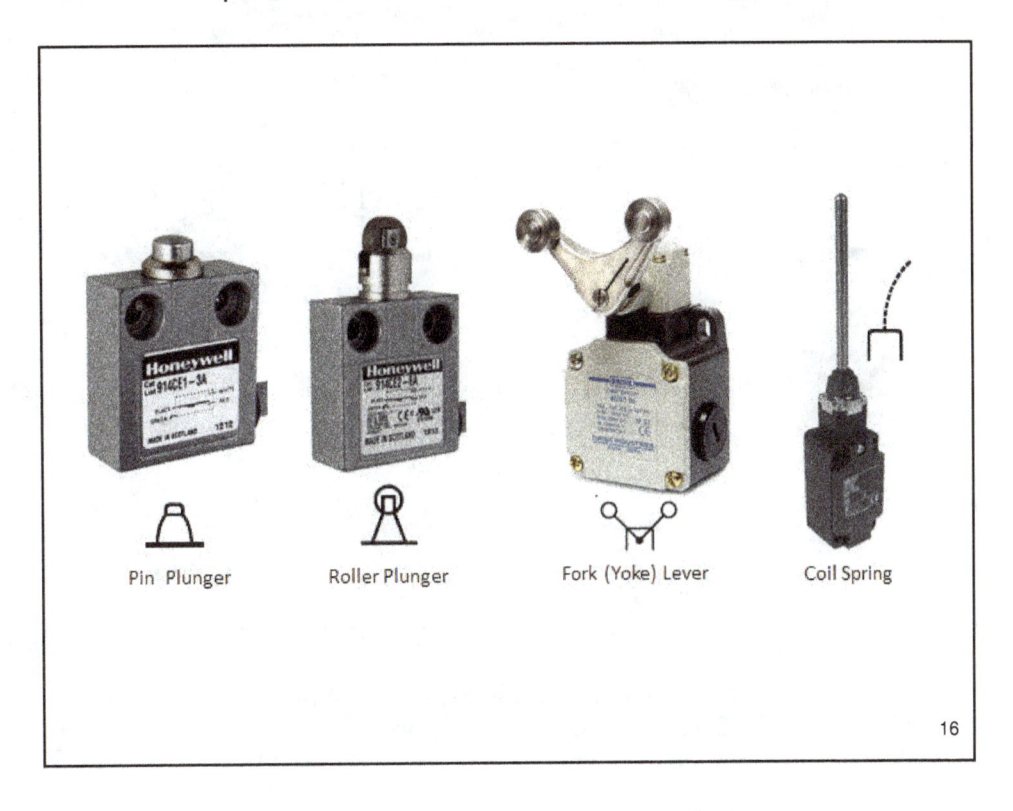

Pin Plunger   Roller Plunger   Fork (Yoke) Lever   Coil Spring

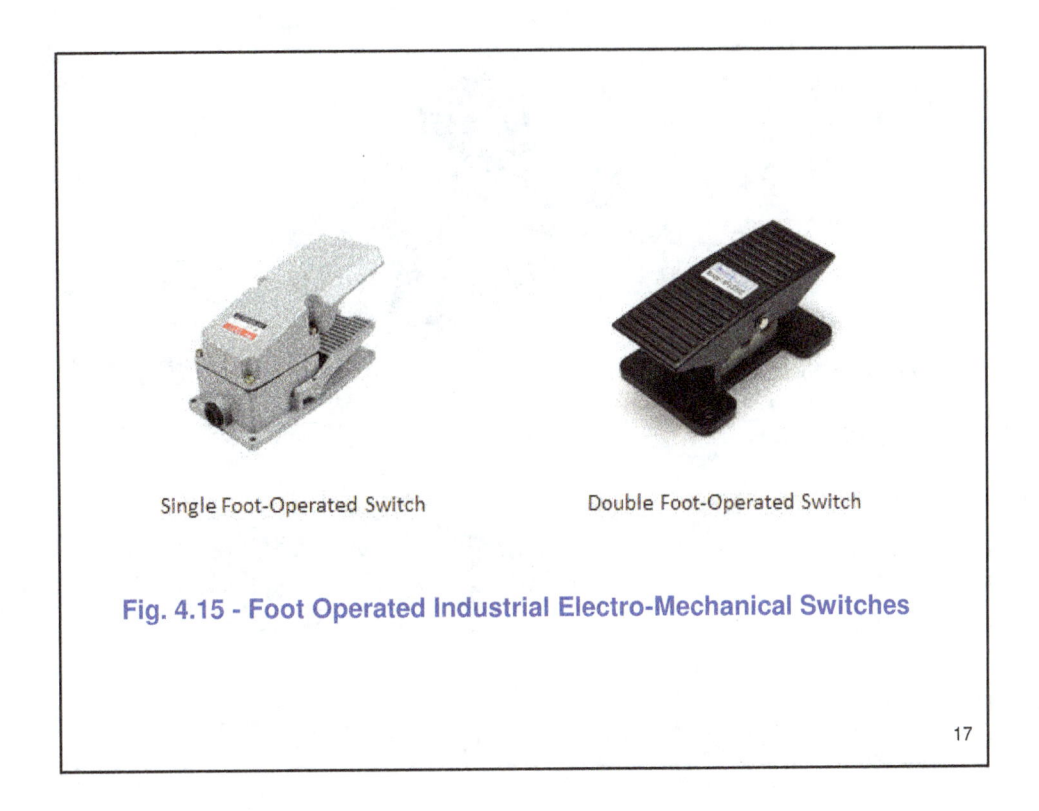

Single Foot-Operated Switch   Double Foot-Operated Switch

Fig. 4.15 - Foot Operated Industrial Electro-Mechanical Switches

4.3.4- Proximity Switches

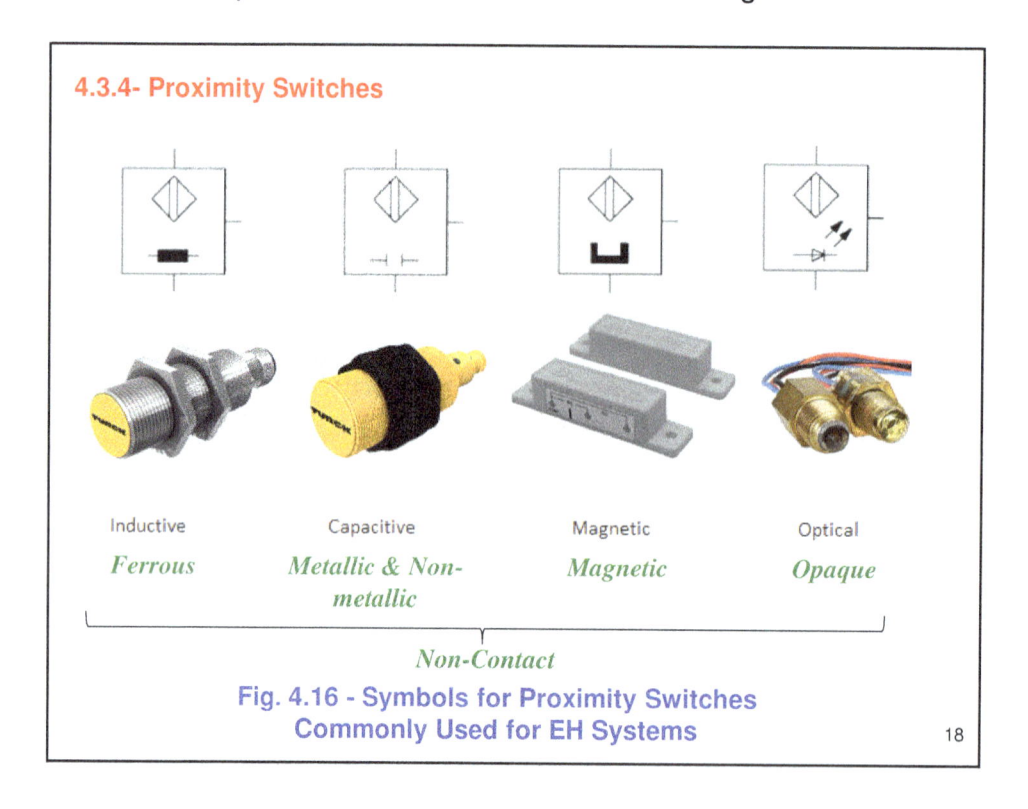

Inductive	Capacitive	Magnetic	Optical
Ferrous	*Metallic & Non-metallic*	*Magnetic*	*Opaque*

Non-Contact

Fig. 4.16 - Symbols for Proximity Switches Commonly Used for EH Systems

4.3.5- Pressure Switches

Pressure Switches vs. Vacuum Switches?

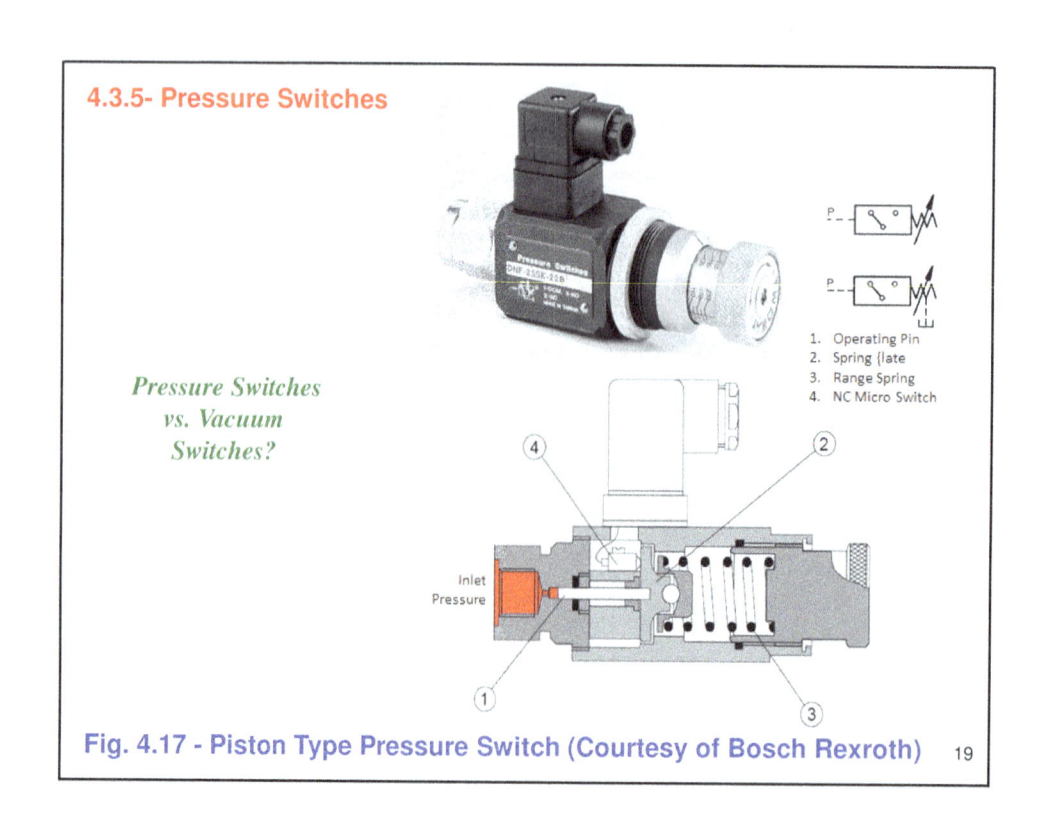

1. Operating Pin
2. Spring Plate
3. Range Spring
4. NC Micro Switch

Fig. 4.17 - Piston Type Pressure Switch (Courtesy of Bosch Rexroth)

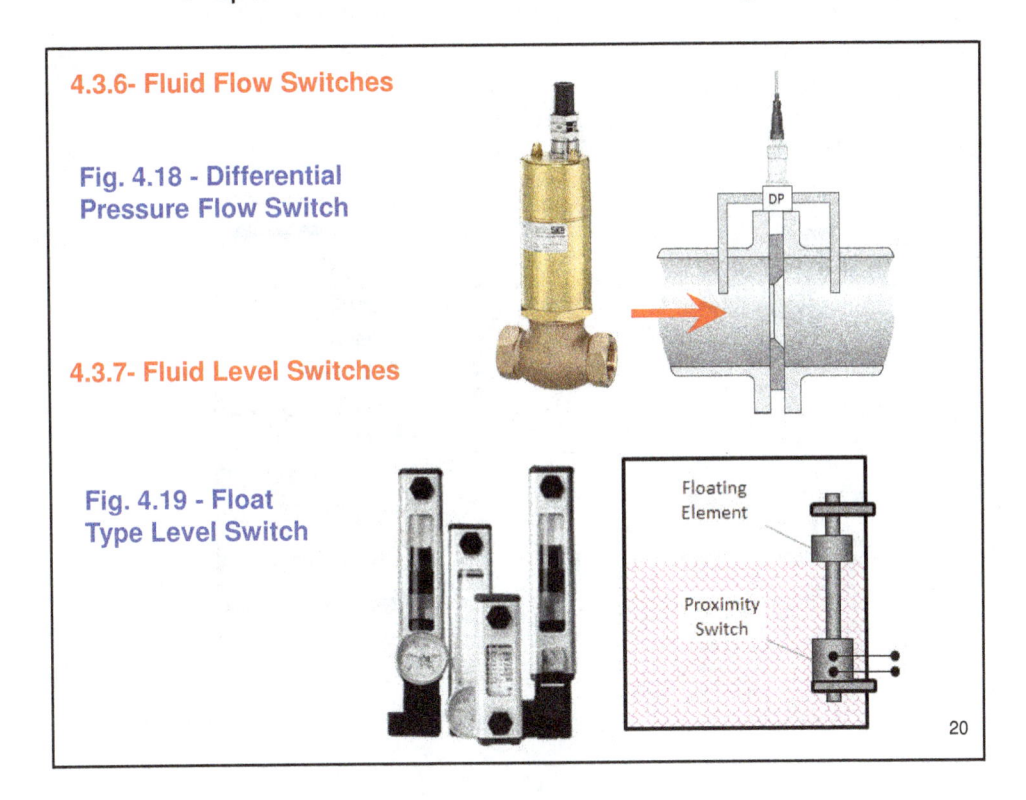

Workbook: Electro-Hydraulic Components and Systems
Chapter 4: Electrical Circuits for Switching Valves

4.3.6- Fluid Flow Switches

Fig. 4.18 - Differential Pressure Flow Switch

4.3.7- Fluid Level Switches

Fig. 4.19 - Float Type Level Switch

4.3.8- Temperature Switches

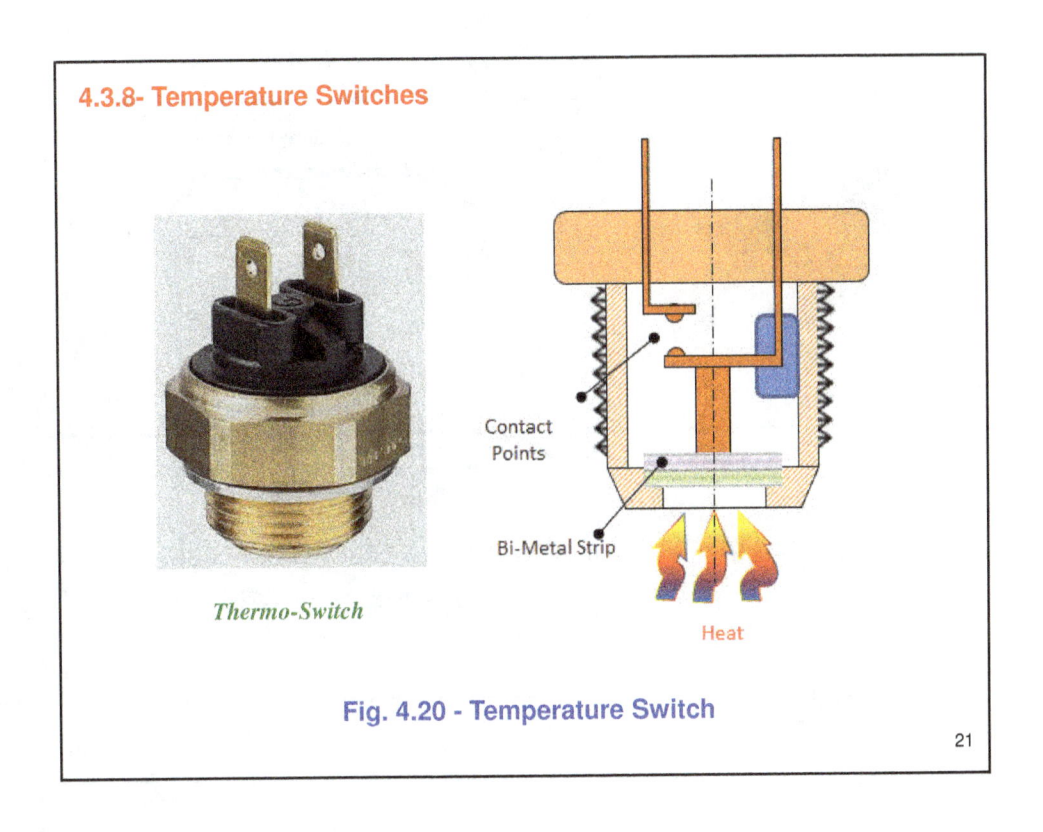

Thermo-Switch

Fig. 4.20 - Temperature Switch

Workbook: Electro-Hydraulic Components and Systems
Chapter 4: Electrical Circuits for Switching Valves

4.3.9- Control Relays

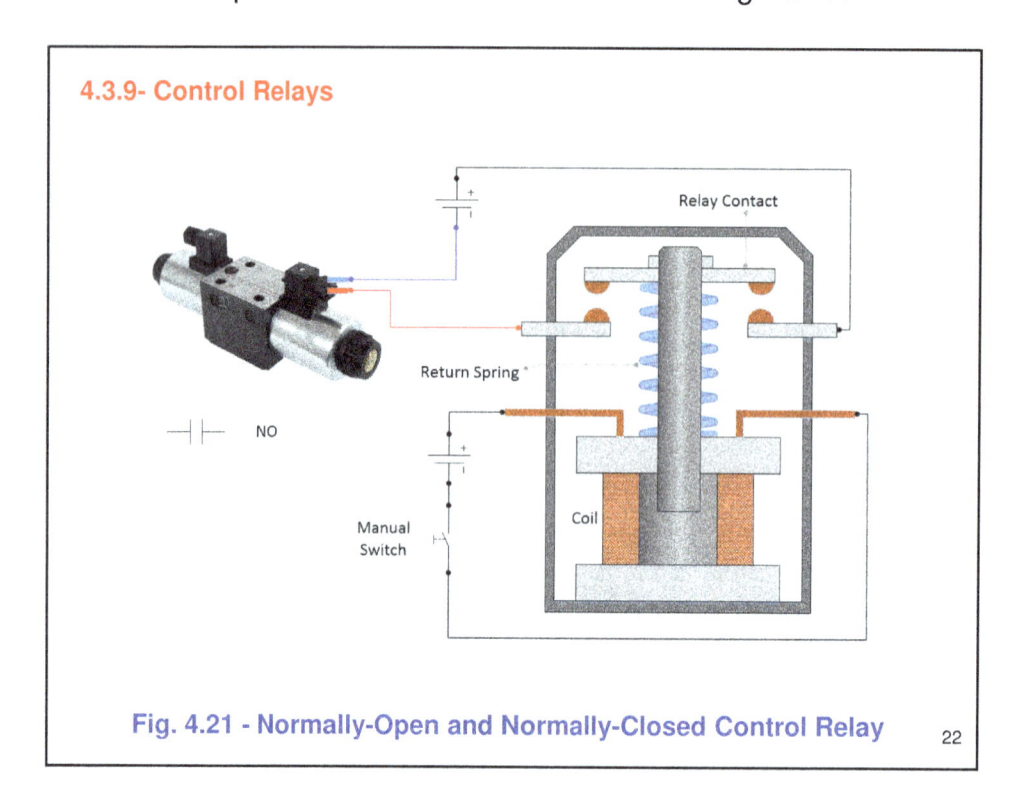

Fig. 4.21 - Normally-Open and Normally-Closed Control Relay

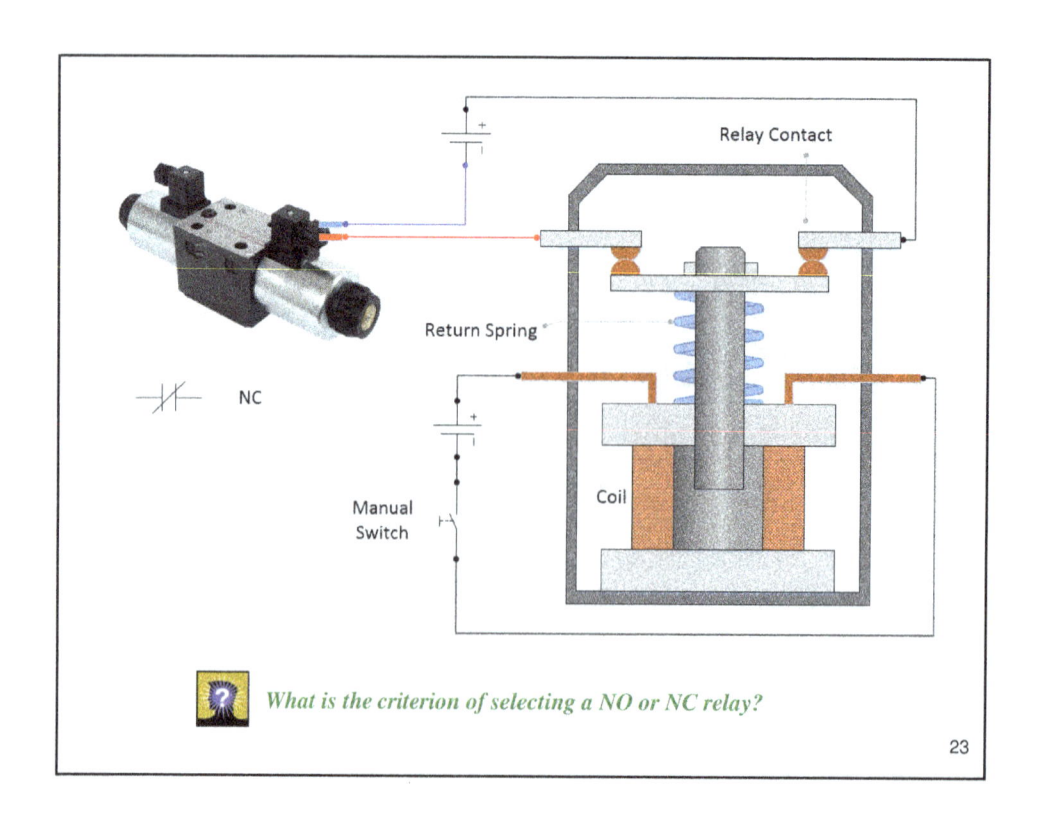

What is the criterion of selecting a NO or NC relay?

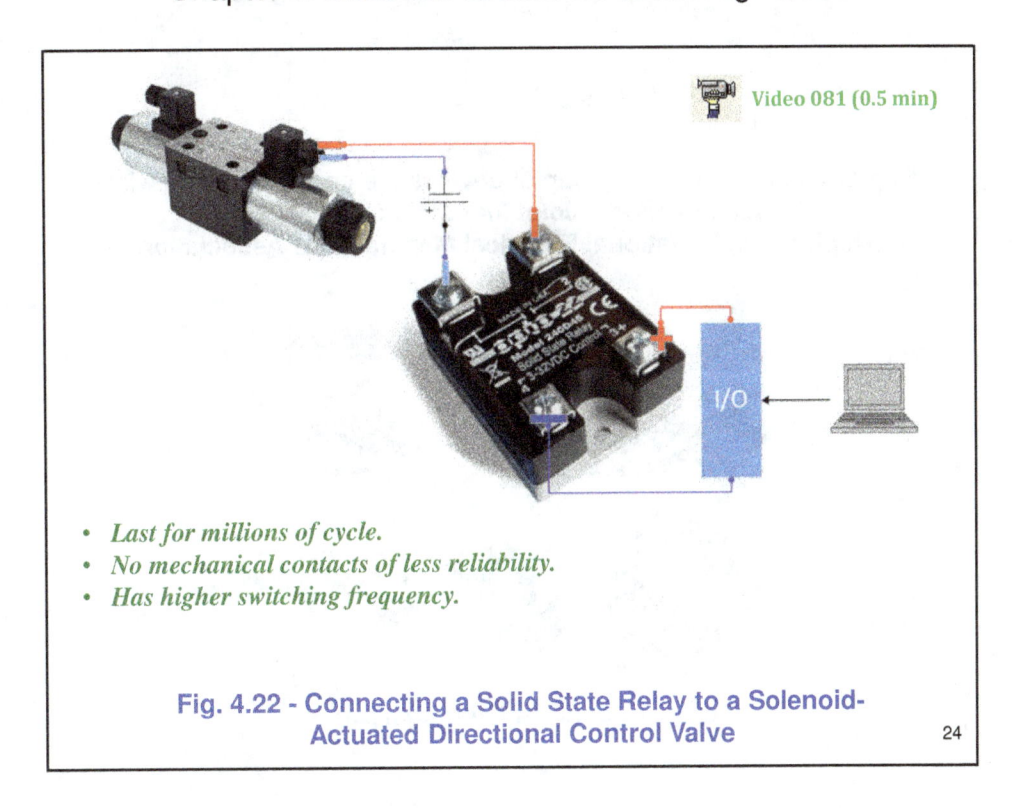

- *Last for millions of cycle.*
- *No mechanical contacts of less reliability.*
- *Has higher switching frequency.*

Fig. 4.22 - Connecting a Solid State Relay to a Solenoid-Actuated Directional Control Valve

4.3.10 - Programmable Logic Controllers (PLCs)

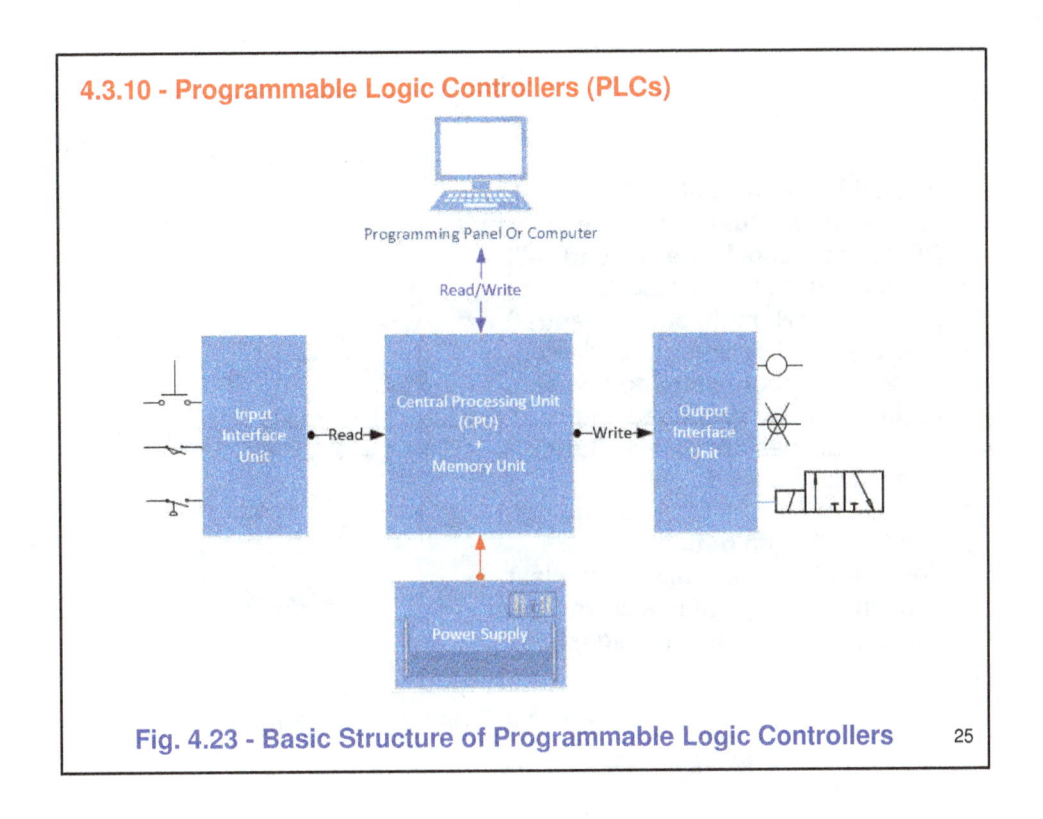

Fig. 4.23 - Basic Structure of Programmable Logic Controllers

4.4- Electrical Schematic Diagrams
4.4.1- JIC Schematic Diagrams

A diagram shows how is a group of devices are connected with wires.
Example: JIC "Joint Industrial Council"
Example: NEMA "National Electrical Manufacturer Association"

Fig. 4.24 - Electrical Terminal Blocks

- Vertical ladder format.
- Vertical lines "Buses" are power lines.
- DC circuits: Labels are "+" and "−".
- AC circuits: "L1" and "L2".
- AC circuits: H (high) and N (neutral).
- Left line "hot" or ungrounded line.
- Switches are connected to this bus.
- Right line is the grounded line.
- Horizontal lines are called "Rungs."
- Rungs labeled R1, R2, etc.
- The diagram is read from left to right and from the top bottom.
- The output device is always the last item on the rung and is always drawn close to the right (low voltage) bus.

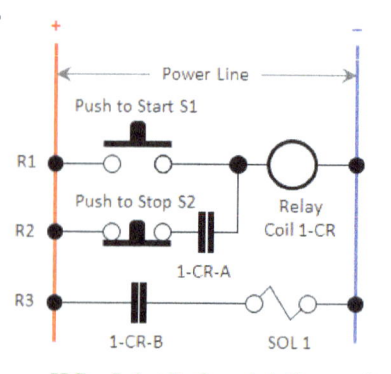

JIC : Joint Industrial Councel

Fig. 4.25 - Example of a JIC Schematic Diagram

Workbook: Electro-Hydraulic Components and Systems
Chapter 4: Electrical Circuits for Switching Valves

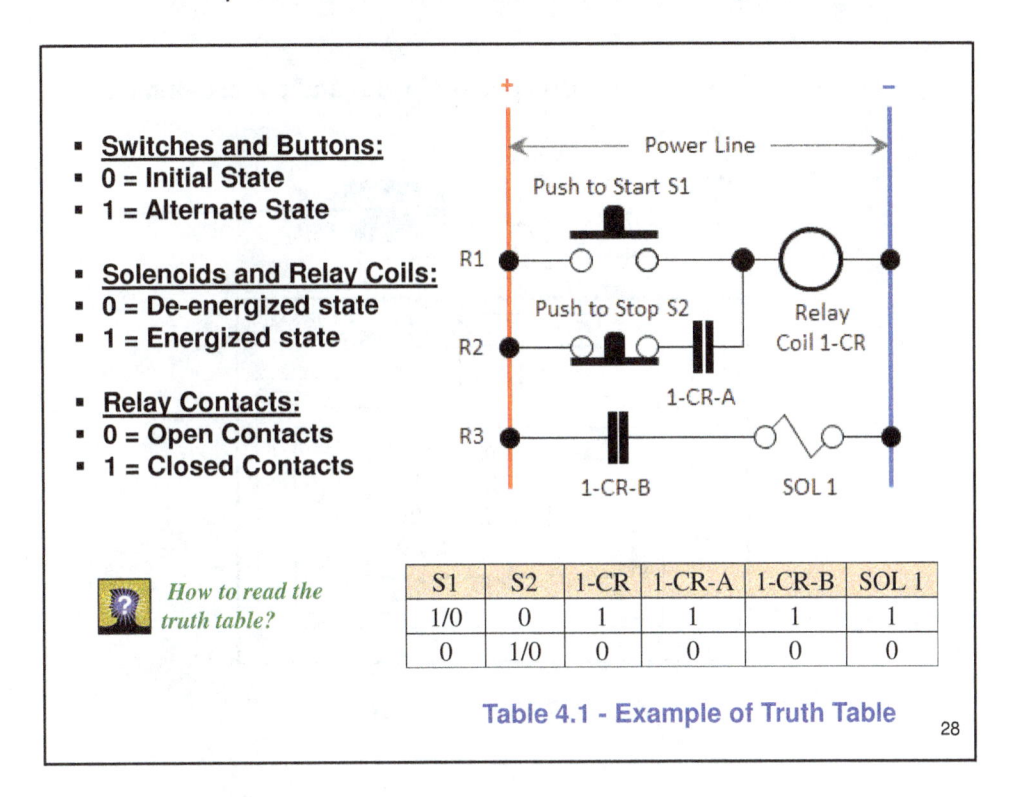

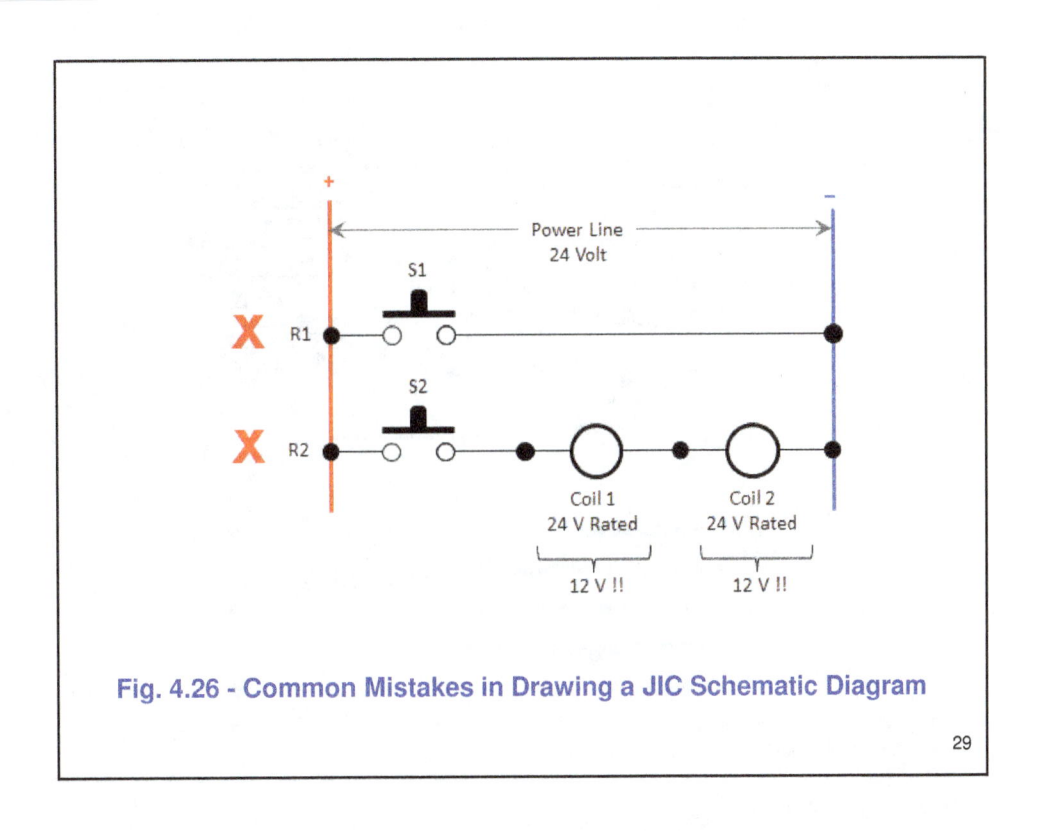

- **Switches and Buttons:**
- 0 = Initial State
- 1 = Alternate State

- **Solenoids and Relay Coils:**
- 0 = De-energized state
- 1 = Energized state

- **Relay Contacts:**
- 0 = Open Contacts
- 1 = Closed Contacts

How to read the truth table?

S1	S2	1-CR	1-CR-A	1-CR-B	SOL 1
1/0	0	1	1	1	1
0	1/0	0	0	0	0

Table 4.1 - Example of Truth Table

Fig. 4.26 - Common Mistakes in Drawing a JIC Schematic Diagram

Workbook: Electro-Hydraulic Components and Systems
Chapter 4: Electrical Circuits for Switching Valves

4.4.2- Wiring Diagrams — Component layout and pin assignment

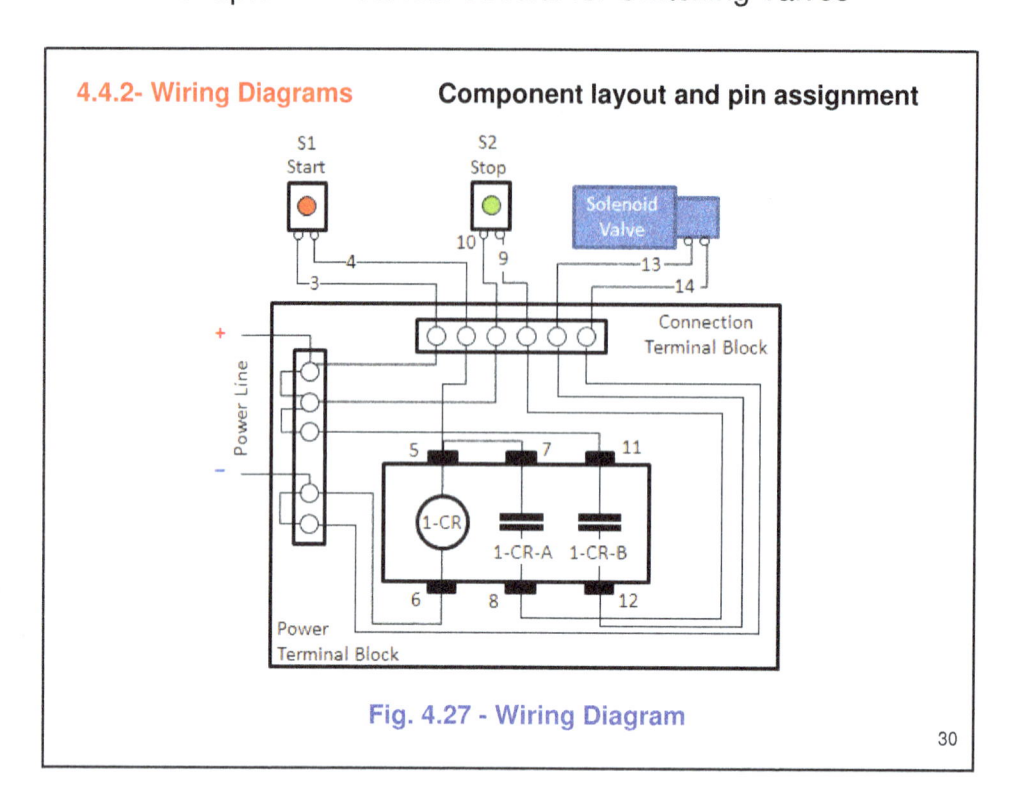

Fig. 4.27 - Wiring Diagram

4.4.3- Sequence Diagrams

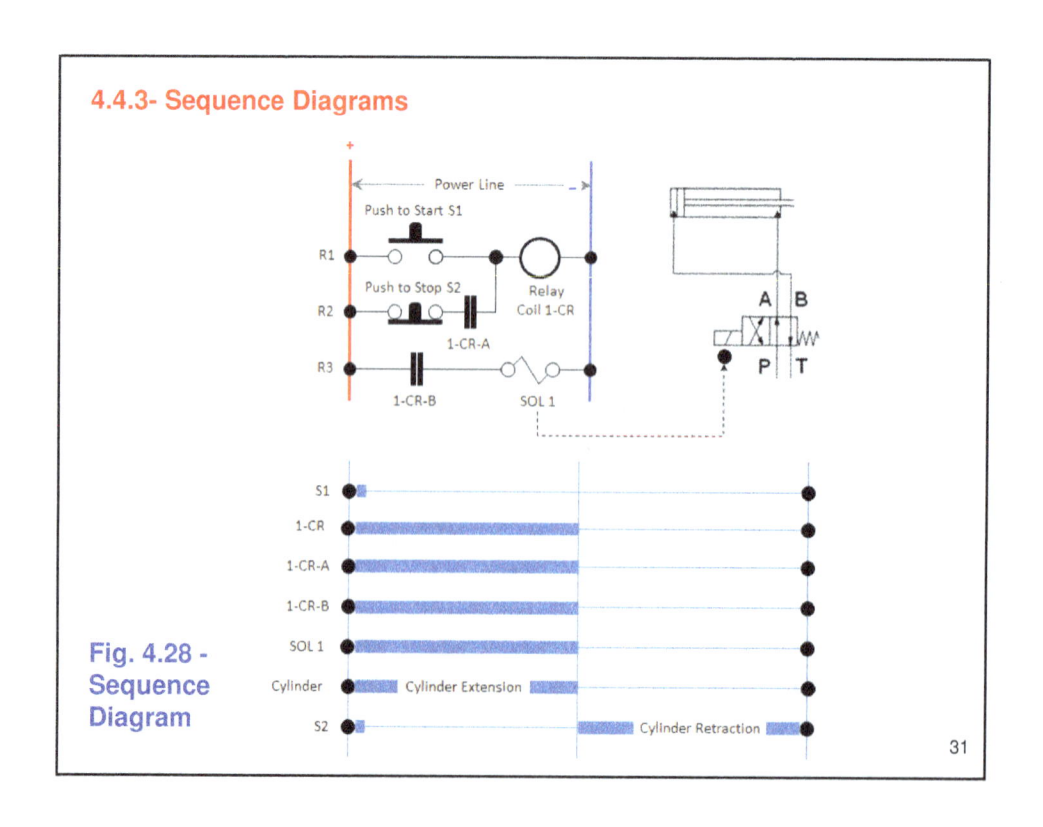

Fig. 4.28 - Sequence Diagram

Workbook: Electro-Hydraulic Components and Systems
Chapter 4: Electrical Circuits for Switching Valves

4.5- Electrical Circuits for Applications of Switching Valves

In this section, the following abbreviations are used:

- **PS: Pressure Switches.**
- **LS: Limit or Position Switches (Mechanical).**
- **TS: Toggle Switches.**
- **PX: Proximity Switches.**
- **S: Pushbuttons.**
- **K1, K2, etc.: Relay Coils.**
- **K1A, K1B, etc.: Relay Contacts.**

4.5.1- Identity (YES Function) by Direct Activation

S1	Y1	Pump
0	0	Loaded
1	1	Unloaded

Fig. 4.29 - Identity (YES Function) by Direct Activation

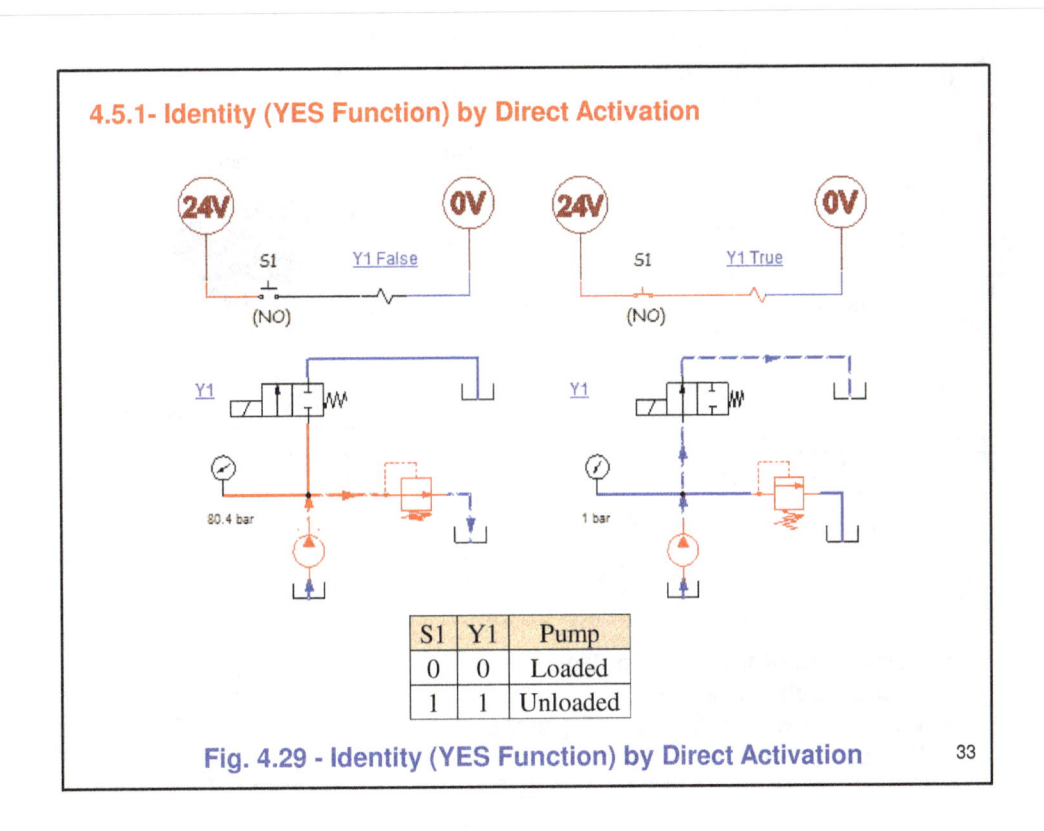

4.5.2 - Negation (NOT Function) by Direct Activation

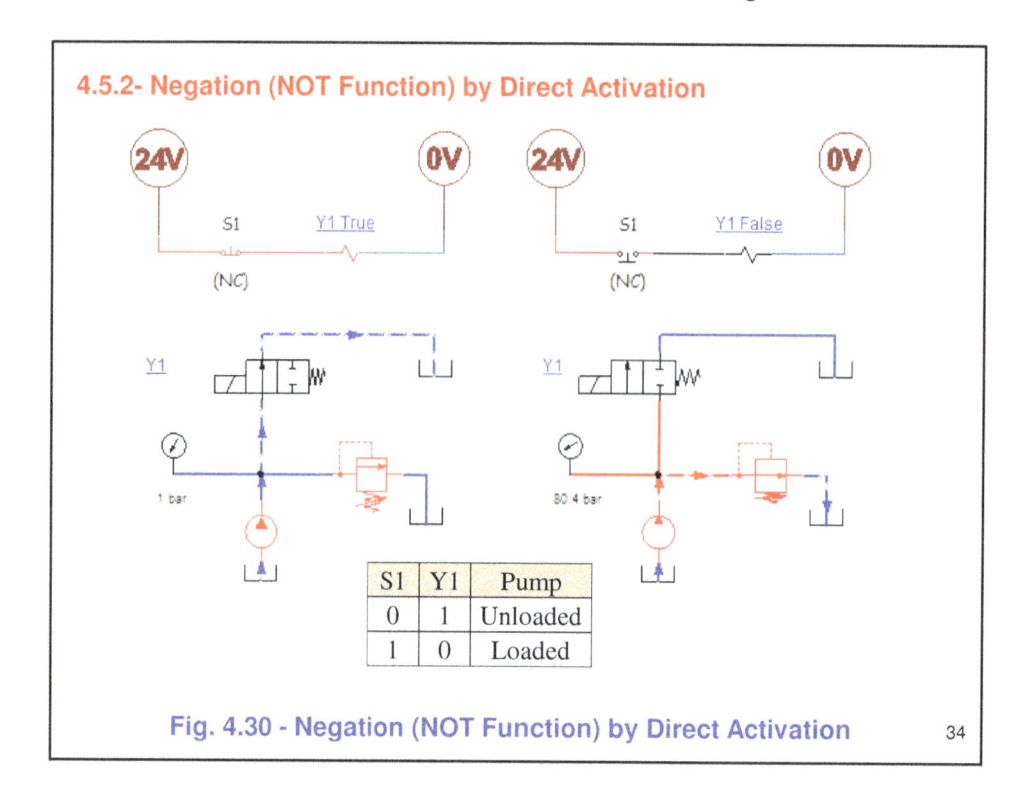

S1	Y1	Pump
0	1	Unloaded
1	0	Loaded

Fig. 4.30 - Negation (NOT Function) by Direct Activation

4.5.3 - Identity (YES Function) by Indirect Activation

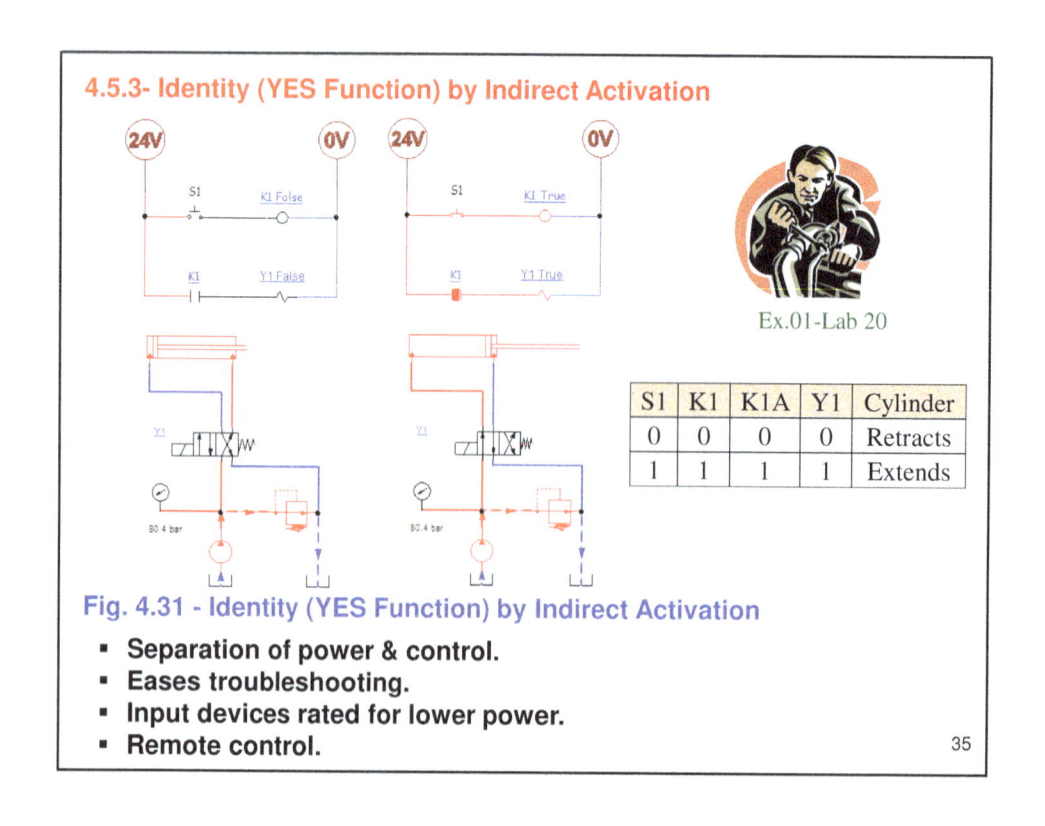

Ex.01-Lab 20

S1	K1	K1A	Y1	Cylinder
0	0	0	0	Retracts
1	1	1	1	Extends

Fig. 4.31 - Identity (YES Function) by Indirect Activation

- **Separation of power & control.**
- **Eases troubleshooting.**
- **Input devices rated for lower power.**
- **Remote control.**

Workbook: Electro-Hydraulic Components and Systems
Chapter 4: Electrical Circuits for Switching Valves

4.5.4- Negation (NOT Function) by Indirect Activation

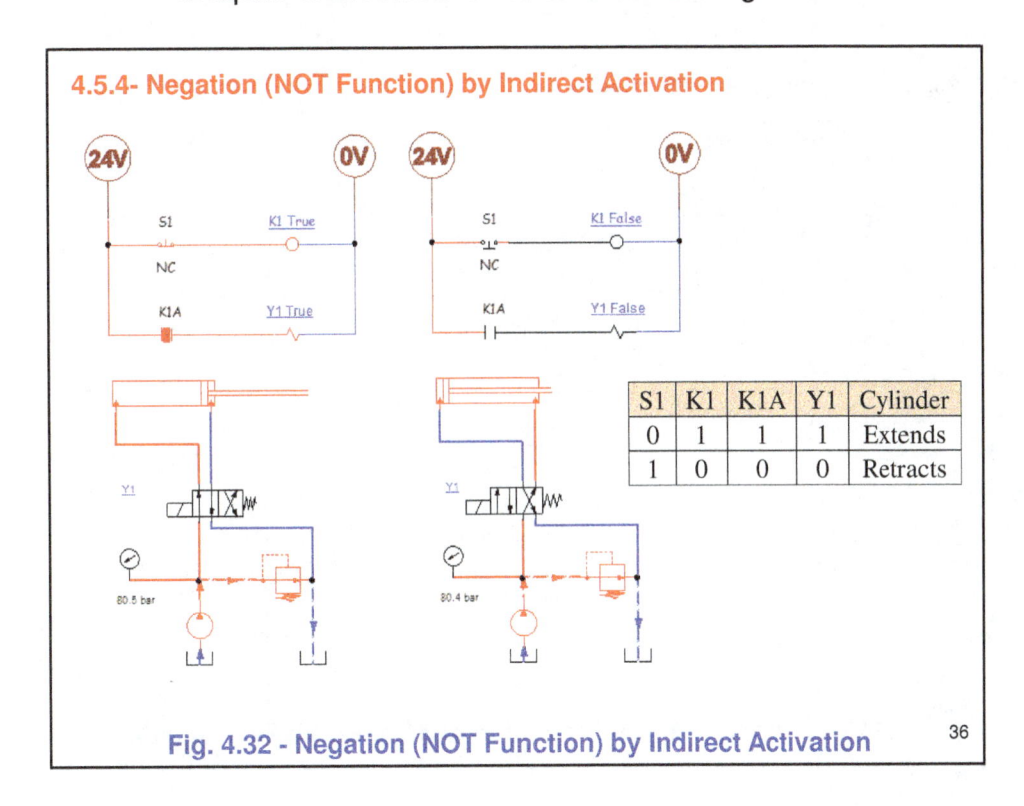

S1	K1	K1A	Y1	Cylinder
0	1	1	1	Extends
1	0	0	0	Retracts

Fig. 4.32 - Negation (NOT Function) by Indirect Activation

4.5.5- Signal Storage by Electrical Latching

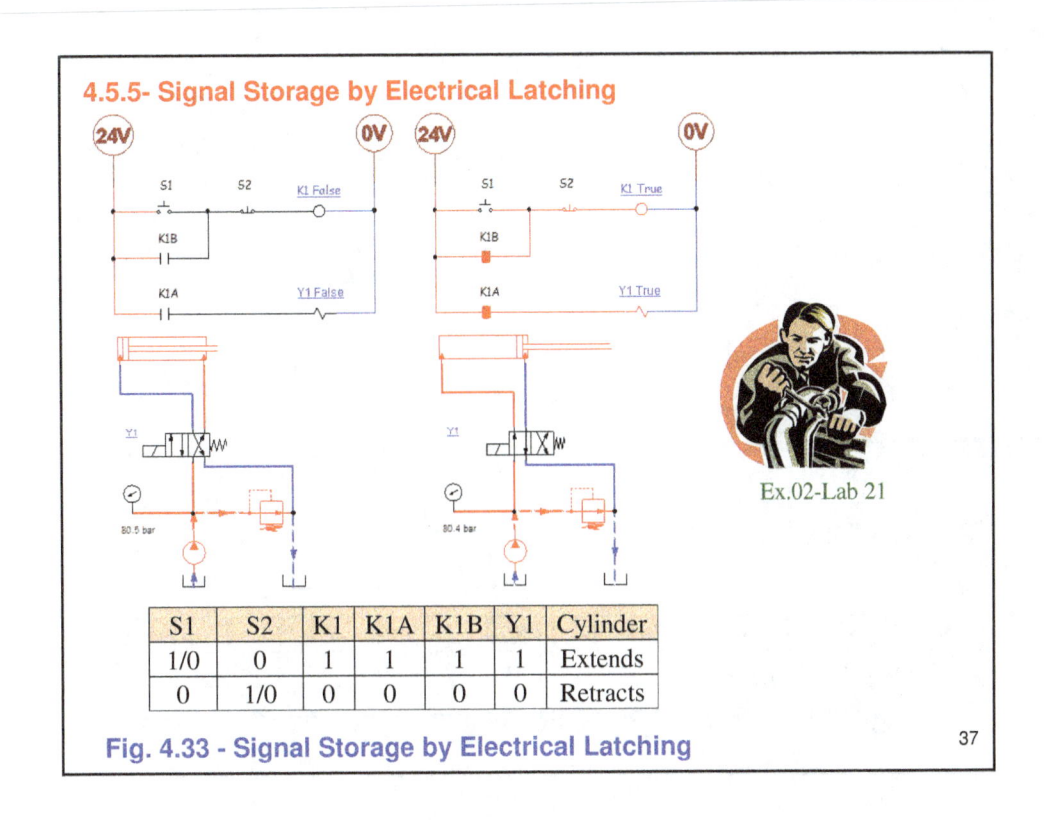

S1	S2	K1	K1A	K1B	Y1	Cylinder
1/0	0	1	1	1	1	Extends
0	1/0	0	0	0	0	Retracts

Ex.02-Lab 21

Fig. 4.33 - Signal Storage by Electrical Latching

4.5.6 - Electromechanical Protection of a Valve with Two Solenoids

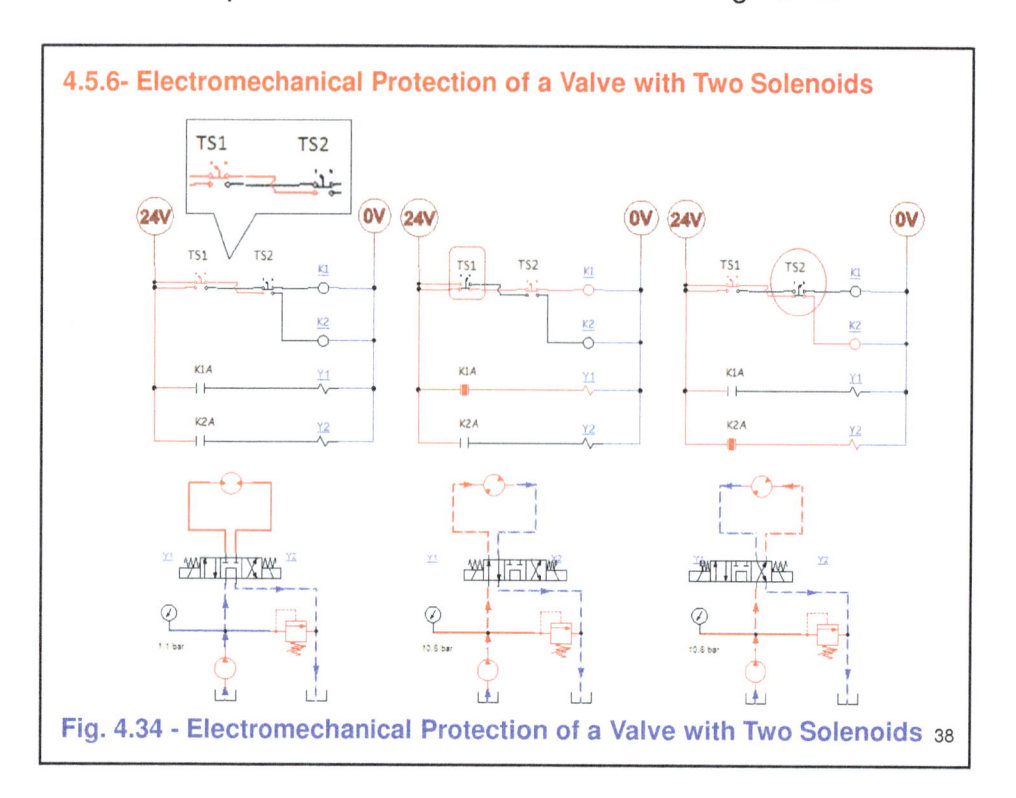

Fig. 4.34 - Electromechanical Protection of a Valve with Two Solenoids

4.5.7 - Electrical Protection of a Valve with Two Solenoids

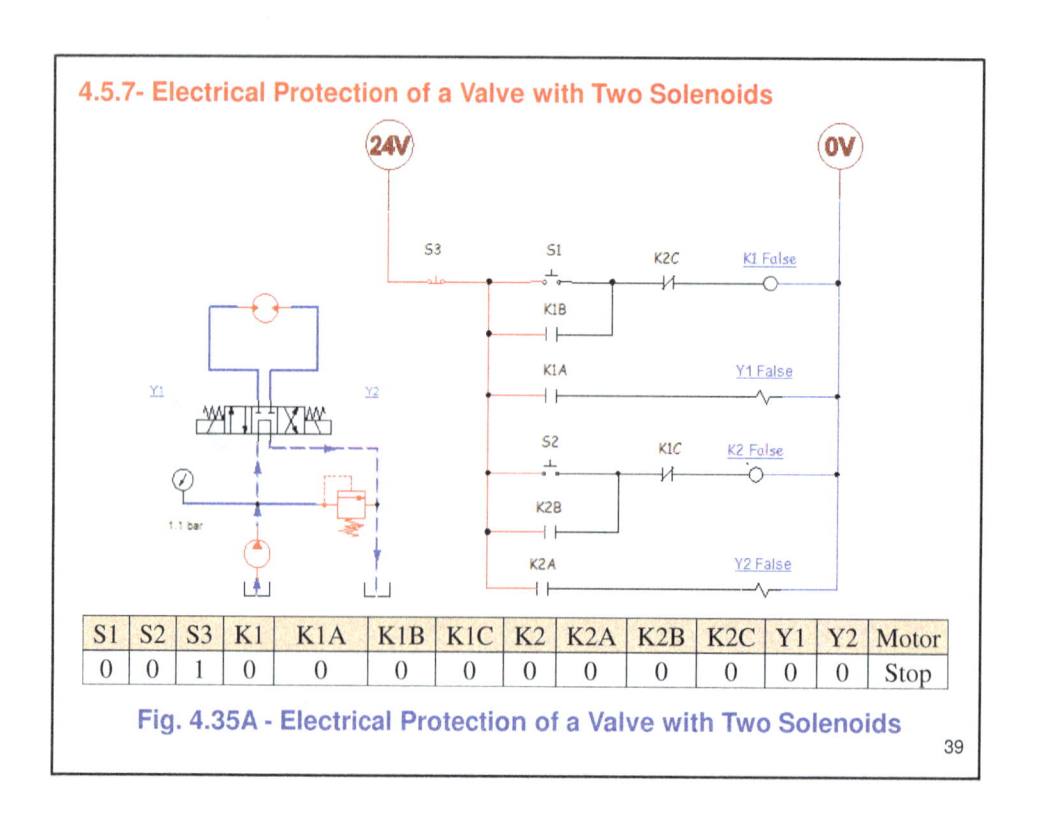

S1	S2	S3	K1	K1A	K1B	K1C	K2	K2A	K2B	K2C	Y1	Y2	Motor
0	0	1	0	0	0	0	0	0	0	0	0	0	Stop

Fig. 4.35A - Electrical Protection of a Valve with Two Solenoids

Workbook: Electro-Hydraulic Components and Systems
Chapter 4: Electrical Circuits for Switching Valves

Ex.03-Lab 22

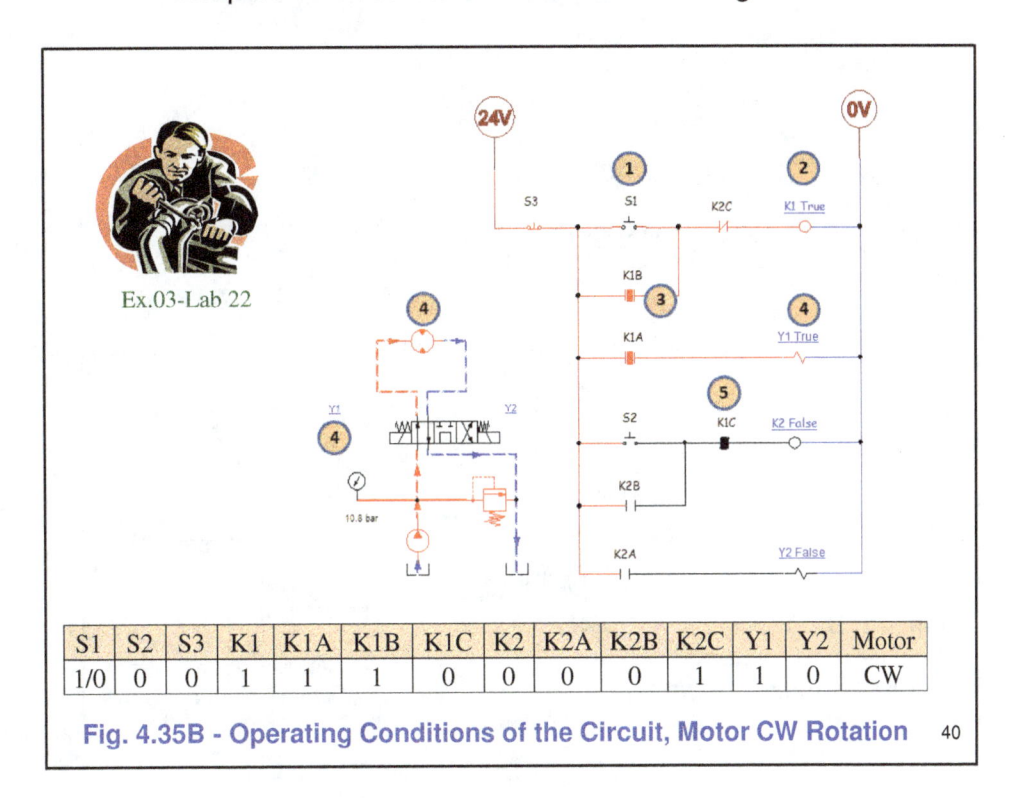

S1	S2	S3	K1	K1A	K1B	K1C	K2	K2A	K2B	K2C	Y1	Y2	Motor
1/0	0	0	1	1	1	0	0	0	0	1	1	0	CW

Fig. 4.35B - Operating Conditions of the Circuit, Motor CW Rotation

---

4.5.8- Position-Dependent Cylinder Deceleration

S1	S2	S3	K1	K1A	K1B	K1C	K2	K2A	K2B
0	0	1	0	0	0	1	0	0	0

K2C	Y1	Y2	PX	K3	K3A	Y3	Cylinder
1	0	0	0	0	0	0	Stop

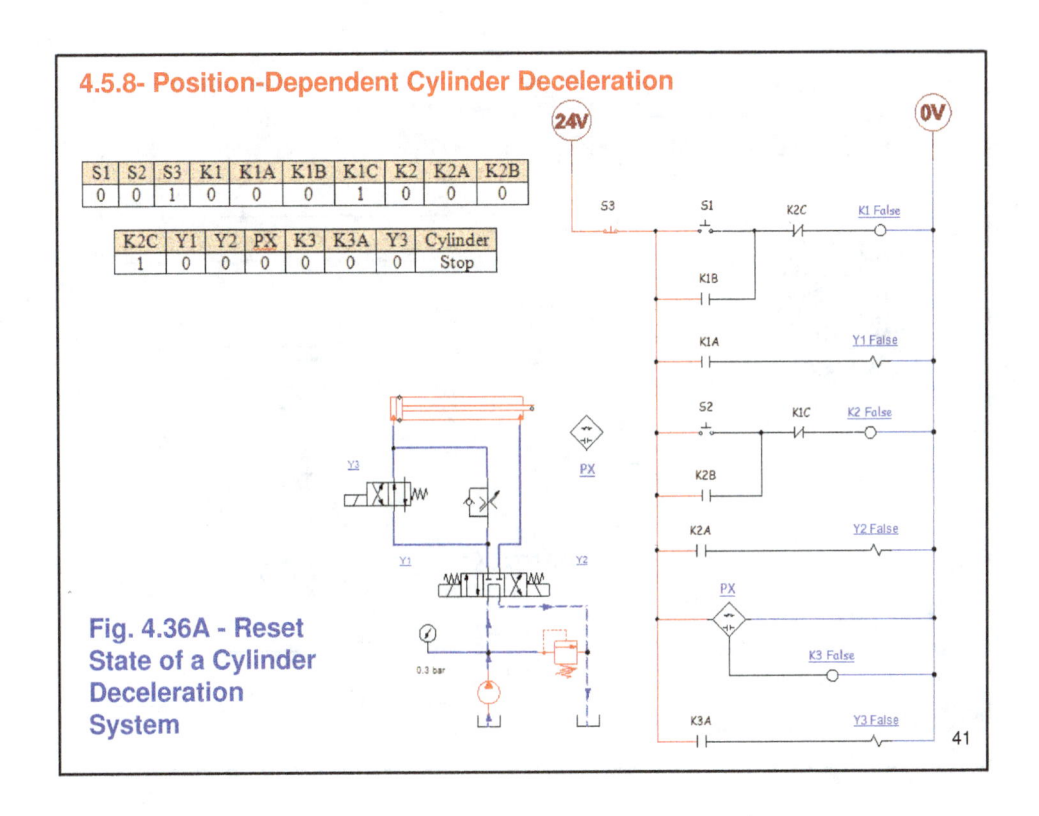

Fig. 4.36A - Reset State of a Cylinder Deceleration System

Workbook: Electro-Hydraulic Components and Systems
Chapter 4: Electrical Circuits for Switching Valves

S1	S2	S3	K1	K1A	K1B	K1C	K2	K2A	K2B
1/0	0	0	1	1	1	1	0	0	0

K2C	Y1	Y2	PX	K3	K3A	Y3	Cylinder
1	1	0	1	1	1	1	Extends

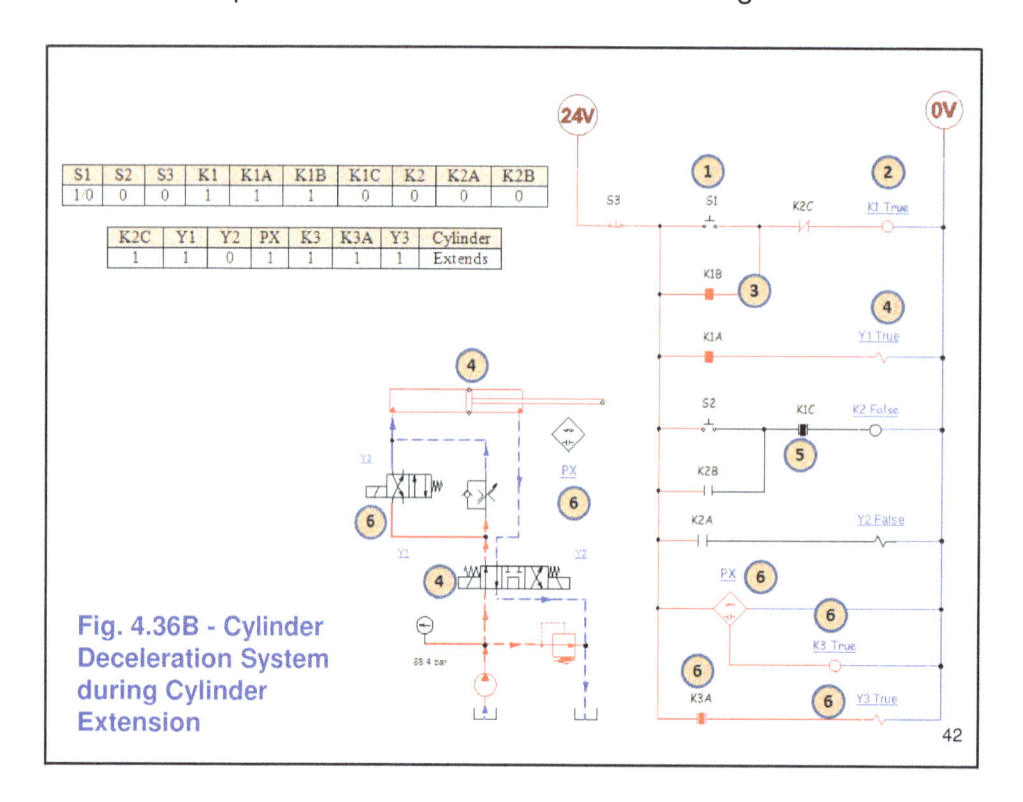

Fig. 4.36B - Cylinder Deceleration System during Cylinder Extension

---

S1	S2	S3	K1	K1A	K1B	K1C	K2	K2A	K2B
0	0/1	0	0	0	0	0	1	1	1

K2C	Y1	Y2	PX	K3	K3A	Y3	Cylinder
0	0	1	0	0	0	0	Extends

Ex.04-Lab 23

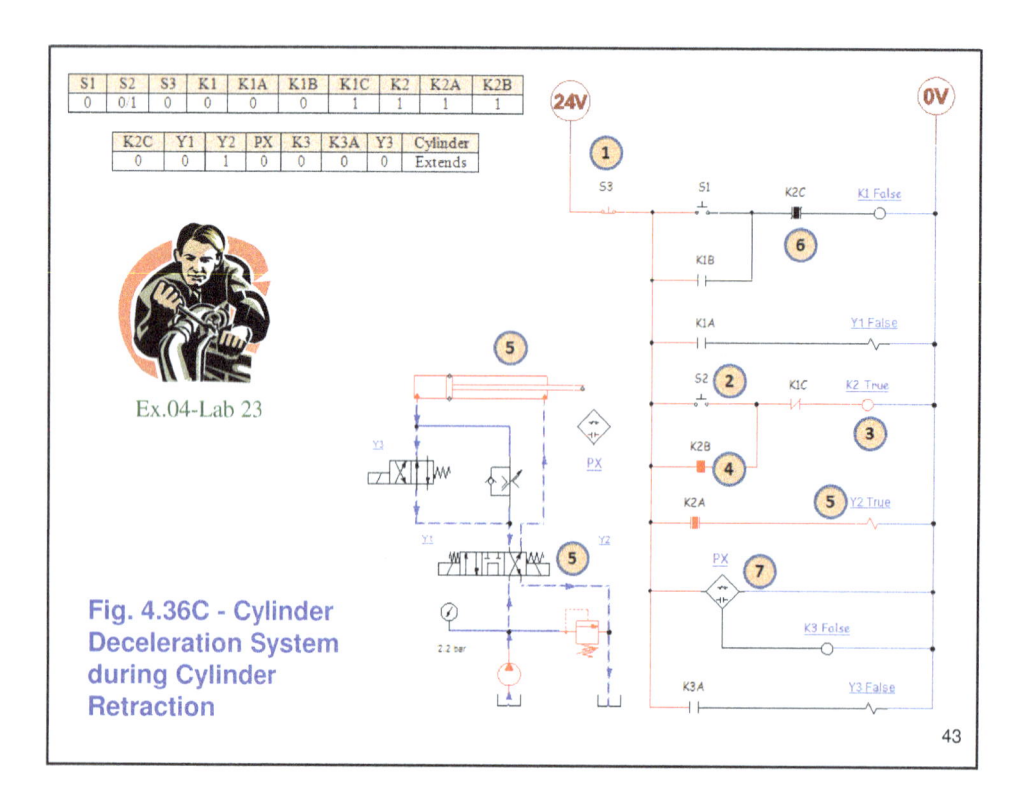

Fig. 4.36C - Cylinder Deceleration System during Cylinder Retraction

4.5.9 - One-Cycle Hydraulic Cylinder Reciprocation

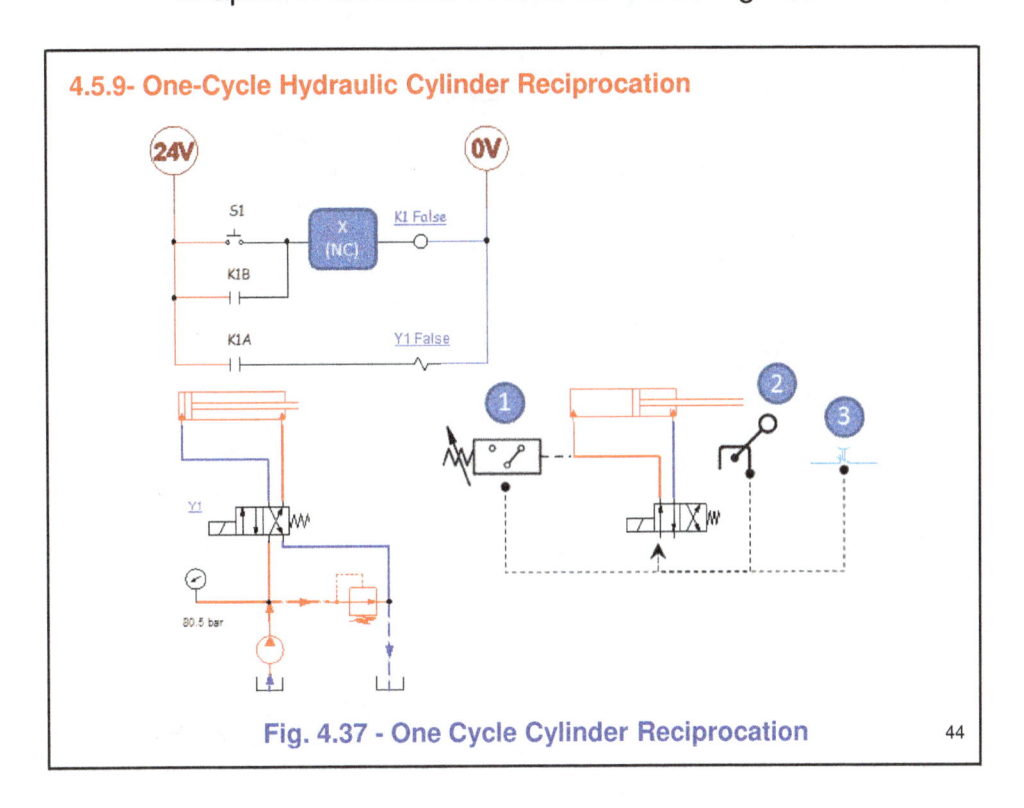

Fig. 4.37 - One Cycle Cylinder Reciprocation

Ex.05-Lab 24

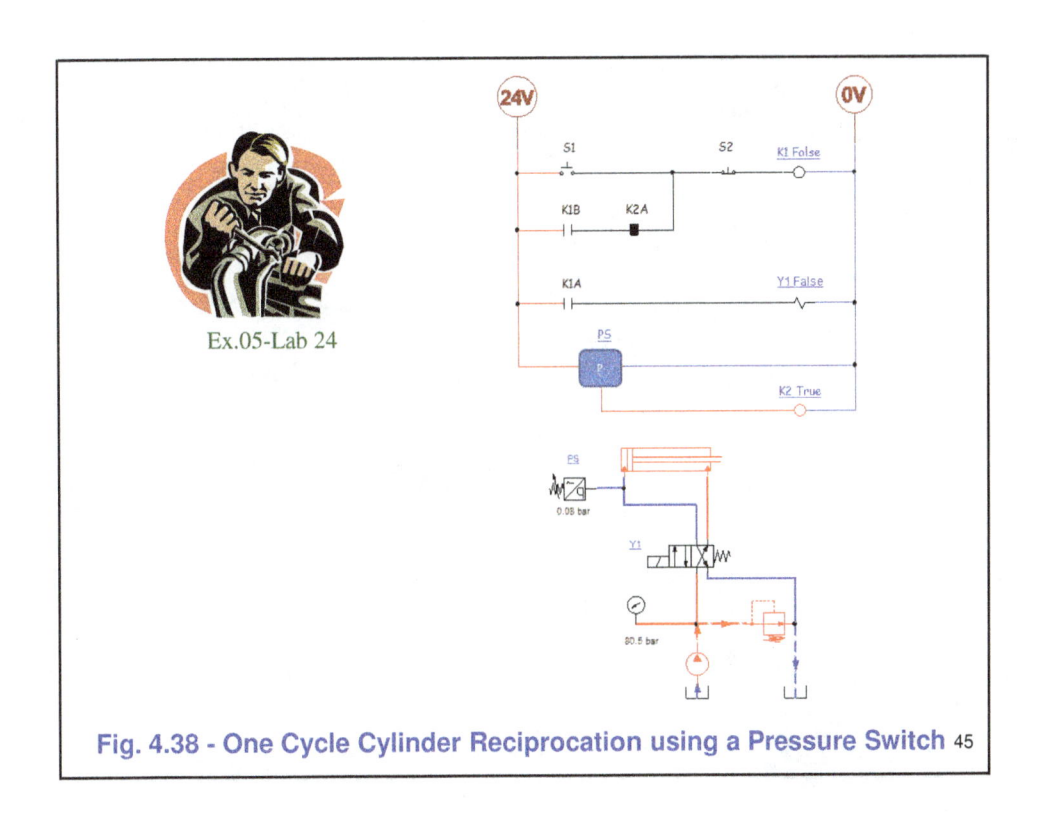

Fig. 4.38 - One Cycle Cylinder Reciprocation using a Pressure Switch

Workbook: Electro-Hydraulic Components and Systems
Chapter 4: Electrical Circuits for Switching Valves

4.5.10- Continuous Cylinder Reciprocation

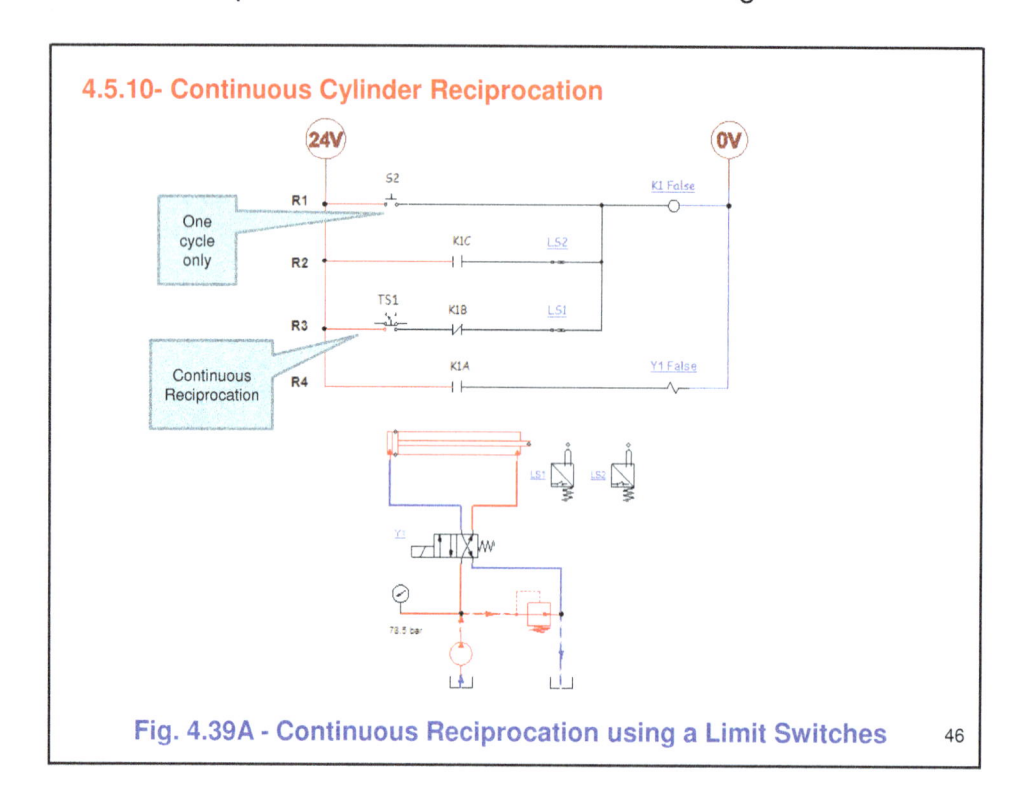

Fig. 4.39A - Continuous Reciprocation using a Limit Switches

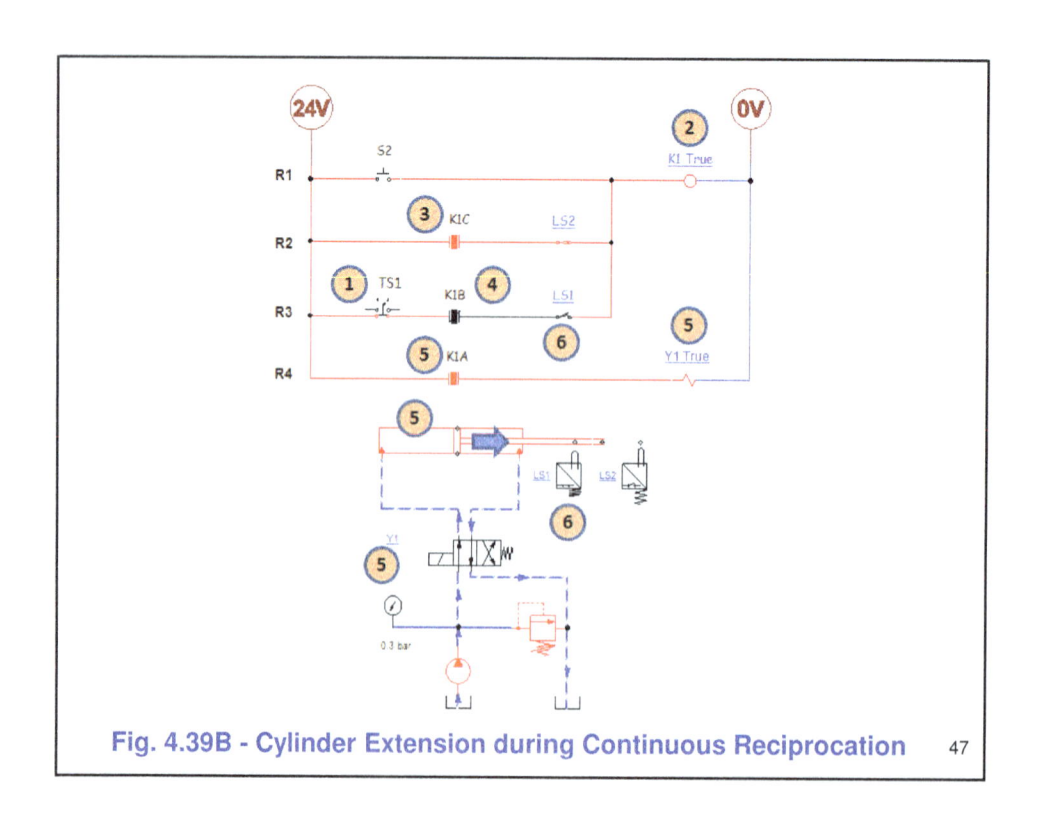

Fig. 4.39B - Cylinder Extension during Continuous Reciprocation

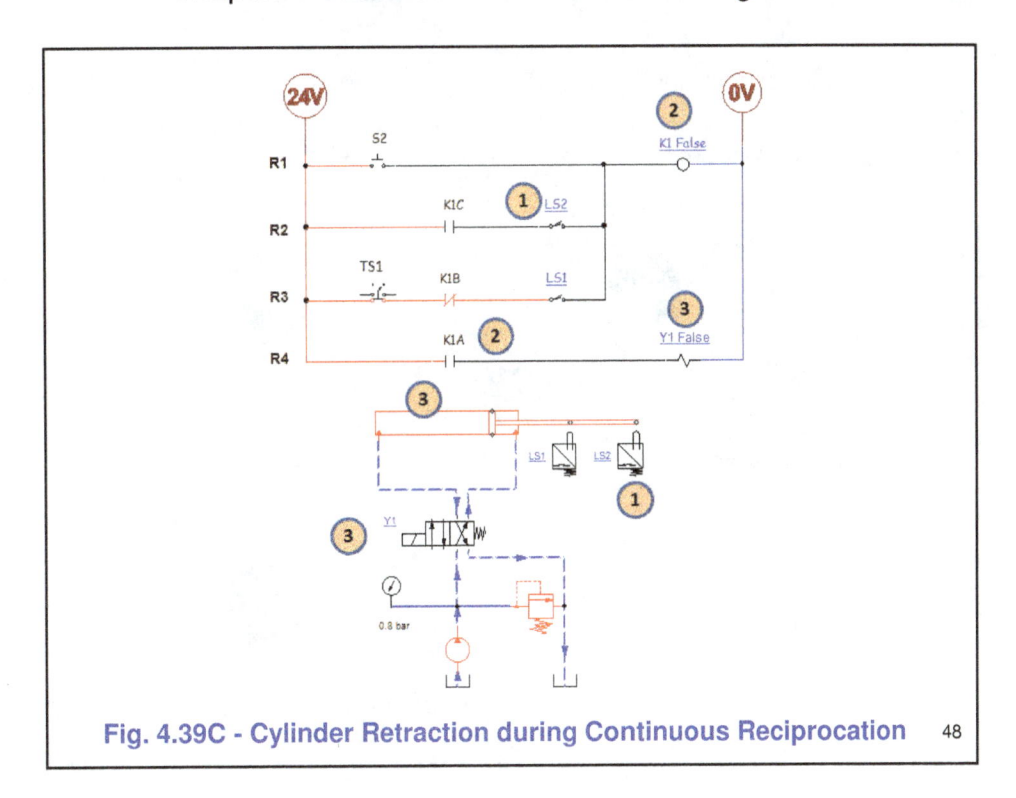

Workbook: Electro-Hydraulic Components and Systems
Chapter 4: Electrical Circuits for Switching Valves

Fig. 4.39C - Cylinder Retraction during Continuous Reciprocation

4.5.11- Panic Circuit

Ex.06-Lab 25

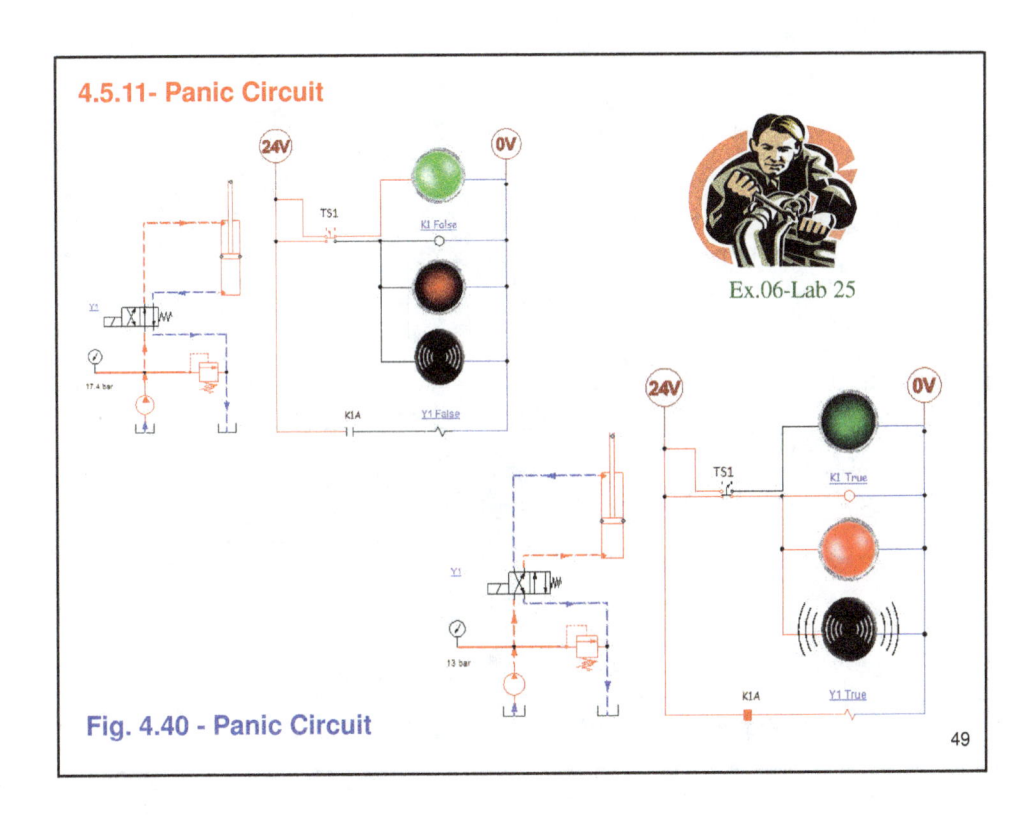

Fig. 4.40 - Panic Circuit

4.5.12- Conjunction Functions

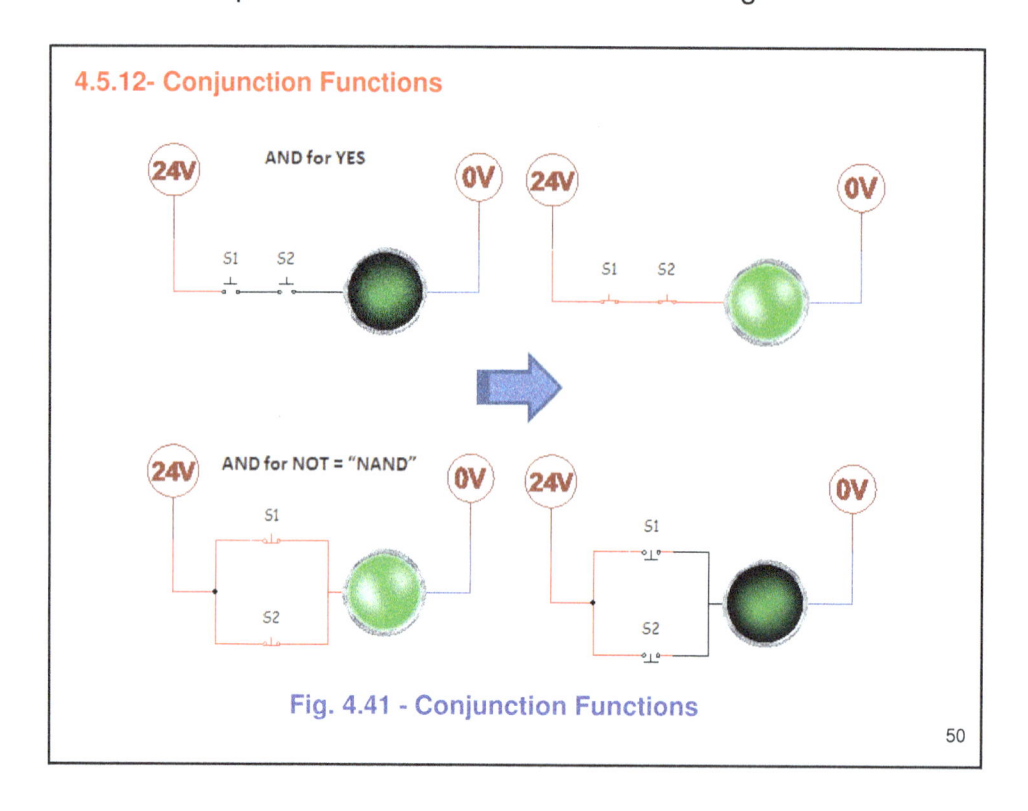

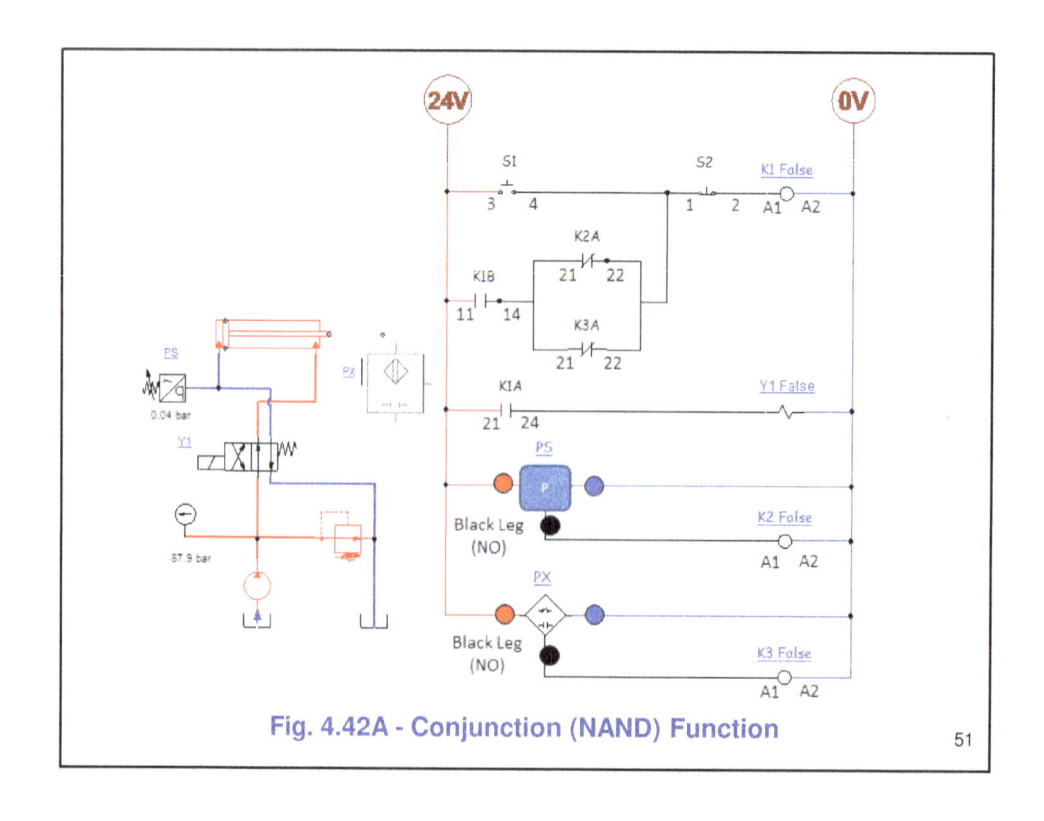

Fig. 4.41 - Conjunction Functions

Fig. 4.42A - Conjunction (NAND) Function

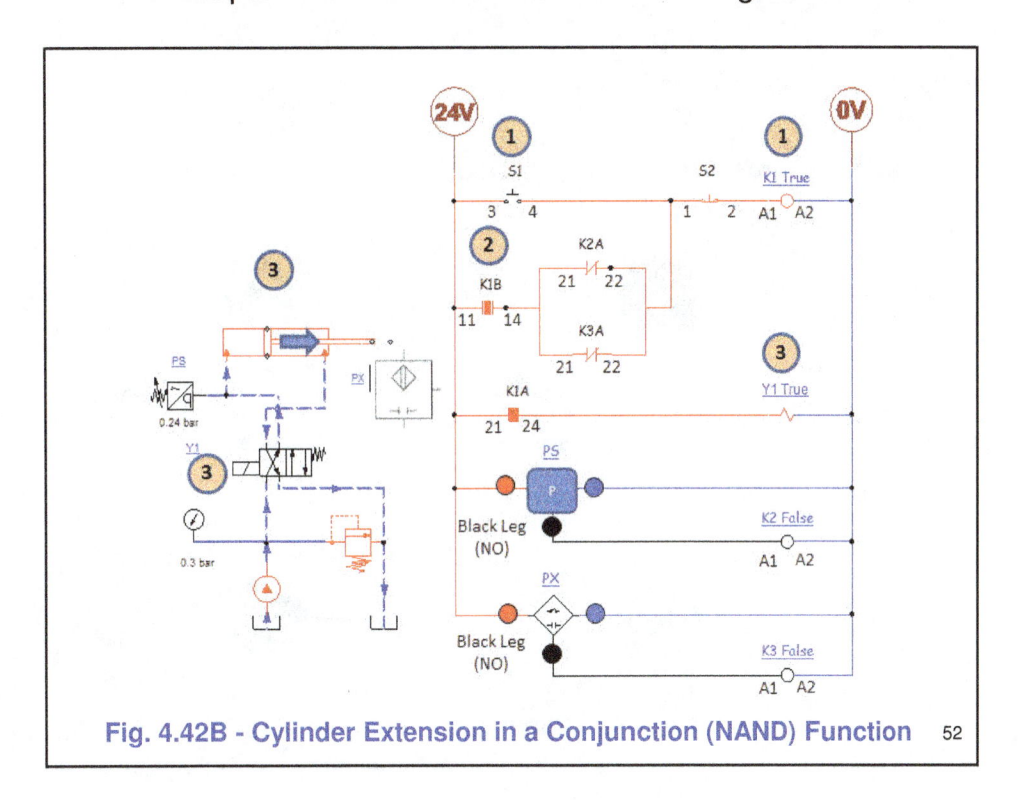

Workbook: Electro-Hydraulic Components and Systems
Chapter 4: Electrical Circuits for Switching Valves

Fig. 4.42B - Cylinder Extension in a Conjunction (NAND) Function 52

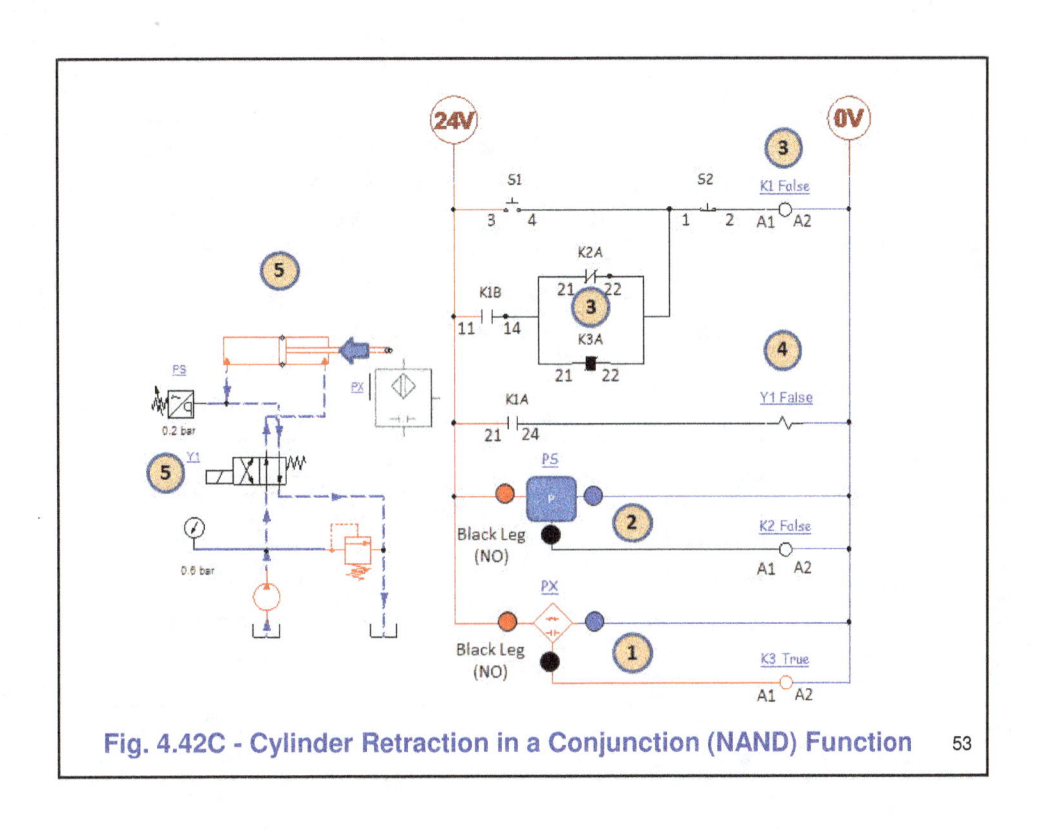

Fig. 4.42C - Cylinder Retraction in a Conjunction (NAND) Function 53

Workbook: Electro-Hydraulic Components and Systems
Chapter 4: Electrical Circuits for Switching Valves

4.5.13- Disjunction Functions

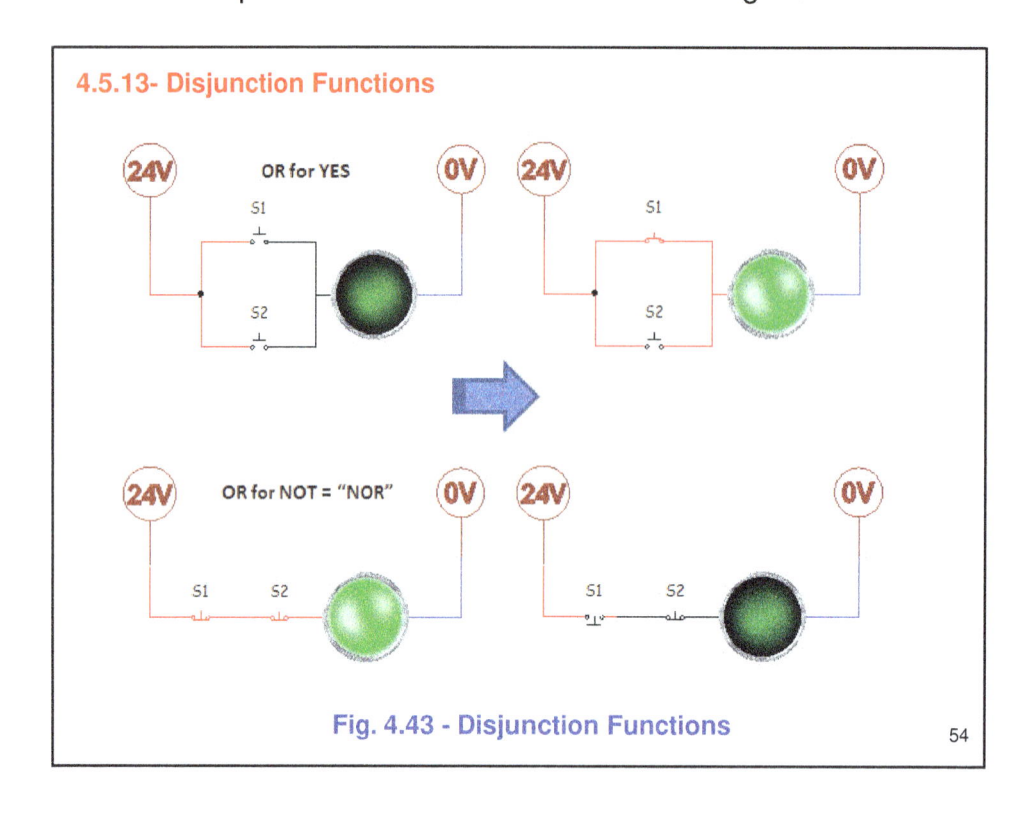

Fig. 4.43 - Disjunction Functions

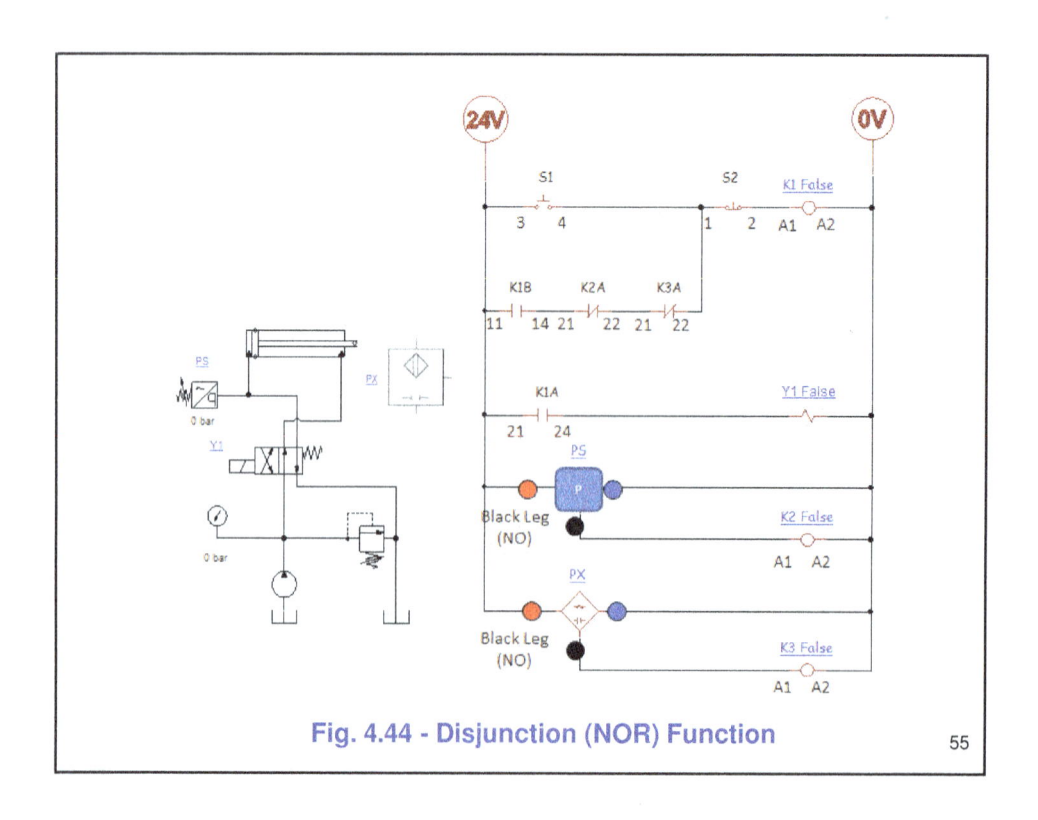

Fig. 4.44 - Disjunction (NOR) Function

Workbook: Electro-Hydraulic Components and Systems
Chapter 4: Electrical Circuits for Switching Valves

4.5.14 - Timer Circuits

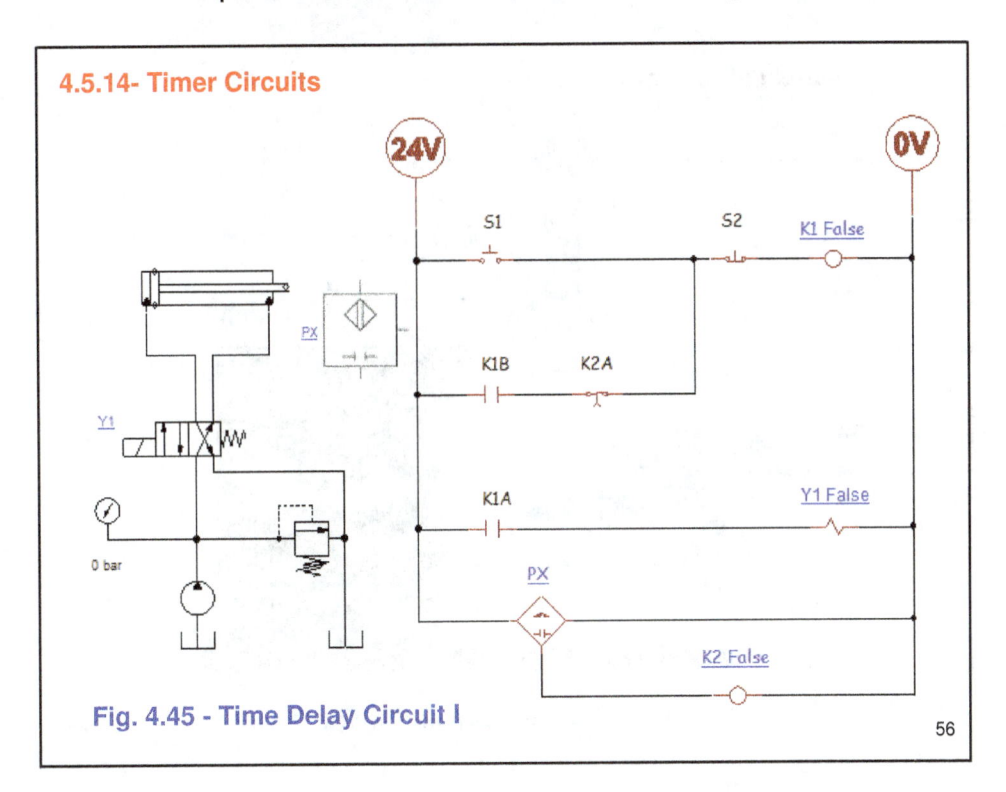

Fig. 4.45 - Time Delay Circuit I

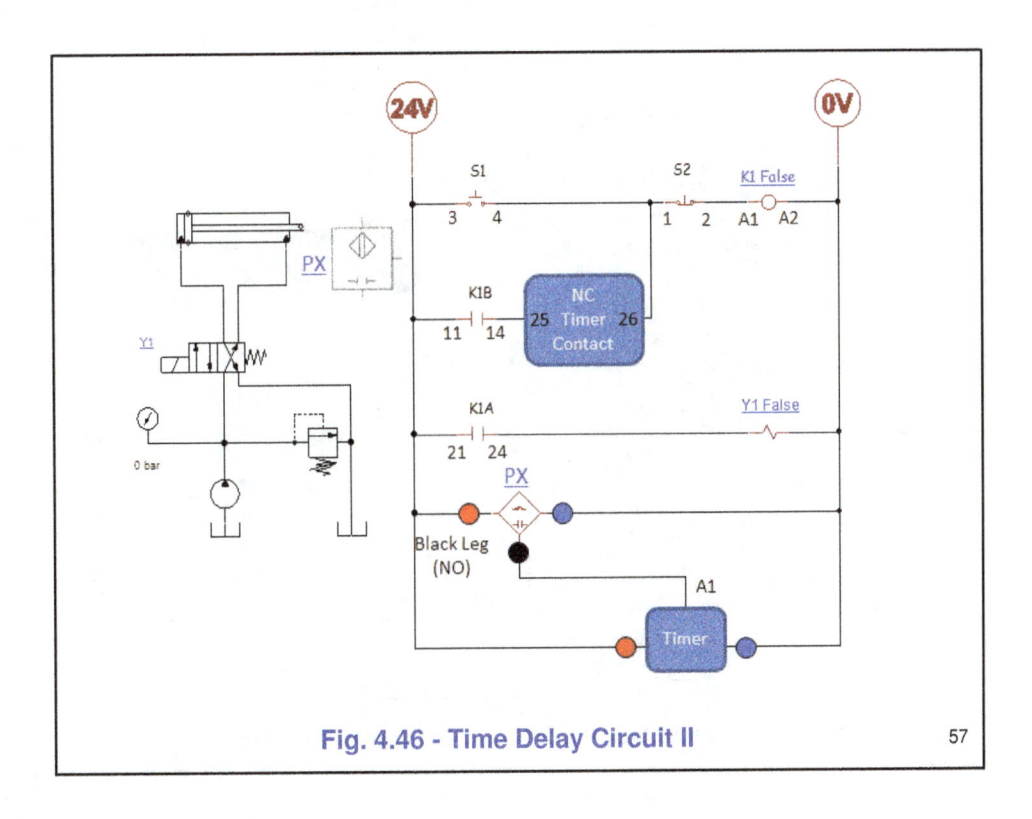

Fig. 4.46 - Time Delay Circuit II

Workbook: Electro-Hydraulic Components and Systems
Chapter 4: Electrical Circuits for Switching Valves

4.5.15- Sequence Control

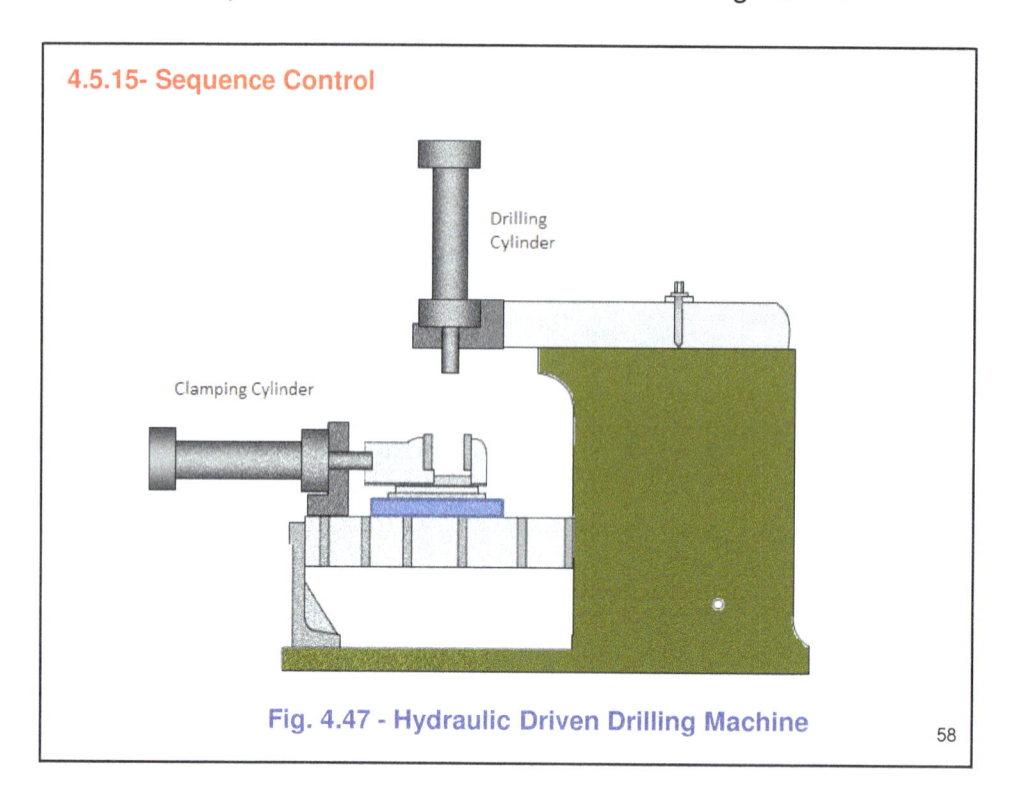

Fig. 4.47 - Hydraulic Driven Drilling Machine

What is the purpose of the pressure switch?

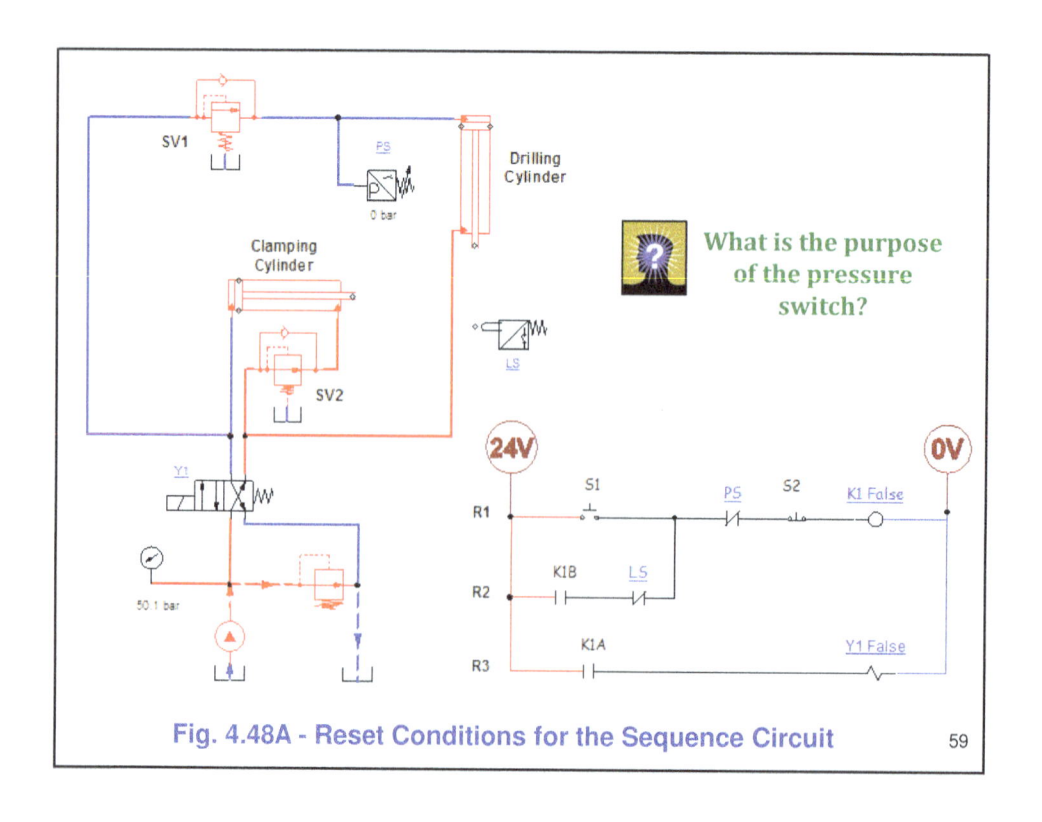

Fig. 4.48A - Reset Conditions for the Sequence Circuit

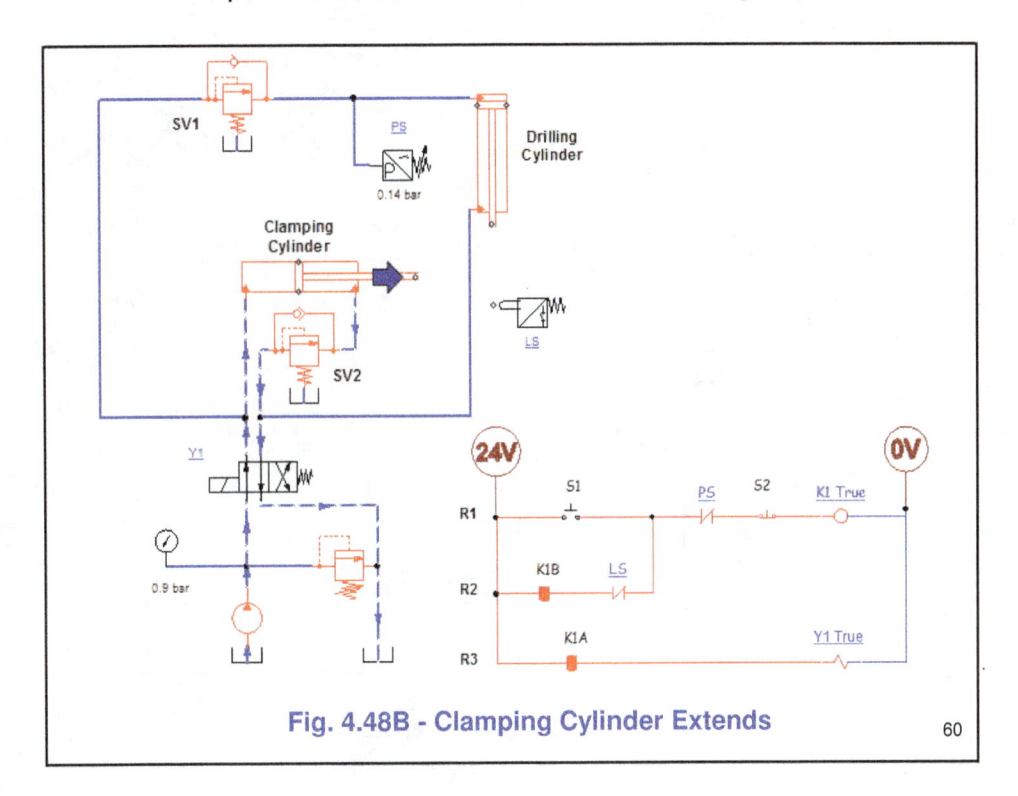

Workbook: Electro-Hydraulic Components and Systems
Chapter 4: Electrical Circuits for Switching Valves

Fig. 4.48B - Clamping Cylinder Extends

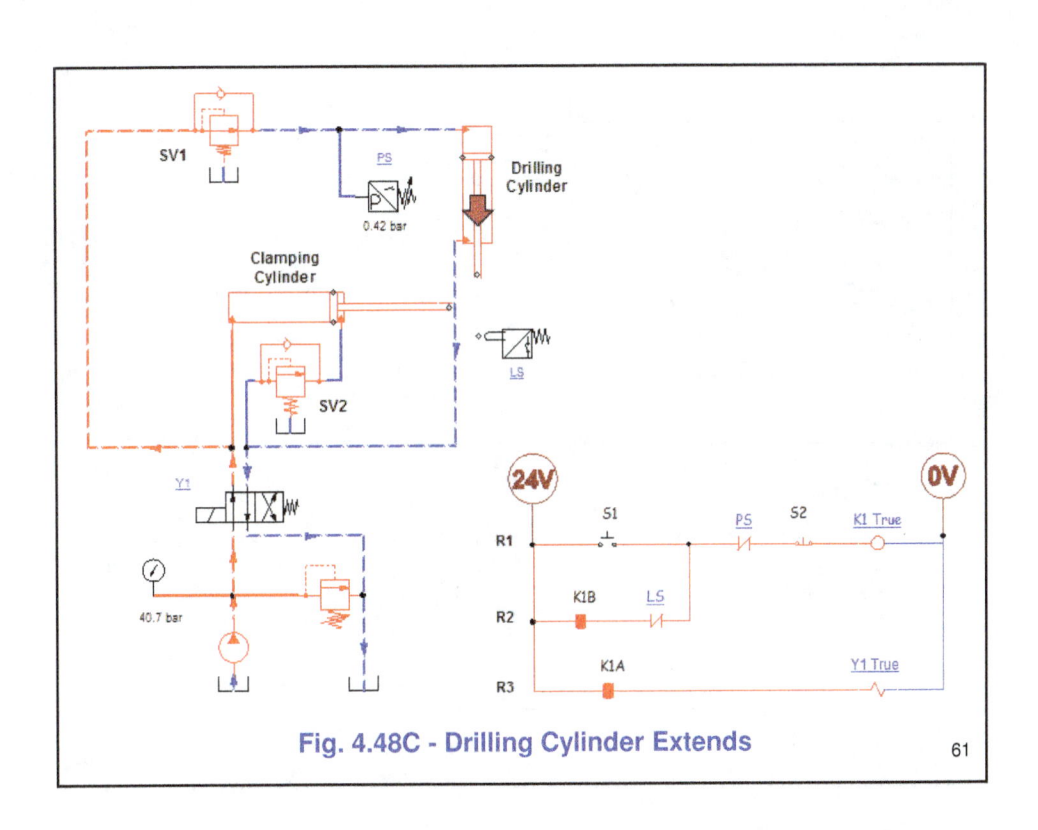

Fig. 4.48C - Drilling Cylinder Extends

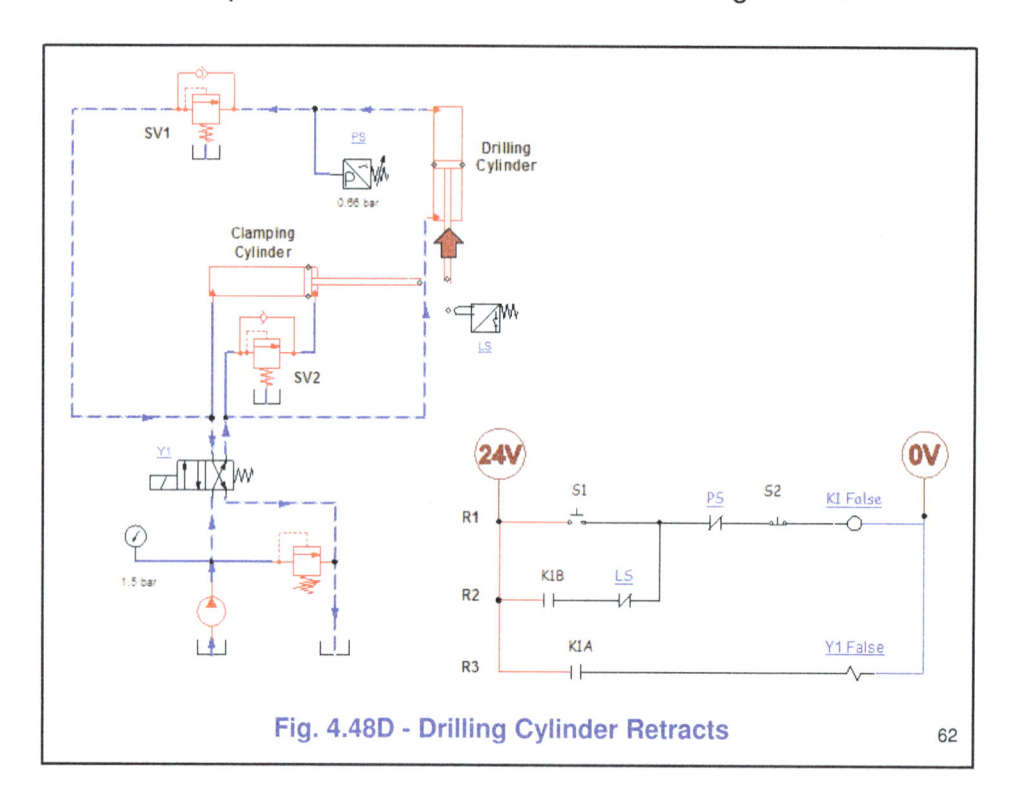

Workbook: Electro-Hydraulic Components and Systems
Chapter 4: Electrical Circuits for Switching Valves

Fig. 4.48D - Drilling Cylinder Retracts

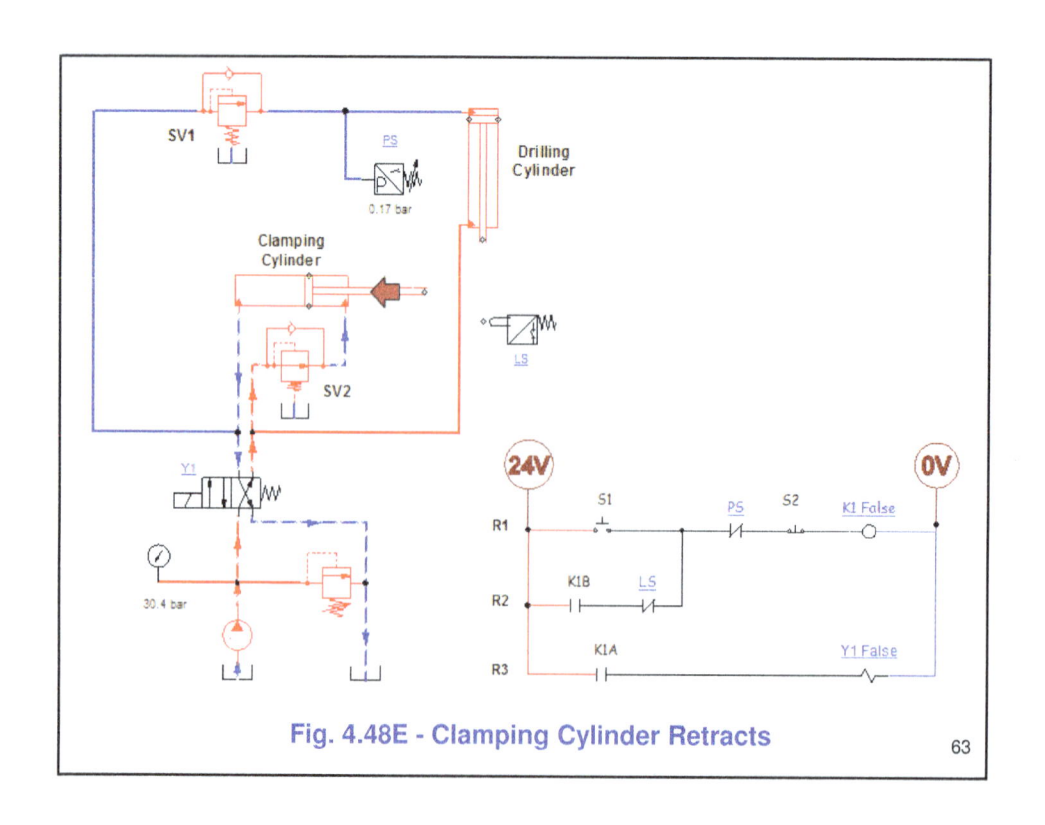

Fig. 4.48E - Clamping Cylinder Retracts

Workbook: Electro-Hydraulic Components and Systems
Chapter 4: Reviews and Assignments

Chapter 4 Reviews

1. Which of the following functions matches the shown below electrical circuit?
   A. Identity by direct activation.
   B. Identity by indirect activation.
   C. Negation by direct activation.
   D. Negation by indirect activation.

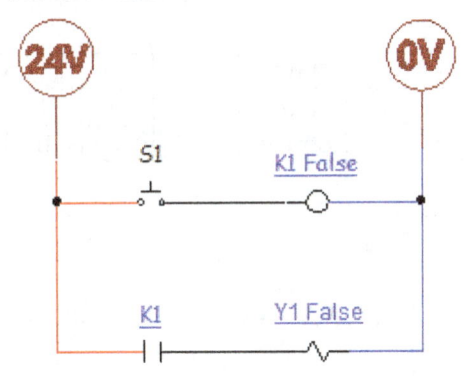

2. Which of the following functions matches the shown below electrical circuit?
   A. Identity by direct activation.
   B. Identity by indirect activation.
   C. Negation by direct activation.
   D. Negation by indirect activation.

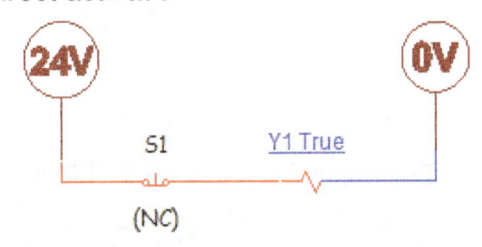

3. Which of the following statements describes the shown below electrical circuit?
   A. Latching circuit to keep K1 energized by pressing/releasing S1.
   B. Pressure switch is used for continuous cylinder reciprocation.
   C. Proximity switch is used for one cycle cylinder reciprocation.
   D. Pressure switch is used for one cycle cylinder reciprocation.

121

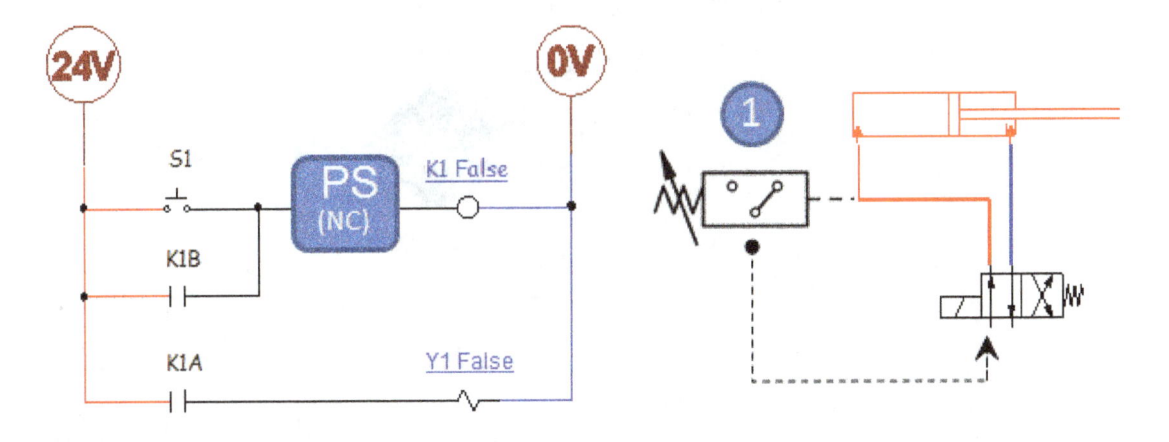

Workbook: Electro-Hydraulic Components and Systems
Chapter 4: Reviews and Assignments

4. Which of the following truth tables describes the shown below electrical circuit?

A

S1	K1	K1A	Y1	Cylinder
0	1	1	1	Extends
1	0	0	0	Retracts

B

S1	K1	K1A	Y1	Cylinder
0	1	1	1	Retracts
1	0	0	0	Extends

C

S1	K1	K1A	Y1	Cylinder
0	1	1	1	Extends
0	1	0	1	Retracts

D

S1	K1	K1A	Y1	Cylinder
0	1	1	1	Extends
1	0	1	1	Retracts

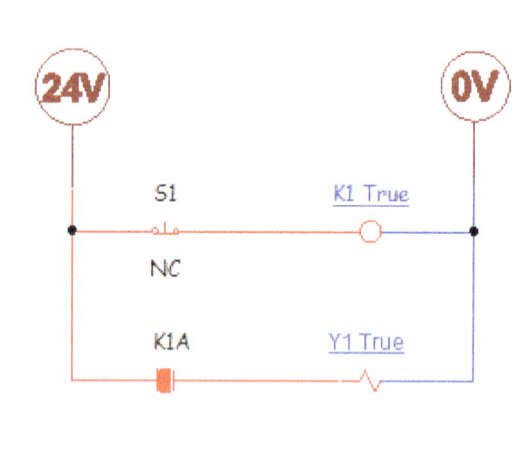

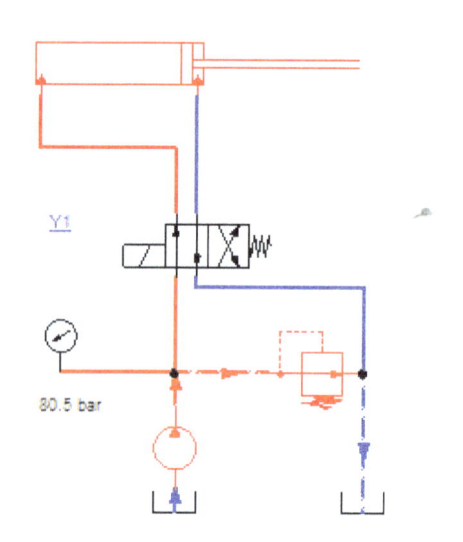

5. Which of the following logic function matches the shown below electrical circuit?
   A. AND.
   B. NAND.
   C. OR.
   D. NOR.

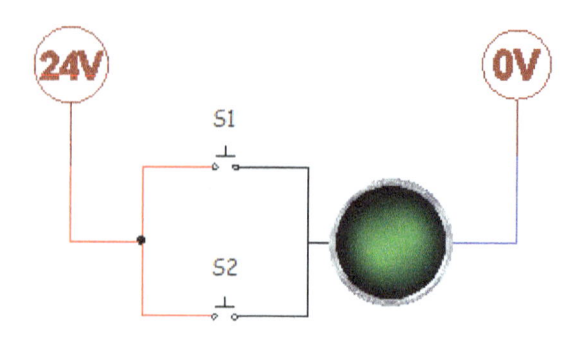

Workbook: Electro-Hydraulic Components and Systems
Chapter 4: Reviews and Assignments

6. Which of the following logic function matches the shown below electrical circuit?
    A. AND.
    B. NAND.
    C. OR.
    D. NOR.

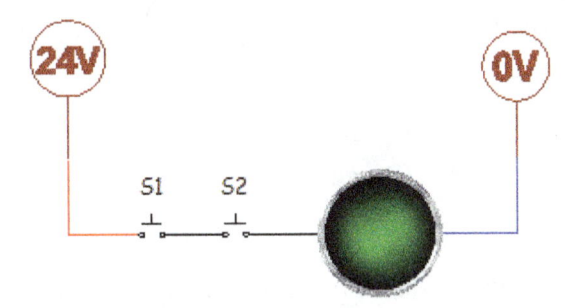

7. Which of the following logic function matches the shown below electrical circuit?
    A. AND.
    B. NAND.
    C. OR.
    D. NOR.

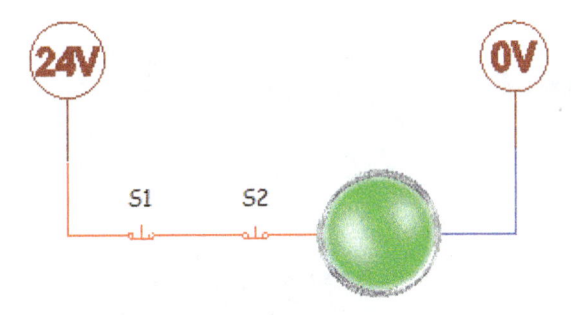

8. Which of the following logic function matches the shown below electrical circuit?
    A. AND.
    B. NAND.
    C. OR.
    D. NOR.

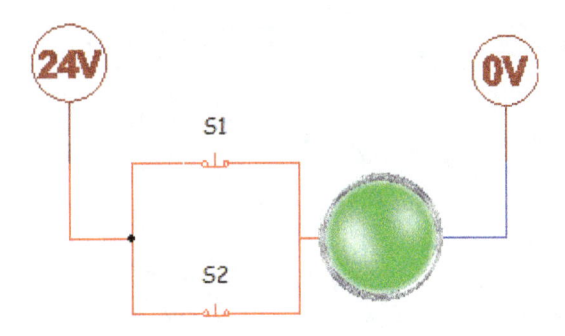

123

Workbook: Electro-Hydraulic Components and Systems
Chapter 4: Reviews and Assignments

9. Which of the following definition matches the shown below symbol?
    A. Relay coil.
    B. Normally-open relay contact.
    C. Normally-closed relay contact.
    D. Proximity switch.

10. Which of the following definition matches the shown below symbol?
    A. Relay coil.
    B. Normally-open relay contact.
    C. Normally-closed relay contact.
    D. Proximity switch.

Workbook: Electro-Hydraulic Components and Systems
Chapter 4: Reviews and Assignments

Chapter 4 Assignment

Student Name: -- Student ID: -------------------

Date: -- Score: ------------------------

Draw a hydraulic and an electrical circuit for continuous cylinder reciprocation using two pressure switches. Develop the corresponding truth table.

Chapter 5
Proportional Valves

Objectives:

This chapter introduces the technology of proportional valves and discusses the construction differences as compared to conventional switching valves. The chapter also introduces the conceptual construction of force-controlled versus stroke-controlled types of proportional valves. Additionally, the chapter presents the control schemes when a proportional valve is used in open-loop and closed-loop control system. More important, the chapter concludes by exploring examples of actual proportional directional, pressure, and flow control valves from various suppliers.

Brief Contents:

5.1- Introduction to Proportional Valves

5.2- Proportional Solenoids

5.3- Proportional Directional Control Valves

5.4- Proportional Pressure Control Valves

5.5- Proportional Flow Control Valves

5.6- Proportional Valves for Mobile Applications

5.1 - Introduction to Proportional Valves

Figure 5.1 - Mechanical Flight Control System

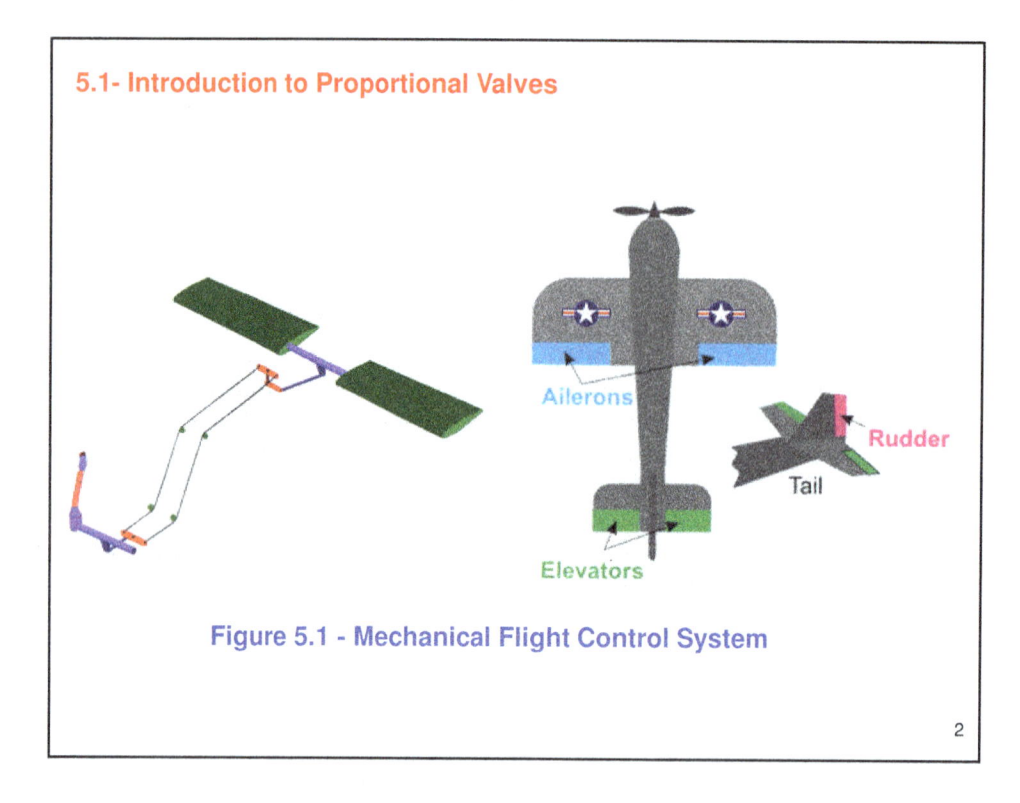

When large jets and powerful propeller planes came into service, the aerodynamic forces that act on the moving surfaces of such powerful planes were considerably increased. Application engineers at the US Air Forces were in a position of having to upgrade the conventional flight control system by a new one that is:

- Power assisted.
- Reliable.
- Offers input-output linear relationship.
- Fast.
- Precise
- Repeatable.
- Automatically compensate the effect of the external disturbances

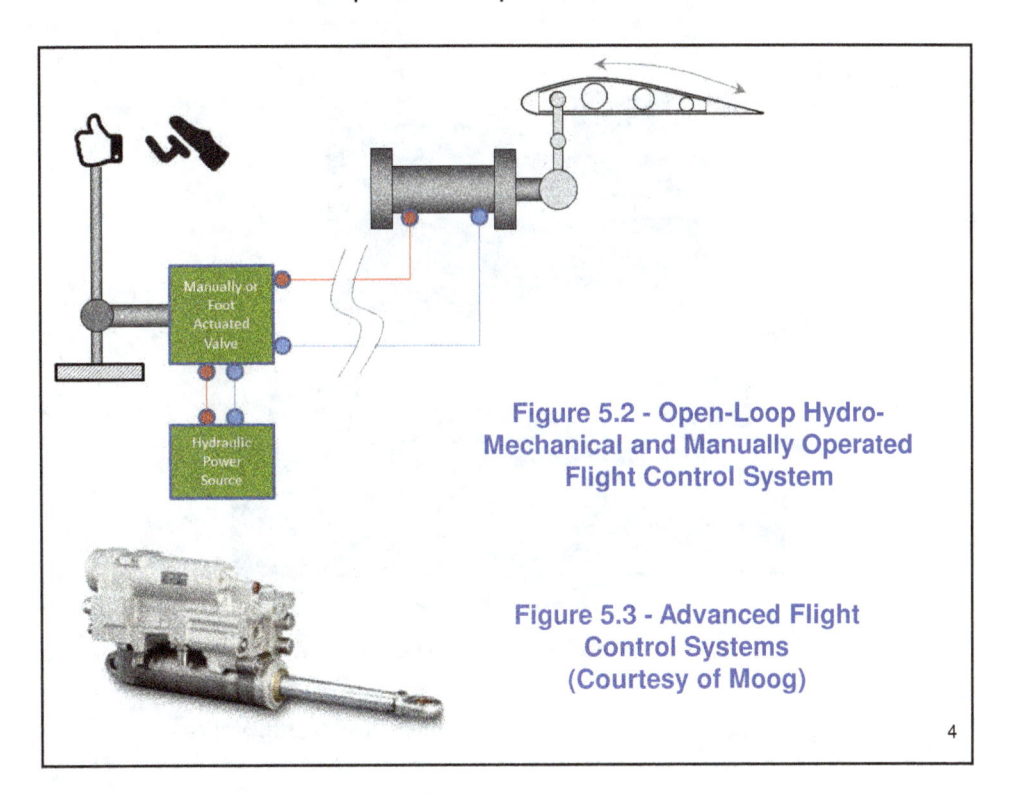

Workbook: Electro-Hydraulic Components and Systems
Chapter 5: Proportional Valves

Figure 5.2 - Open-Loop Hydro-Mechanical and Manually Operated Flight Control System

Figure 5.3 - Advanced Flight Control Systems (Courtesy of Moog)

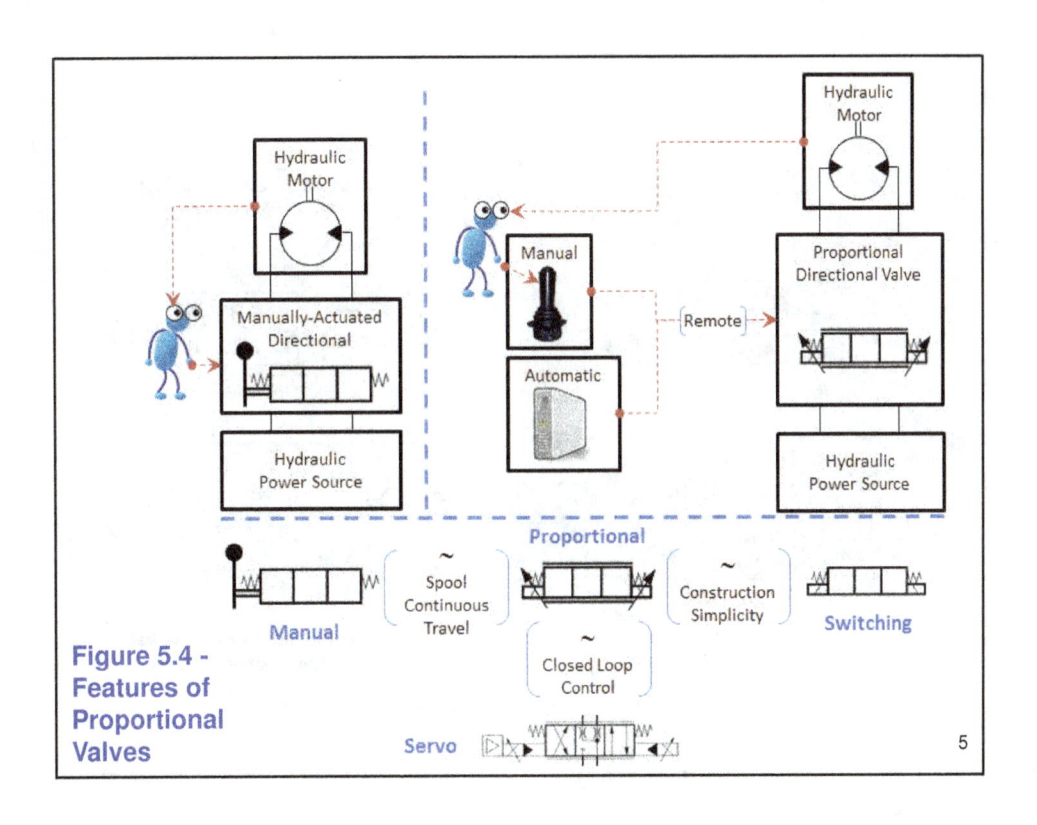

Figure 5.4 - Features of Proportional Valves

Two-Positions

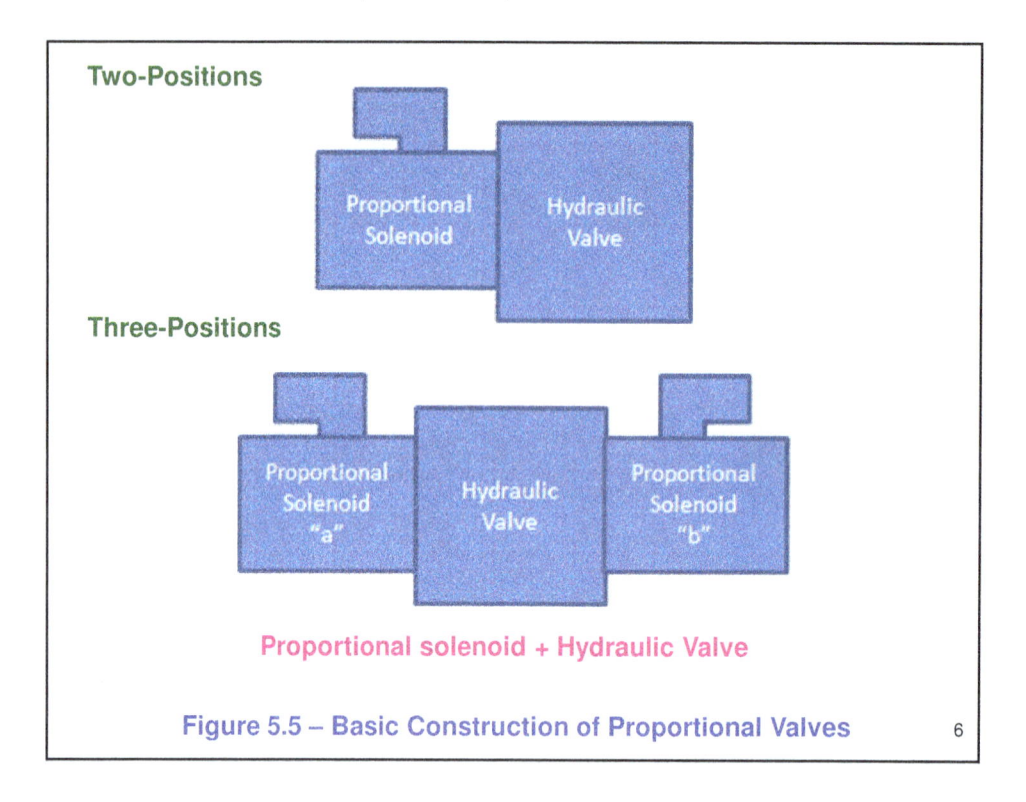

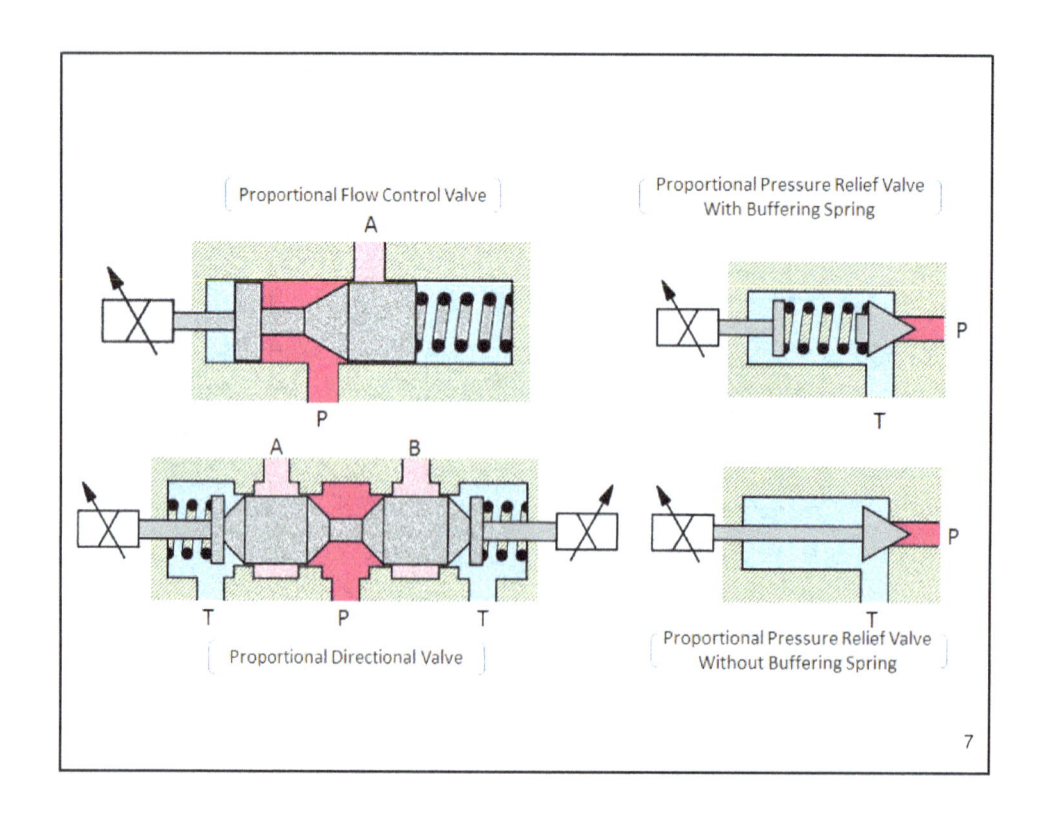

Three-Positions

Proportional solenoid + Hydraulic Valve

Figure 5.5 – Basic Construction of Proportional Valves

Proportional Flow Control Valve

Proportional Directional Valve

Proportional Pressure Relief Valve With Buffering Spring

Proportional Pressure Relief Valve Without Buffering Spring

5.2- Proportional Solenoids

- Interface between the control section and the hydraulic power.
- Only DC-powered.
- Only wet-type.

5.2.1- Force-Controlled Proportional Solenoid

- Spool controlled in an open-loop.
- It can't compensate for the changes in:
  - Spring characteristics.
  - Electrical characteristics.
  - Spool-Sleeve characteristics.

Has the following features:
- Less expensive.
- Simple in construction.
- Compact due to no feedback elements.
- Short stroke, about 1.5 mm.

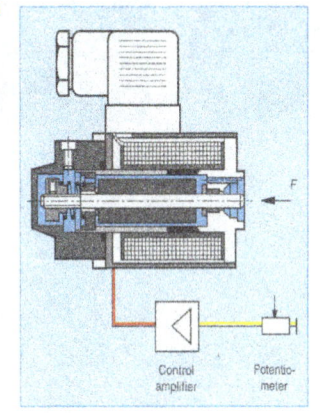

Figure 5.6 - Operation of Force-Controlled Proportional Solenoid (Courtesy of Bosch Rexroth)

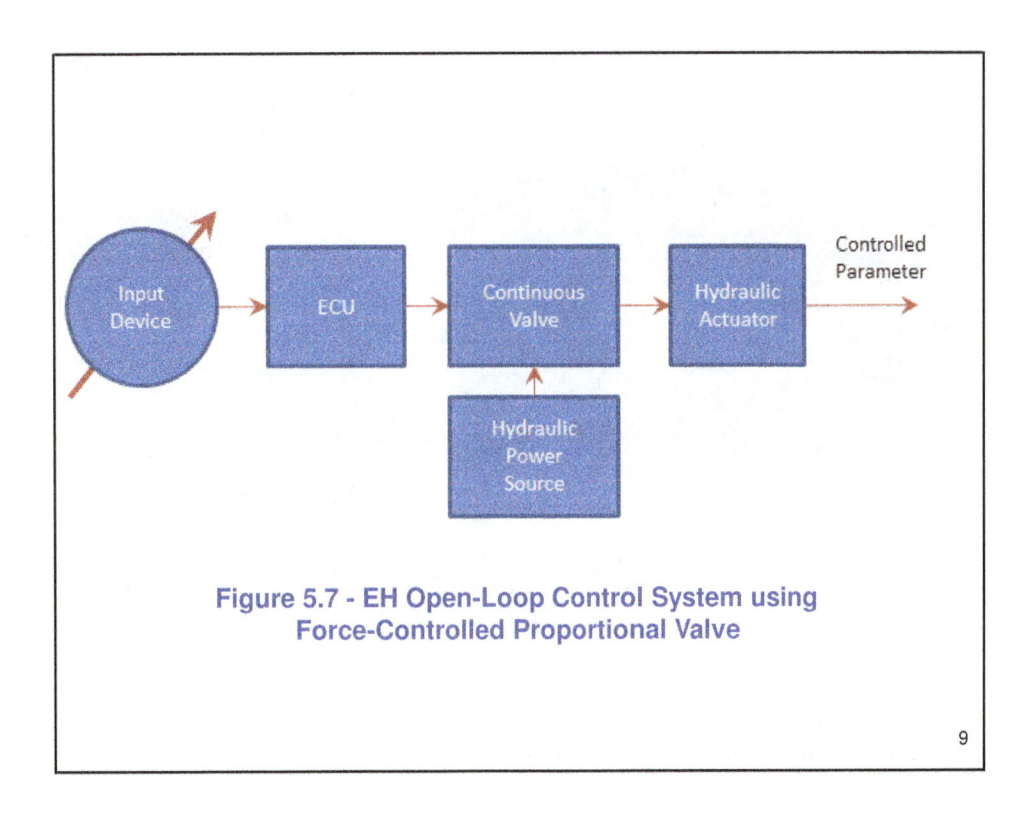

Figure 5.7 - EH Open-Loop Control System using Force-Controlled Proportional Valve

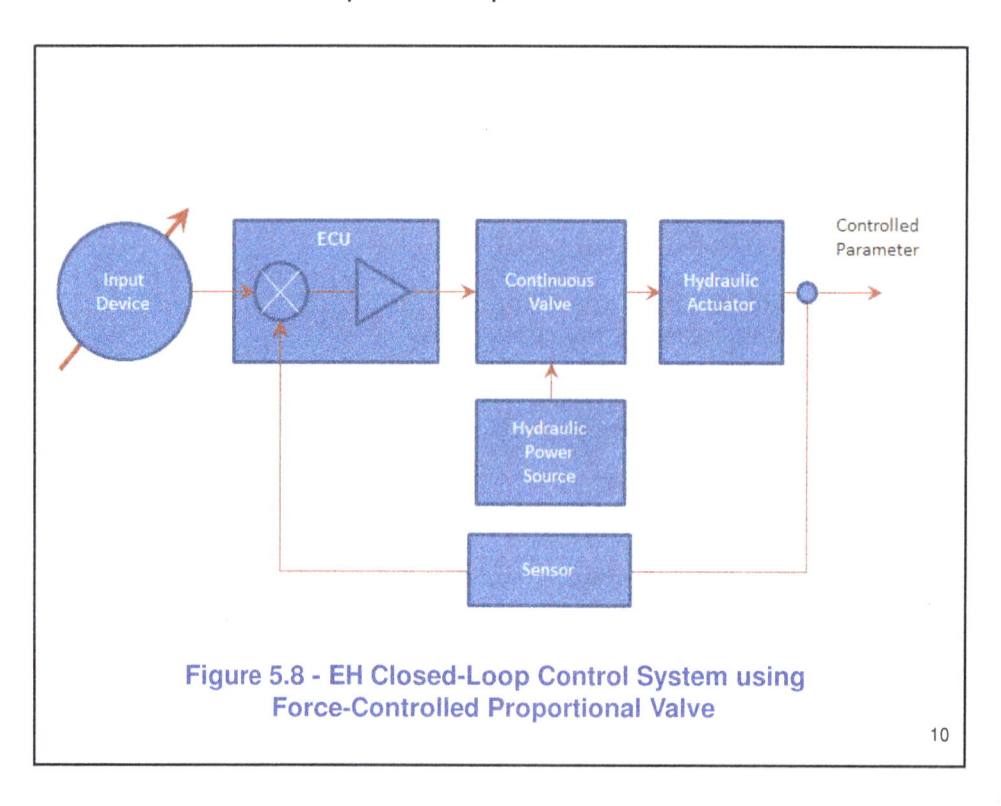

Figure 5.8 - EH Closed-Loop Control System using Force-Controlled Proportional Valve

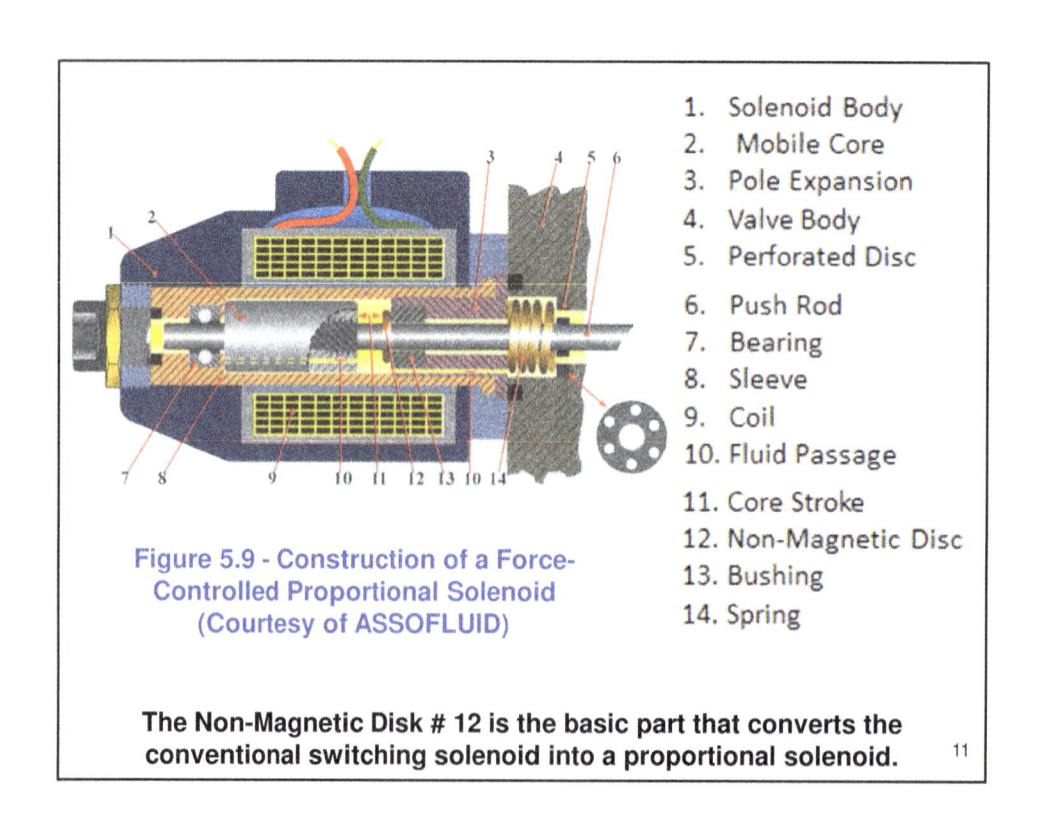

Figure 5.9 - Construction of a Force-Controlled Proportional Solenoid (Courtesy of ASSOFLUID)

1. Solenoid Body
2. Mobile Core
3. Pole Expansion
4. Valve Body
5. Perforated Disc
6. Push Rod
7. Bearing
8. Sleeve
9. Coil
10. Fluid Passage
11. Core Stroke
12. Non-Magnetic Disc
13. Bushing
14. Spring

The Non-Magnetic Disk # 12 is the basic part that converts the conventional switching solenoid into a proportional solenoid.

Proportional Solenoids for Hydraulic Application Type G RC Y
3 sizes: Ø (mm) 37, 45, 63
pressure-proof armature space up to max. 350 bar (= 5075 psi) static
stroke (mm) 2 up to 4
magnetic force (N) 47 up to 112 (at working stroke)
fastening with central thread
pulling design on request

Figure 5.10 - Typical Force-Controlled Proportional Solenoid
(www.magnet-schultz.com)

Figure 5.11 - Characteristic Curves of a Stroke-Controlled Proportional Solenoid

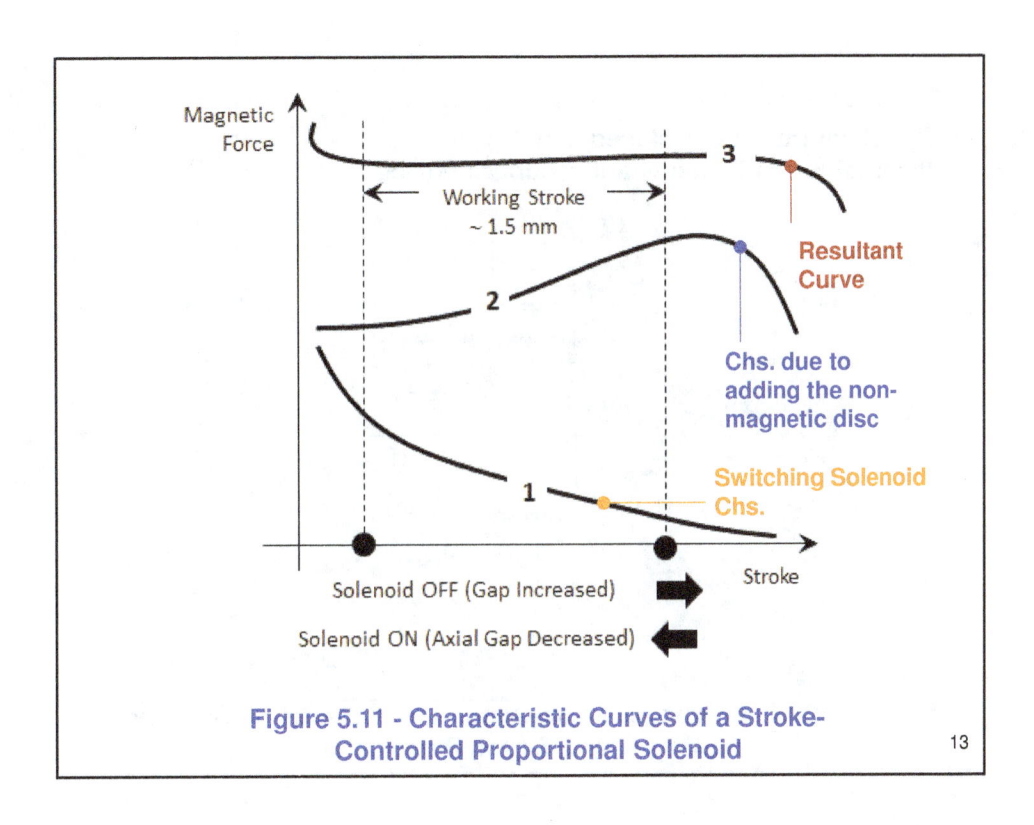

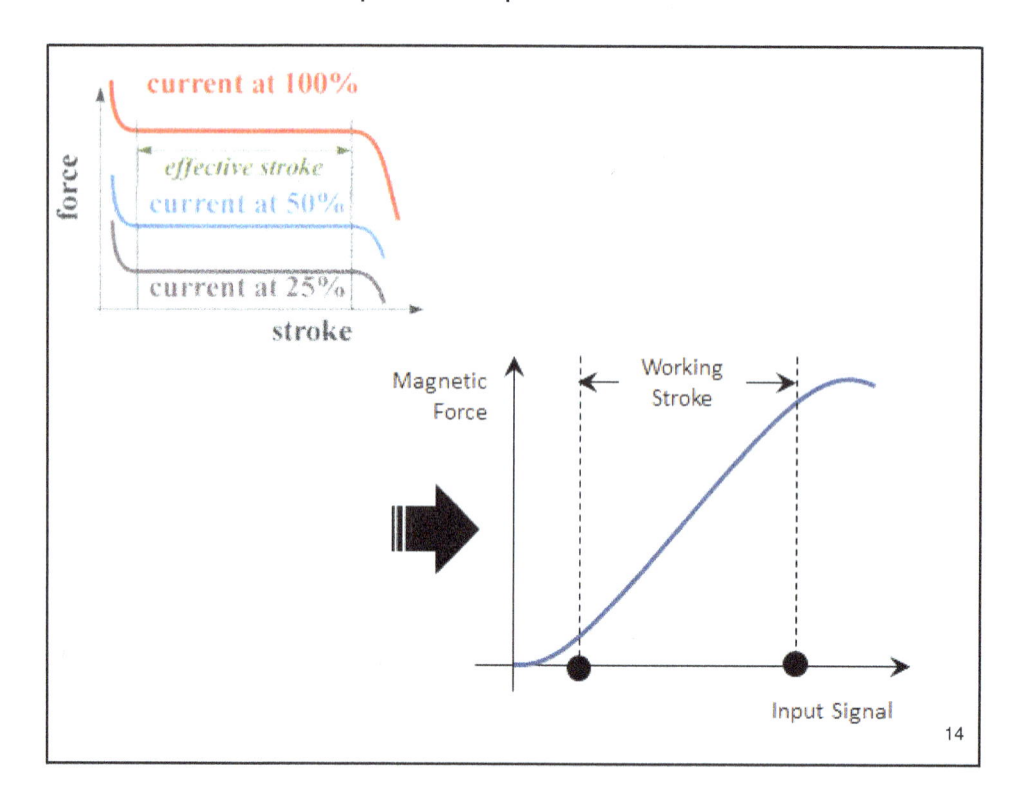

Workbook: Electro-Hydraulic Components and Systems
Chapter 5: Proportional Valves

5.2.2- Stroke-Controlled Proportional Solenoid   Video 227 (2.5 min)
- Spool controlled in a closed-loop.
- It contains a built in position feedback device

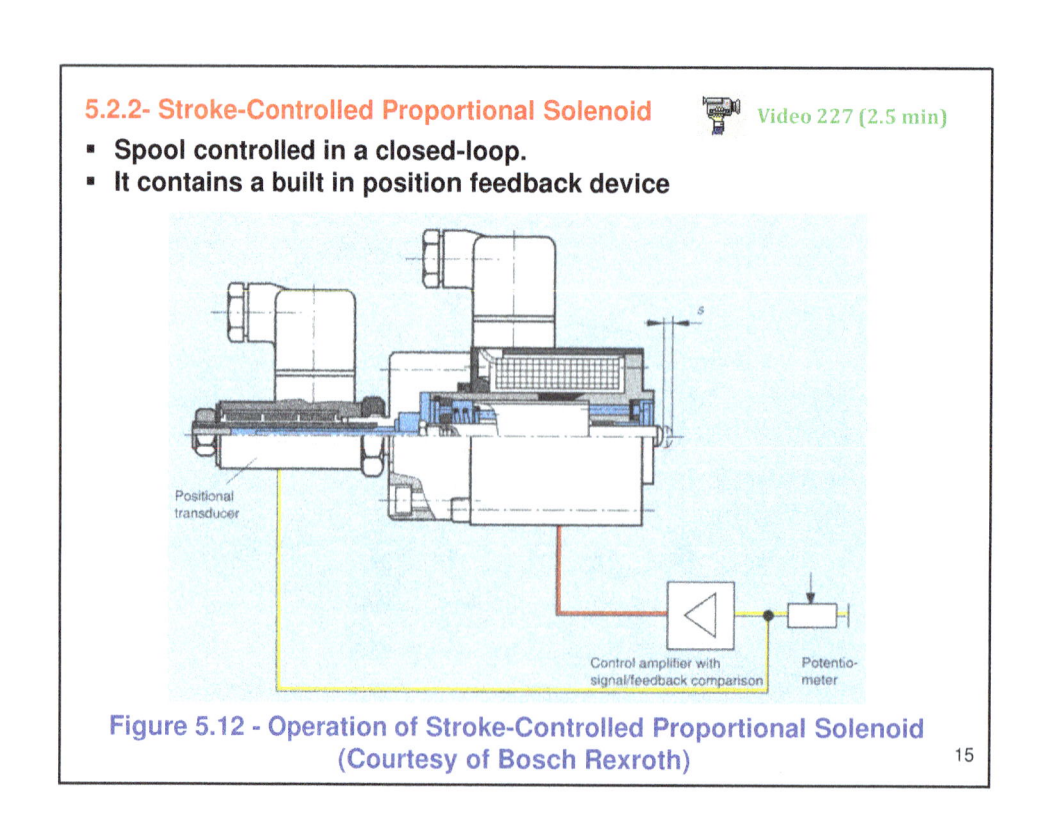

Figure 5.12 - Operation of Stroke-Controlled Proportional Solenoid
(Courtesy of Bosch Rexroth)

Has the following features:

- More expensive.
- More complex in construction.
- Larger in size due to adding feedback elements.
- Longer stroke, about 3-5 mm.
- Less hysteresis.
- Better accuracy in spool positioning.
- To some limit, compensates for the change in spring, spool-sleeve, and electrical characteristics.

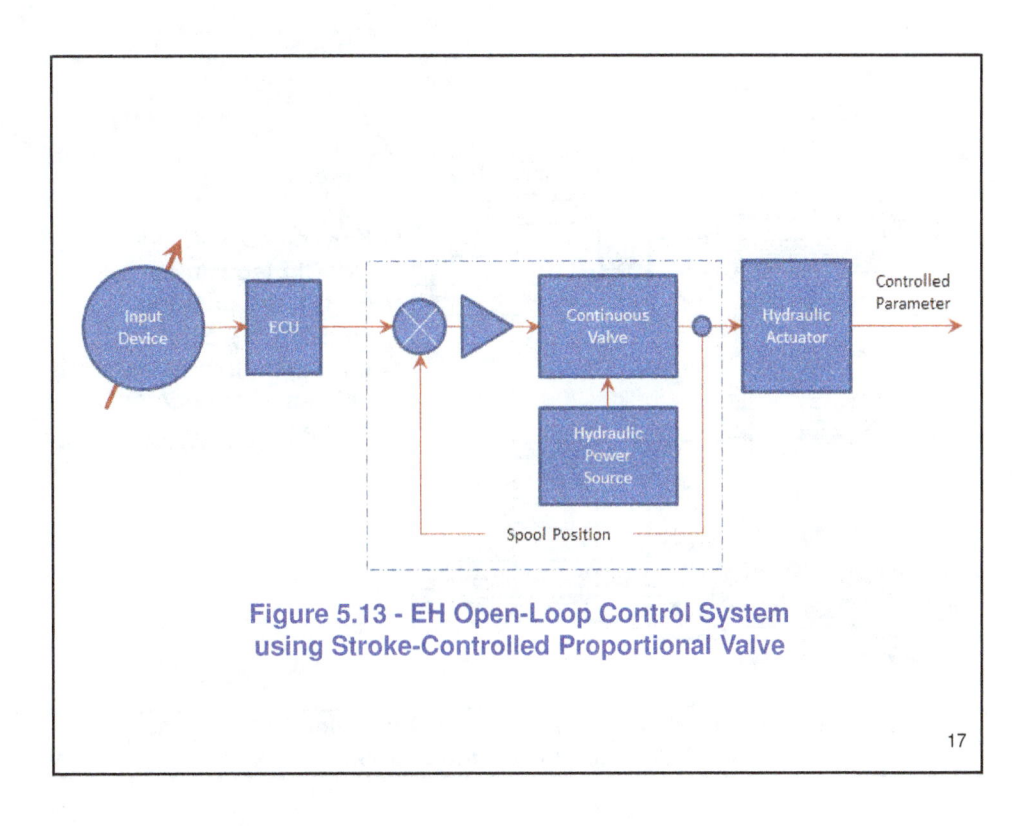

Figure 5.13 - EH Open-Loop Control System using Stroke-Controlled Proportional Valve

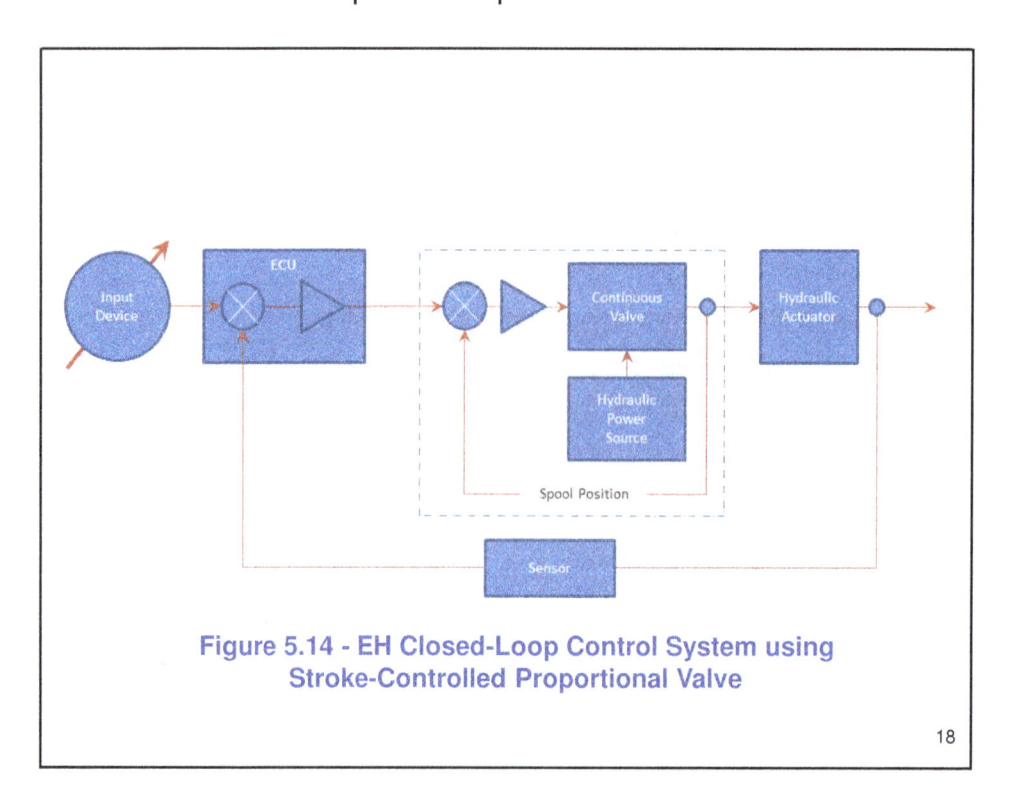

Workbook: Electro-Hydraulic Components and Systems
Chapter 5: Proportional Valves

Figure 5.14 - EH Closed-Loop Control System using Stroke-Controlled Proportional Valve

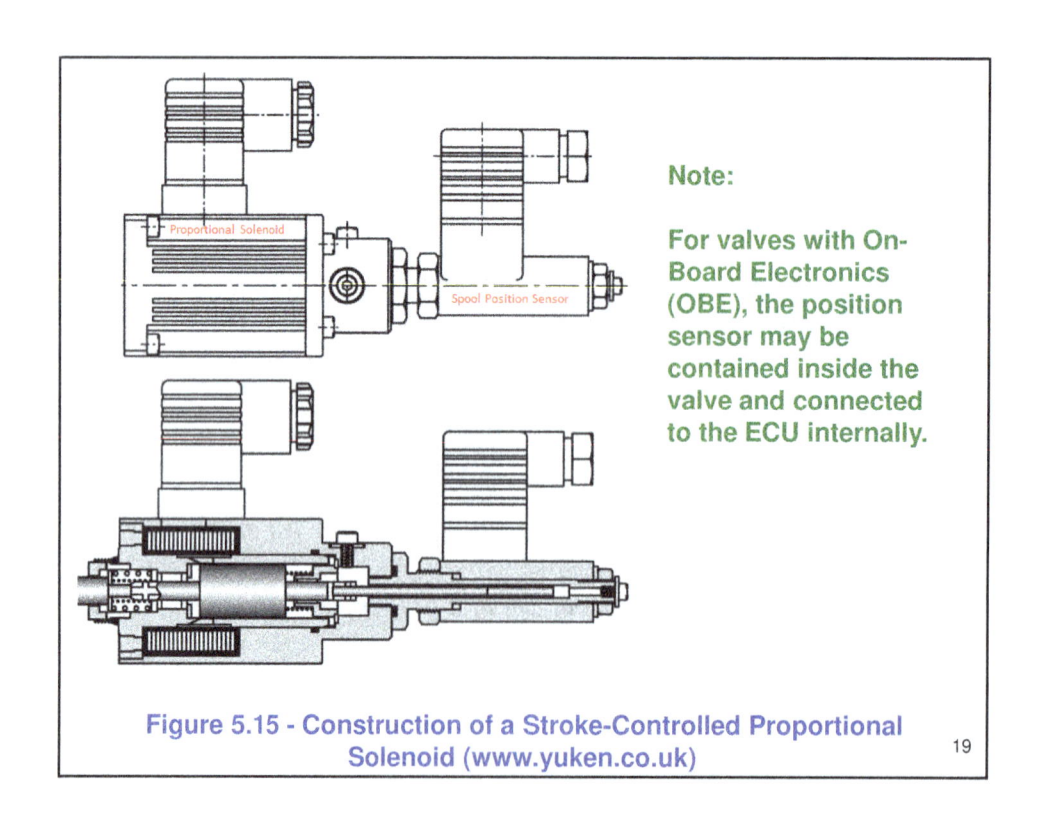

Note:

For valves with On-Board Electronics (OBE), the position sensor may be contained inside the valve and connected to the ECU internally.

Figure 5.15 - Construction of a Stroke-Controlled Proportional Solenoid (www.yuken.co.uk)

Rated Current[A] : 2.5
Normal Resistance[Ω]: 3.7
Rated Attraction Force [N] : 160
Rated Stroke[mm] : 4
Total Stroke [mm]: ≥9
Force Hysteresis [%]: ≤5
Current Hysteresis [%]: ≤3
Repeat Accuracy [%]: ≤1
Operating Oil Temperature[°C]: -90

Figure 5.16 - Typical Stroke-Controlled Proportional Solenoid (http://yfsolenoid.com)

Proportional Solenoids for Hydraulic Application with Transducer type G RC...A62
3 sizes: Ø (mm) 37, 45, 63
pressure-proof armature space
up to max. 320 bar (= 4640 psi-static)
stroke (mm) 5.5 up to 9
magnetic force (N) 43 up to 120
fastening with central thread
extendable to a complete closed
loop control circuit

Figure 5.17 - Typical Stroke-Controlled Proportional Solenoid (www.magnet-schultz.com)

5.3- Proportional Directional Control Valves
5.3.1- Interpretation of Symbols for Proportional Directional Valves

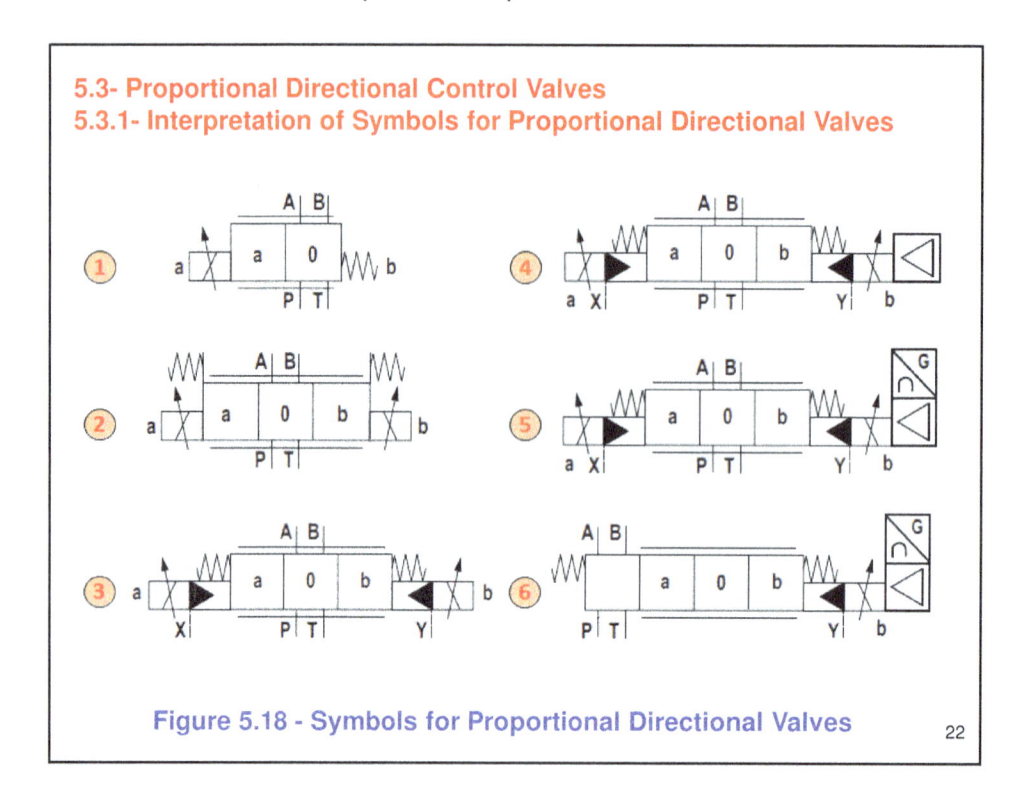

Figure 5.18 - Symbols for Proportional Directional Valves

5.3.2- Hydraulic Static Characteristics of Proportional Directional Valves

Δp = Valve pressure differential

(inlet pressure p_P minus load pressure p_L minus return flow pressure p_T)

25 l/min rated flow with 10 bar valve pressure differential
P → A / B → T
or
P → B / A → T
Type 4WREE 6 V32

1 Δp = 10 bar constant
2 Δp = 20 bar constant
3 Δp = 30 bar constant
4 Δp = 50 bar constant
5 Δp = 100 bar constant

Control spool V
Control spool E- and W
Command value in % →

Type 4WREE (measured with HLP46, ϑ_{Oil} = 40 °C ± 5 °C)

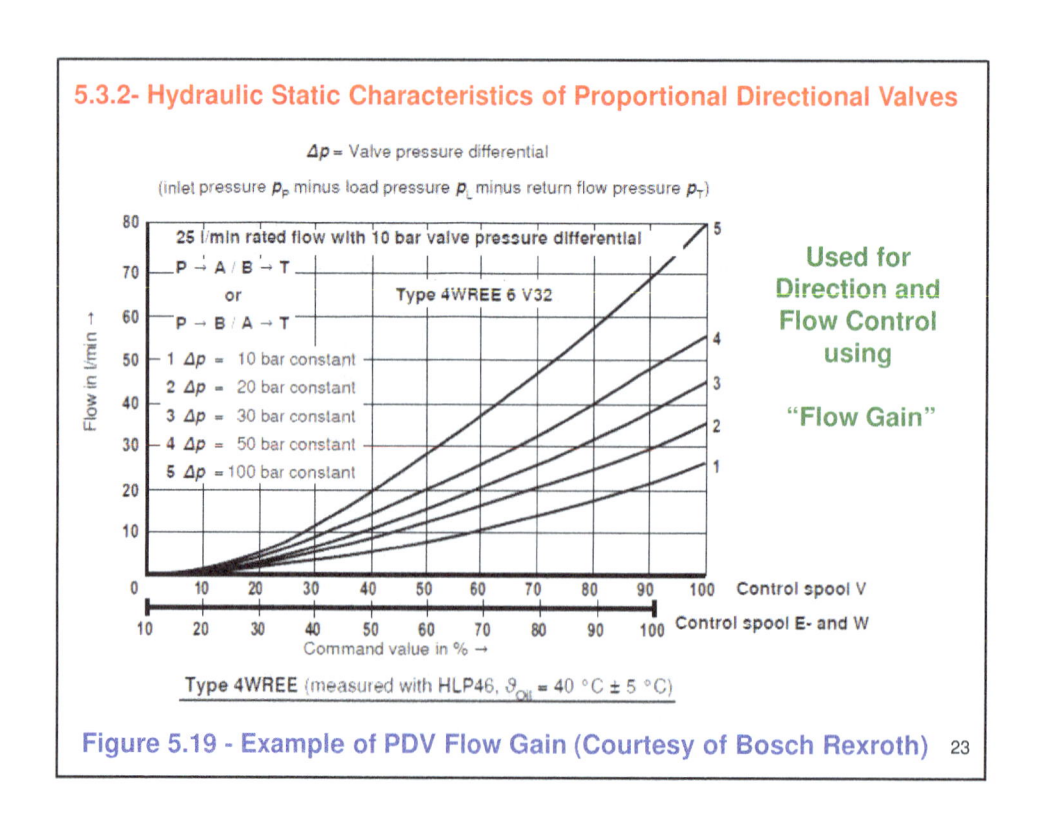

Used for Direction and Flow Control using "Flow Gain"

Figure 5.19 - Example of PDV Flow Gain (Courtesy of Bosch Rexroth)

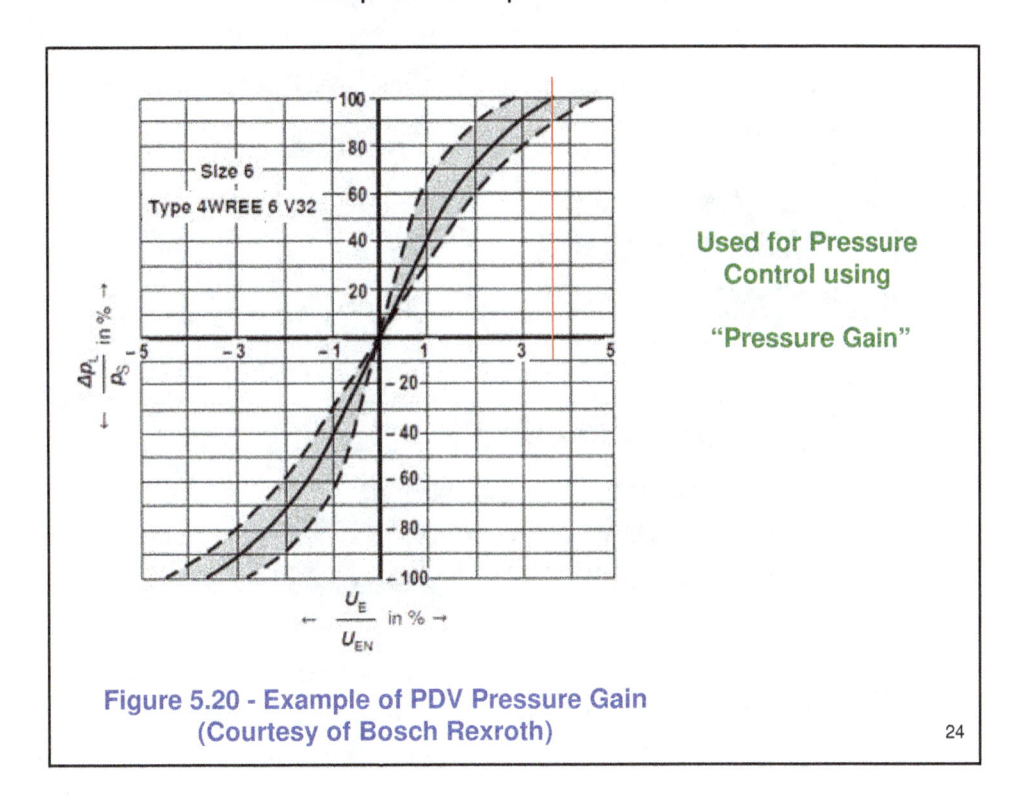

Figure 5.20 - Example of PDV Pressure Gain (Courtesy of Bosch Rexroth)

Used for Pressure Control using

"Pressure Gain"

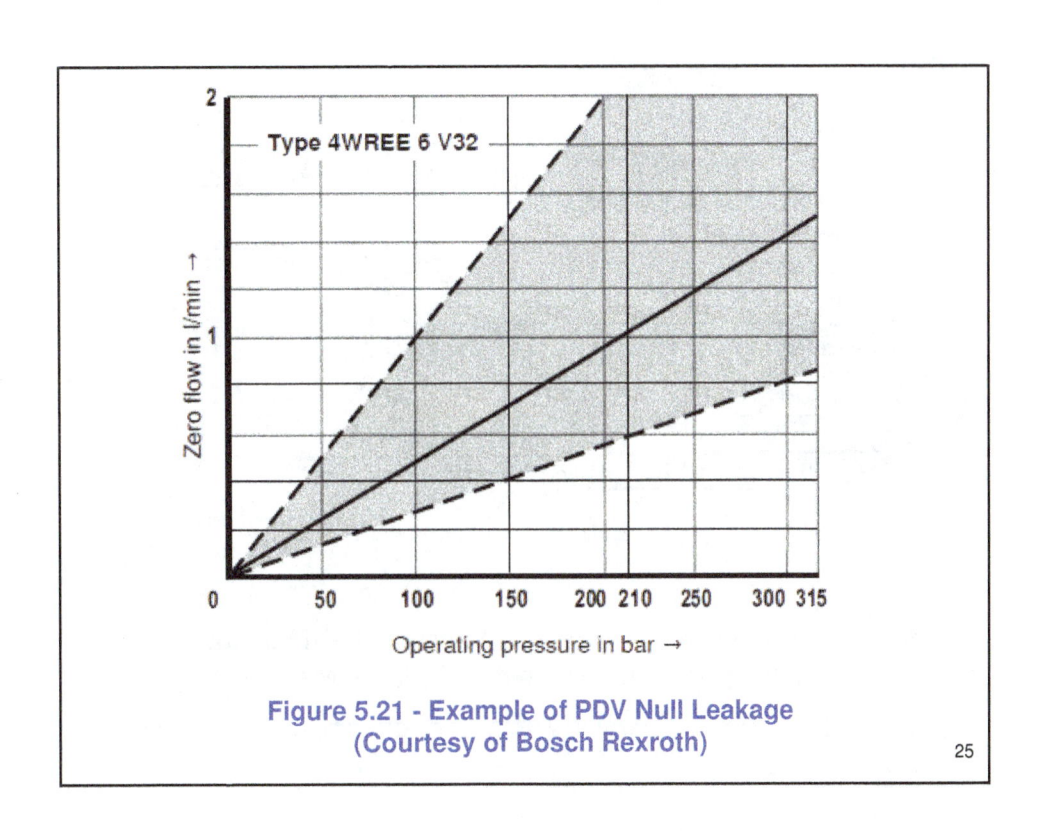

Figure 5.21 - Example of PDV Null Leakage (Courtesy of Bosch Rexroth)

Workbook: Electro-Hydraulic Components and Systems
Chapter 5: Proportional Valves

5.3.3- Direct-Operated Force-Controlled Proportional Directional Valve

1- Valve Body, 2- Spool, 3- Proportional Solenoid, and 4- Electrical Connection

Figure 5.22 - Direct-Operated Force-Controlled Spool-Type PDV with Separate ECU (Courtesy of Atos)

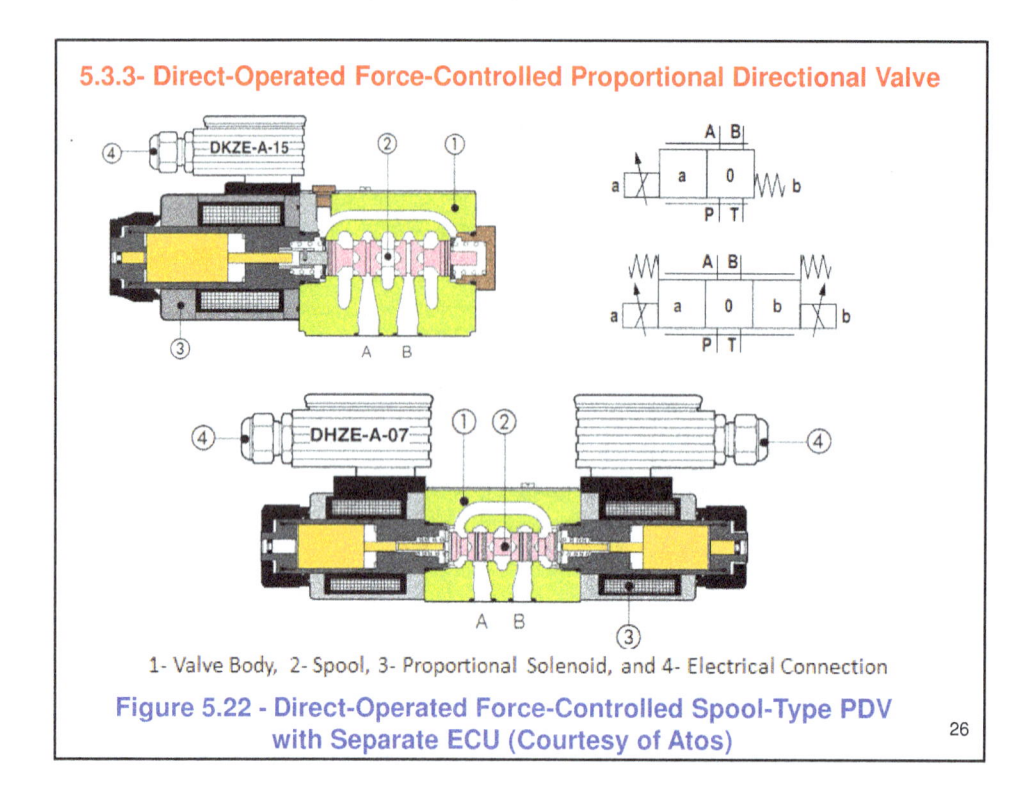

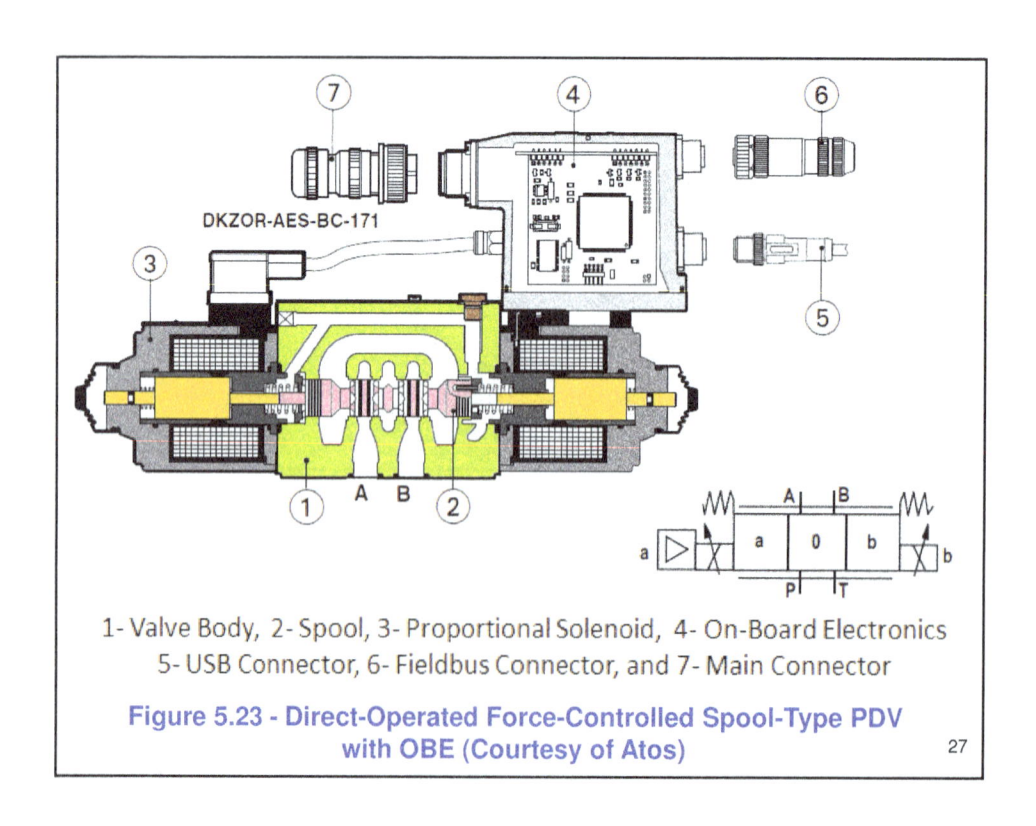

1- Valve Body, 2- Spool, 3- Proportional Solenoid, 4- On-Board Electronics
5- USB Connector, 6- Fieldbus Connector, and 7- Main Connector

Figure 5.23 - Direct-Operated Force-Controlled Spool-Type PDV with OBE (Courtesy of Atos)

Figure 5.24 - Direct-Operated Force-Controlled Cartridge-Type PDV with Separate ECU (Courtesy of Hydraforce)

5.3.4- Direct-Operated Stroke-Controlled Proportional Directional Valve

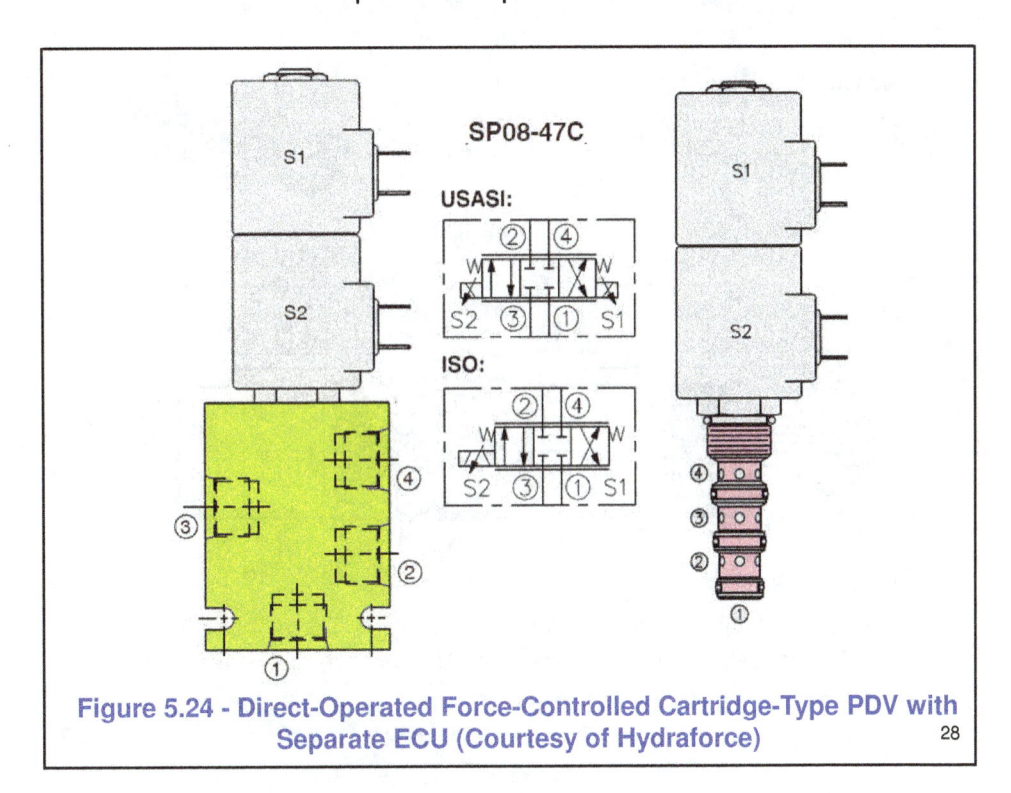

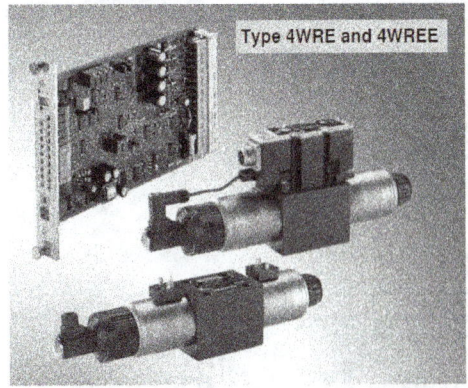

Figure 5.25 - Direct-Operated Stroke-Controlled Spool-Type PDV (Courtesy of Bosch Rexroth)

Duty cycle		%	100
Maximum coil temperature [1]		°C	up to 150
Supply voltage	Nominal voltage	VDC	24
	lower limit value	V	19.4
	upper limit value	V	35
Current consumption of the amplifier	I_{max}	A	< 2
	Pulse current	A	3

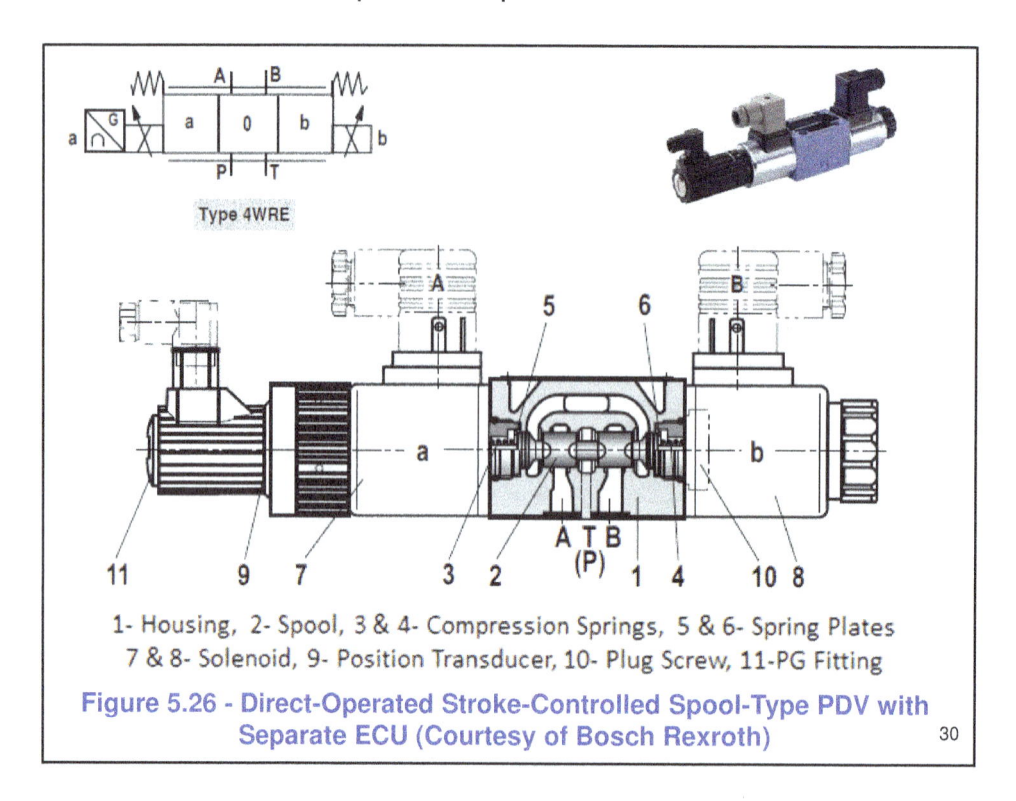

Type 4WRE

1- Housing, 2- Spool, 3 & 4- Compression Springs, 5 & 6- Spring Plates
7 & 8- Solenoid, 9- Position Transducer, 10- Plug Screw, 11-PG Fitting

Figure 5.26 - Direct-Operated Stroke-Controlled Spool-Type PDV with Separate ECU (Courtesy of Bosch Rexroth)

Type 4WREE

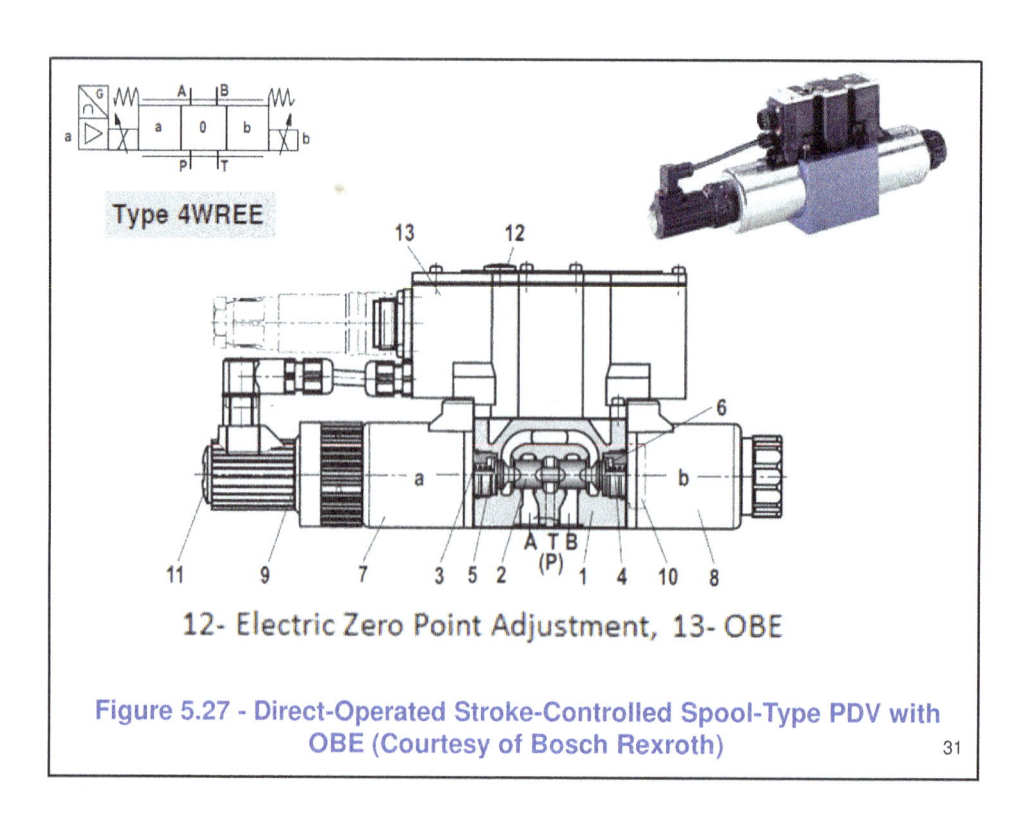

12- Electric Zero Point Adjustment, 13- OBE

Figure 5.27 - Direct-Operated Stroke-Controlled Spool-Type PDV with OBE (Courtesy of Bosch Rexroth)

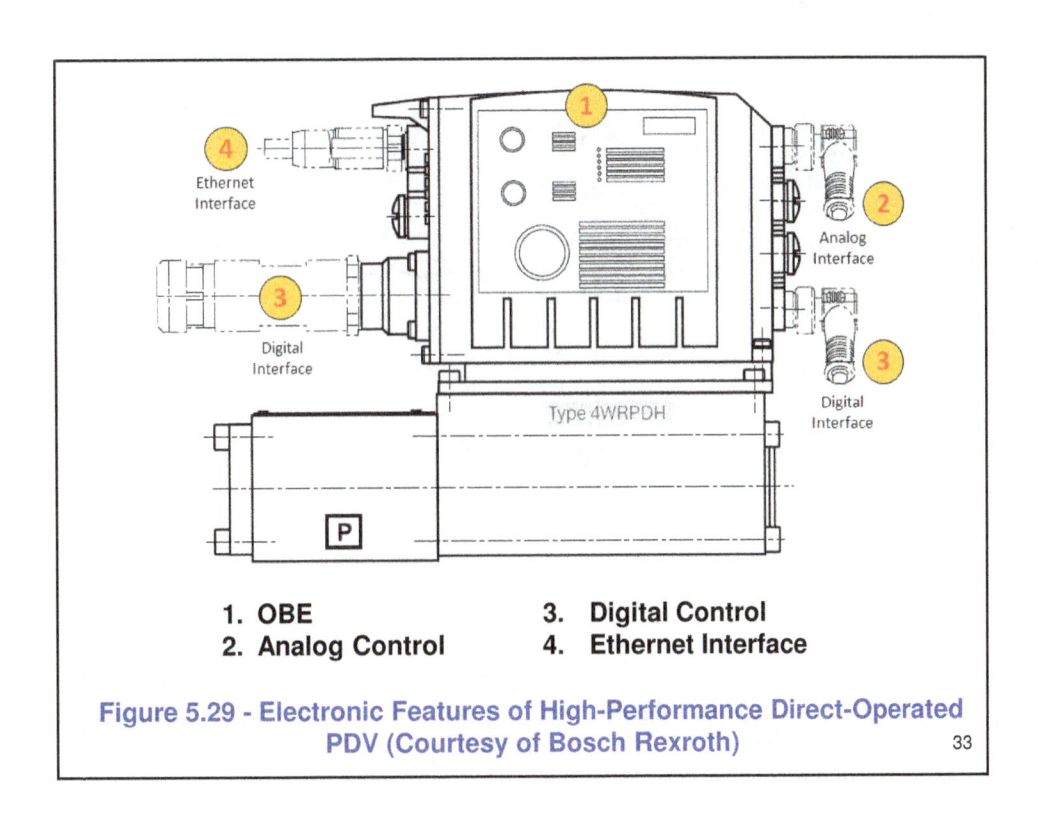

5.3.5- High Performance Proportional Directional Valve

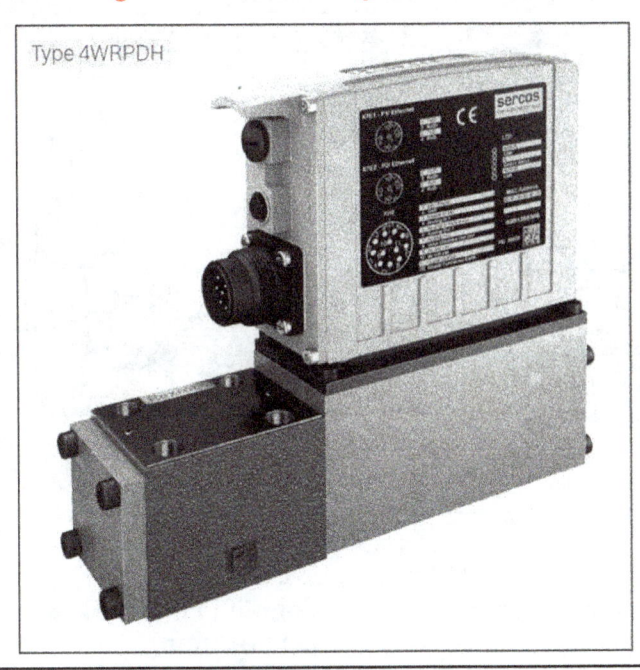

Figure 5.28 - High-Performance Direct-Operated PDV (Courtesy of Bosch Rexroth)

1. OBE
2. Analog Control
3. Digital Control
4. Ethernet Interface

Figure 5.29 - Electronic Features of High-Performance Direct-Operated PDV (Courtesy of Bosch Rexroth)

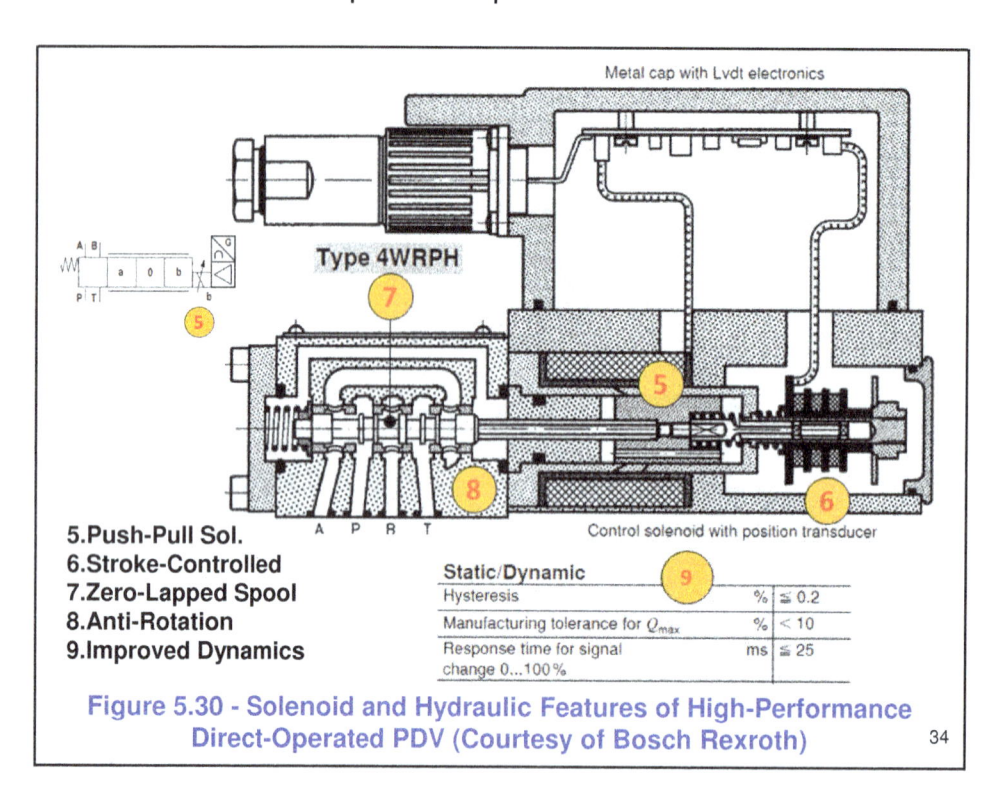

Workbook: Electro-Hydraulic Components and Systems
Chapter 5: Proportional Valves

5. Push-Pull Sol.
6. Stroke-Controlled
7. Zero-Lapped Spool
8. Anti-Rotation
9. Improved Dynamics

Figure 5.30 - Solenoid and Hydraulic Features of High-Performance Direct-Operated PDV (Courtesy of Bosch Rexroth)

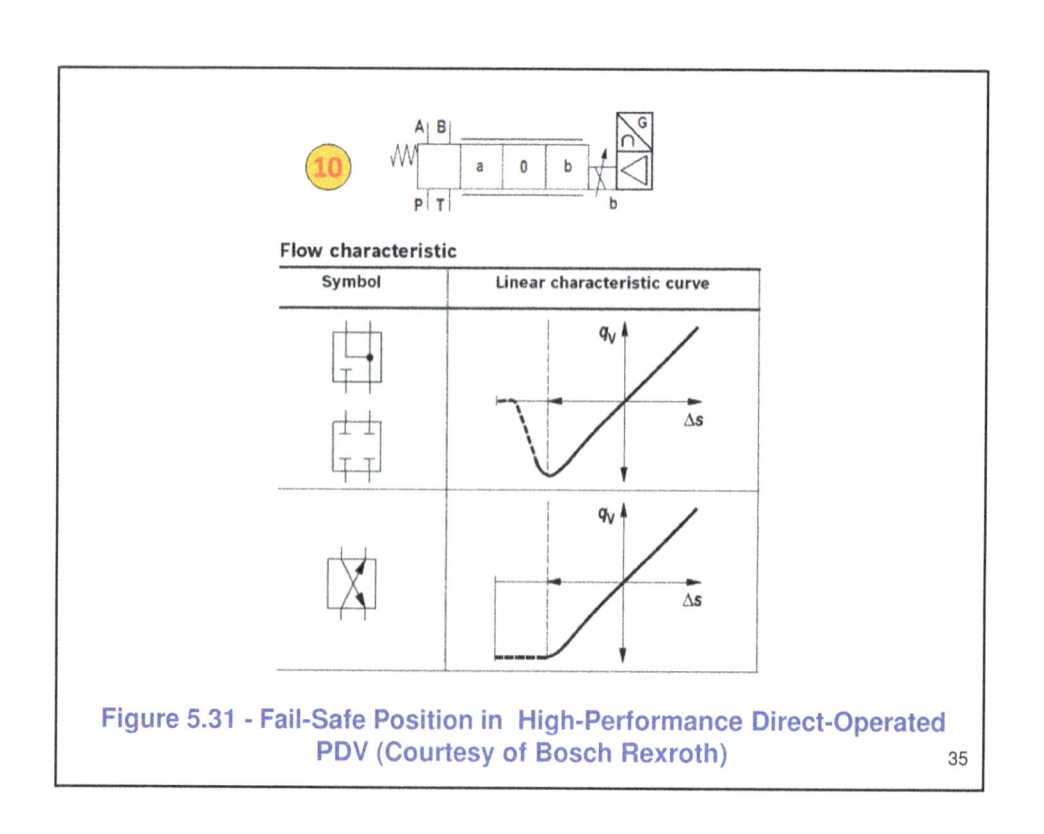

Figure 5.31 - Fail-Safe Position in High-Performance Direct-Operated PDV (Courtesy of Bosch Rexroth)

Such high-performance valves replace servo valves in many applications except where servo valve high dynamic features are needed

Figure 5.32 - Example of using a High-Performance PDV in a Closed-Loop Application (Courtesy of Bosch Rexroth)

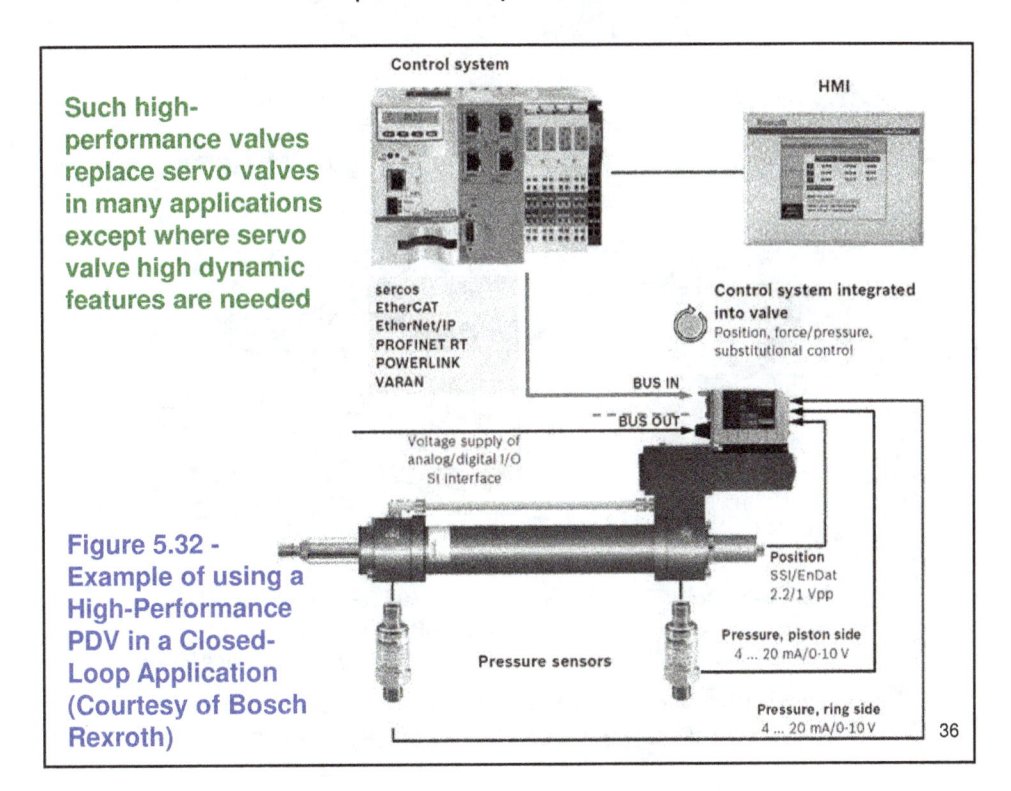

5.3.6 - Pilot-Operated Force-Controlled Proportional Directional Valve

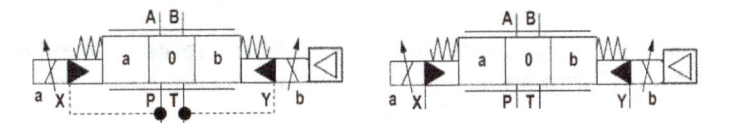

Similar to the pilot-operated switching directional valves:

- It consists of two stages.
- It could be of two positions or three positions.
- Its pilot stage uses the control pressure X to control the main stage.
- It is called Pilot-Controlled if X is sourced externally.
- It is called Direct-Controlled if X is generated from the main pressure line P.
- Control pressure should not exceed a certain value (30 – 100 bar).
- If needed, regardless the source of X, an optional sandwich-type pressure reducing valve may be used to control the proper value of X.
- The return of the pilot pressure Y may be drained externally or merged with the main tank line of the valve T.
- External drainage is recommended if T line is experiencing back pressure.
- If X is external, it is not necessary that Y to be external too.

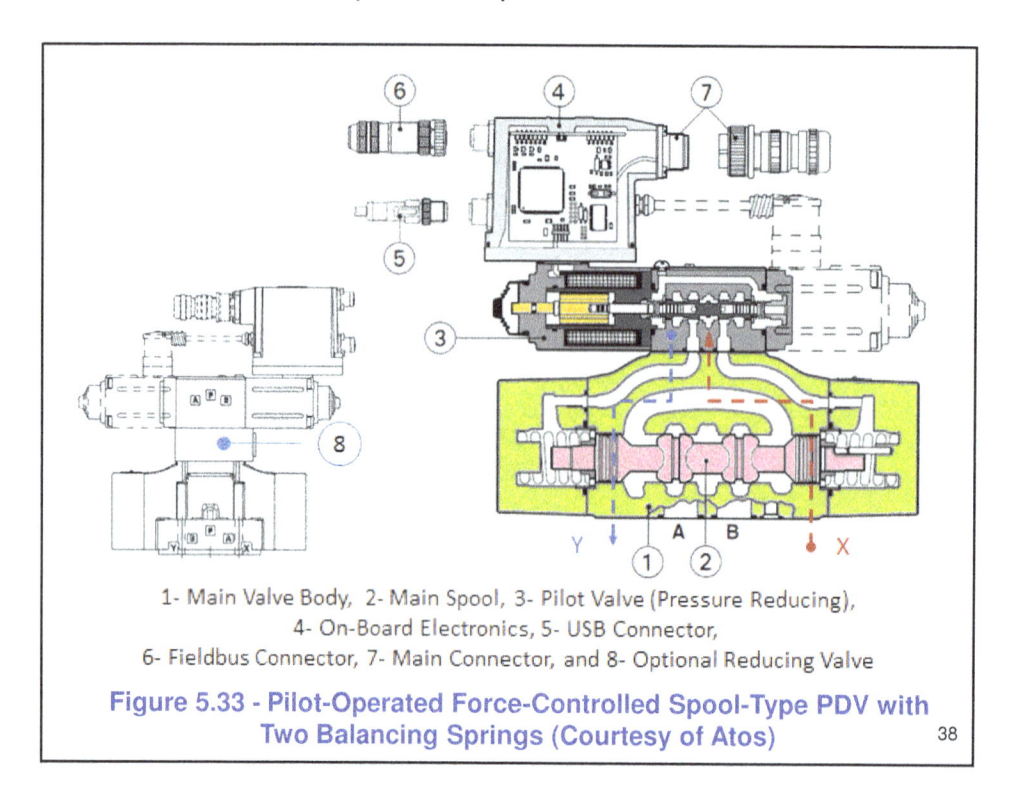

1- Main Valve Body, 2- Main Spool, 3- Pilot Valve (Pressure Reducing),
4- On-Board Electronics, 5- USB Connector,
6- Fieldbus Connector, 7- Main Connector, and 8- Optional Reducing Valve

Figure 5.33 - Pilot-Operated Force-Controlled Spool-Type PDV with Two Balancing Springs (Courtesy of Atos)

Unlike the pilot-operated switching directional valves:

Positioning of the main spool in a pilot-operated force-controlled PDV is converted into an open-loop pressure control solution:

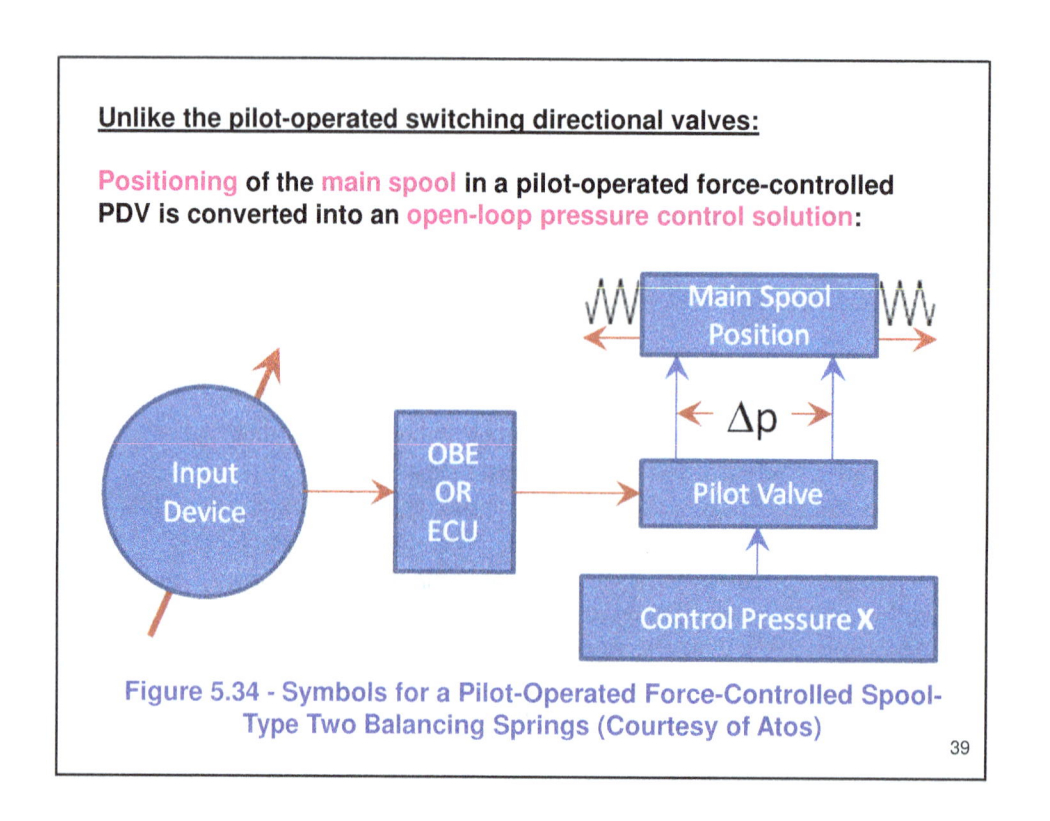

Figure 5.34 - Symbols for a Pilot-Operated Force-Controlled Spool-Type Two Balancing Springs (Courtesy of Atos)

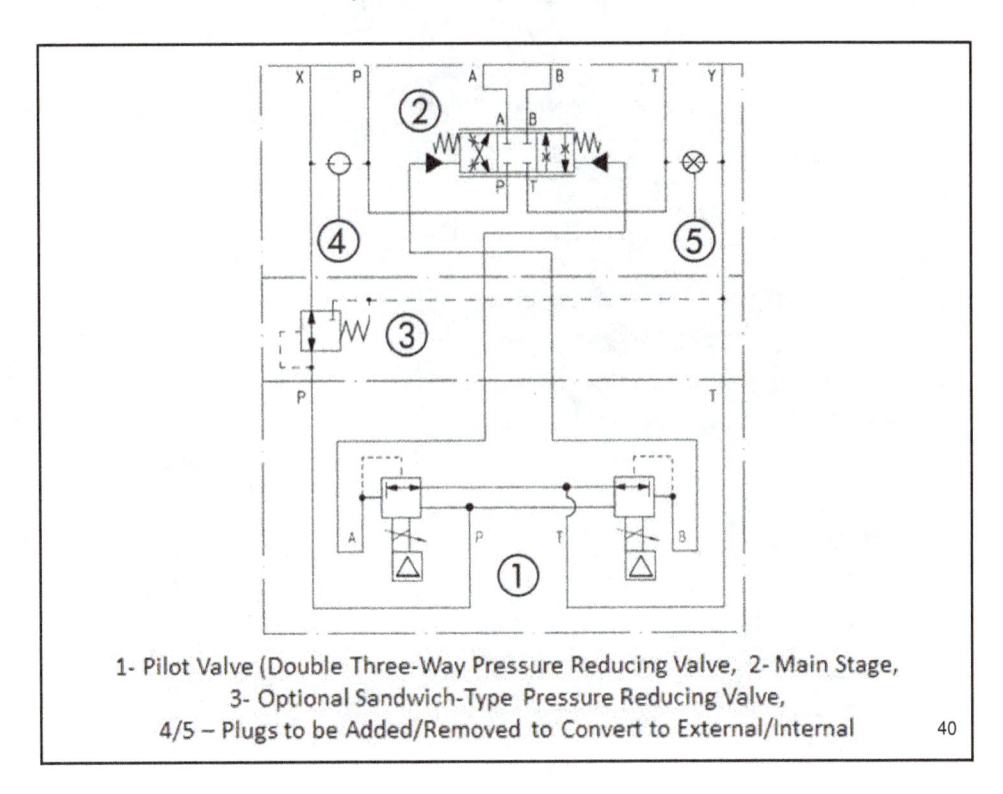

Workbook: Electro-Hydraulic Components and Systems
Chapter 5: Proportional Valves

1- Pilot Valve (Double Three-Way Pressure Reducing Valve, 2- Main Stage,
3- Optional Sandwich-Type Pressure Reducing Valve,
4/5 – Plugs to be Added/Removed to Convert to External/Internal

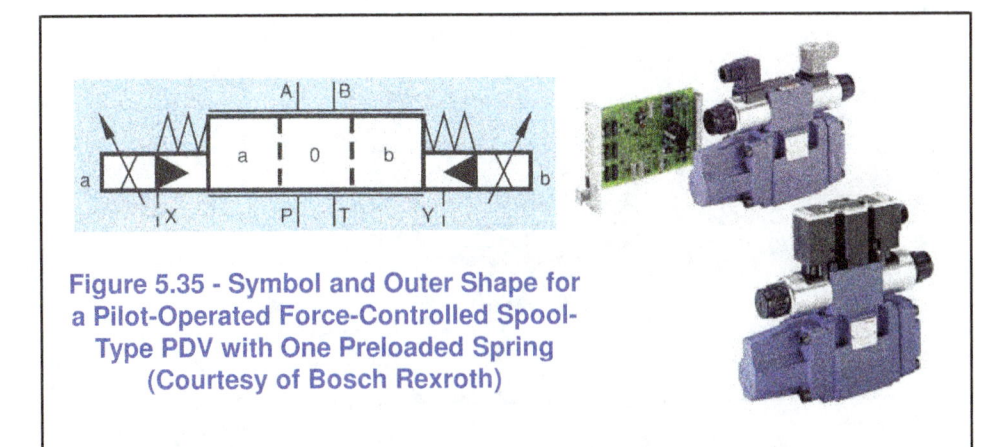

Figure 5.35 - Symbol and Outer Shape for a Pilot-Operated Force-Controlled Spool-Type PDV with One Preloaded Spring (Courtesy of Bosch Rexroth)

Having one preloaded spring are as follows:

- **Ensures, an identical valve reaction in each direction to any given signal, and thus, equal deflection in each direction**

- **The change in spring characteritics will be distributed evenly and will not affect the center position of the main spool.**

147

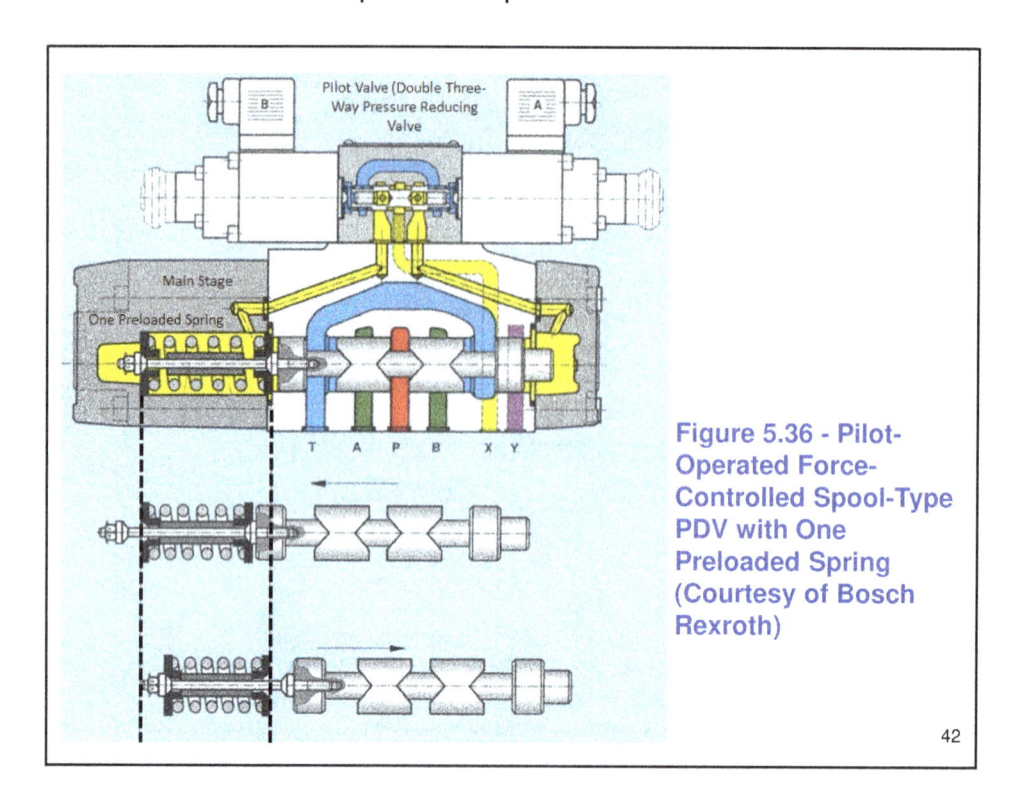

Figure 5.36 - Pilot-Operated Force-Controlled Spool-Type PDV with One Preloaded Spring (Courtesy of Bosch Rexroth)

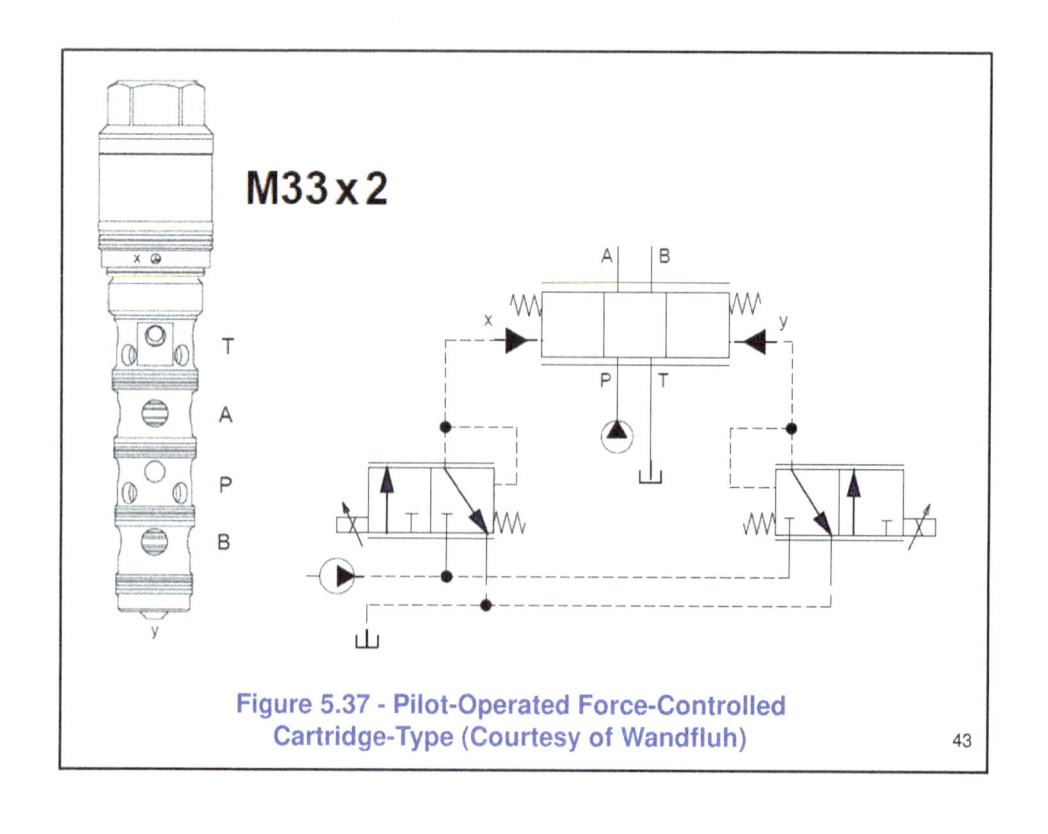

Figure 5.37 - Pilot-Operated Force-Controlled Cartridge-Type (Courtesy of Wandfluh)

- Are called **screw-in** cartridge valves in the market.

- The valves are designed for maximum pressures of up to 400 bar.

- With volume flow of up to 400 l/min,

- Ambient temperatures of up to 70°C can be accepted without any loss of performance.

Figure 5.38 - Cartridge-Type Proportional Valves (Courtesy of Wandfluh)

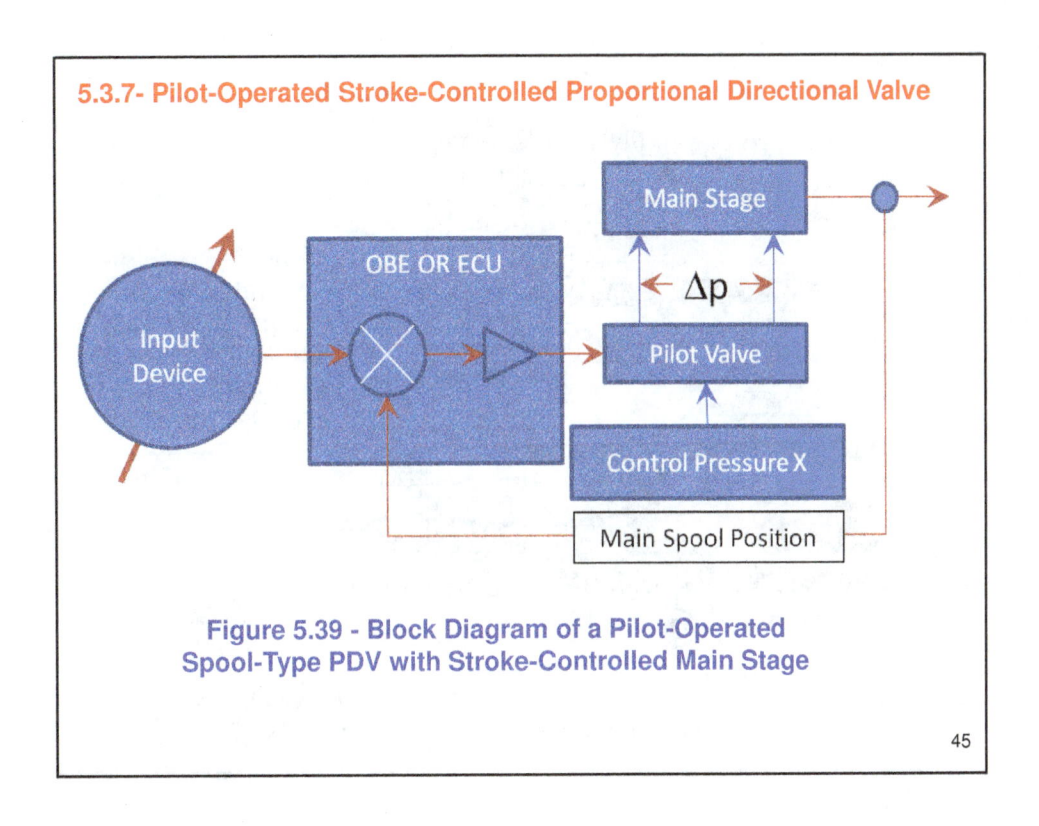

5.3.7- Pilot-Operated Stroke-Controlled Proportional Directional Valve

Figure 5.39 - Block Diagram of a Pilot-Operated Spool-Type PDV with Stroke-Controlled Main Stage

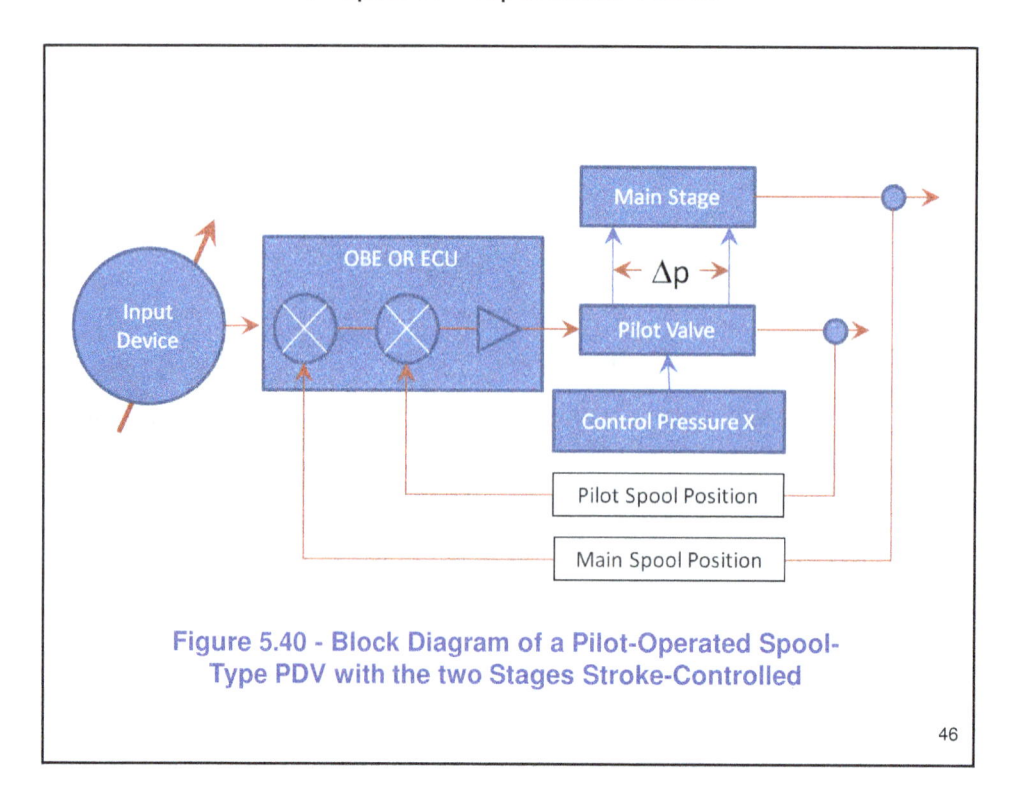

Figure 5.40 - Block Diagram of a Pilot-Operated Spool-Type PDV with the two Stages Stroke-Controlled

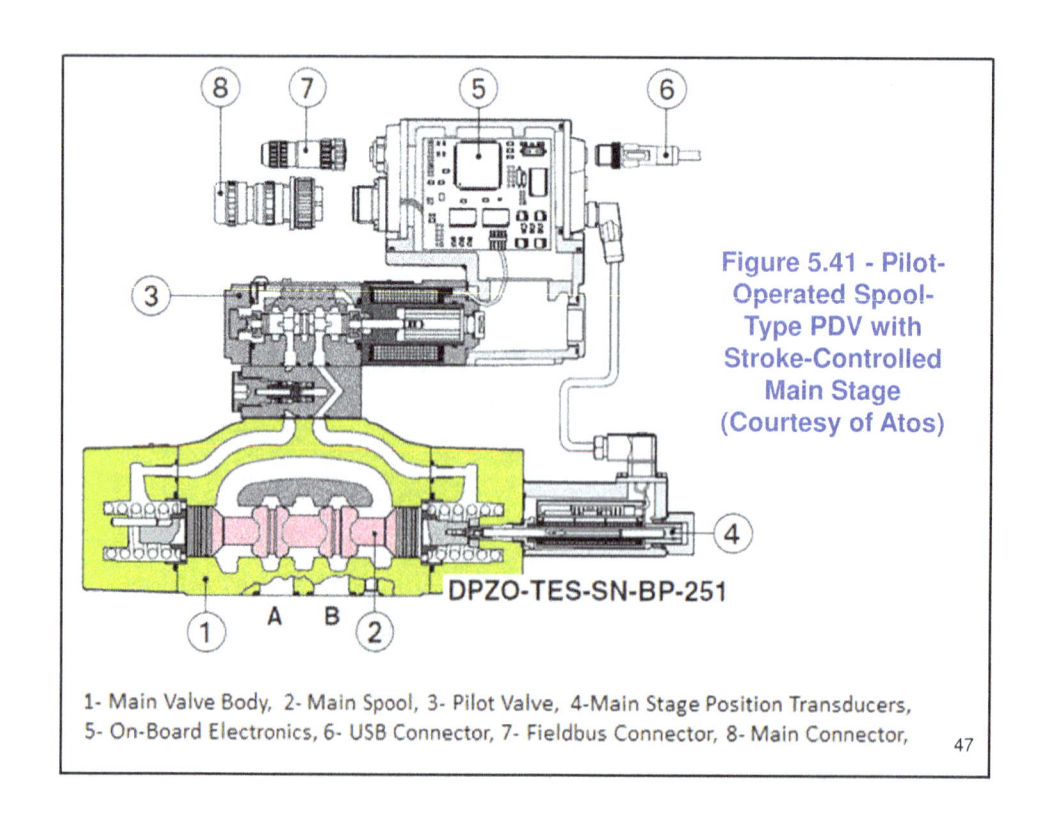

Figure 5.41 - Pilot-Operated Spool-Type PDV with Stroke-Controlled Main Stage (Courtesy of Atos)

1- Main Valve Body, 2- Main Spool, 3- Pilot Valve, 4- Main Stage Position Transducers, 5- On-Board Electronics, 6- USB Connector, 7- Fieldbus Connector, 8- Main Connector,

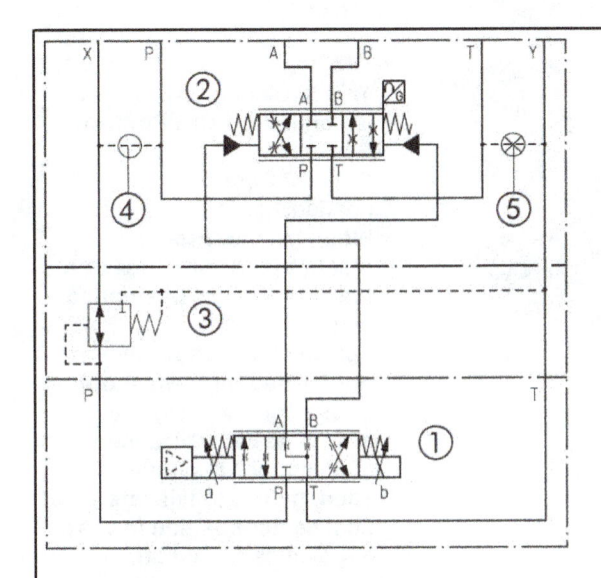

- **Pilot valve float-center**
- This stroke-controlled valve is configured as a Control System **Type 0**.
- **PID** Controller is needed.
- The **pilot valve must be kept energized** so the main spool stays in its controlled position.

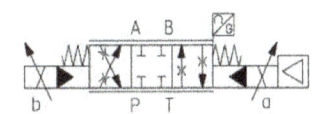

1- Pilot Valve, 2- Stroke-Controlled Main Stage, 3- Optional Sandwich-Type Pressure Reducing Valve, 4/5 – Plugs to be Added/Removed to Convert to External/Internal

Figure 5.42 - Symbols for a Pilot-Operated Spool-Type PDV with Stroke-Controlled Main Stage (Courtesy of Atos)

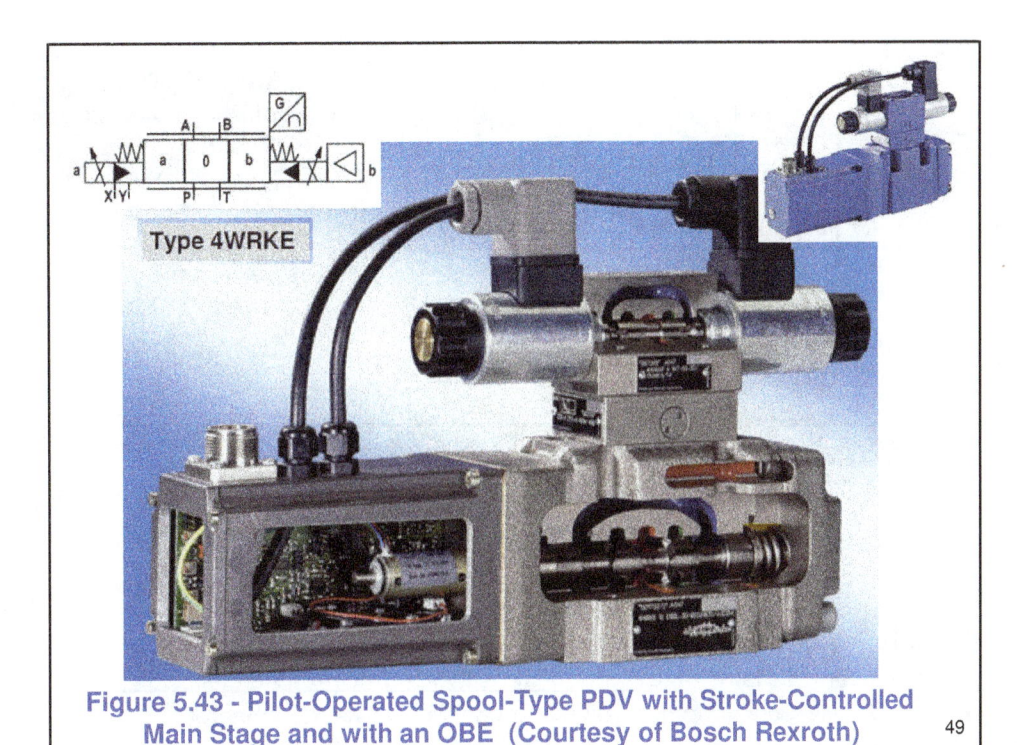

Type 4WRKE

Figure 5.43 - Pilot-Operated Spool-Type PDV with Stroke-Controlled Main Stage and with an OBE (Courtesy of Bosch Rexroth)

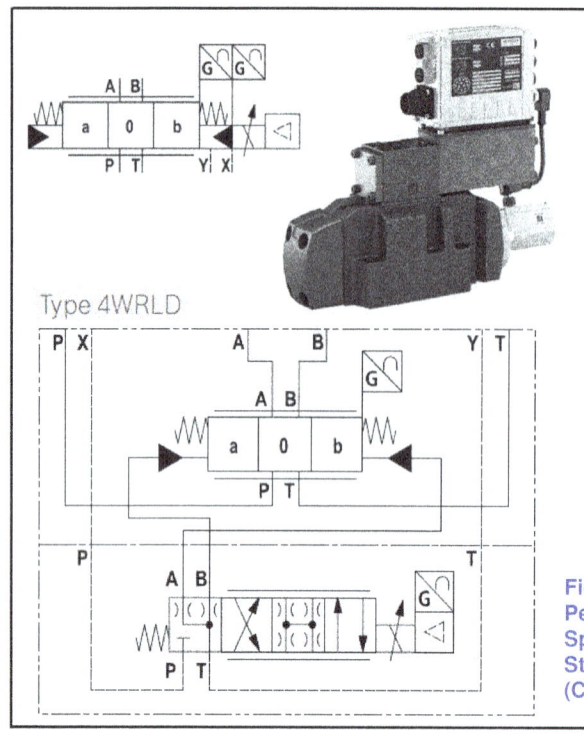

- **Pilot valve critical-center.**
- This stroke-controlled valve is configured as a Control System Type 1.
- Only PD Controller is needed.
- When the main spool reaches its desired position, the pilot valve is positioned in its zero-lapped position.
- Springs on main stage are not needed for main spool stroke control. They are needed to centralize the main spool when the pilot spool moves to fail-safe float center position in case of power is turned off.

Figure 5.44 - High-Performance Pilot-Operated Spool-Type PDV with the two Stages Stroke-Controlled (Courtesy of Bosch Rexroth)

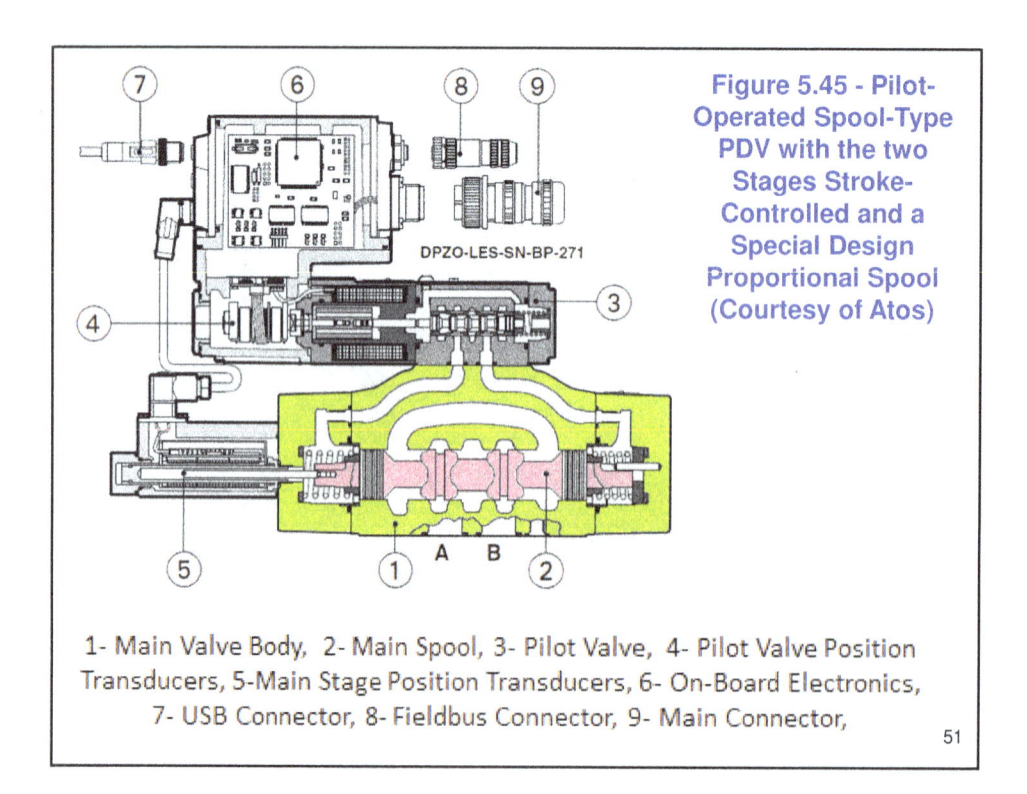

Figure 5.45 - Pilot-Operated Spool-Type PDV with the two Stages Stroke-Controlled and a Special Design Proportional Spool (Courtesy of Atos)

1- Main Valve Body, 2- Main Spool, 3- Pilot Valve, 4- Pilot Valve Position Transducers, 5-Main Stage Position Transducers, 6- On-Board Electronics, 7- USB Connector, 8- Fieldbus Connector, 9- Main Connector,

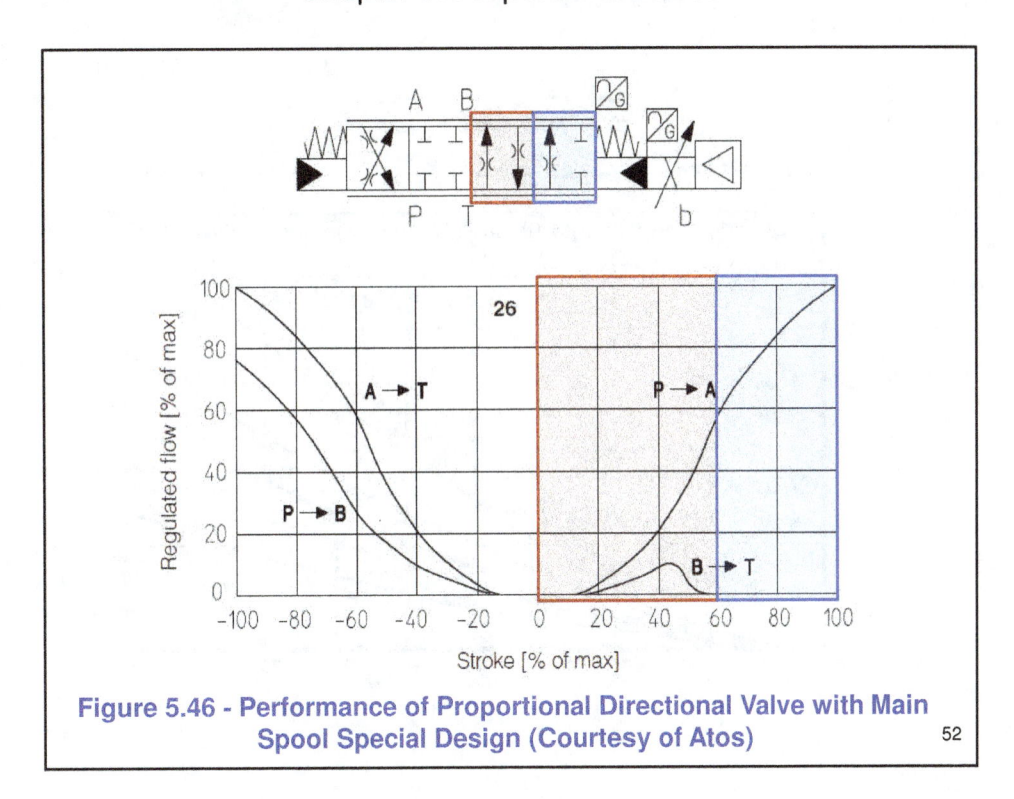

Workbook: Electro-Hydraulic Components and Systems
Chapter 5: Proportional Valves

Figure 5.46 - Performance of Proportional Directional Valve with Main Spool Special Design (Courtesy of Atos)

5.3.8- Proportional Directional Valves with Variable Load Compensator

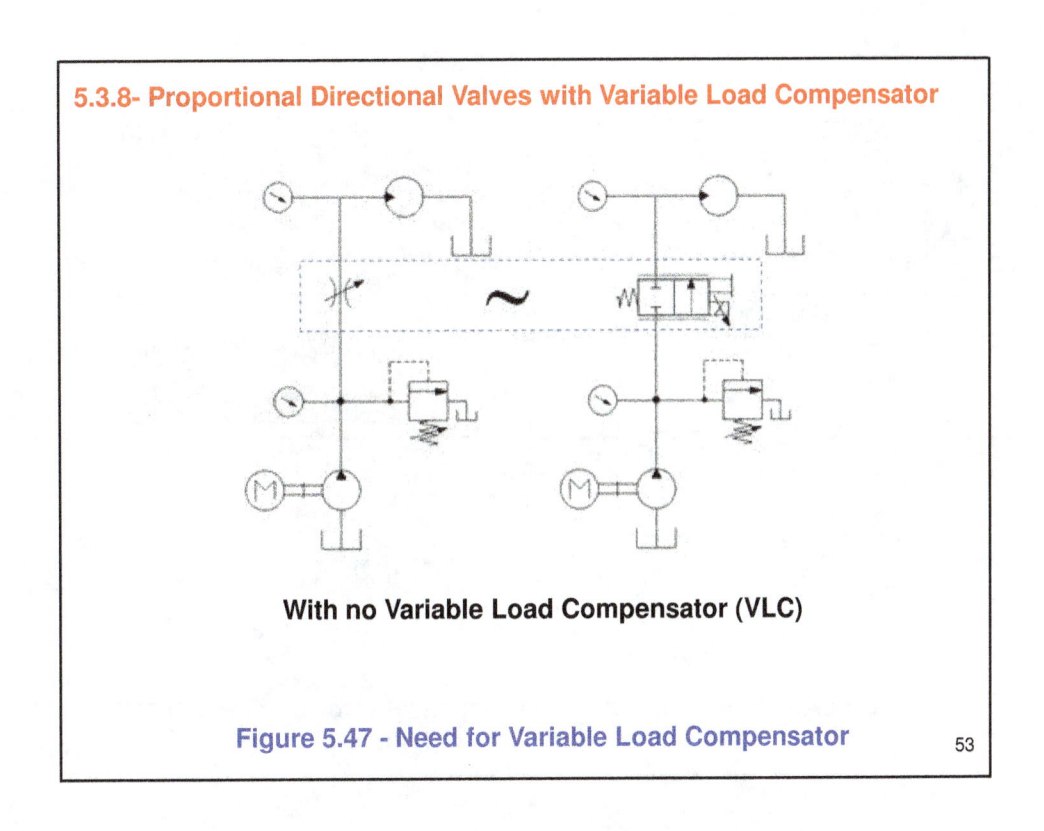

With no Variable Load Compensator (VLC)

Figure 5.47 - Need for Variable Load Compensator

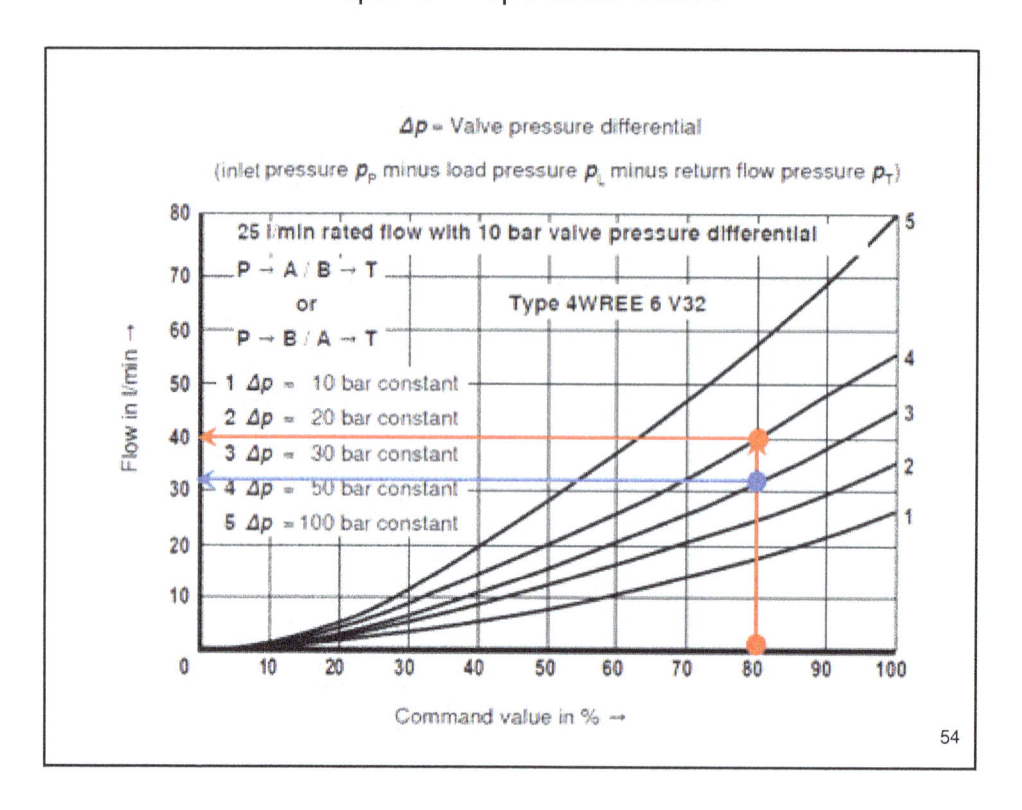

Workbook: Electro-Hydraulic Components and Systems
Chapter 5: Proportional Valves

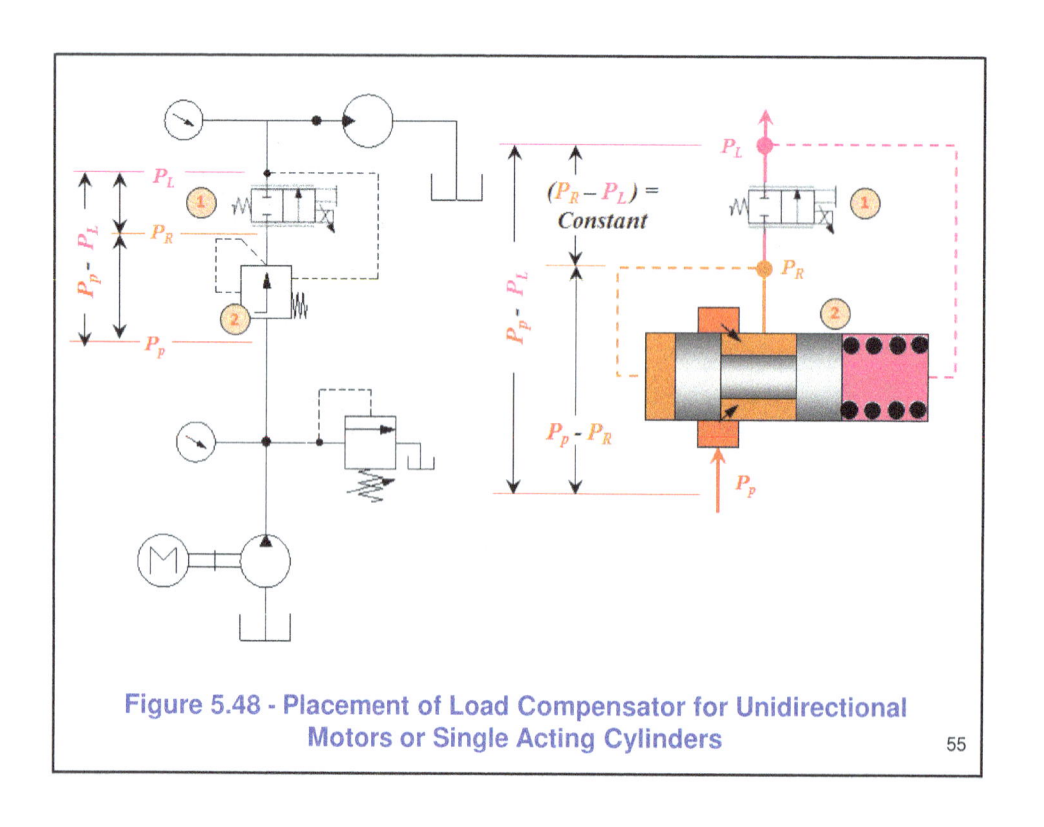

Figure 5.48 - Placement of Load Compensator for Unidirectional Motors or Single Acting Cylinders

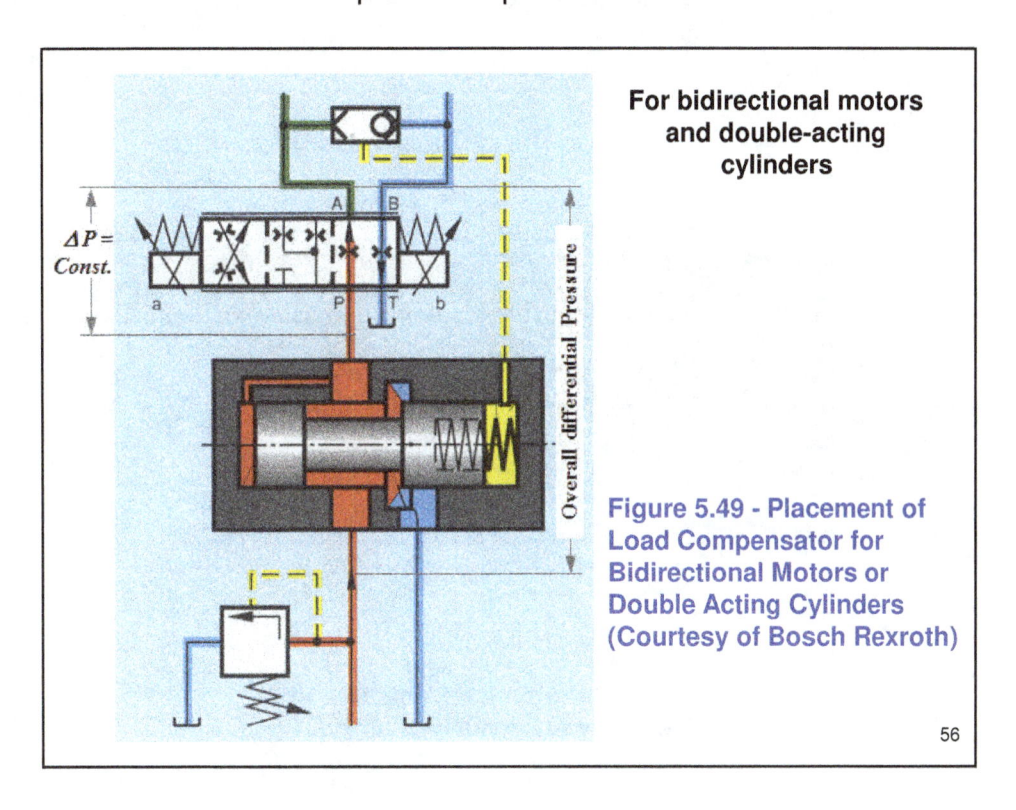

Figure 5.49 - Placement of Load Compensator for Bidirectional Motors or Double Acting Cylinders (Courtesy of Bosch Rexroth)

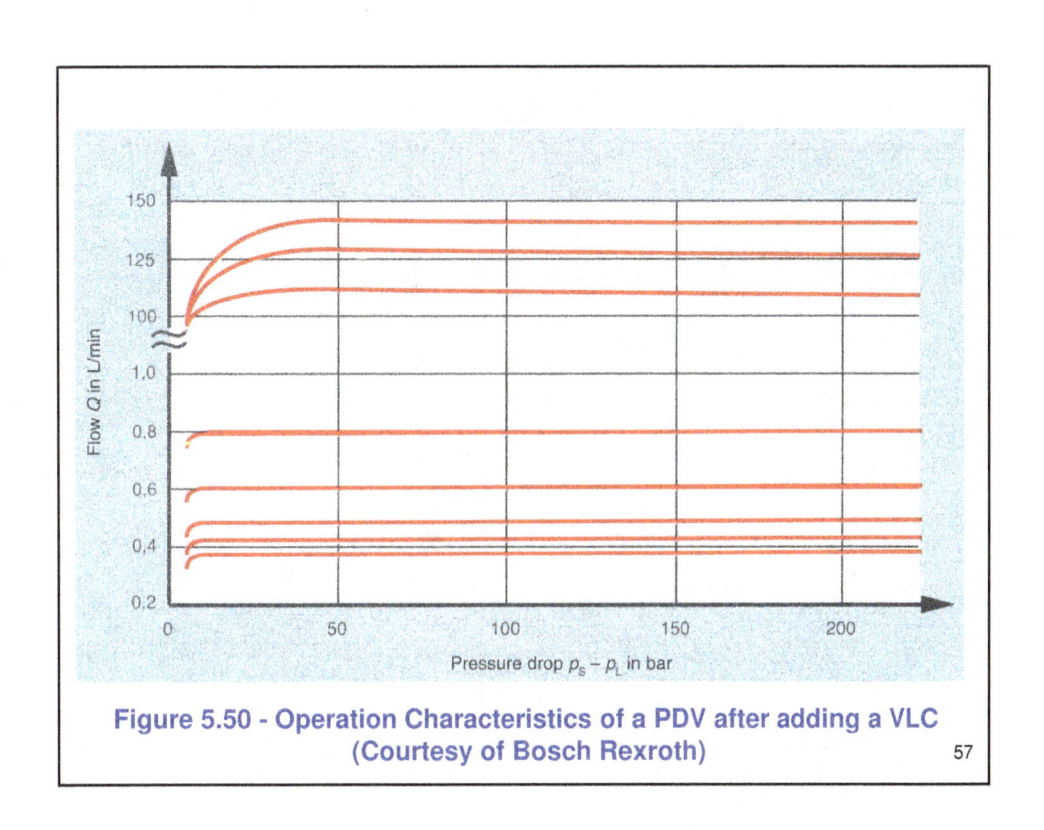

Figure 5.50 - Operation Characteristics of a PDV after adding a VLC (Courtesy of Bosch Rexroth)

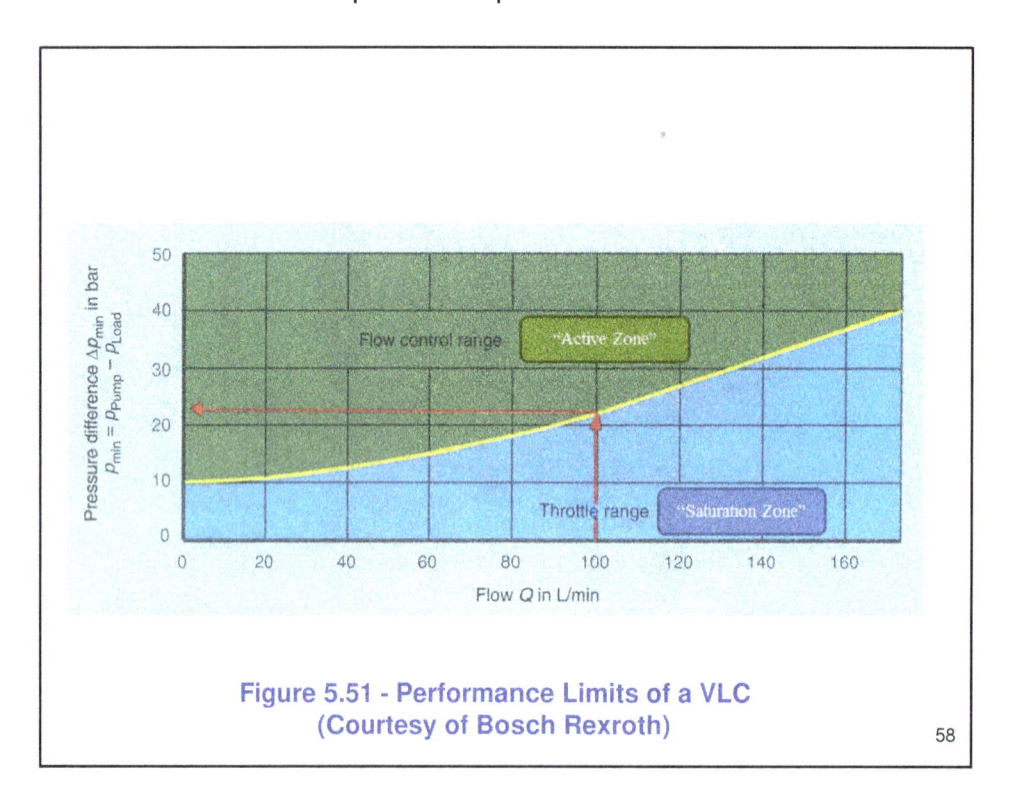

Figure 5.51 - Performance Limits of a VLC (Courtesy of Bosch Rexroth)

5.4- Proportional Pressure Control Valves
5.4.1- Interpretation of Symbols for Proportional Pressure Relief Valves

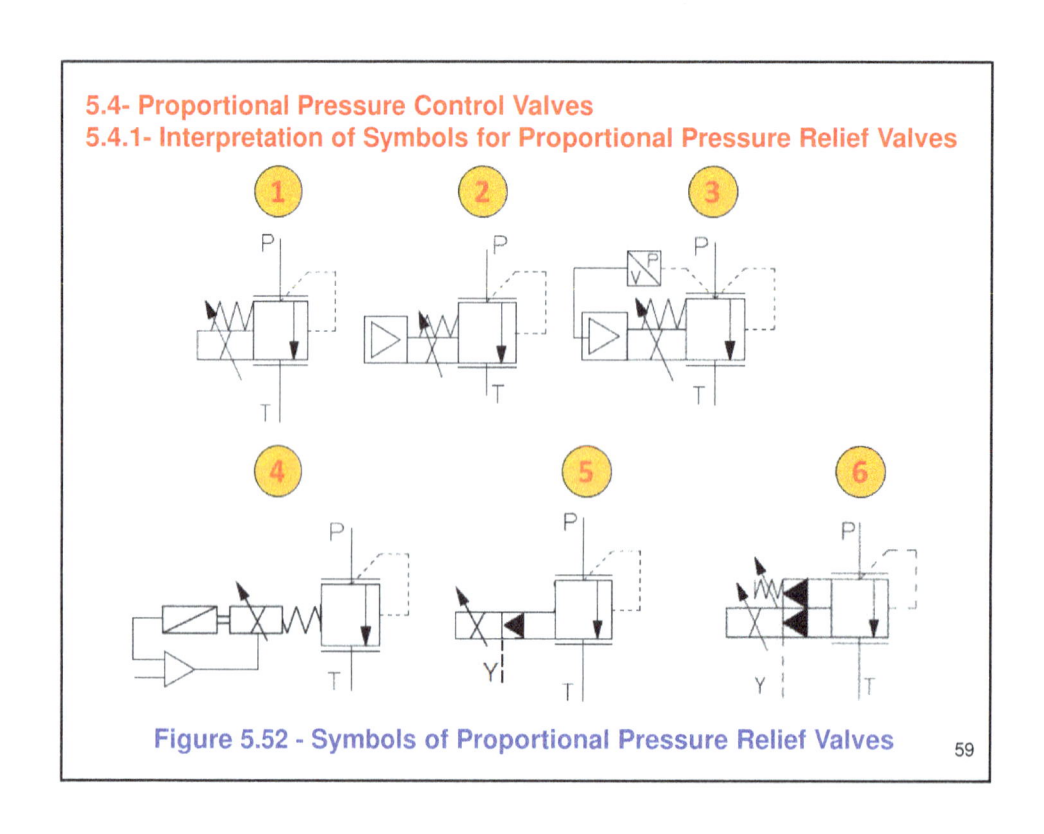

Figure 5.52 - Symbols of Proportional Pressure Relief Valves

5.4.2- Hydraulic Static Characteristics of Proportional Pressure Relief Valves

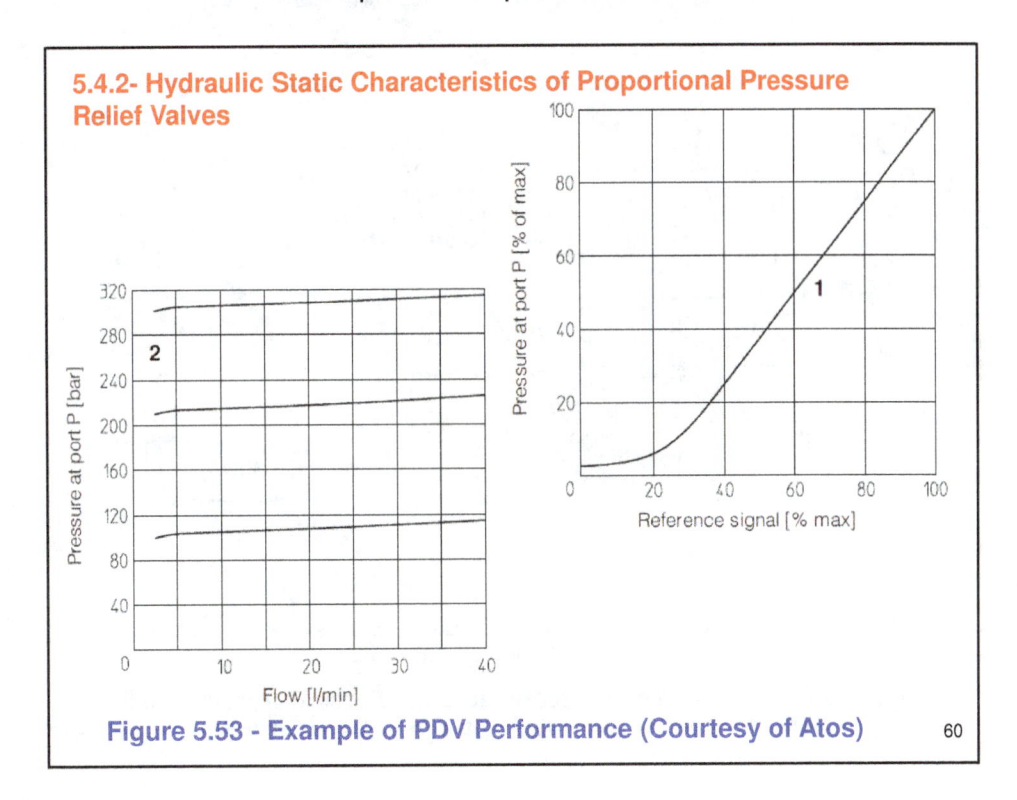

Figure 5.53 - Example of PDV Performance (Courtesy of Atos)

5.4.3- Direct-Operated Force-Controlled Proportional Pressure Relief Valves

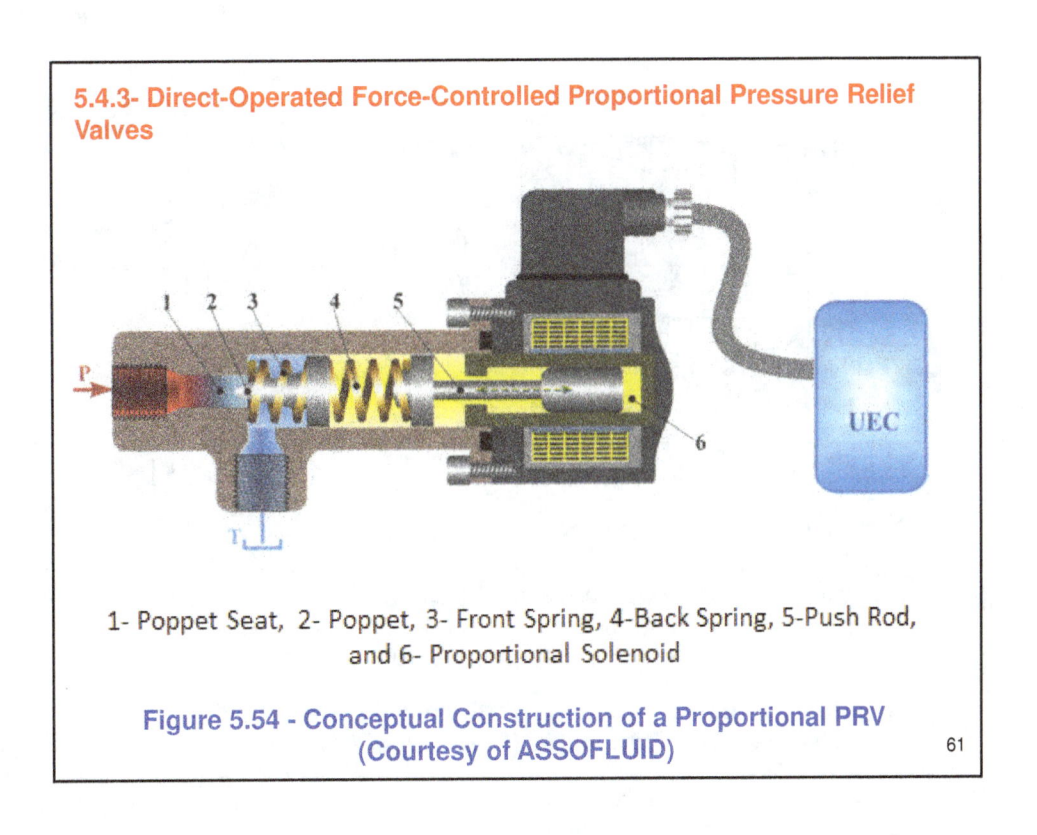

1- Poppet Seat, 2- Poppet, 3- Front Spring, 4- Back Spring, 5- Push Rod, and 6- Proportional Solenoid

Figure 5.54 - Conceptual Construction of a Proportional PRV (Courtesy of ASSOFLUID)

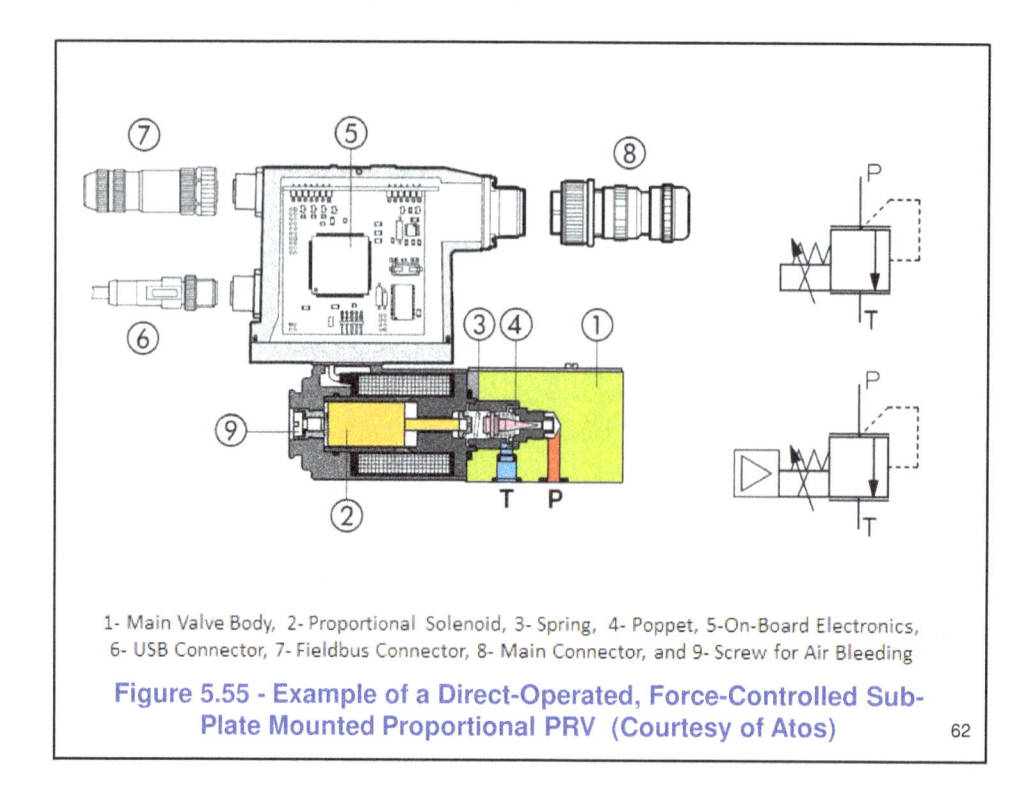

1- Main Valve Body, 2- Proportional Solenoid, 3- Spring, 4- Poppet, 5-On-Board Electronics, 6- USB Connector, 7- Fieldbus Connector, 8- Main Connector, and 9- Screw for Air Bleeding

Figure 5.55 - Example of a Direct-Operated, Force-Controlled Sub-Plate Mounted Proportional PRV (Courtesy of Atos)

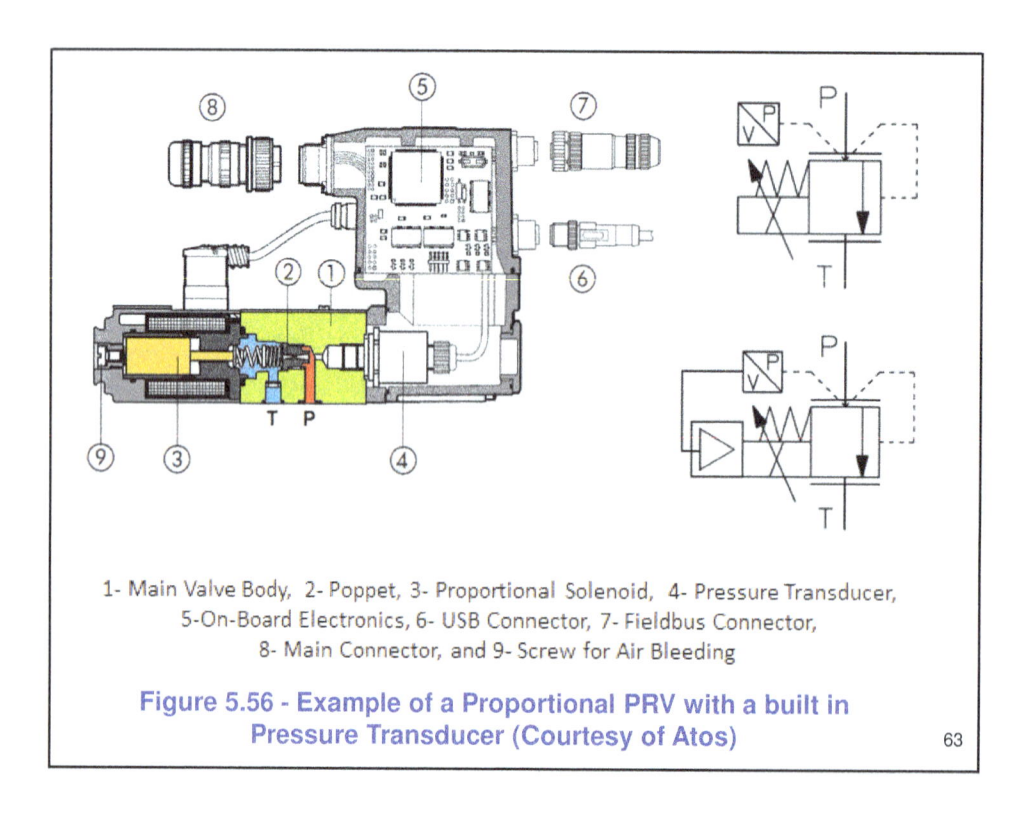

1- Main Valve Body, 2- Poppet, 3- Proportional Solenoid, 4- Pressure Transducer, 5-On-Board Electronics, 6- USB Connector, 7- Fieldbus Connector, 8- Main Connector, and 9- Screw for Air Bleeding

Figure 5.56 - Example of a Proportional PRV with a built in Pressure Transducer (Courtesy of Atos)

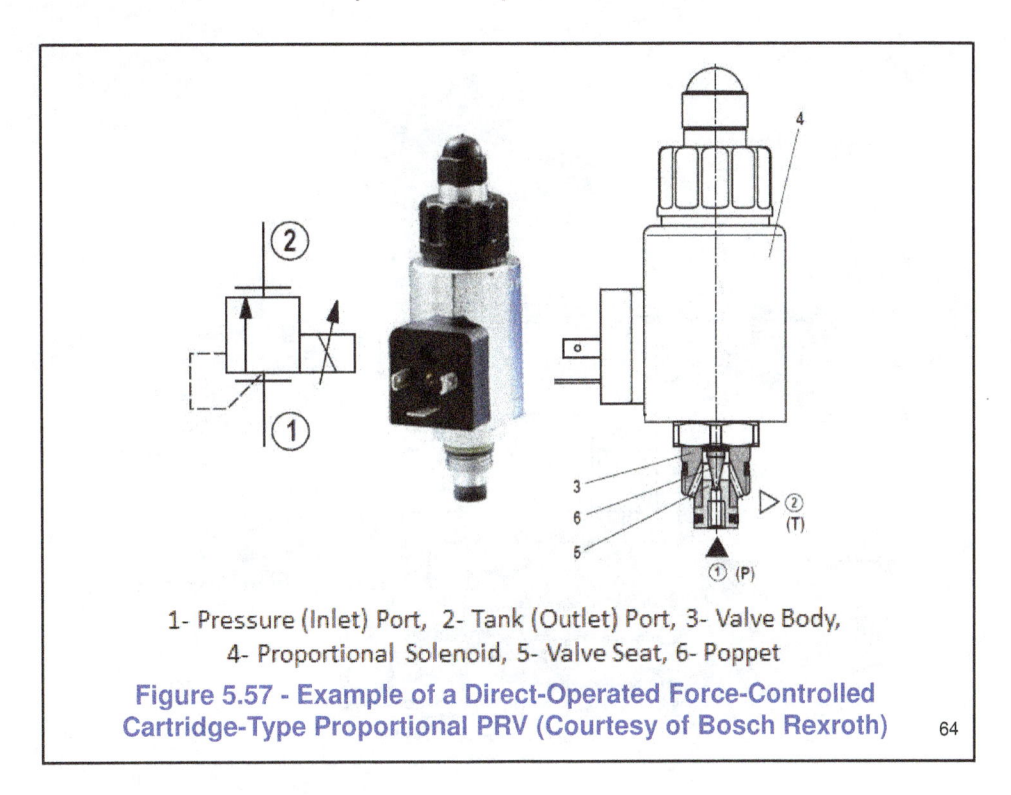

1- Pressure (Inlet) Port, 2- Tank (Outlet) Port, 3- Valve Body,
4- Proportional Solenoid, 5- Valve Seat, 6- Poppet

Figure 5.57 - Example of a Direct-Operated Force-Controlled Cartridge-Type Proportional PRV (Courtesy of Bosch Rexroth)

5.4.4- Direct-Operated Stroke-Controlled Proportional Pressure Relief Valves

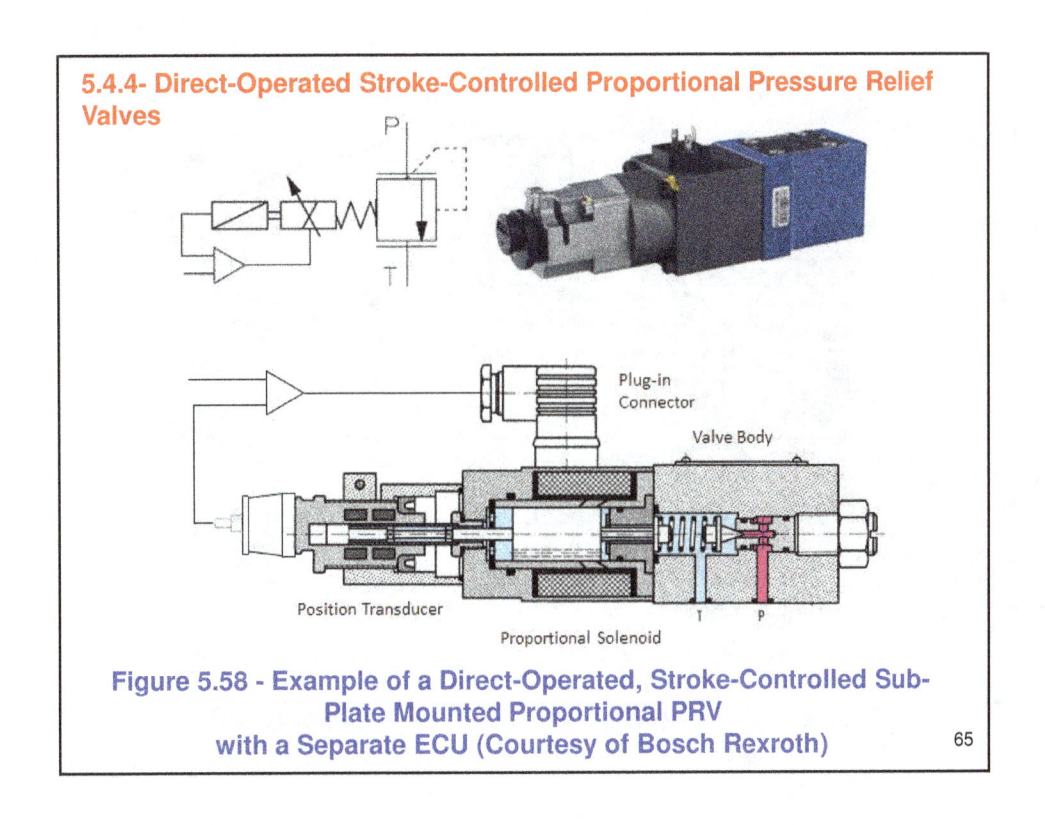

Figure 5.58 - Example of a Direct-Operated, Stroke-Controlled Sub-Plate Mounted Proportional PRV with a Separate ECU (Courtesy of Bosch Rexroth)

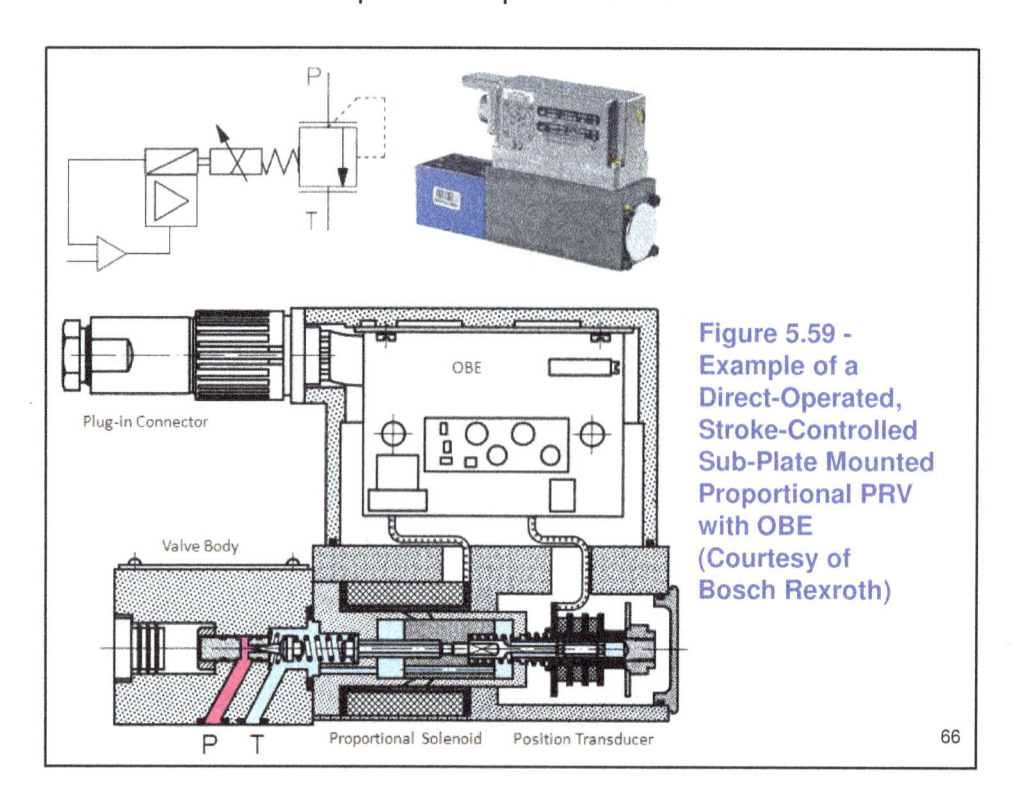

Workbook: Electro-Hydraulic Components and Systems
Chapter 5: Proportional Valves

Figure 5.59 - Example of a Direct-Operated, Stroke-Controlled Sub-Plate Mounted Proportional PRV with OBE (Courtesy of Bosch Rexroth)

5.4.5- Pilot-Operated Force-Controlled Proportional Pressure Relief Valves

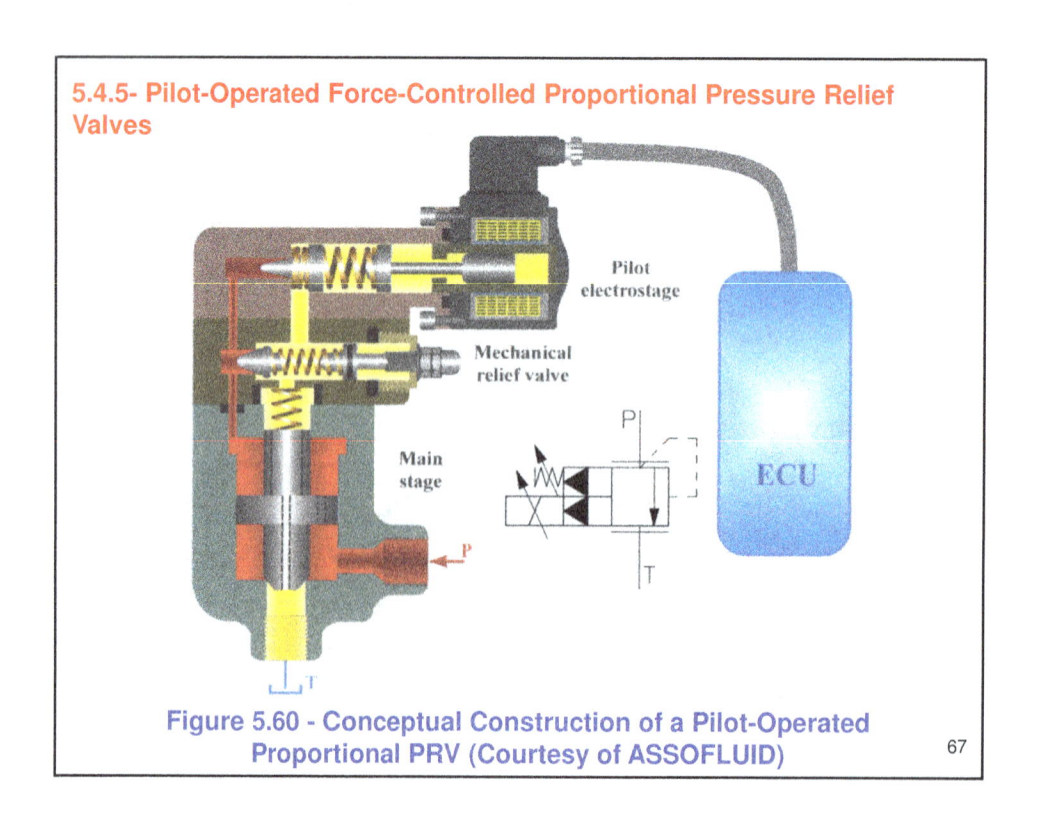

Figure 5.60 - Conceptual Construction of a Pilot-Operated Proportional PRV (Courtesy of ASSOFLUID)

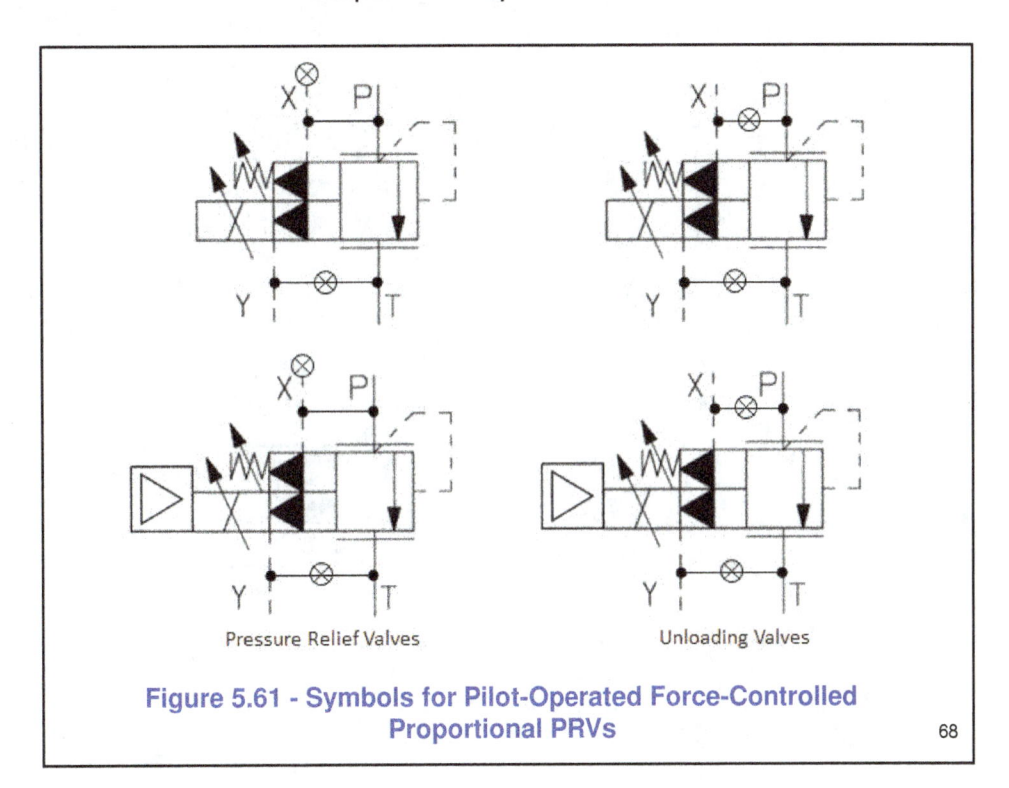

Pressure Relief Valves   Unloading Valves

Figure 5.61 - Symbols for Pilot-Operated Force-Controlled Proportional PRVs

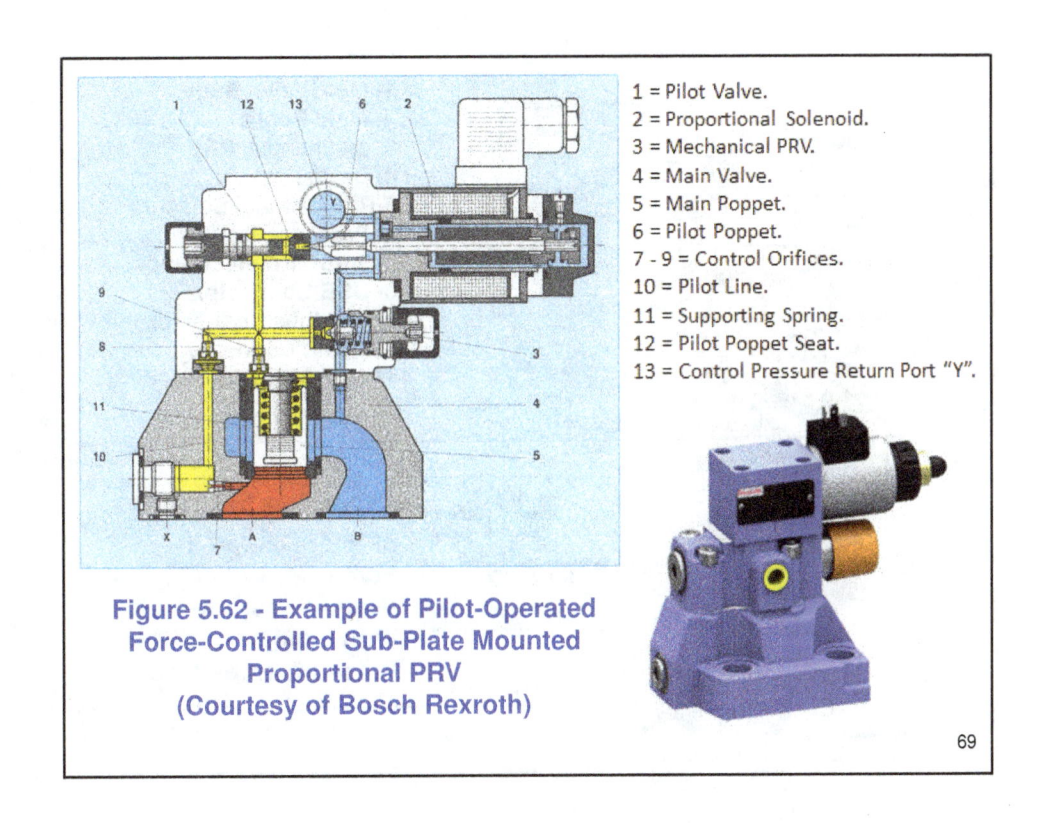

1 = Pilot Valve.
2 = Proportional Solenoid.
3 = Mechanical PRV.
4 = Main Valve.
5 = Main Poppet.
6 = Pilot Poppet.
7 - 9 = Control Orifices.
10 = Pilot Line.
11 = Supporting Spring.
12 = Pilot Poppet Seat.
13 = Control Pressure Return Port "Y".

Figure 5.62 - Example of Pilot-Operated Force-Controlled Sub-Plate Mounted Proportional PRV (Courtesy of Bosch Rexroth)

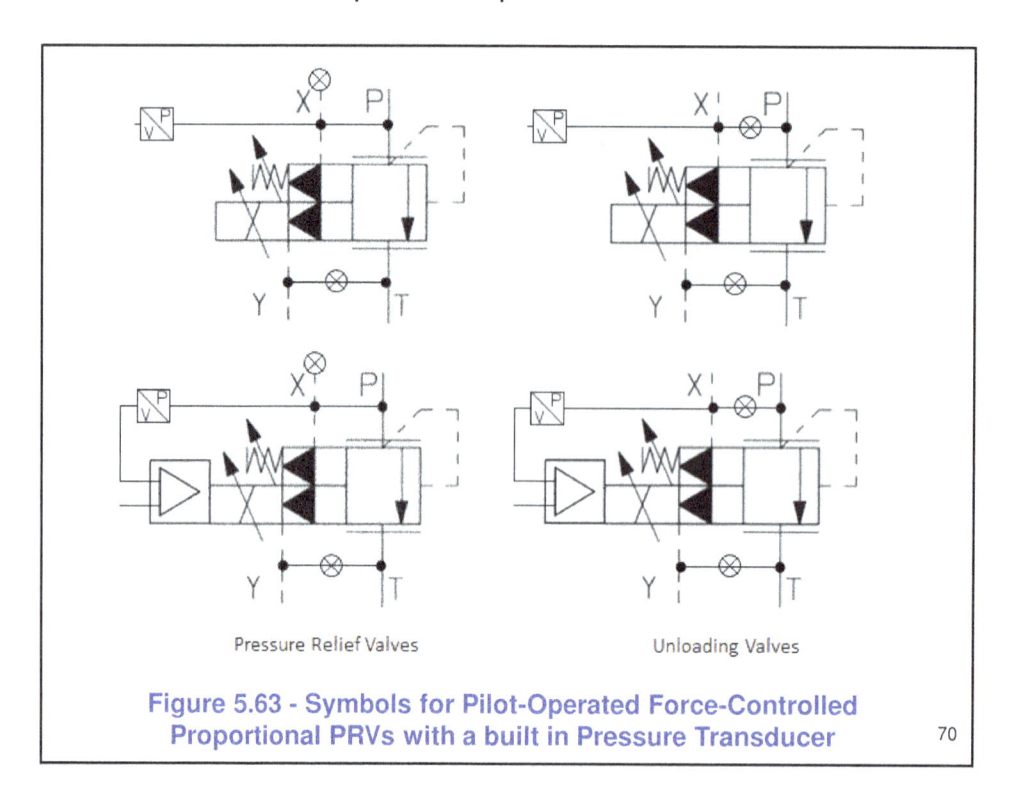

Workbook: Electro-Hydraulic Components and Systems
Chapter 5: Proportional Valves

Pressure Relief Valves   Unloading Valves

Figure 5.63 - Symbols for Pilot-Operated Force-Controlled Proportional PRVs with a built in Pressure Transducer

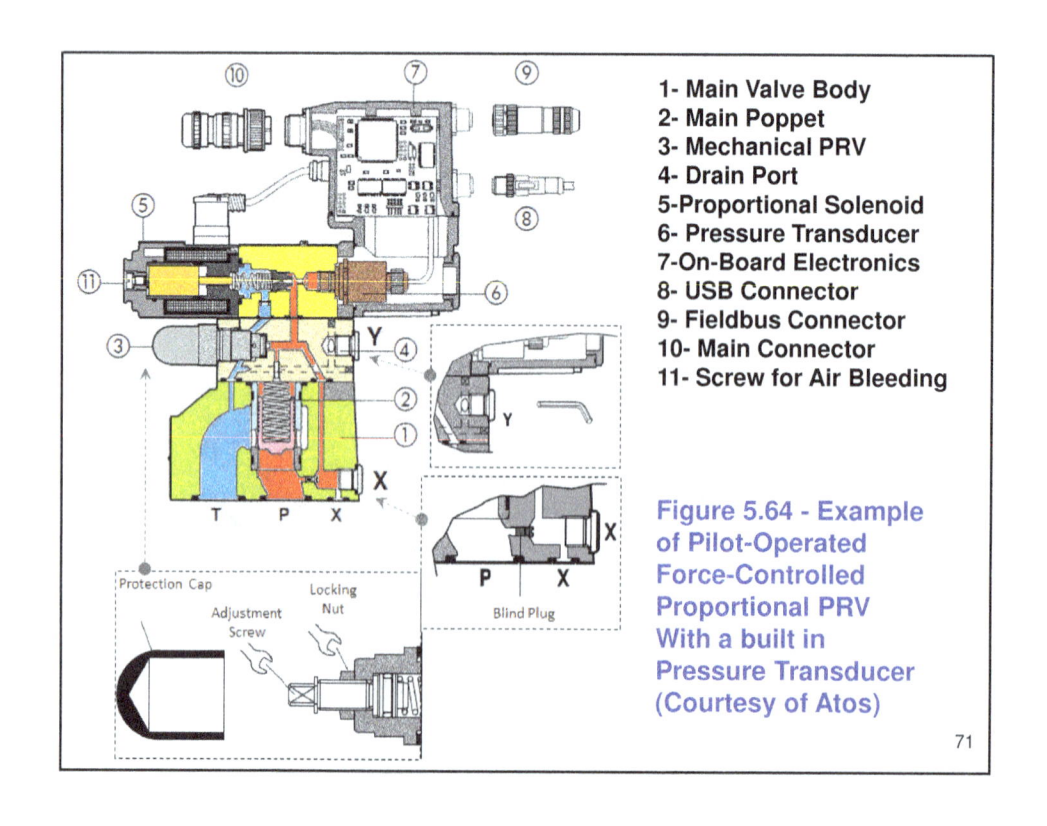

1- Main Valve Body
2- Main Poppet
3- Mechanical PRV
4- Drain Port
5- Proportional Solenoid
6- Pressure Transducer
7- On-Board Electronics
8- USB Connector
9- Fieldbus Connector
10- Main Connector
11- Screw for Air Bleeding

Figure 5.64 - Example of Pilot-Operated Force-Controlled Proportional PRV With a built in Pressure Transducer (Courtesy of Atos)

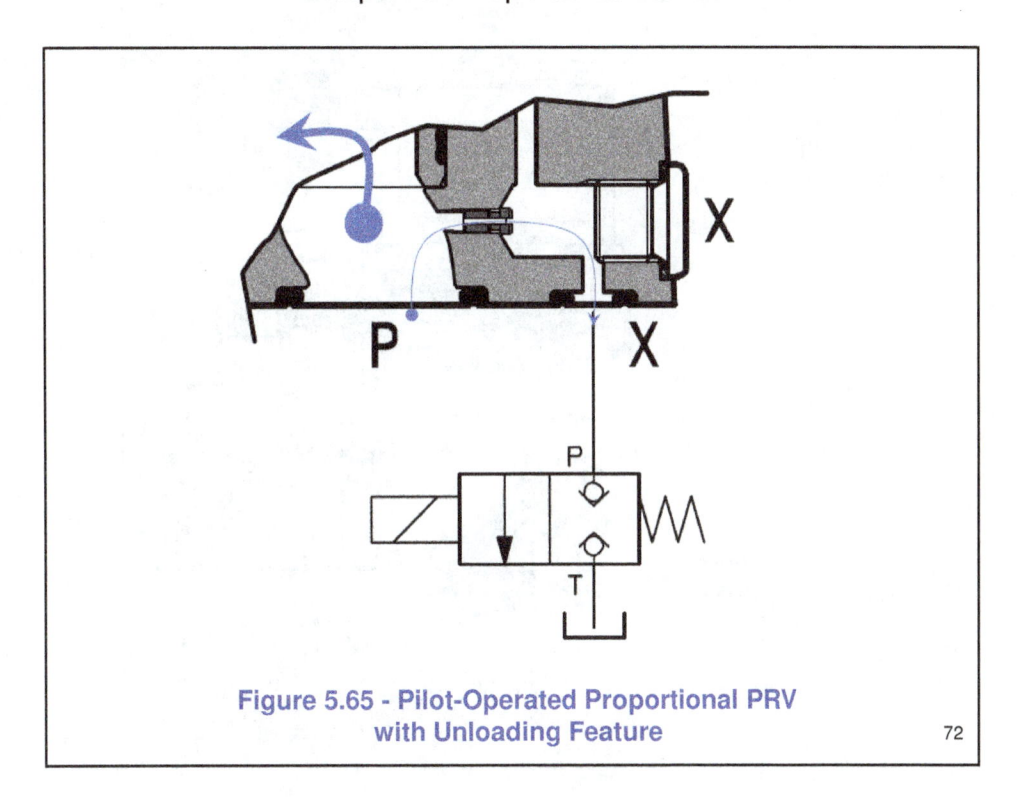

Workbook: Electro-Hydraulic Components and Systems
Chapter 5: Proportional Valves

Figure 5.65 - Pilot-Operated Proportional PRV with Unloading Feature

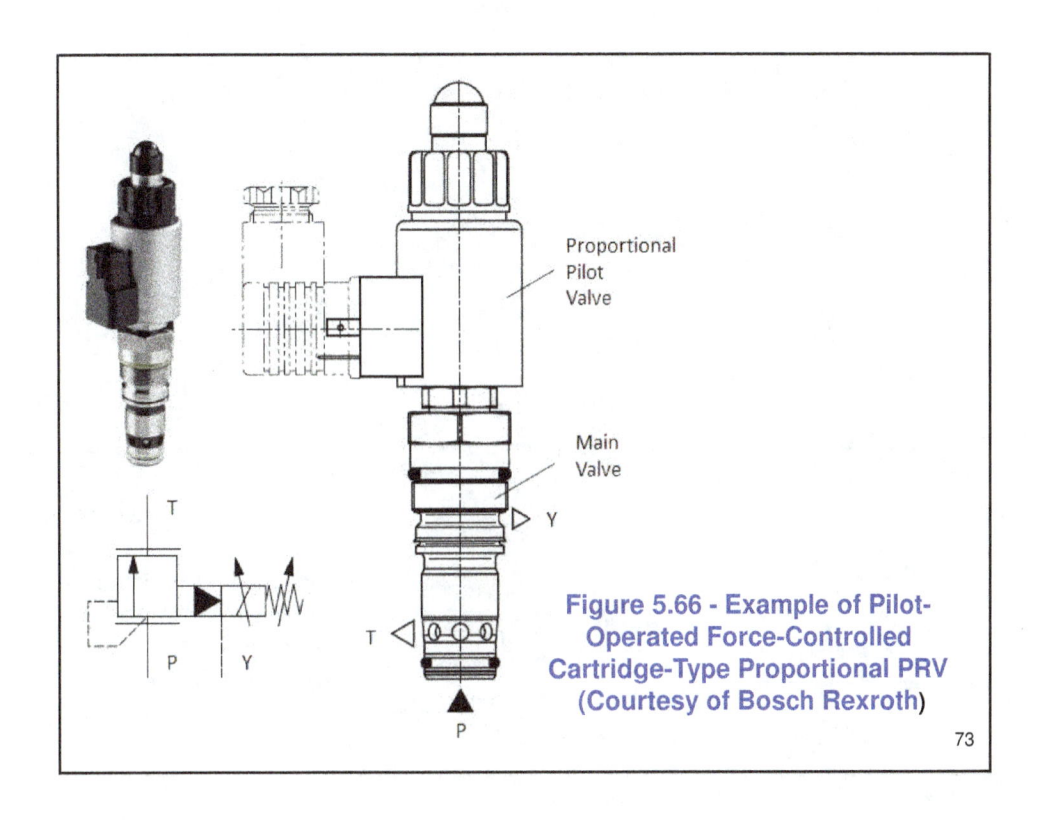

Figure 5.66 - Example of Pilot-Operated Force-Controlled Cartridge-Type Proportional PRV (Courtesy of Bosch Rexroth)

Workbook: Electro-Hydraulic Components and Systems
Chapter 5: Proportional Valves

5.4.6- Pilot-Operated Stroke-Controlled Proportional Pressure Relief Valves

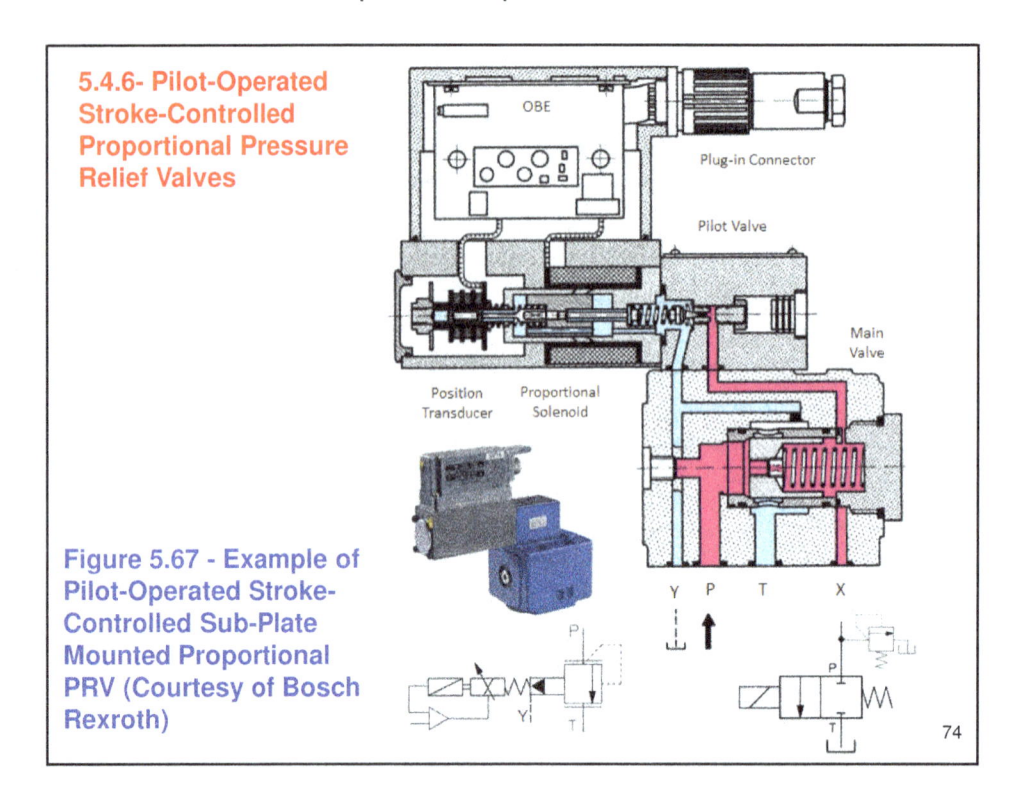

Figure 5.67 - Example of Pilot-Operated Stroke-Controlled Sub-Plate Mounted Proportional PRV (Courtesy of Bosch Rexroth)

5.4.7- Proportional Pressure Reducing Valves

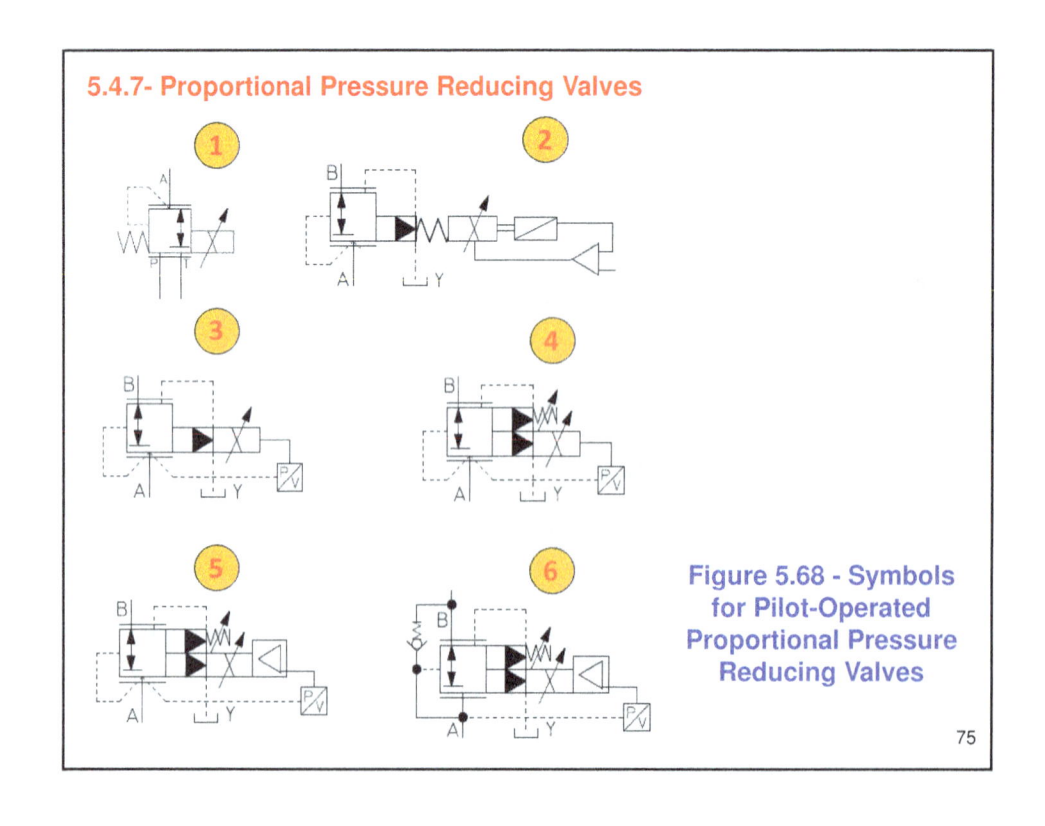

Figure 5.68 - Symbols for Pilot-Operated Proportional Pressure Reducing Valves

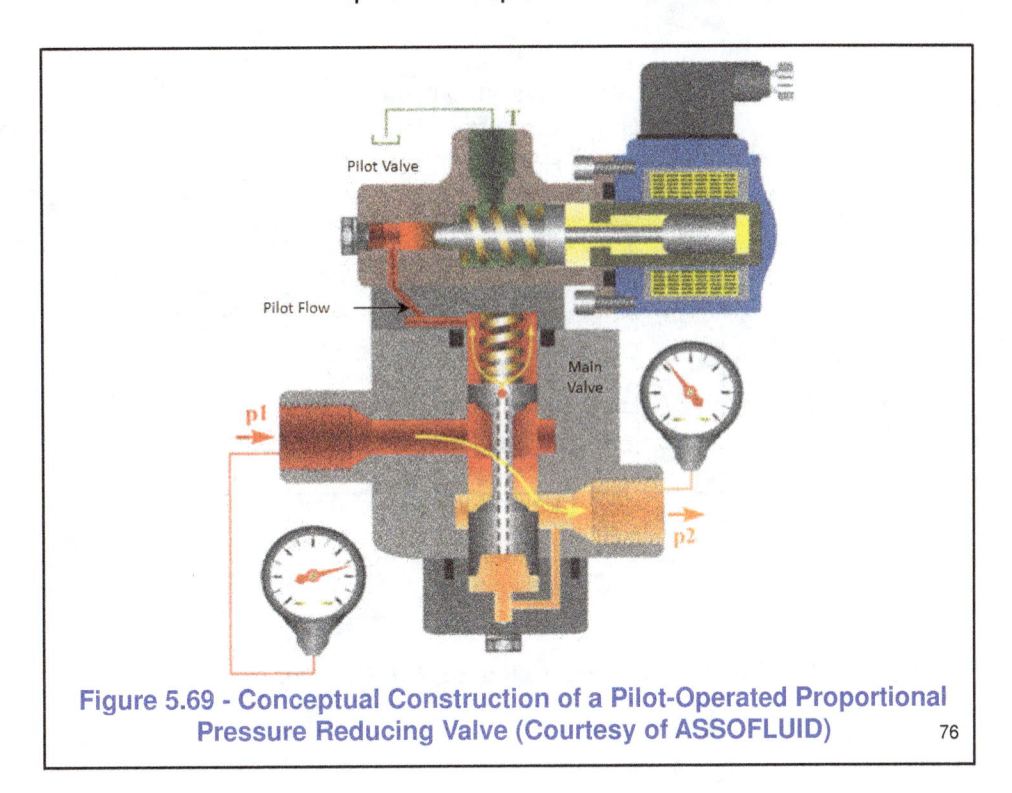

Figure 5.69 - Conceptual Construction of a Pilot-Operated Proportional Pressure Reducing Valve (Courtesy of ASSOFLUID)

5.5- Proportional Flow Control Valves

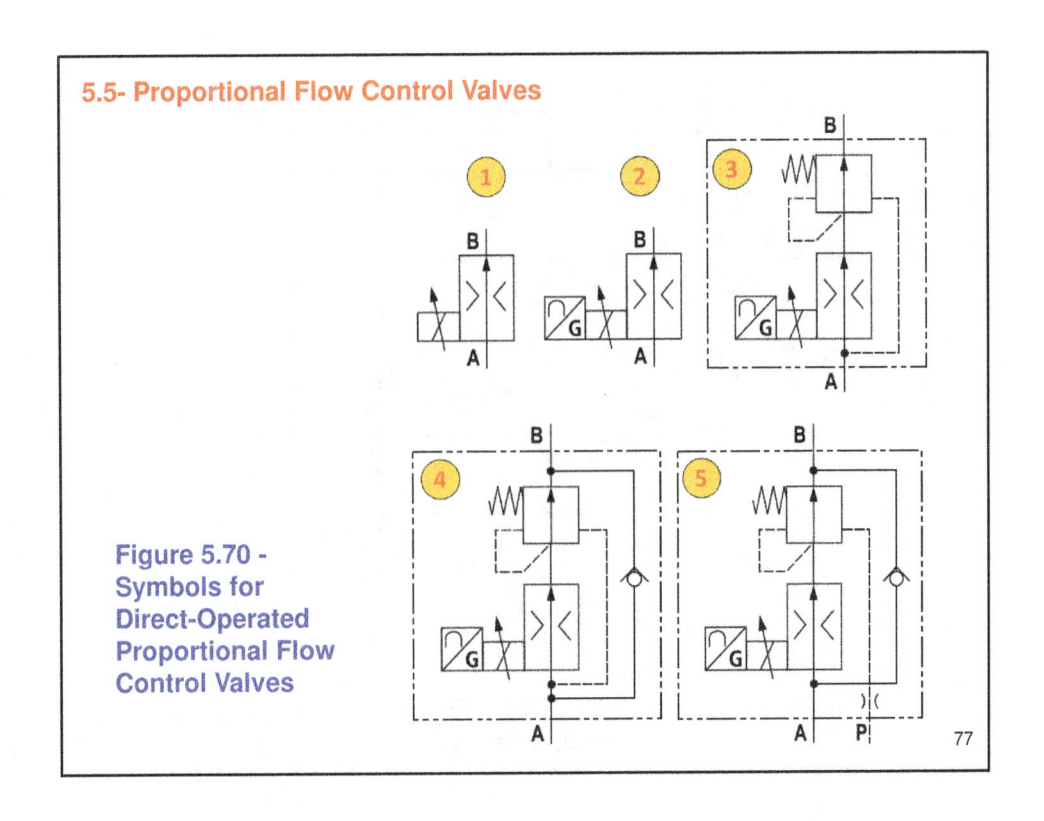

Figure 5.70 - Symbols for Direct-Operated Proportional Flow Control Valves

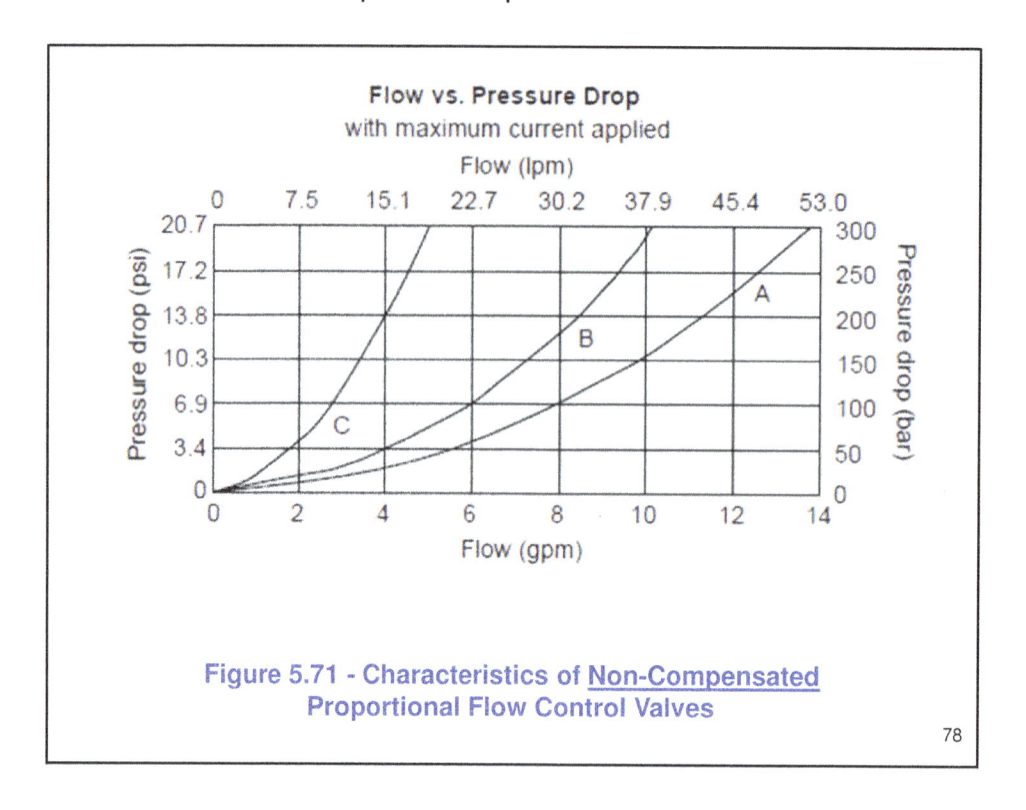

Figure 5.71 - Characteristics of **Non-Compensated** Proportional Flow Control Valves

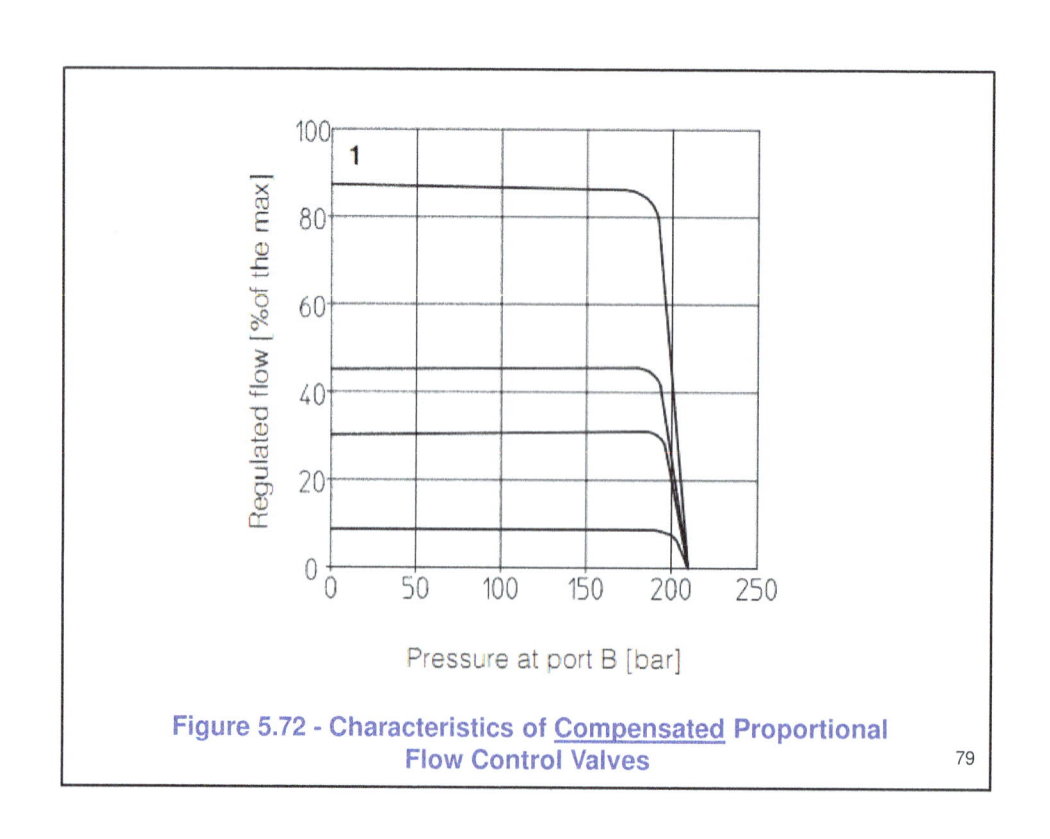

Figure 5.72 - Characteristics of **Compensated** Proportional Flow Control Valves

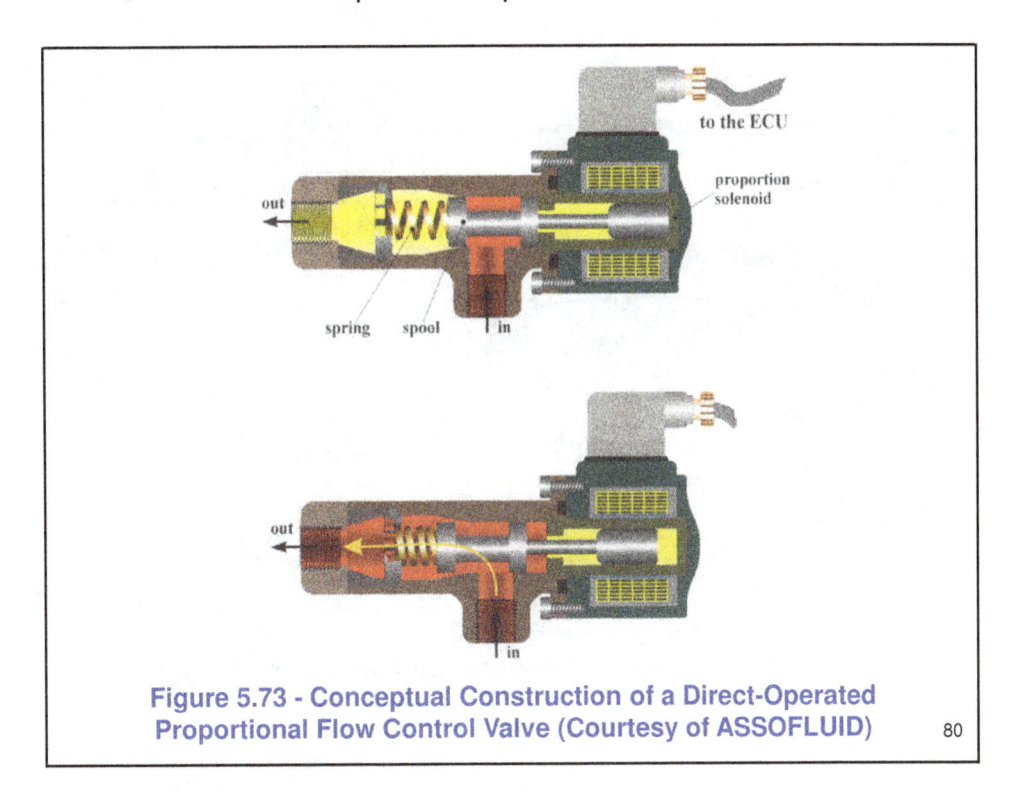

Figure 5.73 - Conceptual Construction of a Direct-Operated Proportional Flow Control Valve (Courtesy of ASSOFLUID)

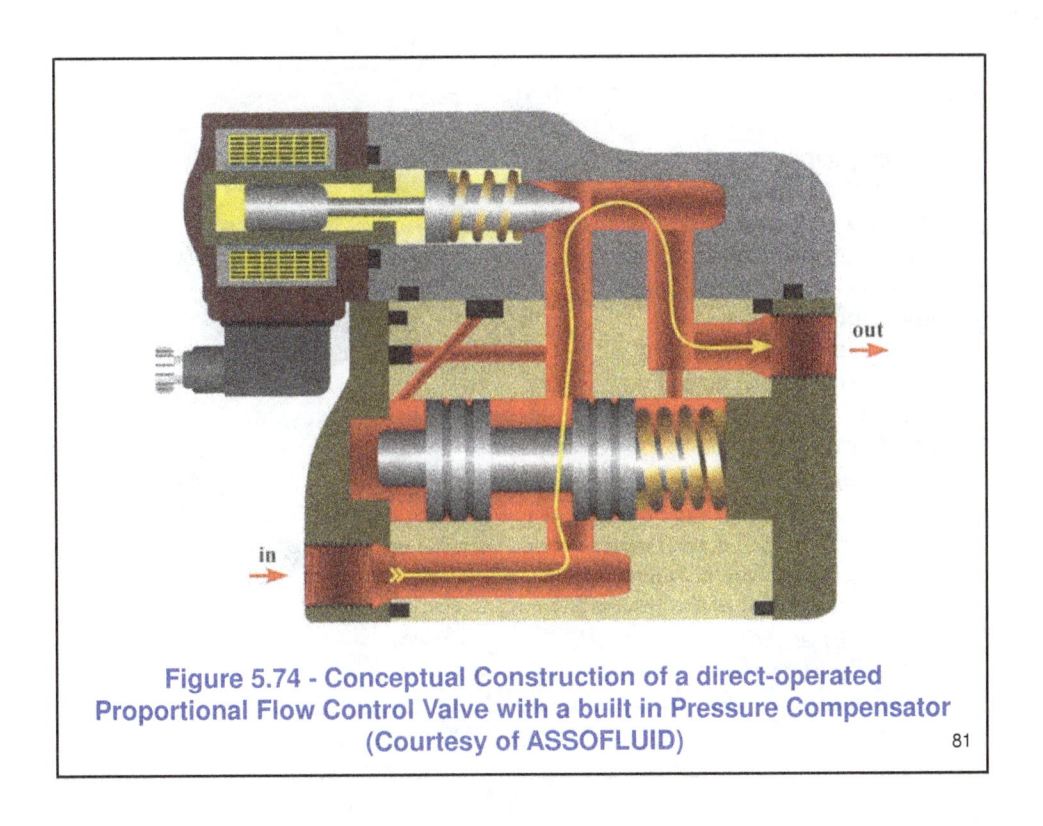

Figure 5.74 - Conceptual Construction of a direct-operated Proportional Flow Control Valve with a built in Pressure Compensator (Courtesy of ASSOFLUID)

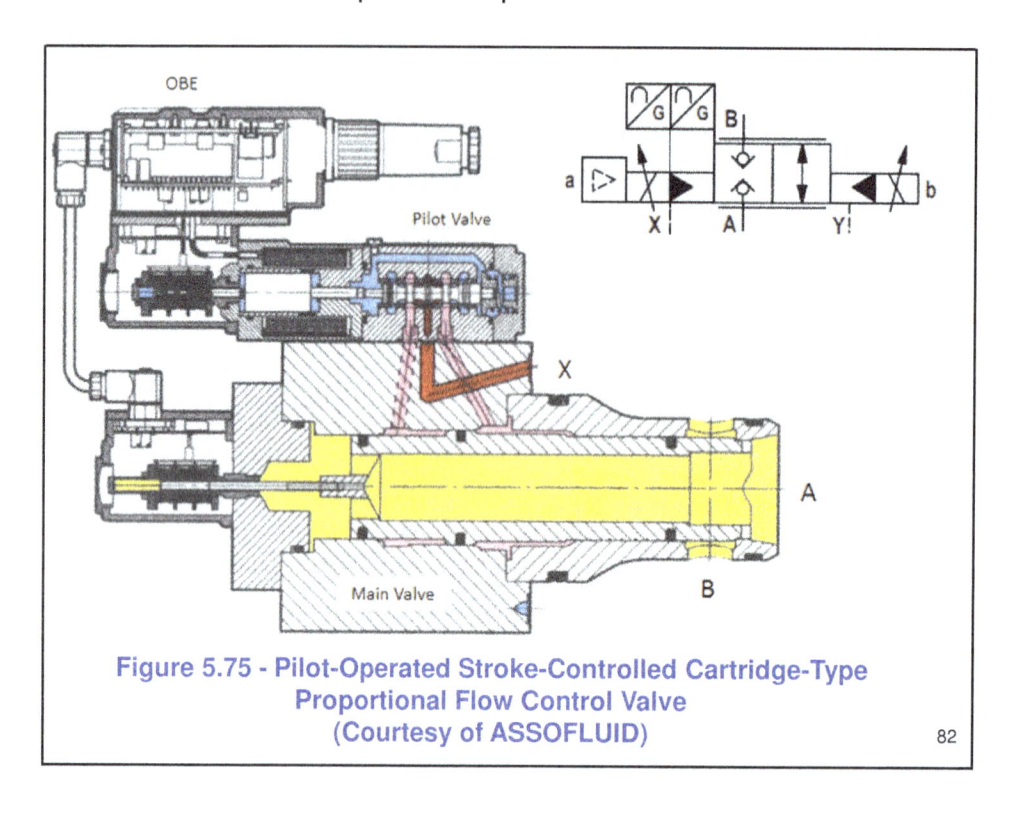

Figure 5.75 - Pilot-Operated Stroke-Controlled Cartridge-Type Proportional Flow Control Valve (Courtesy of ASSOFLUID)

5.6- Proportional Valves for Mobile Applications

Mobile proportional control valves that cover simple to complex machinery needs in agriculture, forestry, earth-moving, paving, road-building, railway, marine, waste handling, crushers, shredders, or any application involving heavy mobile or industrial equipment.

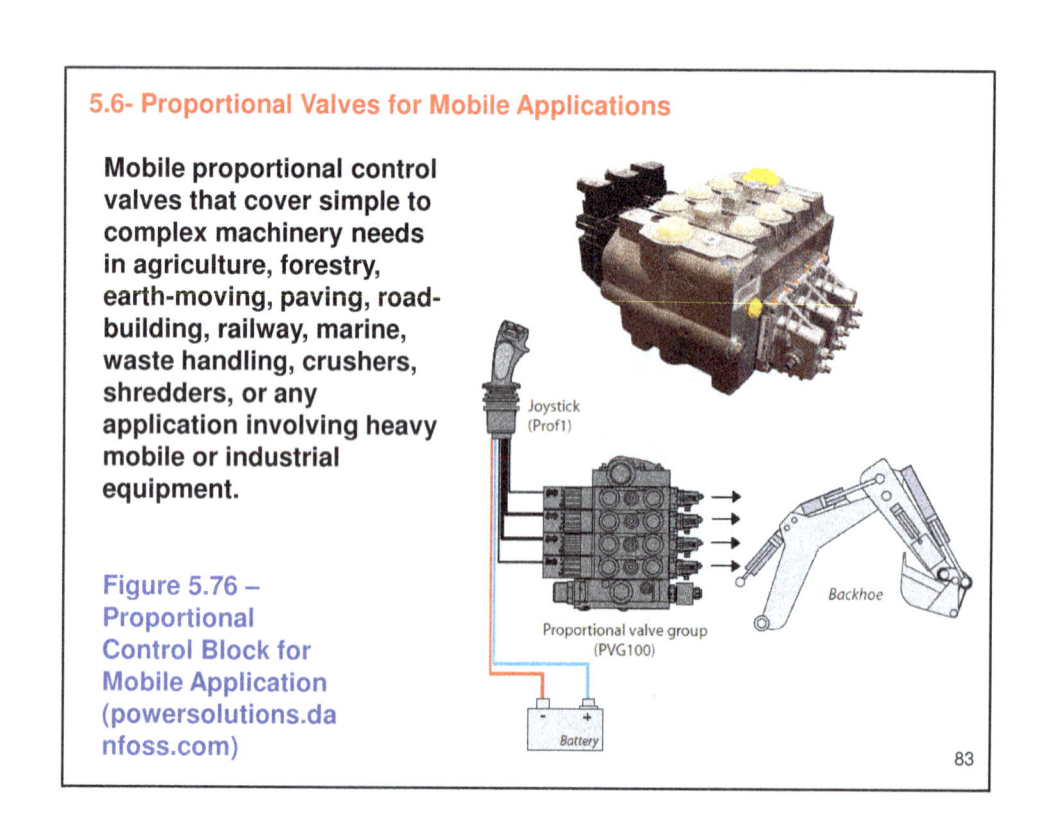

Figure 5.76 – Proportional Control Block for Mobile Application (powersolutions.danfoss.com)

Chapter 5 Reviews

1. A proportional valve shares which of the following features with an ON/OFF valve?
    A. Construction simplicity.
    B. Spool continuous travel.
    C. Can be used to build closed-loop control systems.
    D. No sharing features.

2. A proportional valve shares which of the following features with a manual valve?
    A. Each valve contains a built in solenoid.
    B. Spool continuous travel.
    C. Can be used to build closed-loop control systems.
    D. No sharing features.

3. A proportional valve shares which of the following features with a servo valves?
    A. Construction simplicity.
    B. Each valve contains a built in solenoid.
    C. Can be used to build closed-loop control systems.
    D. No sharing features.

4. Force-controlled proportional valves means?
    A. Proportional valves with feedback on the spool position.
    B. Proportional valves with no feedback on the spool position.
    C. Proportional valves with larger bandwidth.
    D. Proportional valves with two solenoids.

5. Stroke-controlled proportional valves means?
    A. Proportional valves with feedback on the spool position.
    B. Proportional valves with no feedback on the spool position.
    C. Proportional valves with larger bandwidth.
    D. Proportional valves with two solenoids.

6- The system shown below represents?
   A. A Closed-loop control system using a stroke-controlled proportional valve.
   B. A Closed-loop control system using a force-controlled proportional valve.
   C. An open-loop control system using a force-controlled proportional valve.
   D. An open-loop control system using a stroke-controlled proportional valve.

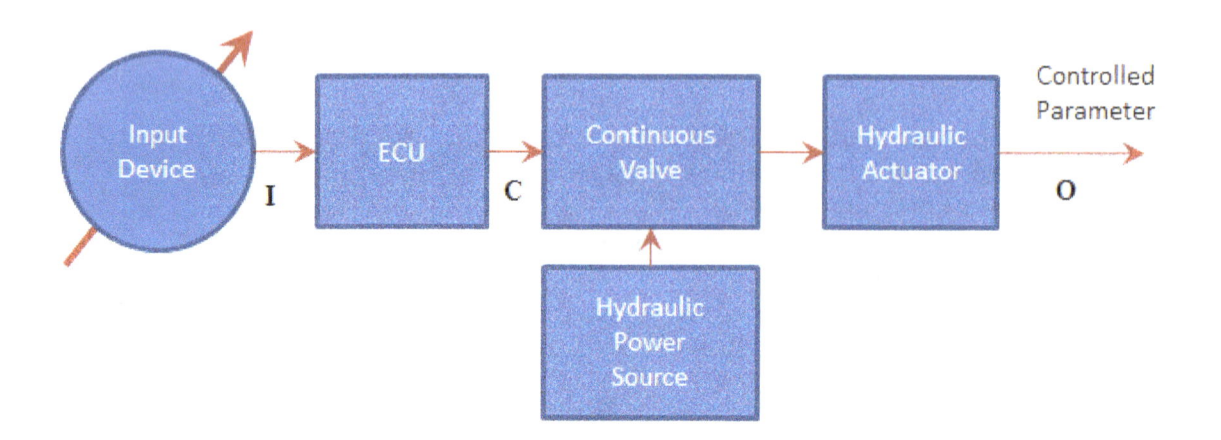

7- The system shown below represents?
   A. A Closed-loop control system using a stroke-controlled proportional valve.
   B. A Closed-loop control system using a force-controlled proportional valve.
   C. An open-loop control system using a force-controlled proportional valve.
   D. An open-loop control system using a stroke-controlled proportional valve.

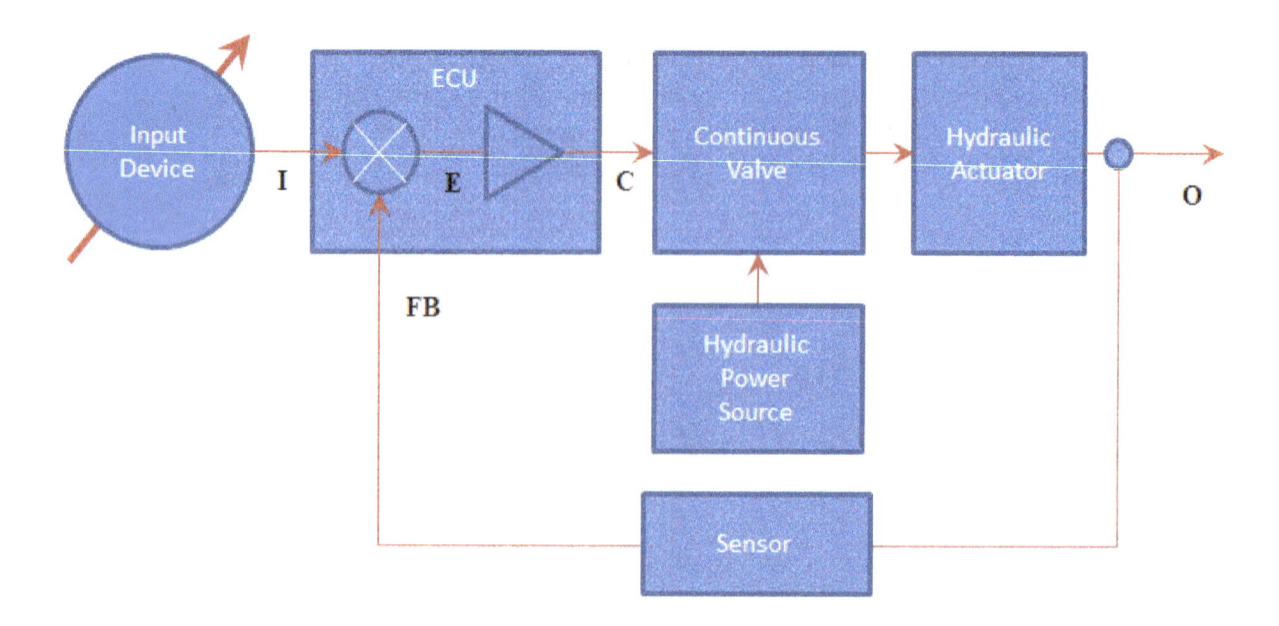

8- The system shown below represents?
   A. A Closed-loop control system using a stroke-controlled proportional valve.
   B. A Closed-loop control system using a force-controlled proportional valve.
   C. An open-loop control system using a force-controlled proportional valve.
   D. An open-loop control system using a stroke-controlled proportional valve.

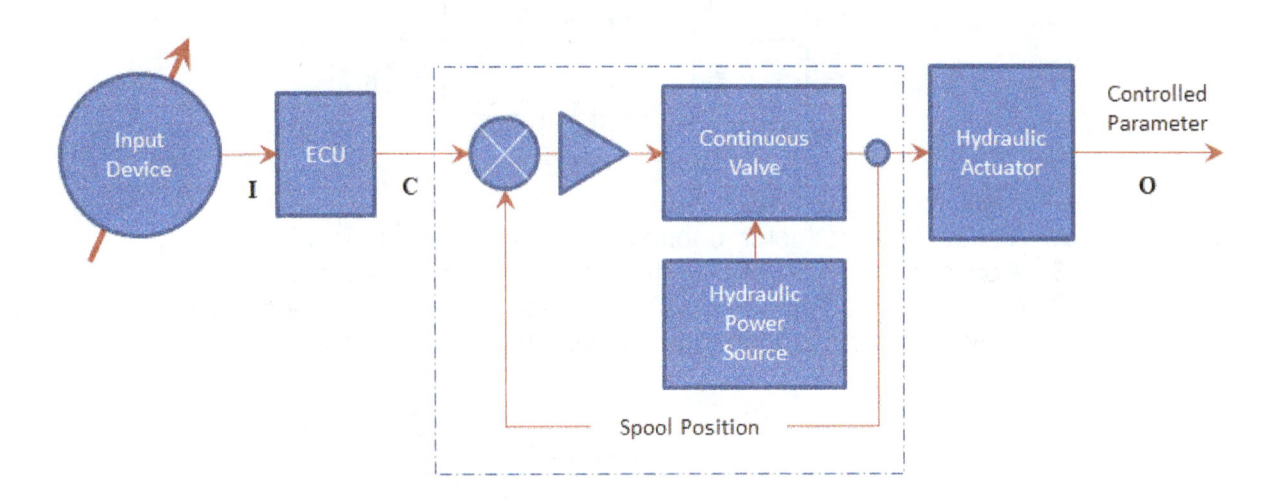

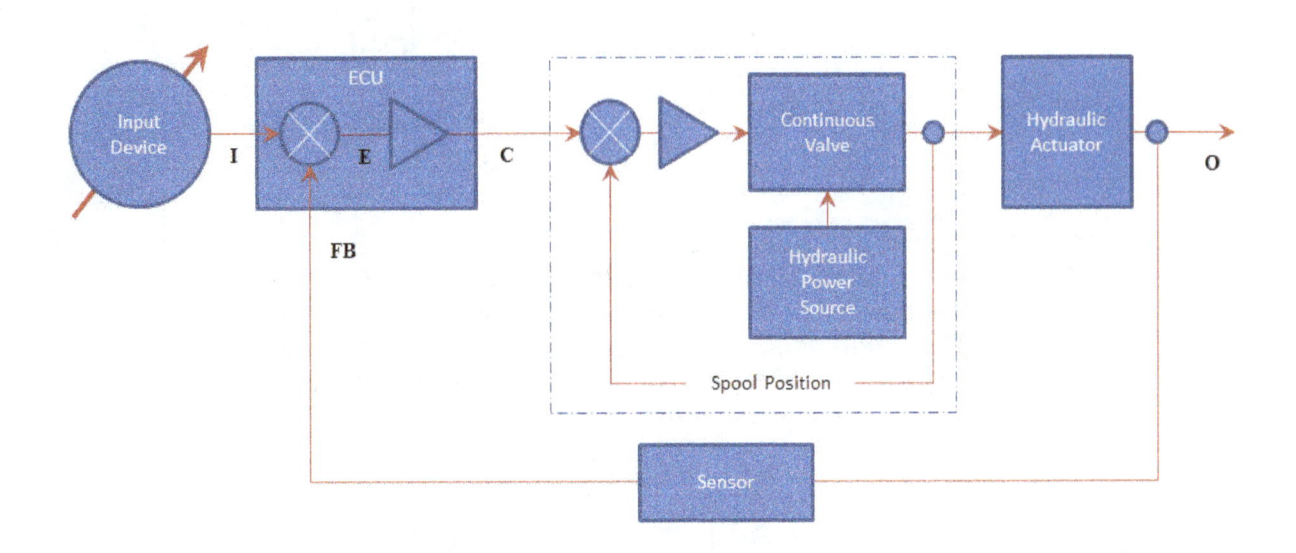

9- The system shown below represents?
   A. A Closed-loop control system using a stroke-controlled proportional valve.
   B. A Closed-loop control system using a force-controlled proportional valve.
   C. An open-loop control system using a force-controlled proportional valve.
   D. An open-loop control system using a stroke-controlled proportional valve.

10- The symbol shown below represents?
   A. A direct-operated force-controlled proportional directional valve.
   B. A direct-operated stroke-controlled proportional directional valve.
   C. A pilot-operated proportional directional valve with separate ECU.
   D. A pilot-operated proportional directional valve with an OBE.

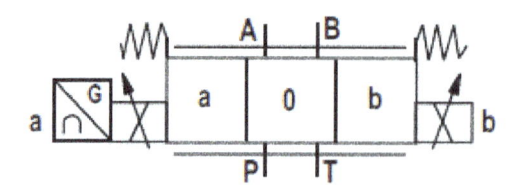

11- The symbol shown below represents?
   A. A direct-operated force-controlled proportional directional valve.
   B. A direct-operated stroke-controlled proportional directional valve.
   C. A pilot-operated proportional directional valve with separate ECU.
   D. A pilot-operated proportional directional valve with an OBE.

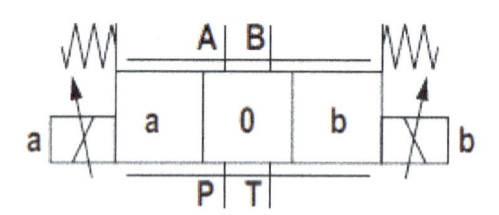

12- The symbol shown below represents?
   A. A direct-operated force-controlled proportional directional valve.
   B. A direct-operated stroke-controlled proportional directional valve.
   C. A pilot-operated proportional directional valve with separate ECU.
   D. A pilot-operated proportional directional valve with an OBE.

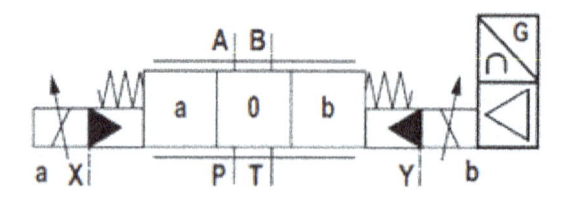

13- The symbol shown below represents?
   A. A direct-operated force-controlled proportional directional valve.
   B. A direct-operated stroke-controlled proportional directional valve.
   C. A pilot-operated proportional directional valve with separate ECU.
   D. A pilot-operated proportional directional valve with an OBE.

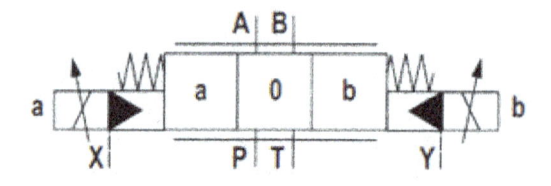

Workbook: Electro-Hydraulic Components and Systems
Chapter 5: Reviews and Assignments

14. The fail-safe position of the valve shown below is of what type?
    A. Open-center type.
    B. Cross-Connection type.
    C. Parallel-Connection type.
    D. Float type.

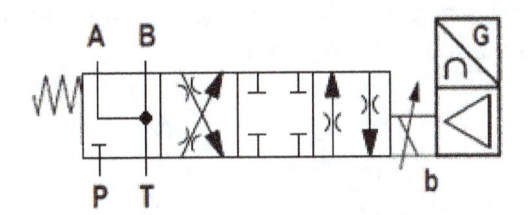

15. The schematic shown below is for a?
    A. Pilot-operated open-center proportional directional valve.
    B. Single-stage open-center proportional directional valve.
    C. Pilot-operated closed-center proportional directional valve
    D. Direct-operated float-center proportional directional valve.

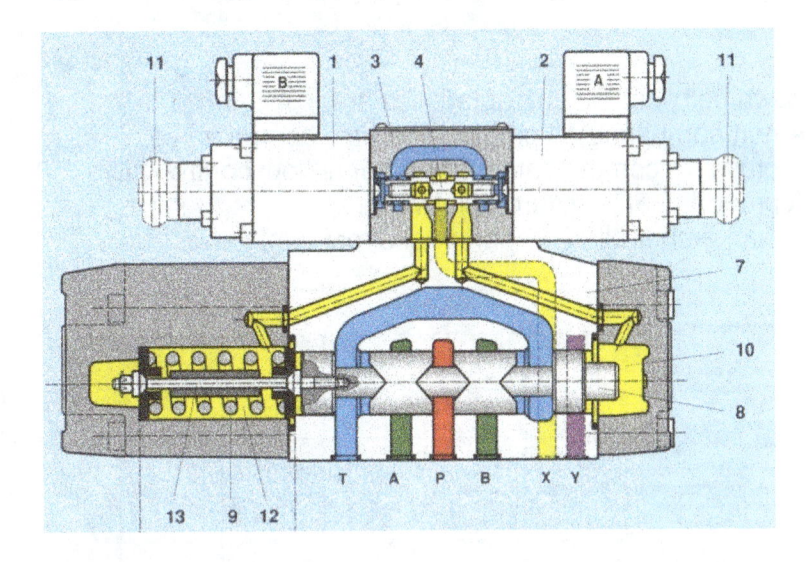

16. In the proportional pressure relief valve shown below, the front spring (3) is for?
    A. Fail-safe conditions so that the valve will open in case of power failure.
    B. Absorbing the pressure fluctuation.
    C. Set the system pressure.
    D. Set the system flow.

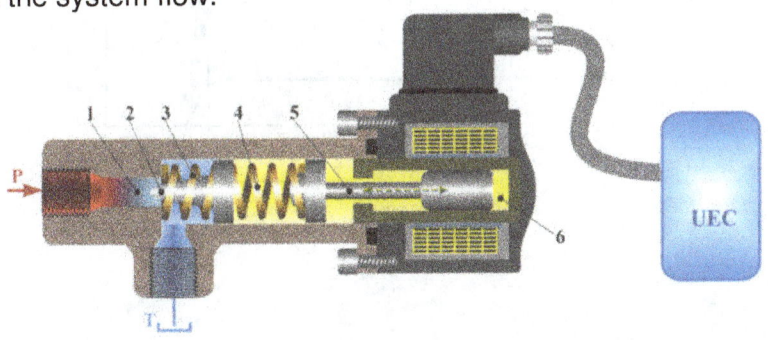

173

17. What is the best description of the symbol shown below?
    A. Pilot-operated pressure relief valve.
    B. Pilot-operated pressure relief valve with mechanical backup PRV.
    C. Pilot-operated pressure relief valve with mechanical backup PRV, and a built-in pressure transducer.
    D. Pilot-operated pressure relief valve with mechanical backup PRV, a built-in pressure transducer, and an OBE.

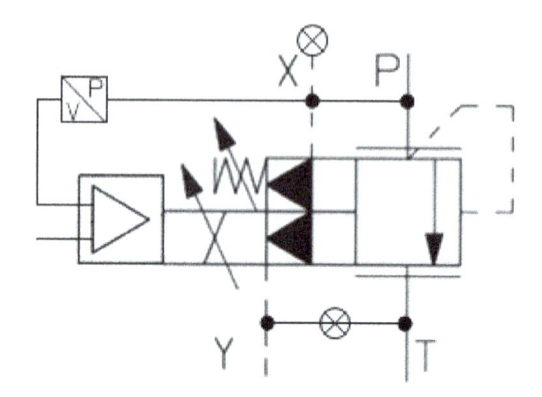

18. The shown below static characteristics are for?
    A. A non-compensated proportional flow control valve.
    B. A pressure compensated proportional flow control valve
    C. A proportional pressure relief valve.
    D. A stroke-controlled proportional directional valve.

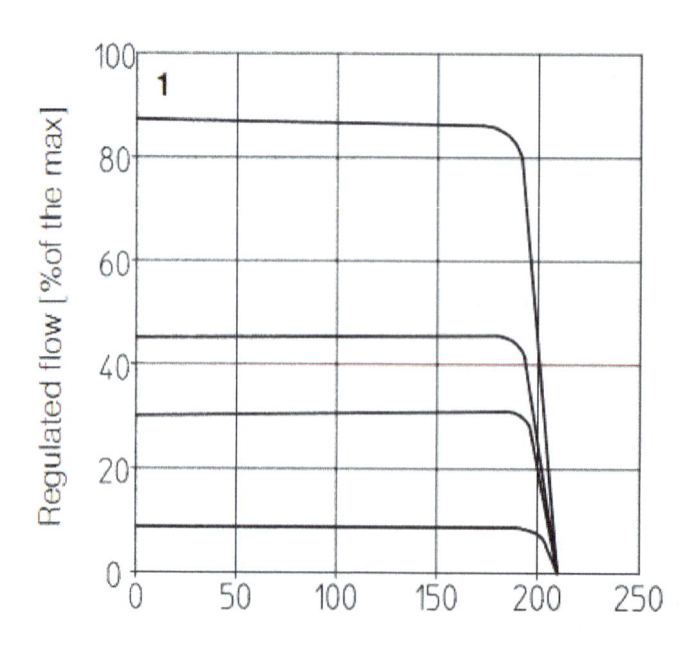

Pressure at port B [bar]

19. The shown below static characteristics are for?
    A. A non-compensated proportional flow control valve.
    B. A pressure compensated proportional flow control valve
    C. A pilot-operated proportional pressure relief valve.
    D. A direct-operated proportional pressure relief valve.

Flow vs. Pressure Drop
with maximum current applied

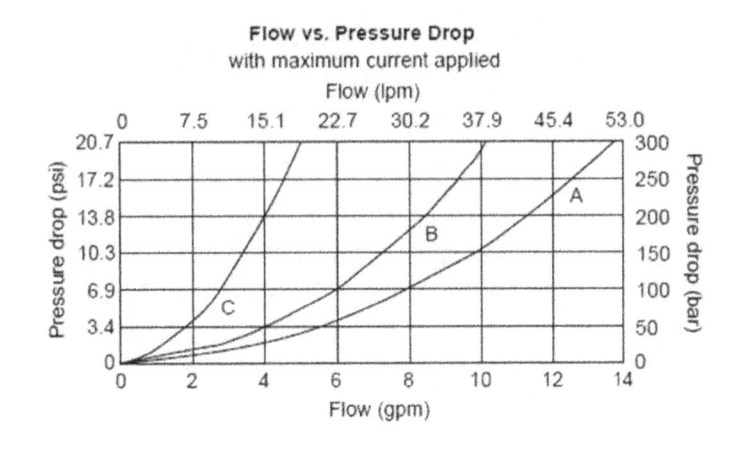

20. The shown below static characteristics are for?
    A. A non-compensated proportional flow control valve.
    B. A pressure compensated proportional flow control valve
    C. A proportional pressure relief valve.
    D. A direct-operated proportional directional valve.

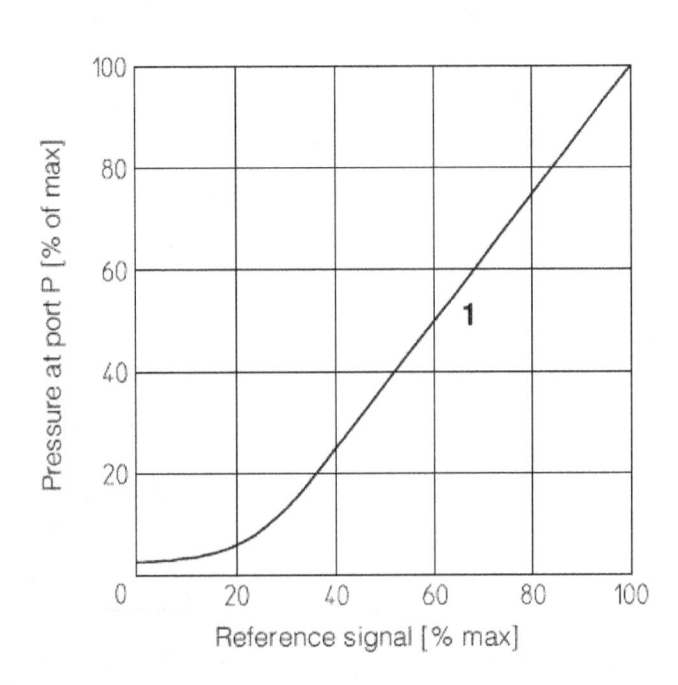

Workbook: Electro-Hydraulic Components and Systems
Chapter 5: Reviews and Assignments

Chapter 5 Assignment

Student Name: -- Student ID: ------------------

Date: -- Score: -------------------------

In the circuit shown below, find out how much flow will pass through the valve if the valve received 60% command and subjected to a 10 bar differential pressure.

Δp = Valve pressure differential

(inlet pressure p_P minus load pressure p_L minus return flow pressure p_T)

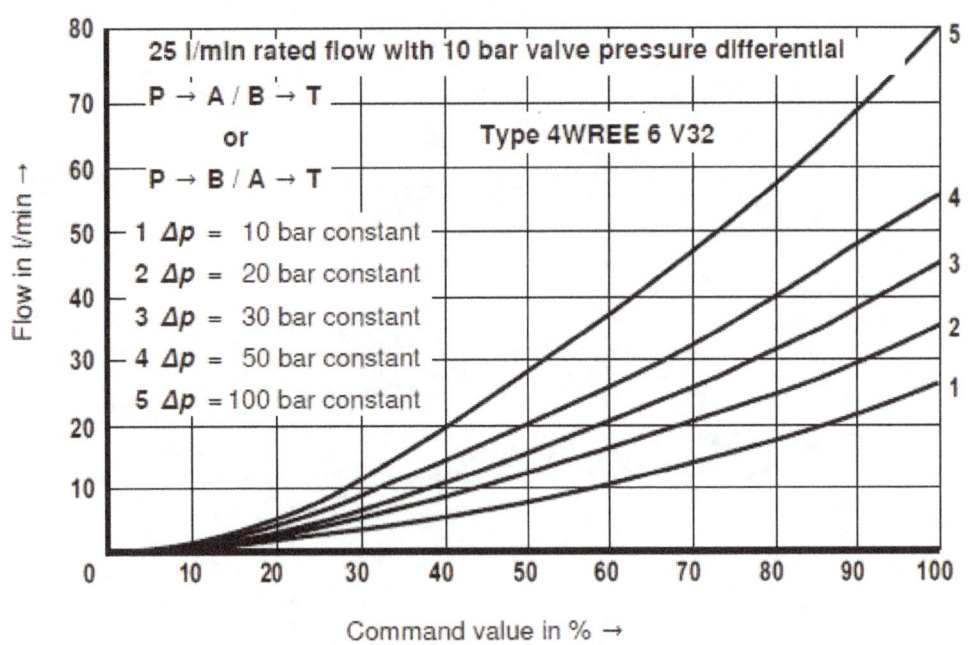

25 l/min rated flow with 10 bar valve pressure differential
P → A / B → T
or
P → B / A → T
Type 4WREE 6 V32

1 Δp = 10 bar constant
2 Δp = 20 bar constant
3 Δp = 30 bar constant
4 Δp = 50 bar constant
5 Δp = 100 bar constant

Command value in % →

Type 4WREE (measured with HLP46, ϑ_{Oil} = 40 °C ± 5 °C)

Chapter 6
Servo Valves

Objectives:

This chapter covers the construction and wiring methods of the main electric components in servo valves including torque motors. The chapter also covers the conceptual construction and the operation of flapper-nozzle and jet-pipe servo valves. The chapter introduces examples of typical valves with mechanical or electrical feedback.

Brief Contents:

6.1- Introduction to Servo Valves
6.2- Torque Motors
6.3- Servo Valves Configurations

Workbook: Electro-Hydraulic Components and Systems
Chapter 6: Servo Valves

6.1- Introduction to Servo Hydraulics
6.1.1- Historical Background

- **Prior WWII**
- Jet Pipe.
- Half Bridge Flapper-Nozzle.
- Full Bridge Flapper-Nozzle.

- **Post WWII.**
- Torque motor instead of a solenoid.
- First 2 stage valve with sliding spool moving inside a sleeve.
- Stroke-controlled valve.
- Pilot-operated three-stages servo valves.
- Enhanced spool-sleeve assembly for high performance.
- Onboard electronics.
- Digital control.

6.1.2- Servo Valve Applications

Force-Controlled Proportional Valves → Stroke-Controlled Proportional Valve → High Performance Proportional Valves → Servo Valves

Dynamics and Precision →

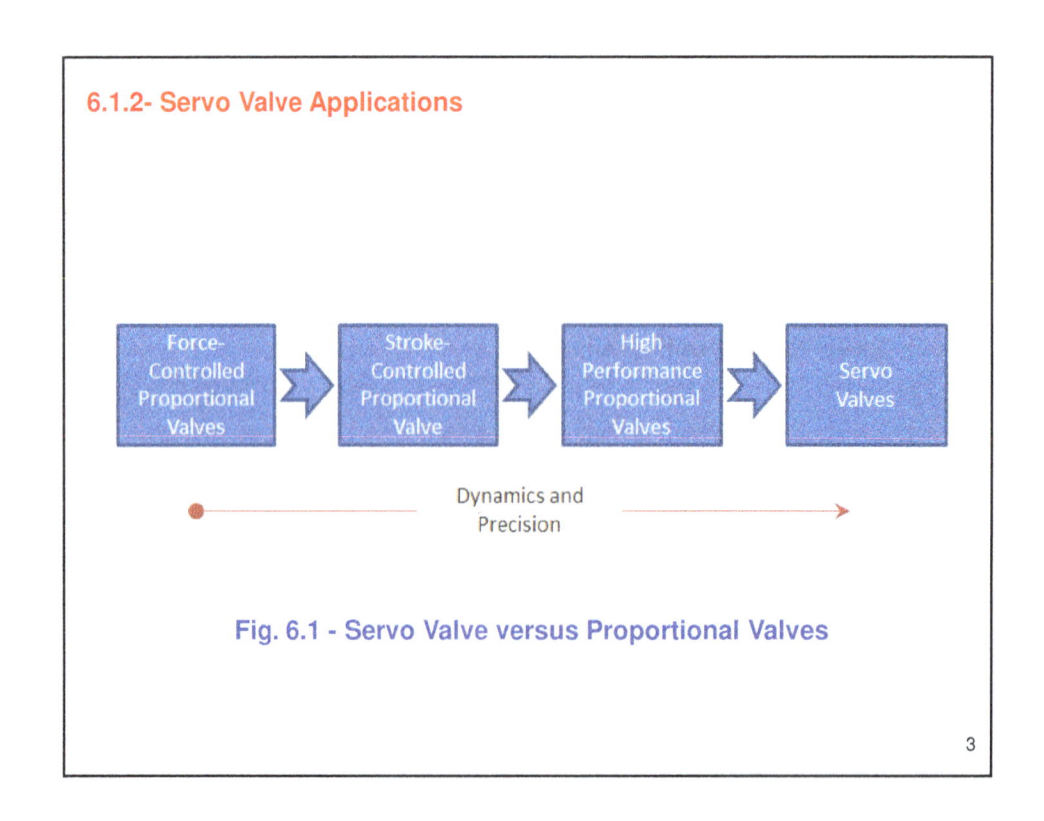

Fig. 6.1 - Servo Valve versus Proportional Valves

- **As a matter of fact, as compared to a proportional valve:**
  - A servo valve has <u>faster</u> dynamic response.
  - A servo valve has <u>higher accuracy</u>.
- **Technically both** servo and proportional valves can be used in open or closed-loop.
- **A proportional** valve is **more cost effective** and convenient for **open loop** applications.
- **Higher performance proportional** valves can be used for closed-loop of faster dynamics.
- **There is no sense to use a servo valve unless** there is a good reason such as a closed-loop with **faster dynamics and high accuracy**.

- A servo valve can be built-in with a hydraulic actuator to form a standalone single-axis system to control **direction, speed or load** of a hydraulic actuator.

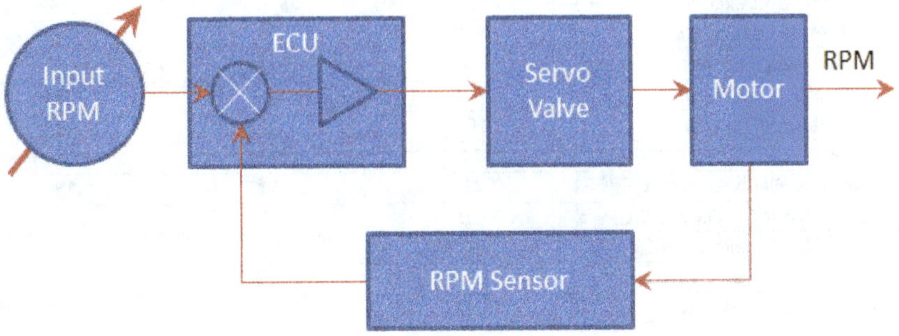

Fig. 6.2 - A Servo Valve in a Single-Axis System

- It can also be a part of a large automated system

Fig. 6.3 - A Servo Valve in a Large Automated System

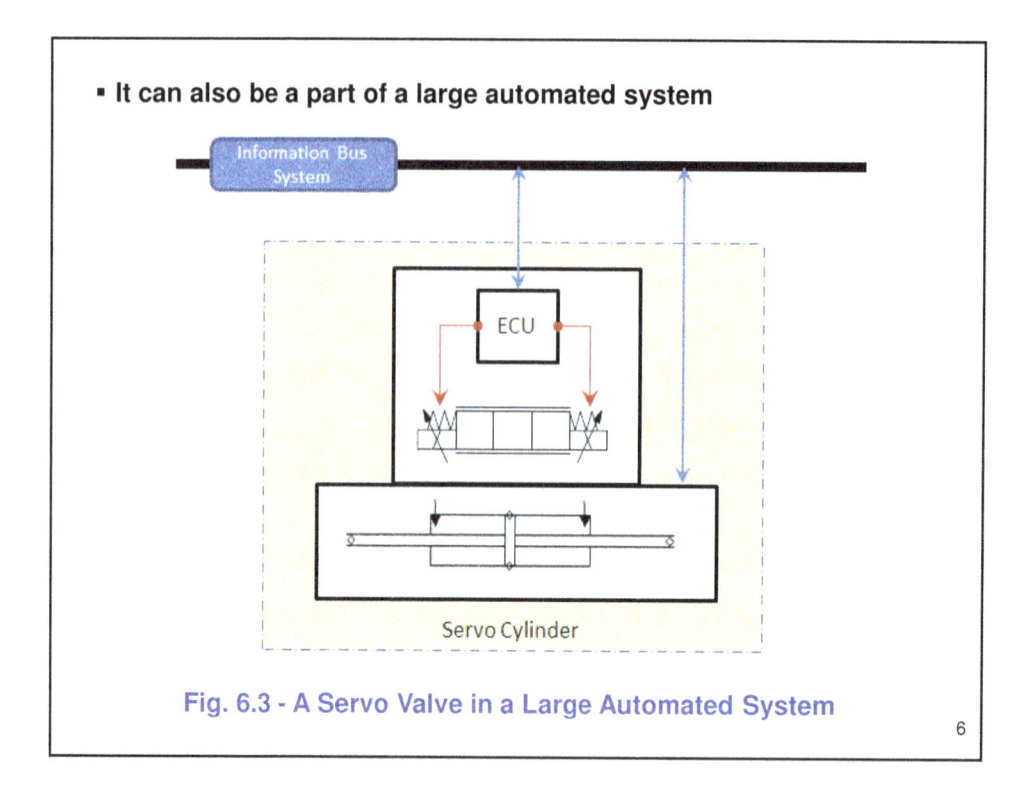

- **The following table** shows examples of using servo valves in industrial and mobile applications.
- Note that **mobile applications are slower** than industrial ones.

Field	Frequency (Hz)	Flow (l/min)	Power (kW)
Vibration Exciters	600	4	1.5
Missile Fin Control	400	4	1.5
Seekers Antenna	300	2	0.75
Oil Exploration Hammers	200	450	190
Fatigue Testing	100	115	40
Flight Simulators	50	190	75
Airplane Flight Controls	40	115	45
Metal Forming Rolling Mills	30	570	225
Agricultural Equipment	15	25	10
Cranes	7	75	30
Process Control	5	7.5	3

Table 6.1 - Servo Valve Application Examples

❑ **Why is the servo valve is faster than the proportional valve?**

1. **Electrically:** A torque motor is much faster than the proportional solenoid and consumes much less power.

2. **Mechanically:** Servo valves have built-in internal feedback mechanisms that assures the correct position of the spool relative to the input command.

3. **Hydraulically:** The pilot stage creates differential pressure against the spool of the main stage much faster.

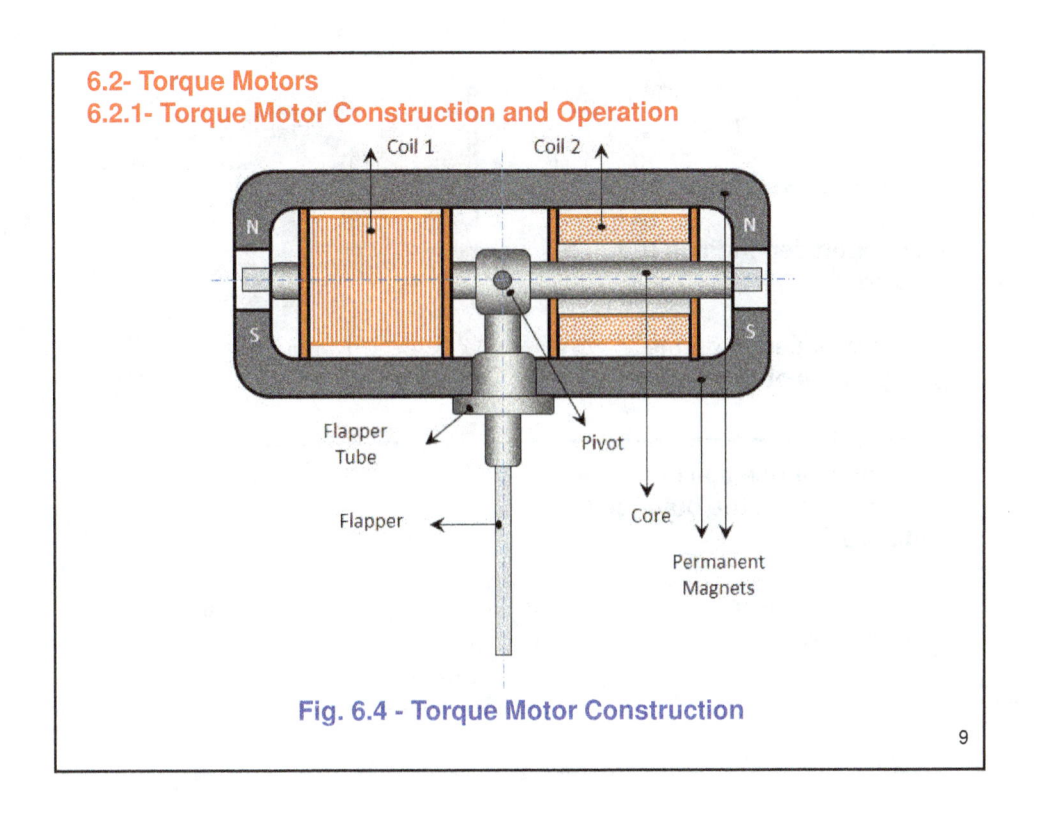

6.2- Torque Motors
6.2.1- Torque Motor Construction and Operation

Fig. 6.4 - Torque Motor Construction

Torque α Current Magnitude

Direction α Current Direction

The core will continue to rotate until the electromagnetic torque is balanced:
- Mechanically by a spring.
- Electrically by wiring style of the coils.

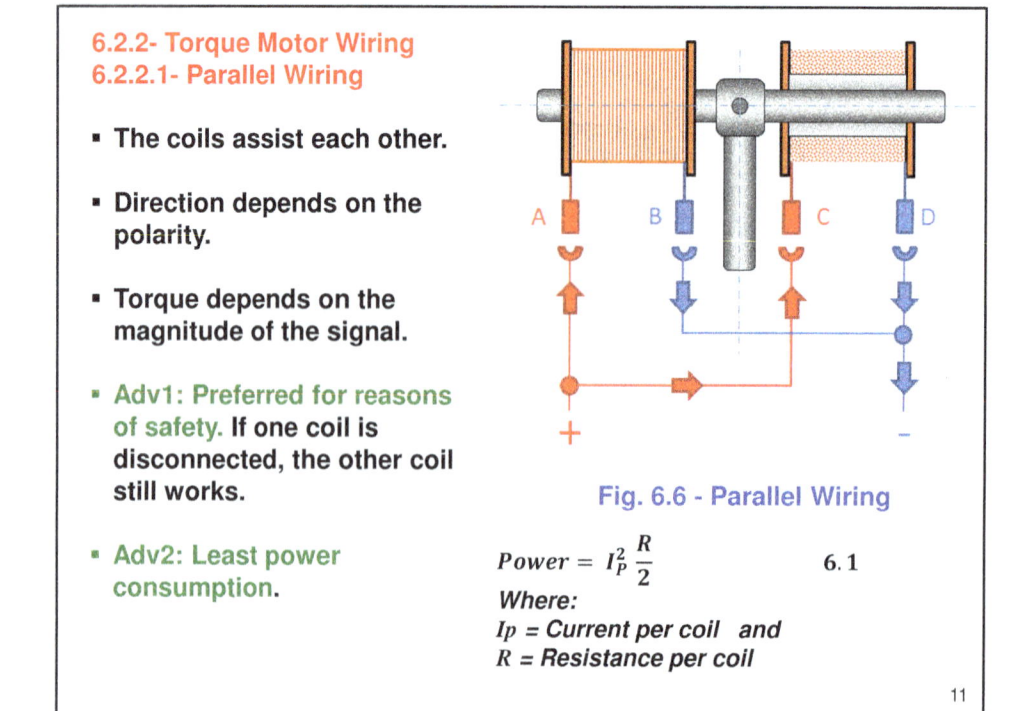

Fig. 6.5 - Torque Motor Operation

6.2.2- Torque Motor Wiring
6.2.2.1- Parallel Wiring

- **The coils assist each other.**

- **Direction depends on the polarity.**

- **Torque depends on the magnitude of the signal.**

- **Adv1: Preferred for reasons of safety.** If one coil is disconnected, the other coil still works.

- **Adv2: Least power consumption.**

Fig. 6.6 - Parallel Wiring

$$Power = I_P^2 \frac{R}{2} \qquad 6.1$$

Where:
I_P = Current per coil   and
R = Resistance per coil

6.2.2.2 - Series Wiring

- The coils assist each other.
- Direction depends on the polarity.
- Torque depends on the magnitude of the signal.
- Operated at double the voltage and half the current (as compared to Parallel Wiring)
- Adv.: Simplest wiring

Fig. 6.7 - Series Wiring

$$Power = I_P^2\,(2R) \qquad 6.2$$

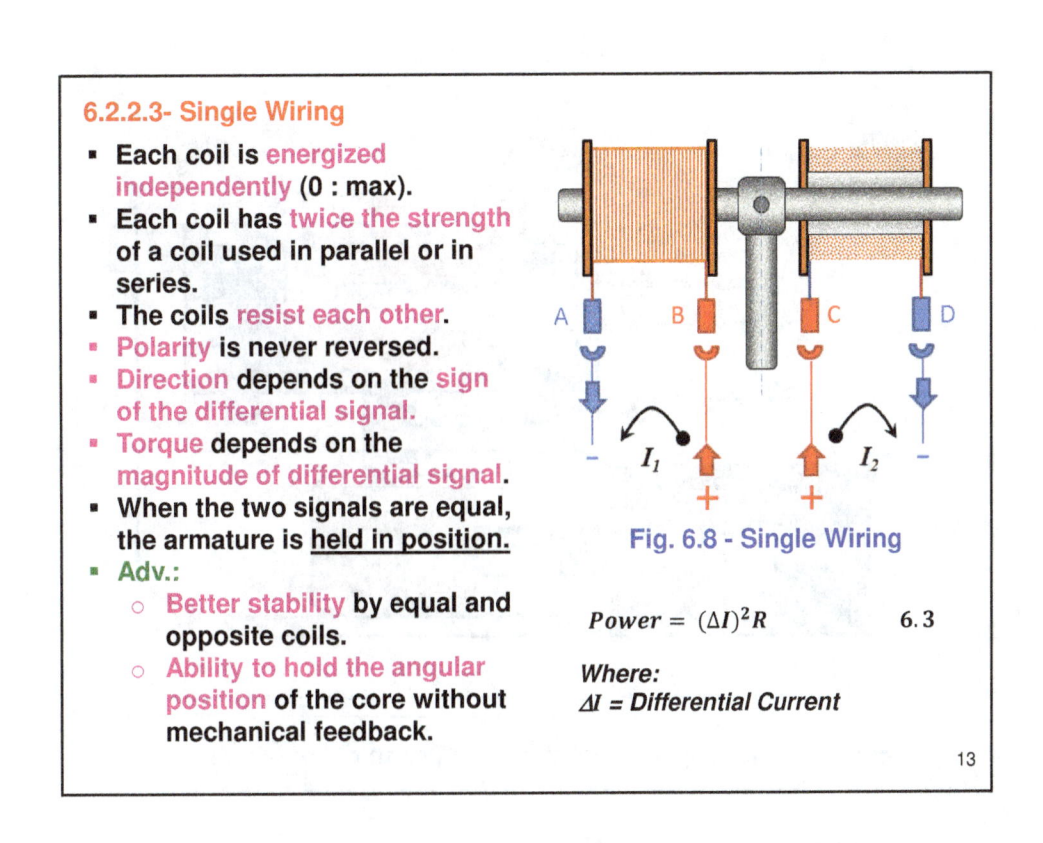

6.2.2.3 - Single Wiring

- Each coil is energized independently (0 : max).
- Each coil has twice the strength of a coil used in parallel or in series.
- The coils resist each other.
- Polarity is never reversed.
- Direction depends on the sign of the differential signal.
- Torque depends on the magnitude of differential signal.
- When the two signals are equal, the armature is held in position.
- Adv.:
  - Better stability by equal and opposite coils.
  - Ability to hold the angular position of the core without mechanical feedback.

Fig. 6.8 - Single Wiring

$$Power = (\Delta I)^2 R \qquad 6.3$$

Where:
ΔI = Differential Current

6.3- Servo Valve Configurations
6.3.1- Power Assistance using Mechanical Servo Valve

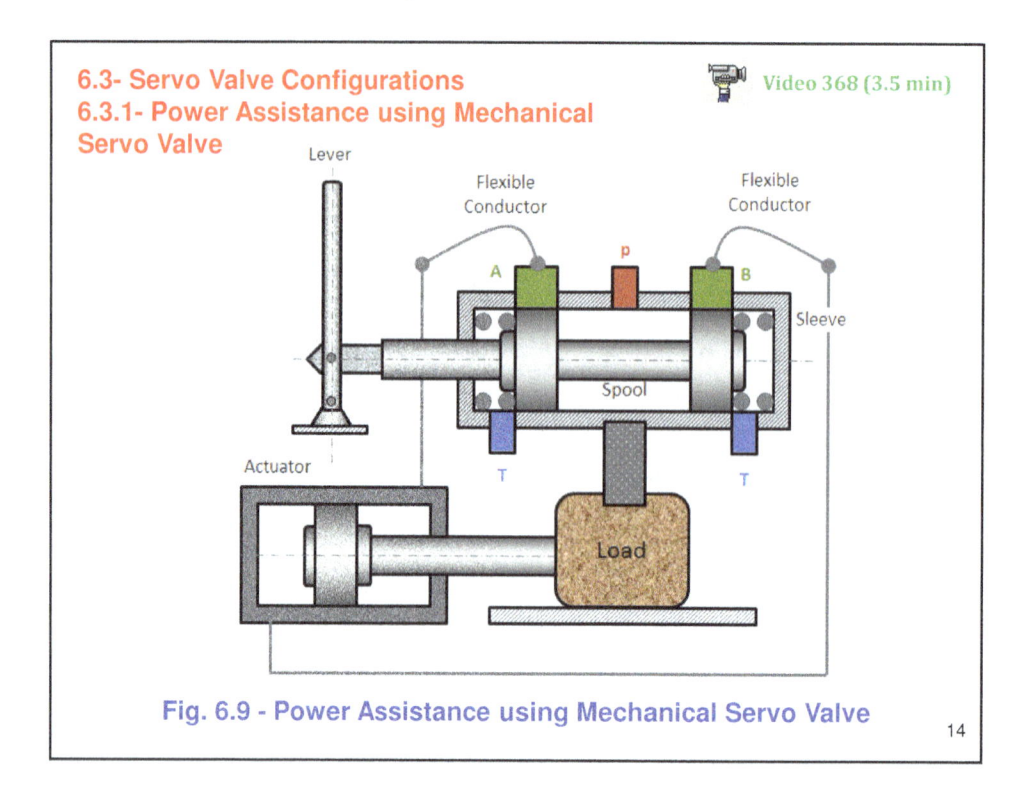

Fig. 6.9 - Power Assistance using Mechanical Servo Valve

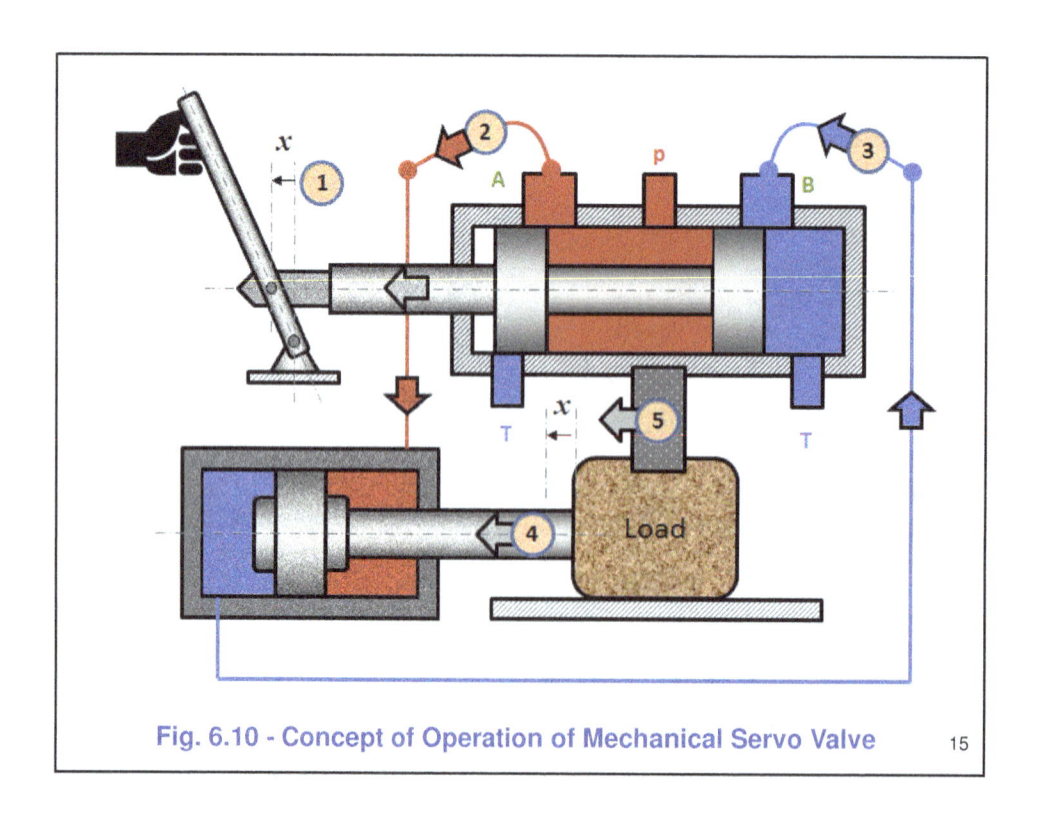

Fig. 6.10 - Concept of Operation of Mechanical Servo Valve

6.3.2- Single-Stage Servo Valve

- Spool movement α input signal to the TM.
- TM is to overcome: spool friction + flow forces + spring forces.
- TM Wiring: Parallel and series are applicable. No need for single wiring.
- Force Feedback: torque from TM balances with control springs
- Good for small flow.

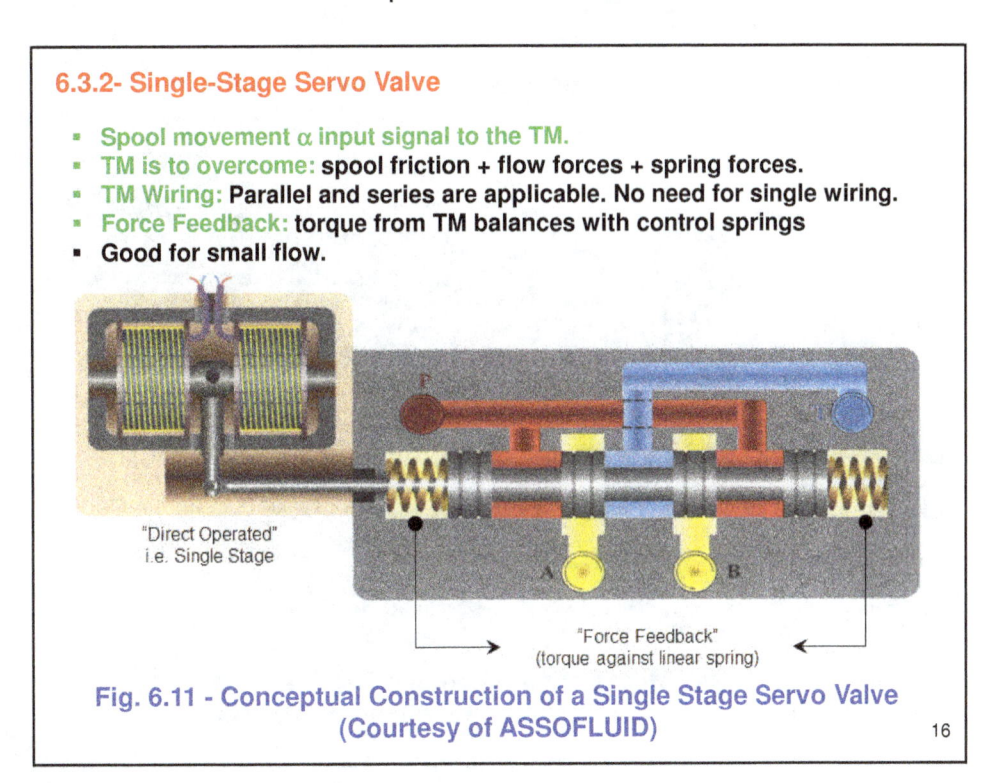

"Direct Operated" i.e. Single Stage

"Force Feedback" (torque against linear spring)

Fig. 6.11 - Conceptual Construction of a Single Stage Servo Valve
(Courtesy of ASSOFLUID)

- TM Wiring: single wiring method.

1. Torque Motor
2. Control Spool
3. Spool Stage
4. Pin
5. Armature
6. Torque Tube
7. Tie Rod
8. Null Adjustment Screw
9. Housing
10. Sleeve

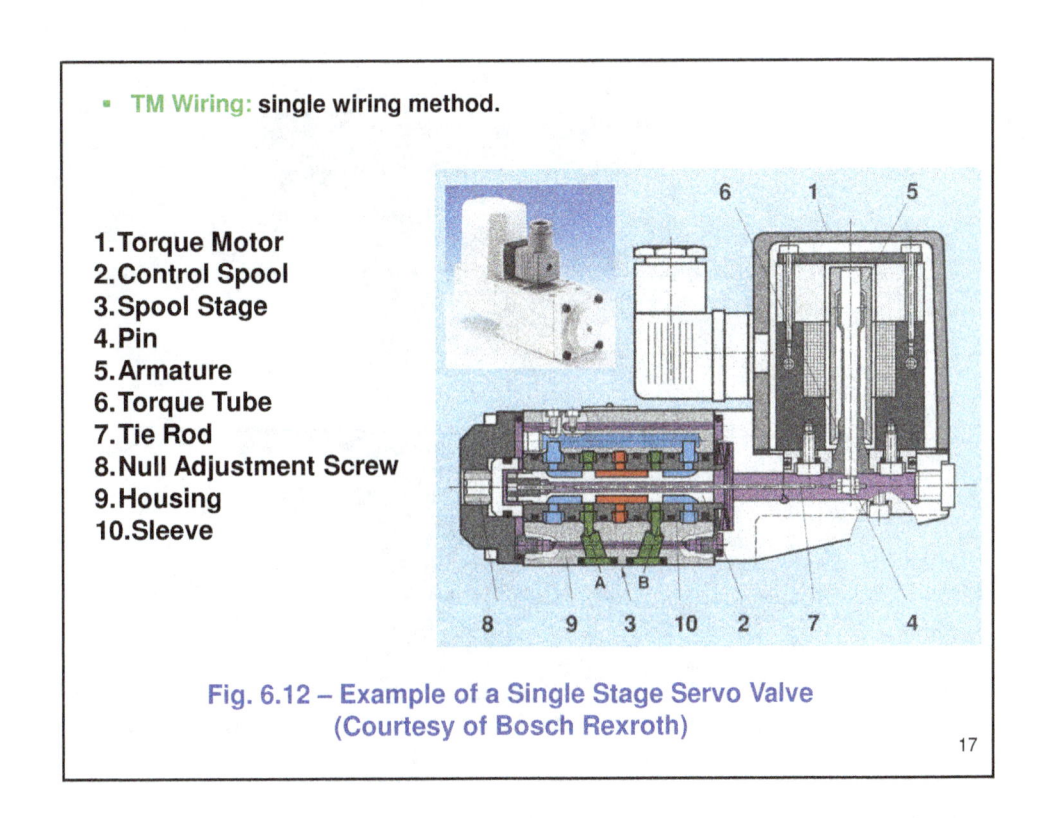

Fig. 6.12 – Example of a Single Stage Servo Valve
(Courtesy of Bosch Rexroth)

6.3.3- Two-Stage Servo Valve with Tracking Sleeve

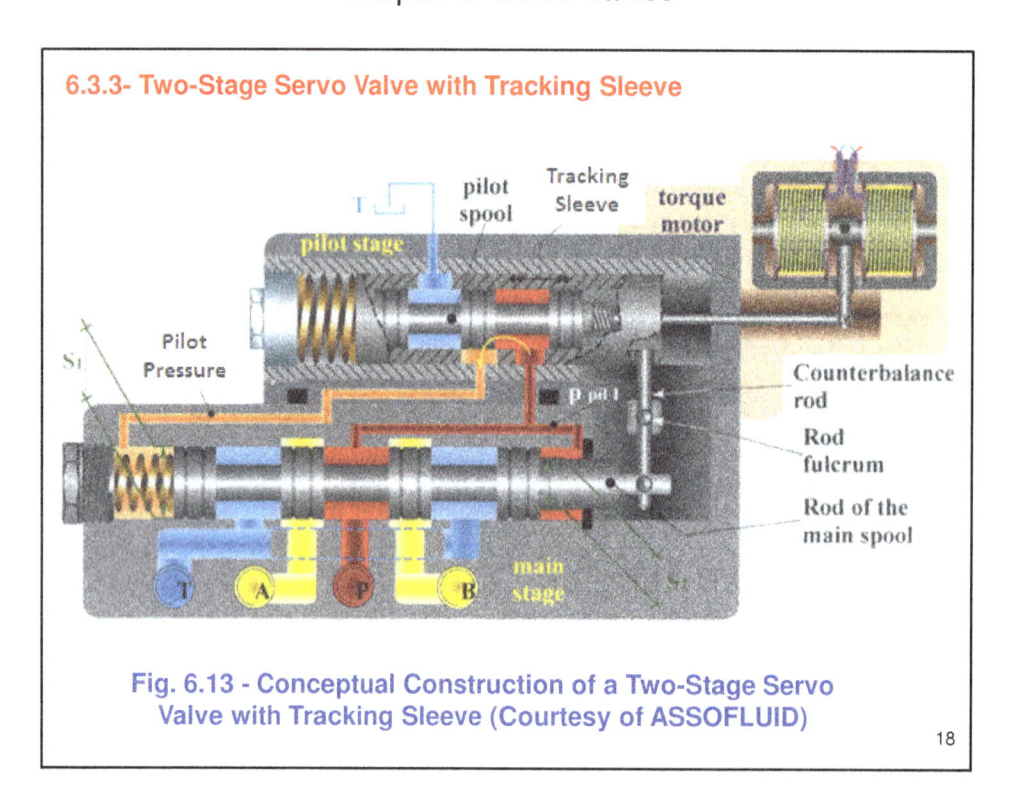

Fig. 6.13 - Conceptual Construction of a Two-Stage Servo Valve with Tracking Sleeve (Courtesy of ASSOFLUID)

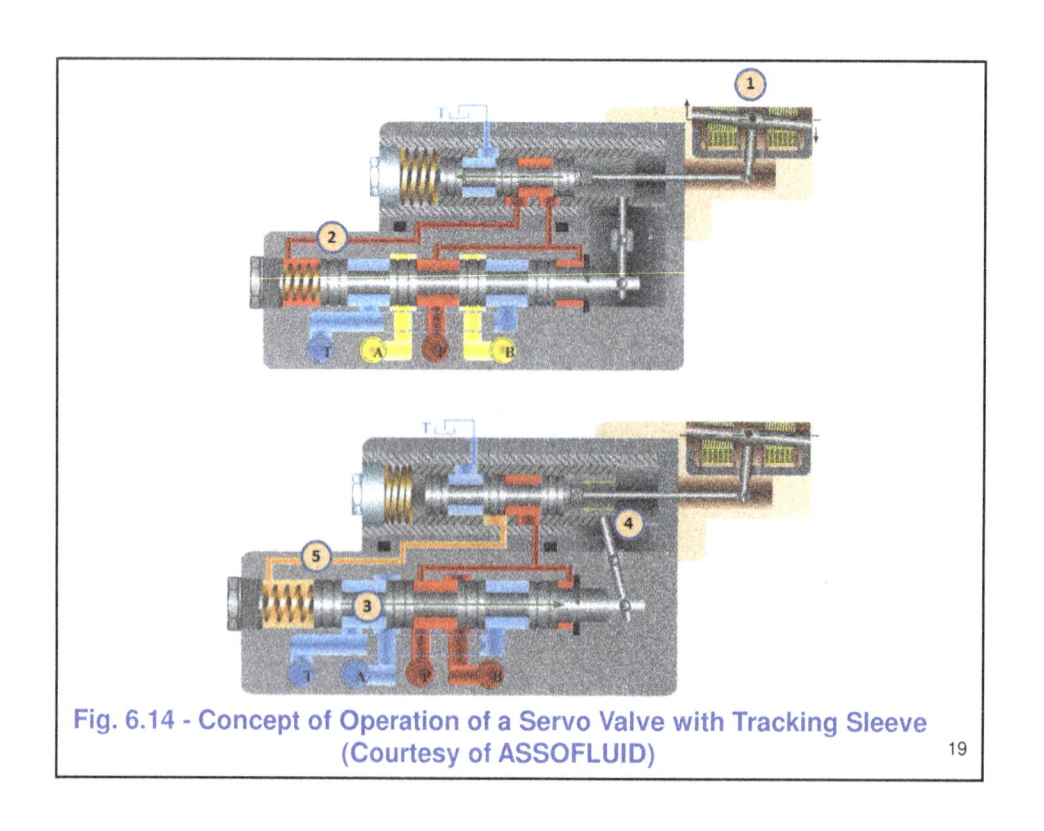

Fig. 6.14 - Concept of Operation of a Servo Valve with Tracking Sleeve (Courtesy of ASSOFLUID)

6.3.4- Flapper–Nozzle Concept of Operation
6.3.4.1- Half-Bridge Flapper-Nozzle Concept

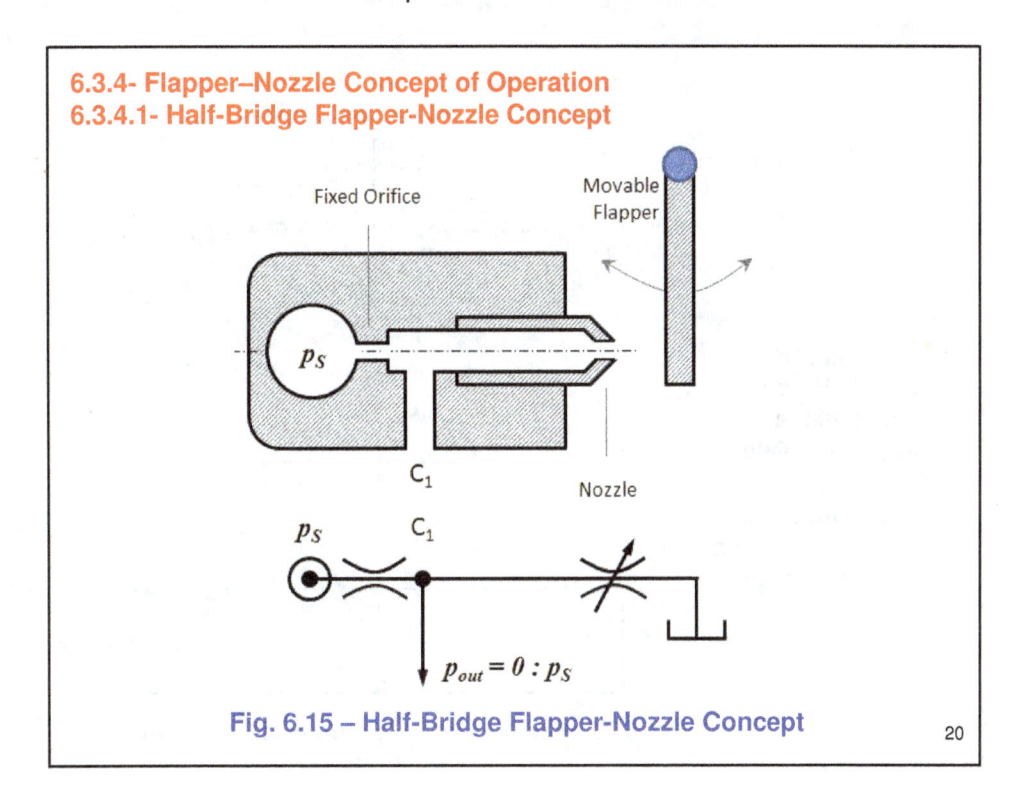

Fig. 6.15 – Half-Bridge Flapper-Nozzle Concept

Ignoring losses and potential head difference between points 1 and 2

$$\left(\frac{p}{\rho g}+\frac{v^2}{2g}\right)_1 = \left(\frac{p}{\rho g}+\frac{v^2}{2g}\right)_2 \quad 6.4$$

$(A\uparrow \to v\downarrow \to p\uparrow)_1 : (A\downarrow \to v\uparrow \to p\downarrow)_2$

Curtain Area << Nozzle Area

$$x_f \cdot \pi d_n \ll \frac{\pi}{4} d_n^2 \quad 6.5$$

$$x_f = 0.1 d_n$$

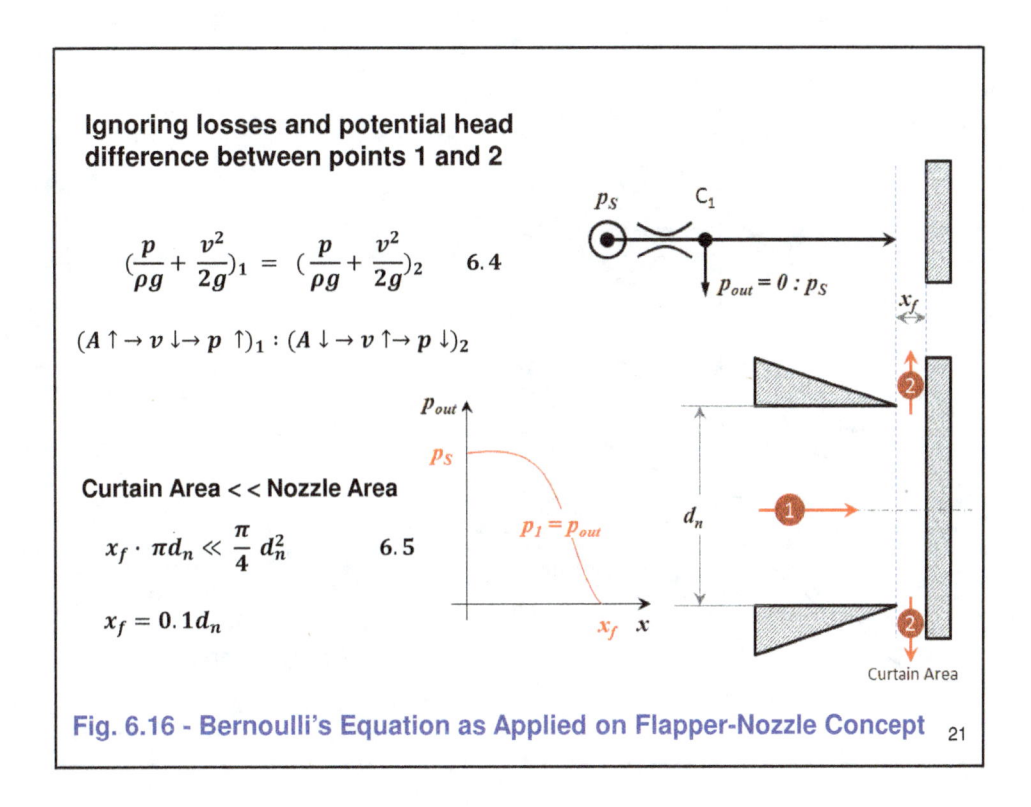

Fig. 6.16 - Bernoulli's Equation as Applied on Flapper-Nozzle Concept

6.3.4.2 - Full-Bridge Flapper-Nozzle Concept

- **The Flapper-Nozzle Stage works like a Pressure Divider**

- **Continuous flow through this stage causes energy losses.**

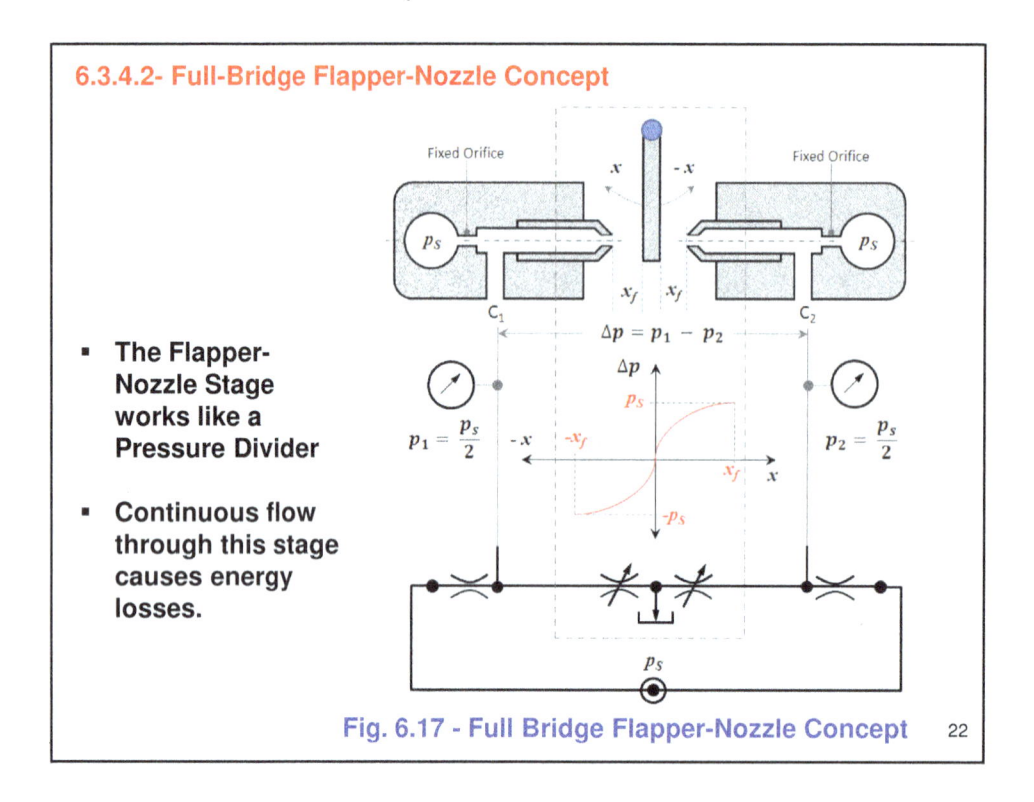

$\Delta p = p_1 - p_2$

$p_1 = \dfrac{p_s}{2}$; $p_2 = \dfrac{p_s}{2}$

Fig. 6.17 - Full Bridge Flapper-Nozzle Concept

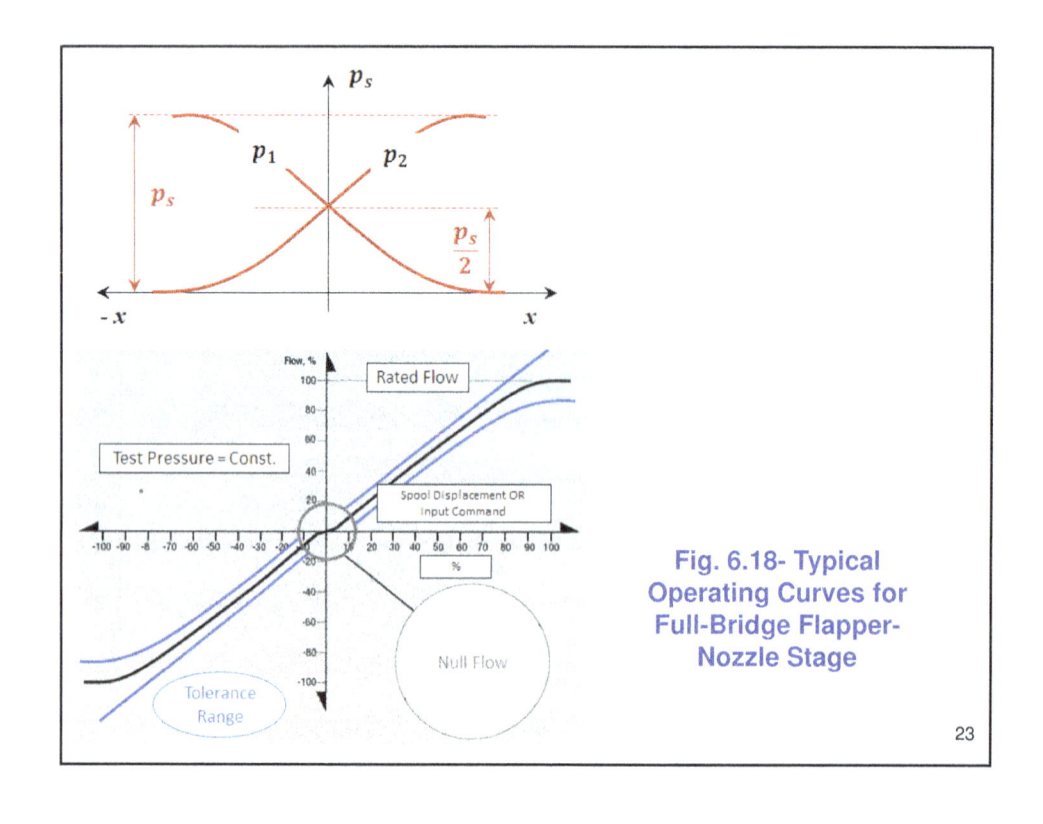

Fig. 6.18 - Typical Operating Curves for Full-Bridge Flapper-Nozzle Stage

6.3.4.3- Power Gain of Servo Valve

Torque Motor Power Gain:
= 100/0.1 = 1000

Pilot Stages Power Gain:
= 1000/100 = 10

Overall Power Gain:
= 1000 x 10 = 10000

Remember:
It is not power amplification!!

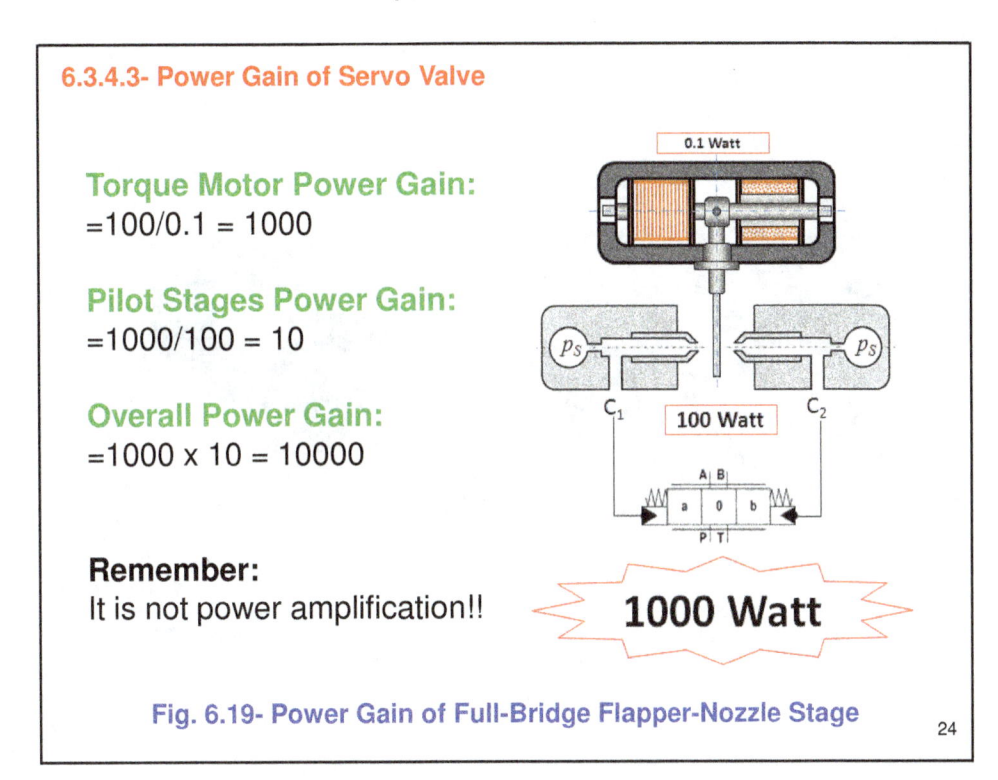

Fig. 6.19- Power Gain of Full-Bridge Flapper-Nozzle Stage

6.3.5- Two-Stage Flapper–Nozzle Servo Valve with Barometric Feedback

- Spool movement α input signal to TM.
- TM is to overcome: **flow forces at the nozzles.**
- TM Wiring: **single wiring to hold the flapper angular position.**
- Barometric Feedback: **differential pressure balances with the control spring.**

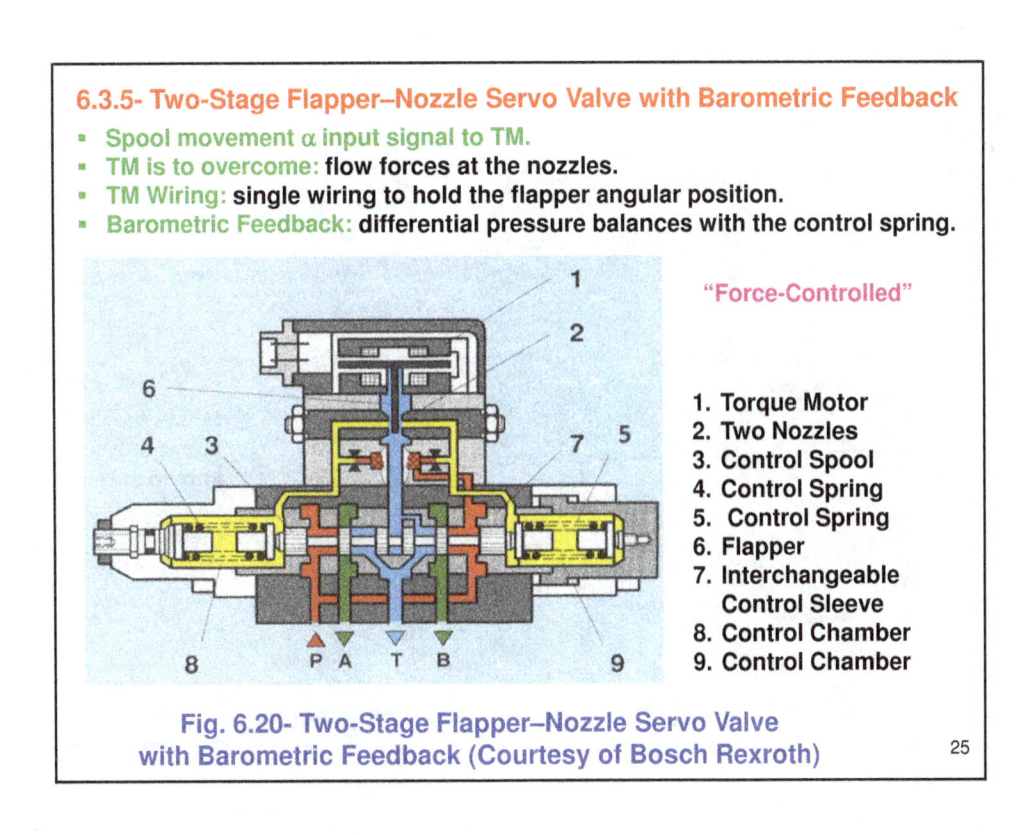

"Force-Controlled"

1. Torque Motor
2. Two Nozzles
3. Control Spool
4. Control Spring
5. Control Spring
6. Flapper
7. Interchangeable Control Sleeve
8. Control Chamber
9. Control Chamber

Fig. 6.20- Two-Stage Flapper–Nozzle Servo Valve with Barometric Feedback (Courtesy of Bosch Rexroth)

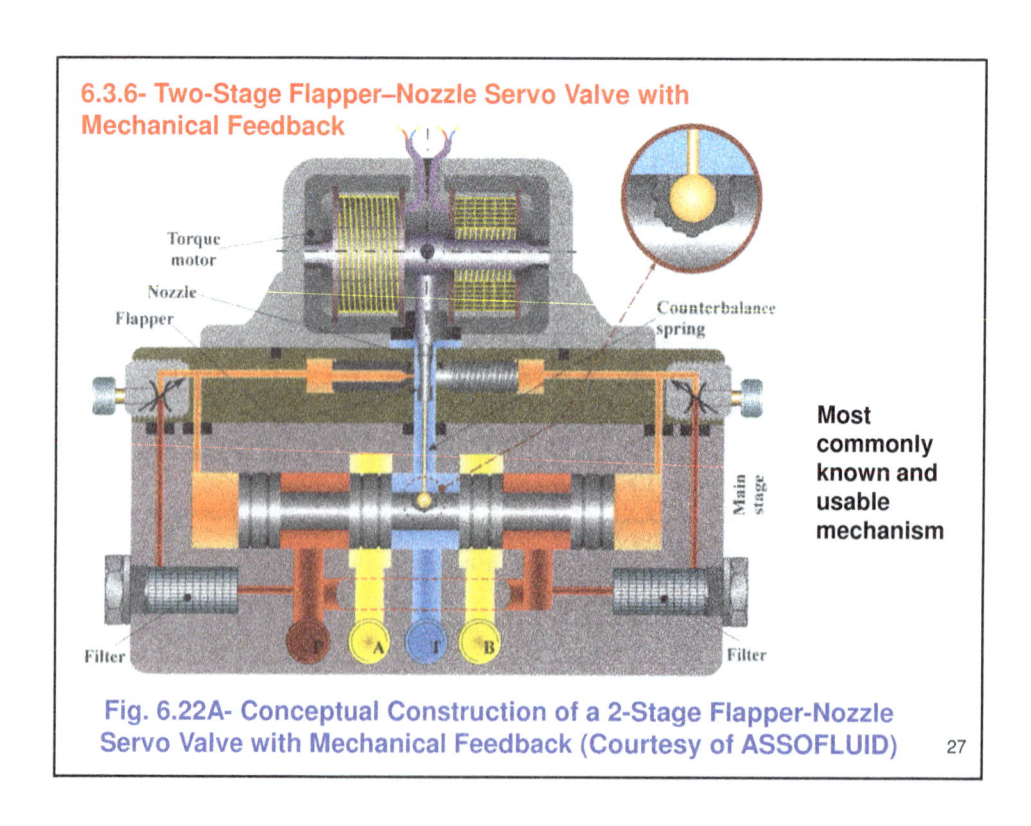

Fig. 6.21- Servo Valve Interchangeable Spool and Sleeve (Courtesy of Moog)

6.3.6- Two-Stage Flapper–Nozzle Servo Valve with Mechanical Feedback

Most commonly known and usable mechanism

Fig. 6.22A- Conceptual Construction of a 2-Stage Flapper-Nozzle Servo Valve with Mechanical Feedback (Courtesy of ASSOFLUID)

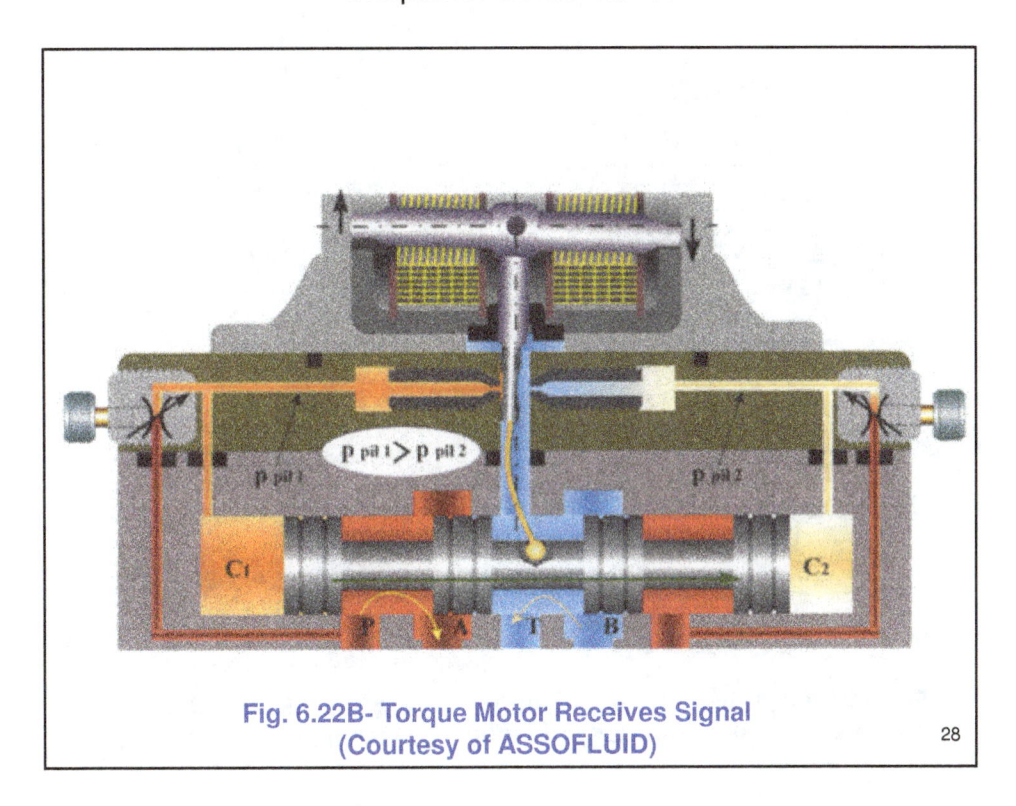

Workbook: Electro-Hydraulic Components and Systems
Chapter 6: Servo Valves

Fig. 6.22B- Torque Motor Receives Signal (Courtesy of ASSOFLUID)

- Spool movement α input signal to the TM.
- TM is to overcome: **spool friction + feedback spring.**
- TM Wiring: **all methods are applicable.**
- Mechanical Feedback: **torque from TM balances with torque from feedback spring.**

"Mechanically Stroke-Controlled"

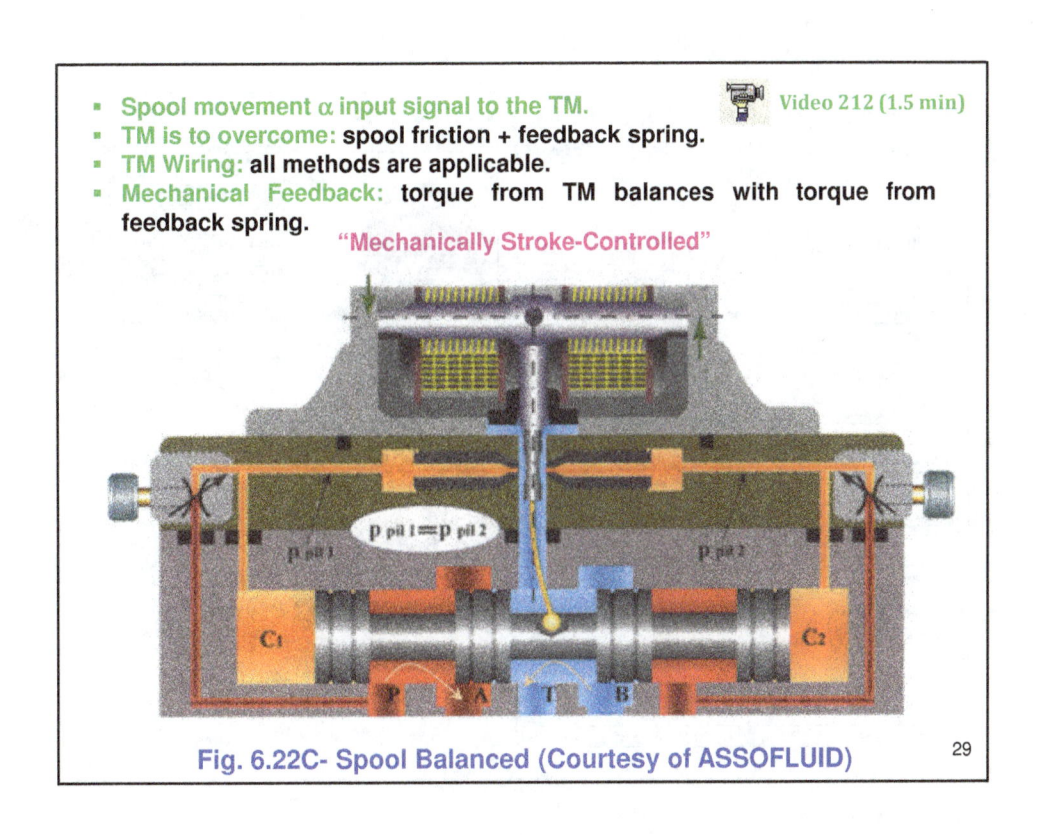

Video 212 (1.5 min)

Fig. 6.22C- Spool Balanced (Courtesy of ASSOFLUID)

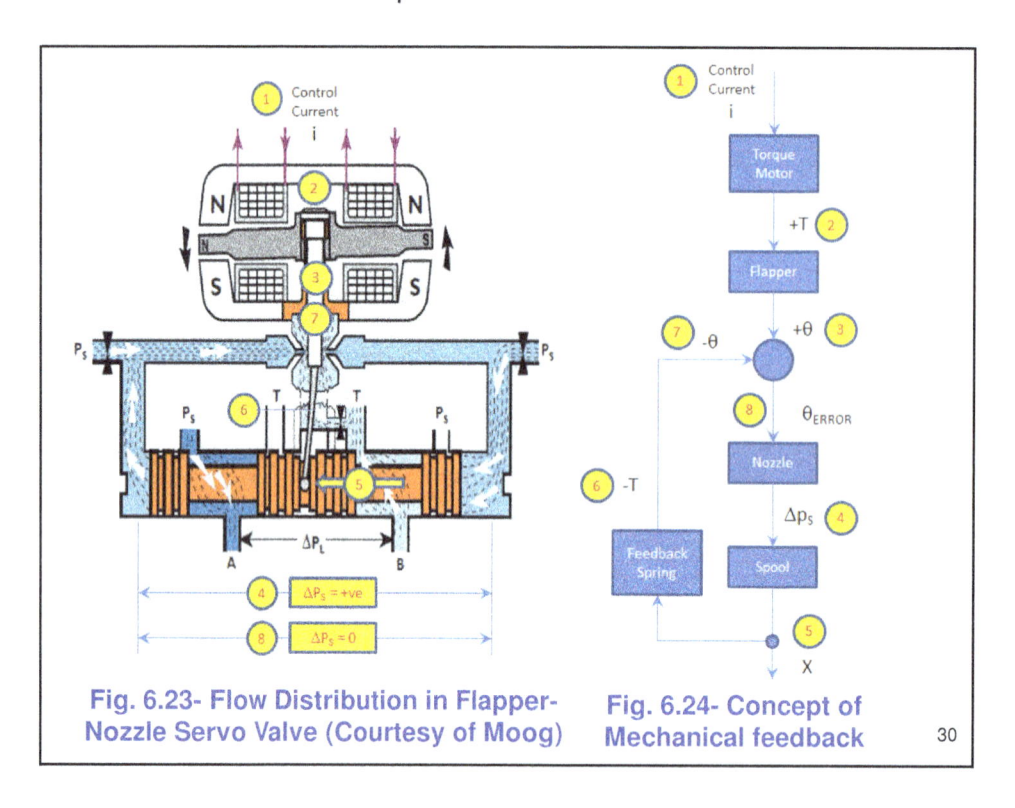

Fig. 6.23- Flow Distribution in Flapper-Nozzle Servo Valve (Courtesy of Moog)

Fig. 6.24- Concept of Mechanical feedback

Fig. 6.25 - Mini Flapper-Nozzle Servo Valve (Courtesy of Parker)

Series SEMT

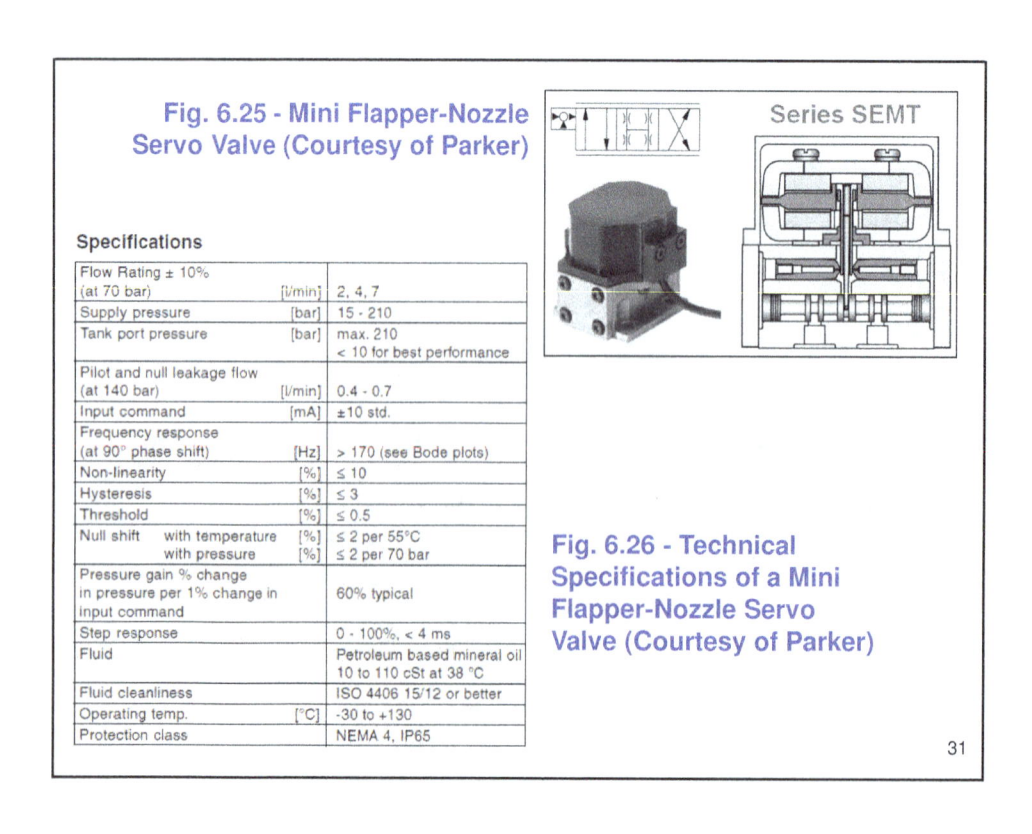

Specifications

Flow Rating ± 10% (at 70 bar)	[l/min]	2, 4, 7
Supply pressure	[bar]	15 - 210
Tank port pressure	[bar]	max. 210 < 10 for best performance
Pilot and null leakage flow (at 140 bar)	[l/min]	0.4 - 0.7
Input command	[mA]	±10 std.
Frequency response (at 90° phase shift)	[Hz]	> 170 (see Bode plots)
Non-linearity	[%]	≤ 10
Hysteresis	[%]	≤ 3
Threshold	[%]	≤ 0.5
Null shift with temperature	[%]	≤ 2 per 55°C
with pressure	[%]	≤ 2 per 70 bar
Pressure gain % change in pressure per 1% change in input command		60% typical
Step response		0 - 100%, < 4 ms
Fluid		Petroleum based mineral oil 10 to 110 cSt at 38 °C
Fluid cleanliness		ISO 4406 15/12 or better
Operating temp.	[°C]	-30 to +130
Protection class		NEMA 4, IP65

Fig. 6.26 - Technical Specifications of a Mini Flapper-Nozzle Servo Valve (Courtesy of Parker)

Valve Features:

- 2-Stage 4/3 servo valve with mechanical feedback.
- 1st stage is nozzle-flapper amplifier.
- For sub-plate mounting.
- Valve to control position, pressure or velocity.
- External electrical control units ECU.
- Valve is factory set and tested.
- Filter of 1st stage freely assembled from outside.

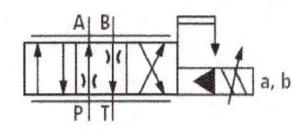

Type 4WS2EM ...XN...-102

Fig. 6.27A-Example of a 2-Stage Flapper–Nozzle Servo Valve with Mechanical Feedback (Courtesy of Bosch Rexroth)

1. Torque Motor
2. Control Chamber
3. Feedback Spring
4. Interchangeable Control Sleeve
5. Control Spool
6. Flapper
7. Two Nozzles
8. Two Adjusting Screws
9. Two Covers

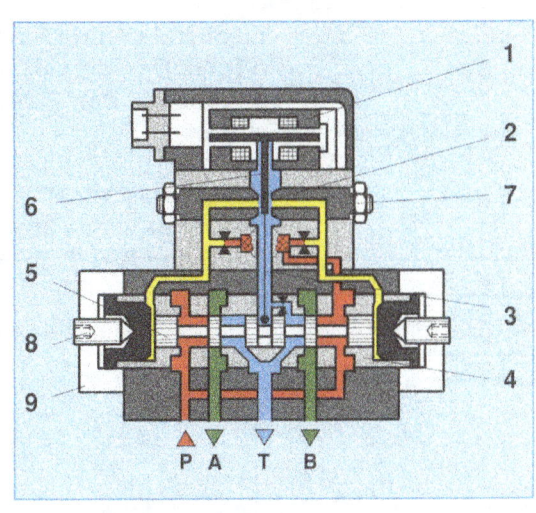

Fig. 6.27B- Detailed Valve Construction (Courtesy of Bosch Rexroth)

hydraulic (measured with HLP32, ϑ_{oil} = 40 °C ± 5 °C)

Operating pressure	Ports P, A, B	bar	10 ... 210 or 10 ... 315
Return flow pressure	Port T	bar	Pressure peaks < 100 static < 10
Hydraulic fluid			Mineral oil (HL, HLP) according to DIN 51524 Ignition temperature > 150 °C
Hydraulic fluid temperature range		°C	–15 ... +70; preferably +40 ... +50
Viscosity range		mm²/s	15 ... 380; preferably 30 ... 45
Cleanliness class according to ISO 4406 (c)			Class 18/16/13 [1]
Rated flows $q_{v\,nom}$, tolerance ±10 % with valve pressure differential Δp = 70 bar		l/min	2; 5; 10; 15; 20; 25

Fig. 6.27C- Valve Hydraulic Specifications (Courtesy of Bosch Rexroth)

---

electric

Type of signal		Analog
Rated current per coil	mA	30
Resistance per coil	Ω	85
Inductivity with 60 Hz and 100% rated current	Serial connection	H 1.0
	Parallel connection	H 0.25

- **Electrical control from A (+) to D (−) provides direction of flow P→A and B → T.**

- **Reverse electrical control provides flow P→ B and A → T.**

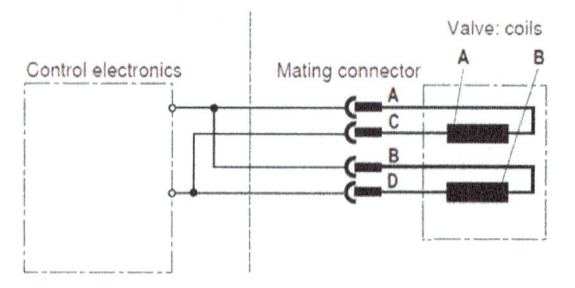

Fig. 6.27D- Valve Electrical Specifications (Courtesy of Bosch Rexroth)

Valve Features:

- 100% factory tested to ensure critical performance specification are met.
- 2-satge design
- Dual coil torque motor.
- Emergency fail-safe position.
- Field replaceable pilot stage filter.
- External null bias adjustment.
- Hardened bushing and spool.
- Carbide ball-in-hall feedback mechanism.

G761/761 Series, Size 04

Fig. 6.28A-Example of a 2-Stage Flapper–Nozzle Servo Valve with Mechanical Feedback (Courtesy of Moog)

General Technical Data

Valve design	2-stage, with spool and busing and dry torque motor
Pilot stage	Nozzle Flapper
Mounting pattern	ISO 10372-04-04-0-92
Installation position	Any orientation, fixed or movable
Weight	1.08 kg (2.4 lb)
Storage temperature range	-40 to +60 °C (-40 to +140 °F)
Ambient temperature range	-40 to +135 °C (-40 to +275 °F)
Vibration resistance	30 g, 3 axis, 10 Hz to 2 kHz
Shock resistance	30 g, 3 axis
Seal material	Fluorocarbon (FKM) 85 SHORE D Others upon request

Fig. 6.28B- Valve General Specifications (Courtesy of Moog)

Hydraulic Data

Maximum operating pressure to ports P, T, A, B, X	315 bar (4,500 psi)
Rated flow at p_N 35 bar/spool land (500 psi/spool land)	4/10/19/38/63 l/min (1/2.5/5/10/16.5 gpm)
Maximum main stage leakage flow rate (zero lap)	2.3 l/min (0.60 gpm)
Null adjust authority	Greater than 10% of rated flow
Hydraulic fluid	Hydraulic oil as per DIN 51524 parts 1 to 3 and ISO 11158. Other fluids on request.
Temperature range	-40 to +60 °C (-40 to +140 °F)
Recommended viscosity range	10 to 97 mm^2/s (cSt)
Maximum permissible viscosity range	5 to 1,250 mm^2/s (cSt)
Recommended cleanliness class as per ISO 4406	
For functional safety	17/14/11
For longer life	15/13/10
Recommended filter rating	
For functional safety	$ß_{10} \leq 75$ (10µm absolute)
For longer life	$ß_5 \leq 75$ (5 µm absolute)

Fig. 6.28C- Valve Hydraulic Specifications (Courtesy of Moog)

6.3.7- Two-Stage Flapper–Nozzle Servo Valve with Electrical Feedback

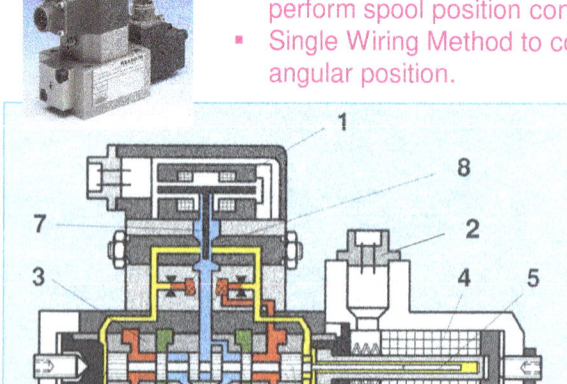

- System type 1 - used with a PD controller to perform spool position control.
- Single Wiring Method to control flapper angular position.

"Stroke-Controlled"

1. Torque Motor
2. Feedback Output
3. Interchangeable Control Sleeve
4. inductive position transducer
5. Core
6. Control Spool
7. Flapper
8. Two Nozzles
9. Two Control Chambers

Fig. 6.29- Example of a 2-Stage Flapper–Nozzle Servo Valve with Electrical Feedback (Courtesy of Bosch Rexroth)

6.3.8- Two-Stage Flapper–Nozzle Servo Valve with Mechanical and Electrical Feedback

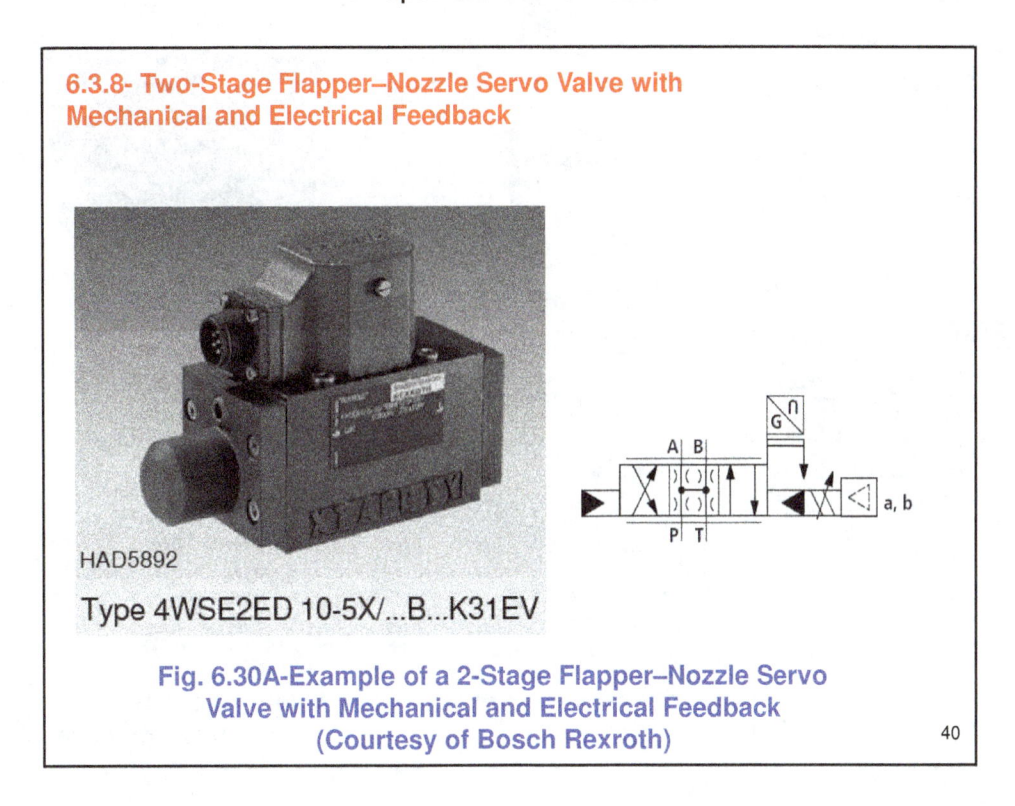

Fig. 6.30A-Example of a 2-Stage Flapper–Nozzle Servo Valve with Mechanical and Electrical Feedback (Courtesy of Bosch Rexroth)

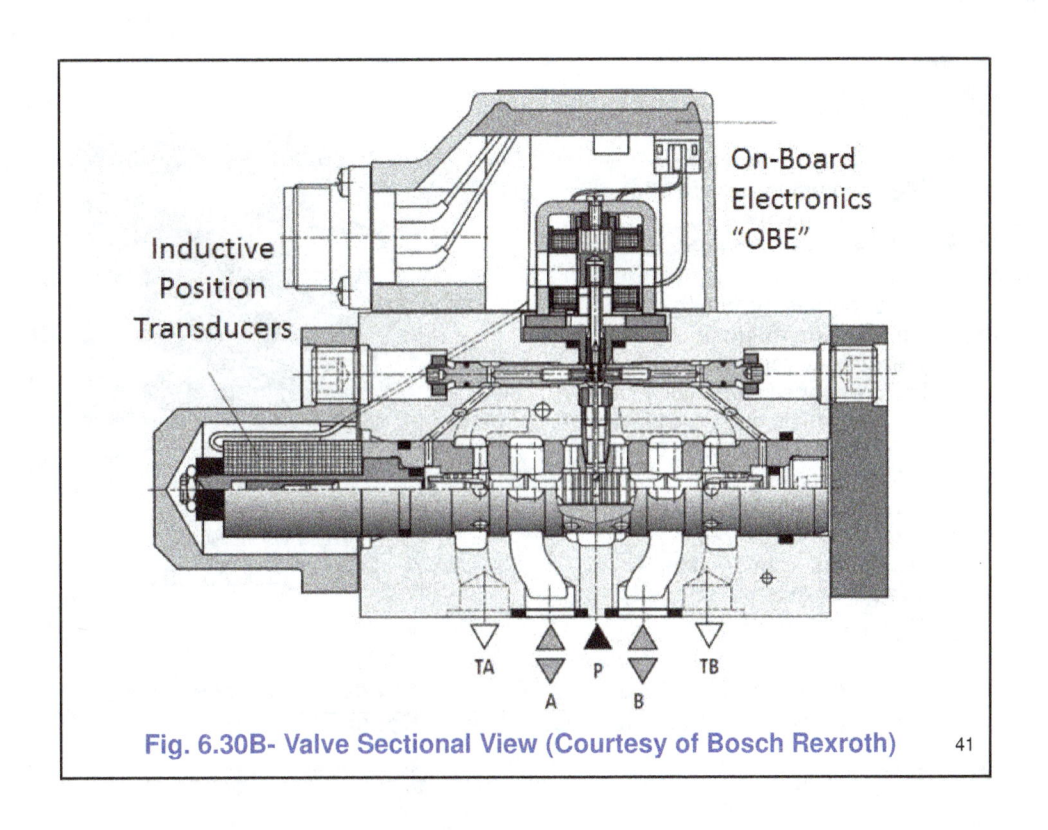

Fig. 6.30B- Valve Sectional View (Courtesy of Bosch Rexroth)

Workbook: Electro-Hydraulic Components and Systems
Chapter 6: Servo Valves

6.3.9- Three-Stage Flapper–Nozzle Servo Valve

"Stage 1"
"Stage 2"
"Stage 3"

1. Main Spool
2. Main Sleeve
3. Two Control Chambers
4. inductive position transducer
5. Core

X T A P B Y

Fig. 6.31-Conceptual Construction of a 3-Stage Flapper–Nozzle Servo Valve with Electrical Feedback (Courtesy of Bosch Rexroth)

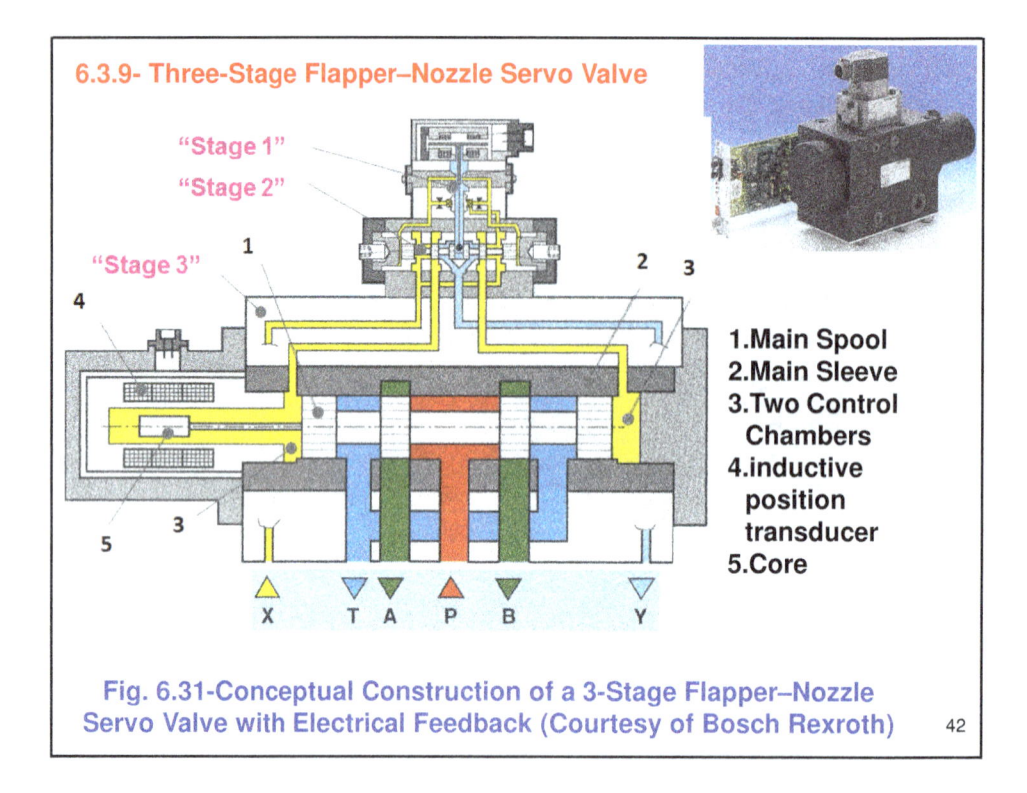

Valve Features:

- 3-Stage 4/3 servo valve.
- High dynamics 2-stage pilot control valve of size 6
- Main stage with electrical feedback.
- For sub-plate mounting.
- Valve to control position, pressure or velocity.
- High response sensitivity, very low hysteresis and zero point drift
- External electrical control units ECU.
- Valve is factory set and tested.
- Filter of 1st stage freely assembled from outside.
- Internal or external pilot oil supply and return

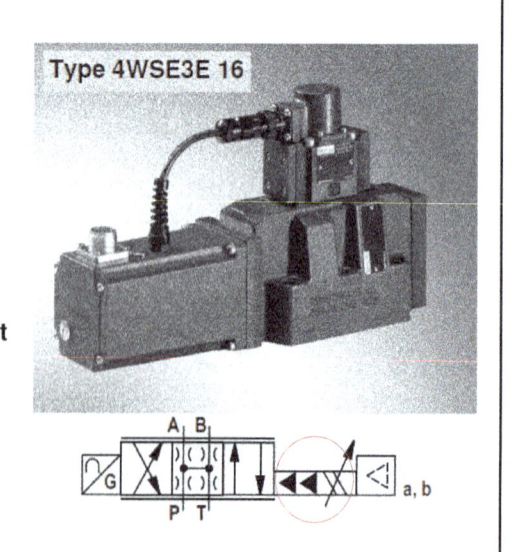

Type 4WSE3E 16

Fig. 6.32A-Example of a 3-Stage Flapper–Nozzle Servo Valve with Electrical Feedback (Courtesy of Bosch Rexroth)

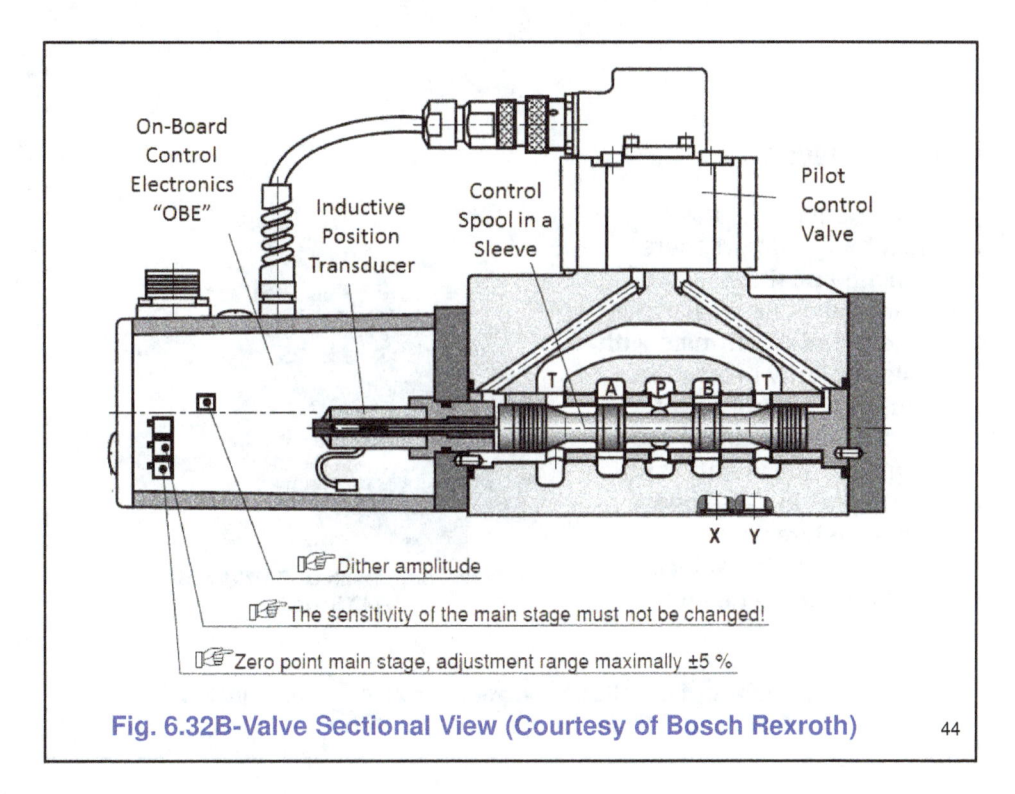

Workbook: Electro-Hydraulic Components and Systems
Chapter 6: Servo Valves

Fig. 6.32B-Valve Sectional View (Courtesy of Bosch Rexroth)

☞ Dither amplitude

☞ The sensitivity of the main stage must not be changed!

☞ Zero point main stage, adjustment range maximally ±5 %

Labels: On-Board Control Electronics "OBE"; Inductive Position Transducer; Control Spool in a Sleeve; Pilot Control Valve; T A P B T; X Y

☞ **Notice!**

Changes in the zero point and/or the dither amplitude may result in damage to the system and may only be implemented by instructed specialists.

The pilot control valve may only be maintained by Bosch Rexroth employees. An exception to this is the replacement of the filter element – see data sheet 29564.

Valve Features:

- Electrical position feedback with pressure isolated position transducer, no wear.
- Integrated electronics with false polarity protection.
- Optional external pilot supply and return connections via fifth and sixth port in valve body.
- Low hysteresis.
- Excellent null stability.
- Pre-adjusted at factory.

Servo Valves with OBE D791 and D792 Series

Fig. 6.33A-Example of a 3-Stage Flapper–Nozzle Servo Valve with Electrical Feedback (Courtesy of Moog)

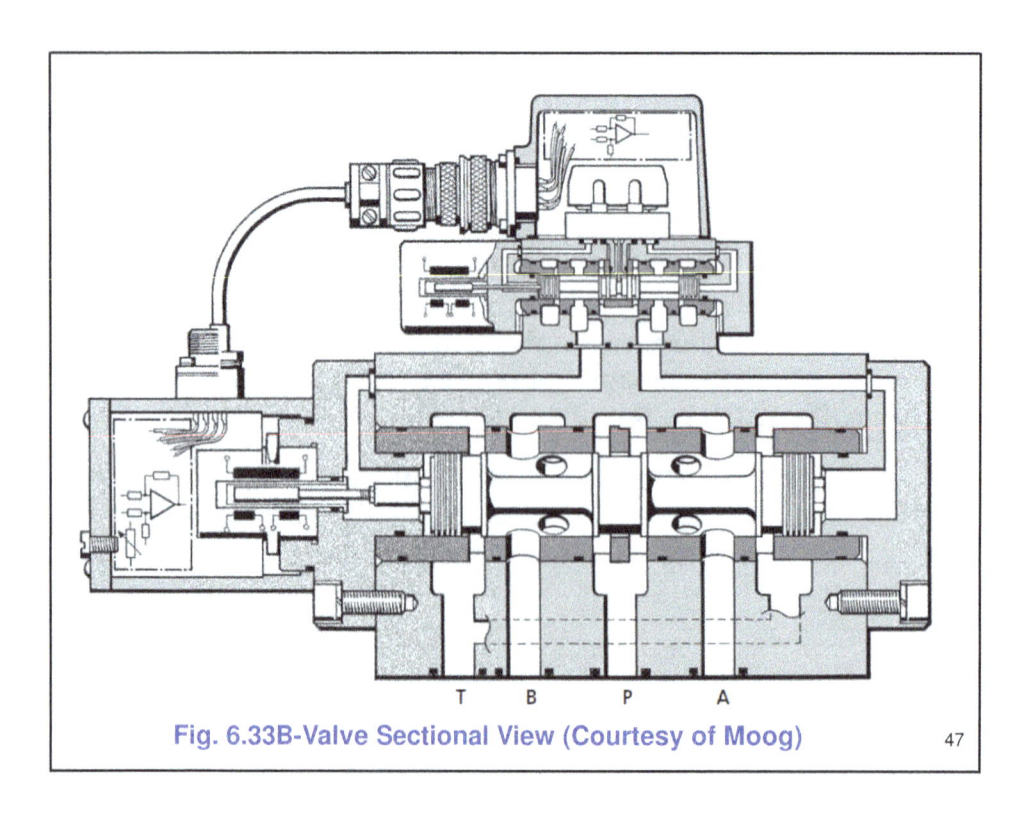

T B P A

Fig. 6.33B-Valve Sectional View (Courtesy of Moog)

Workbook: Electro-Hydraulic Components and Systems
Chapter 6: Servo Valves

6.3.10 - Jet Pipe Concept of Operation

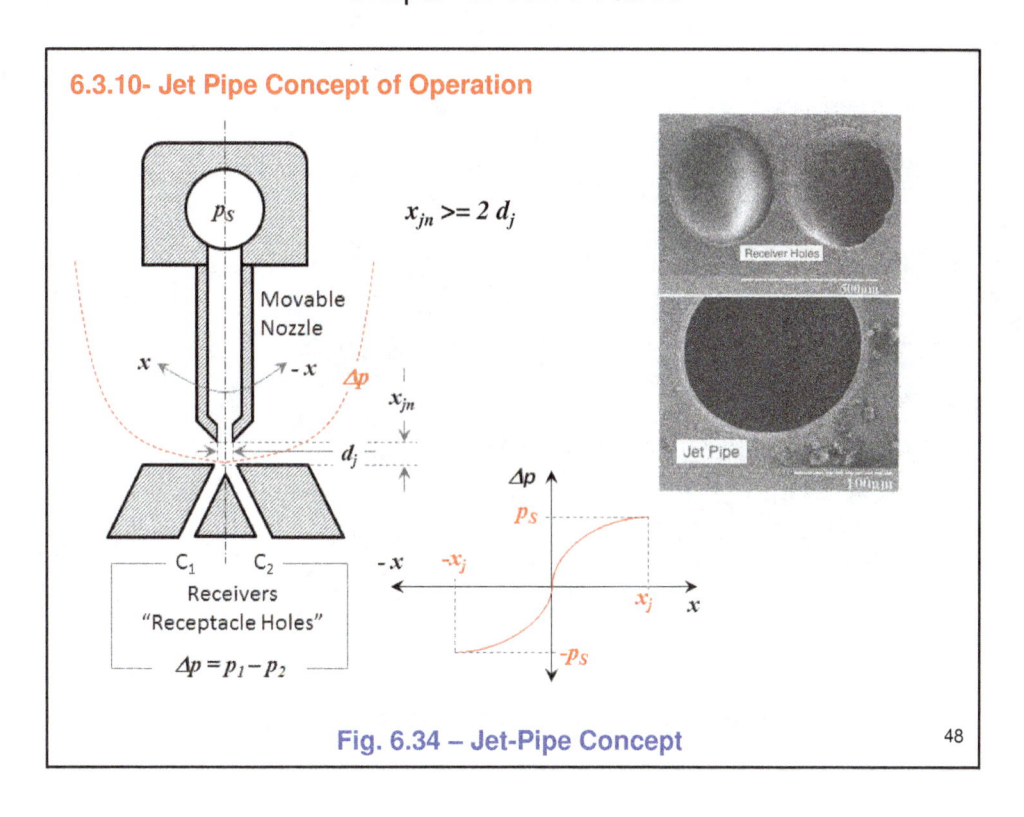

$x_{jn} >= 2\, d_j$

$\Delta p = p_1 - p_2$

Fig. 6.34 – Jet-Pipe Concept

6.3.11 - Two-Stage Jet-Pipe Servo Valve with Mechanical Feedback

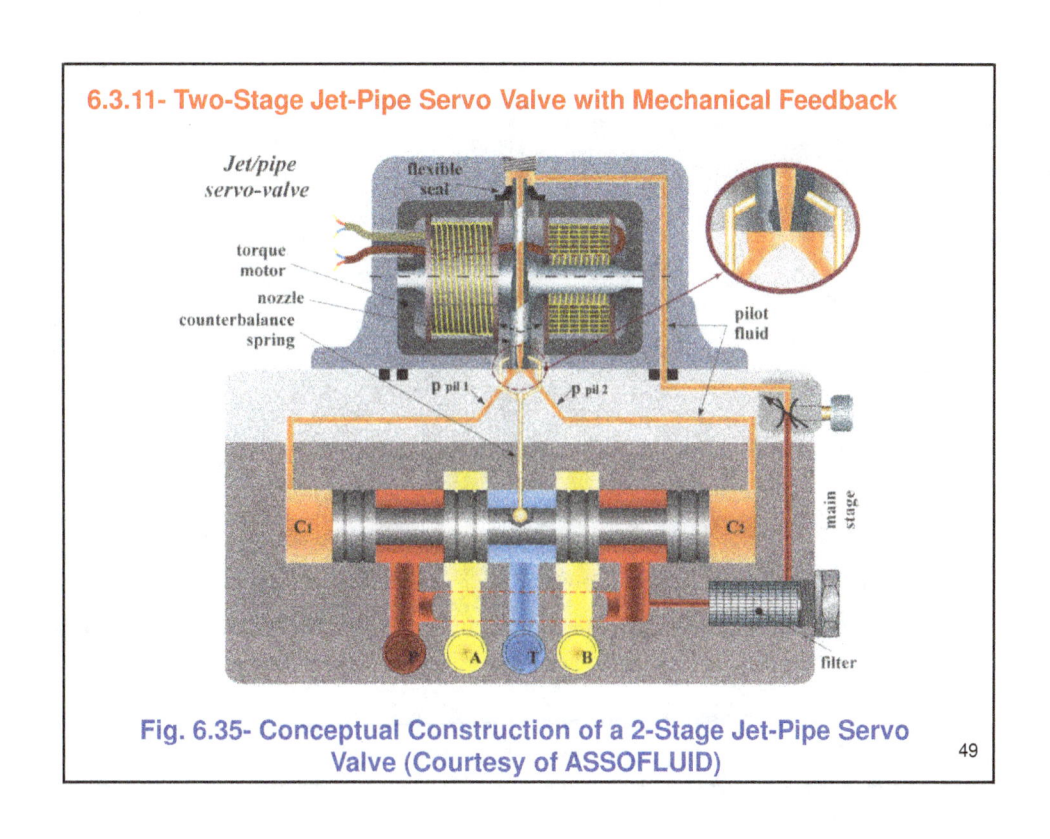

Fig. 6.35 - Conceptual Construction of a 2-Stage Jet-Pipe Servo Valve (Courtesy of ASSOFLUID)

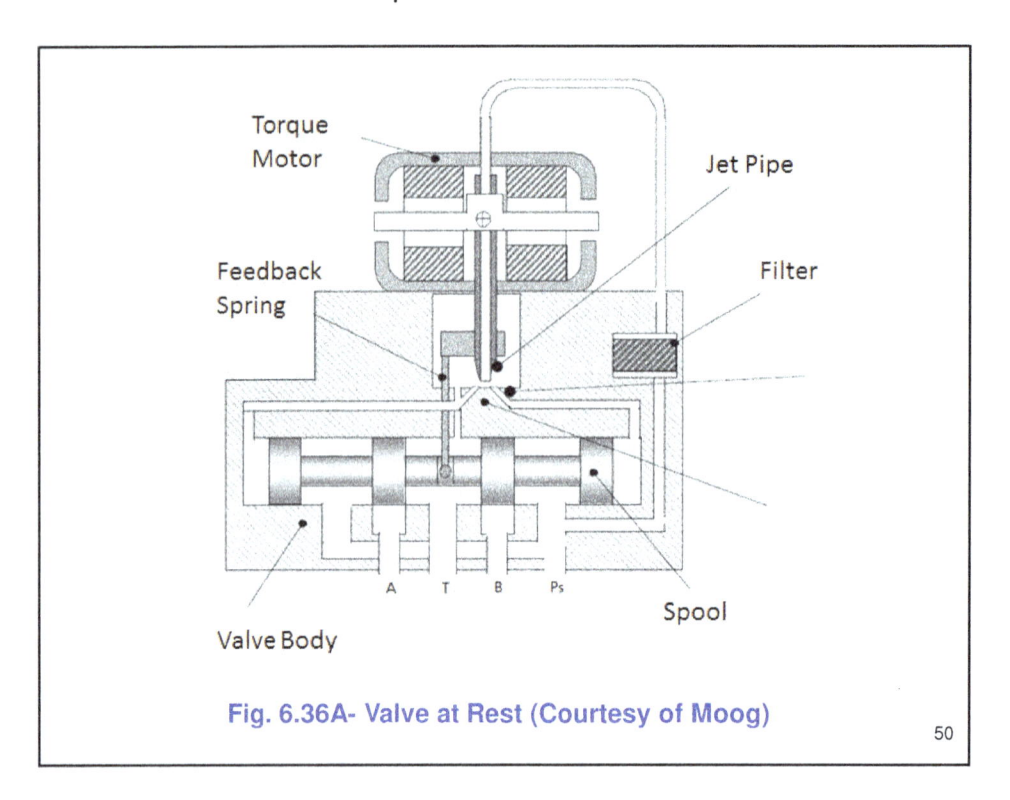

Workbook: Electro-Hydraulic Components and Systems
Chapter 6: Servo Valves

Fig. 6.36A- Valve at Rest (Courtesy of Moog)

Labels: Torque Motor, Jet Pipe, Feedback Spring, Filter, Valve Body, Spool, A T B Ps

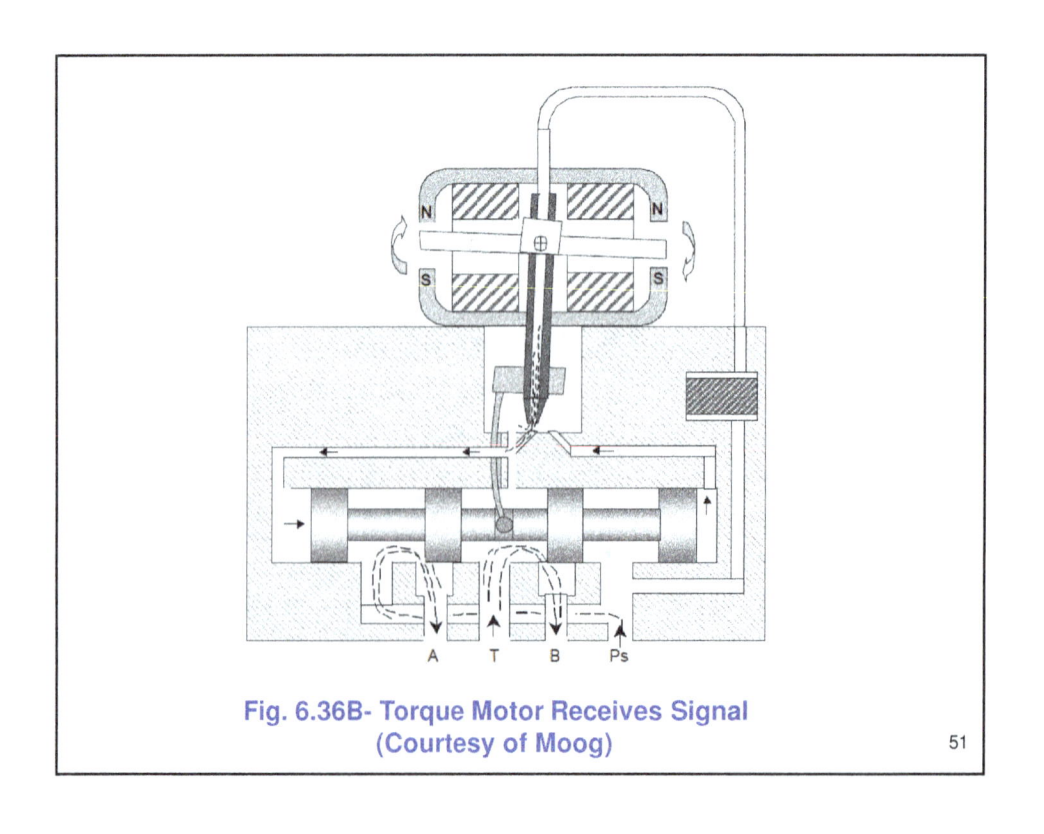

Fig. 6.36B- Torque Motor Receives Signal (Courtesy of Moog)

A T B Ps

Workbook: Electro-Hydraulic Components and Systems
Chapter 6: Servo Valves

Fig. 6.36C- Spool Balanced (Courtesy of Moog)

Specifications

Hydraulic Data

Rated Flow	0 to 225 l/min (0 to 60 gpm) @ Δp 70 bar (1000 psi)
Maximum Operating Pressure	350 bar (5,000 psi)
Mounting Pattern	ISO 10372
100% Step Response	4 to 32 ms @ 210 bar (3,000 psi)

Electrical Data

Supply Voltage	N/A
Rated Signal Options	+/-10, +/-20 and +/-50 mA

Legacy Product Series

Fig. 6.37-Example of a 2-Stage Jet-Pipe Servo Valve with Mechanical Feedback (Courtesy of Moog)

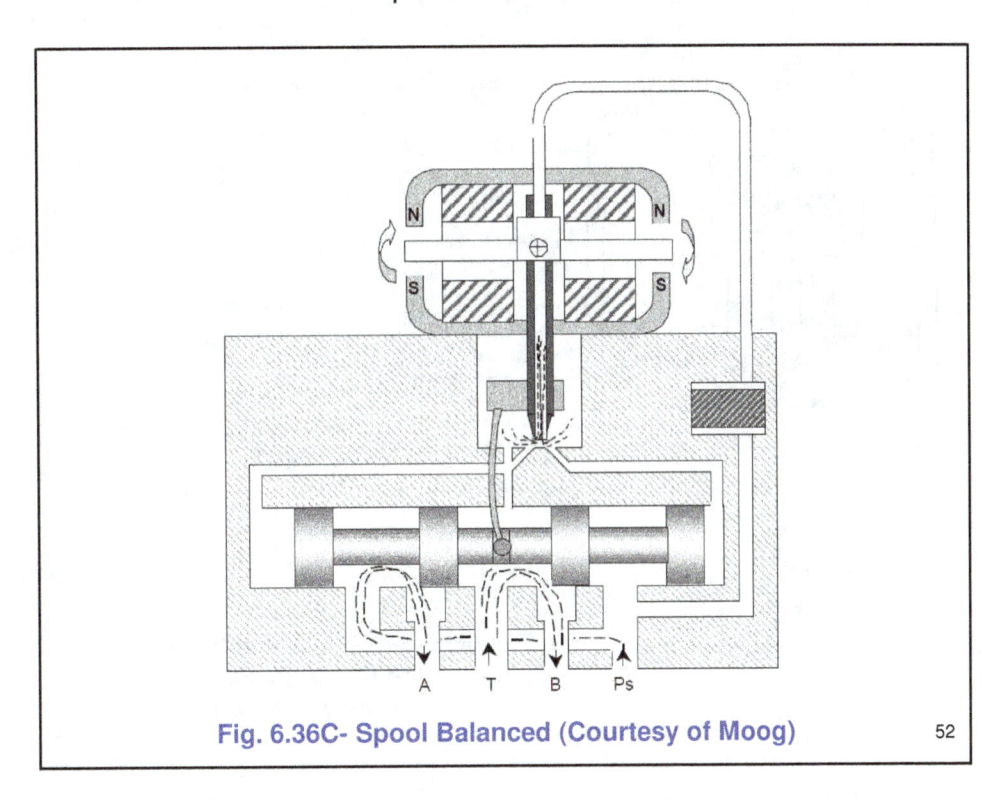

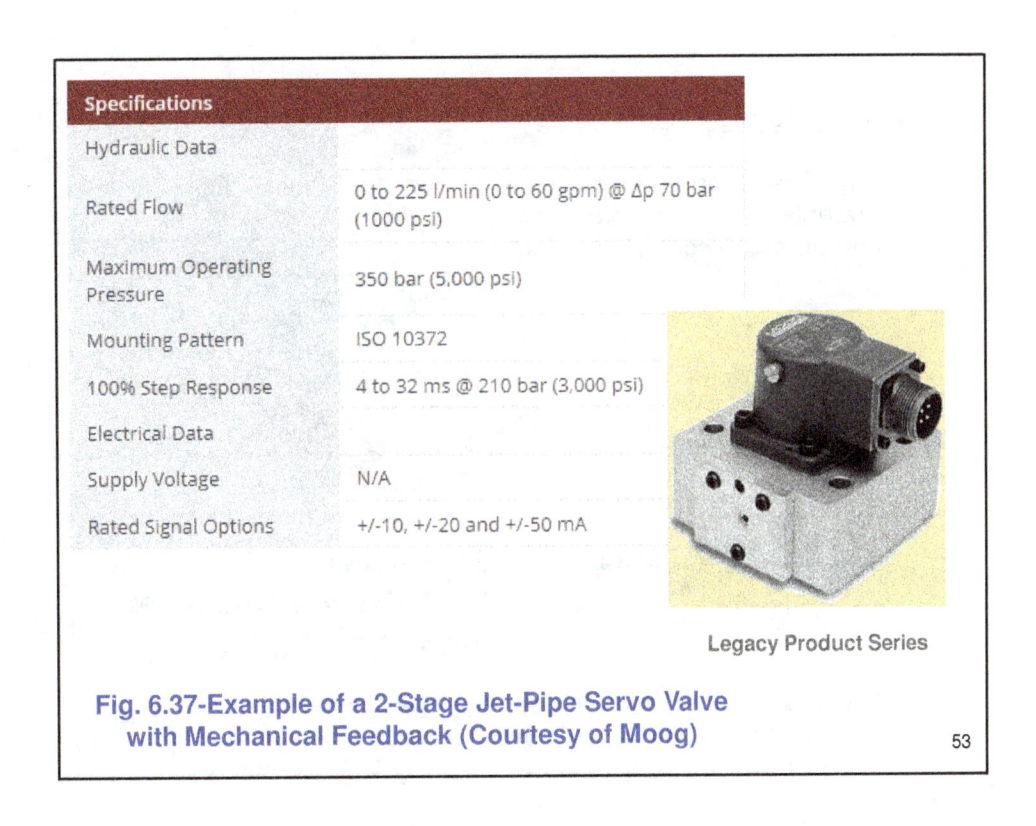

6.3.12- Two-Stage Jet Pipe Servo Valve with Electrical Feedback

1. Coil
2. Armature
3. Jet pipe
4. Annular space below the nozz
5. Pilot ports
6. Receiver holes
7. Nozzle
8. Receiver
9. Drain

"Servo Jet"

Fig. 6.38- Jet-Pipe Stage for Servo Valve with Electrical Feedback (Courtesy of Moog)

Valve Features:

- High Response.
- Suitable for position, velocity, and pressure control.

Servovalve D661 Highresponse Series with SERVOJET pilot stage

Fig. 6.39A-Example of a 2-Stage Jet-Pipe Servo Valve with Electrical Feedback (Courtesy of Moog)

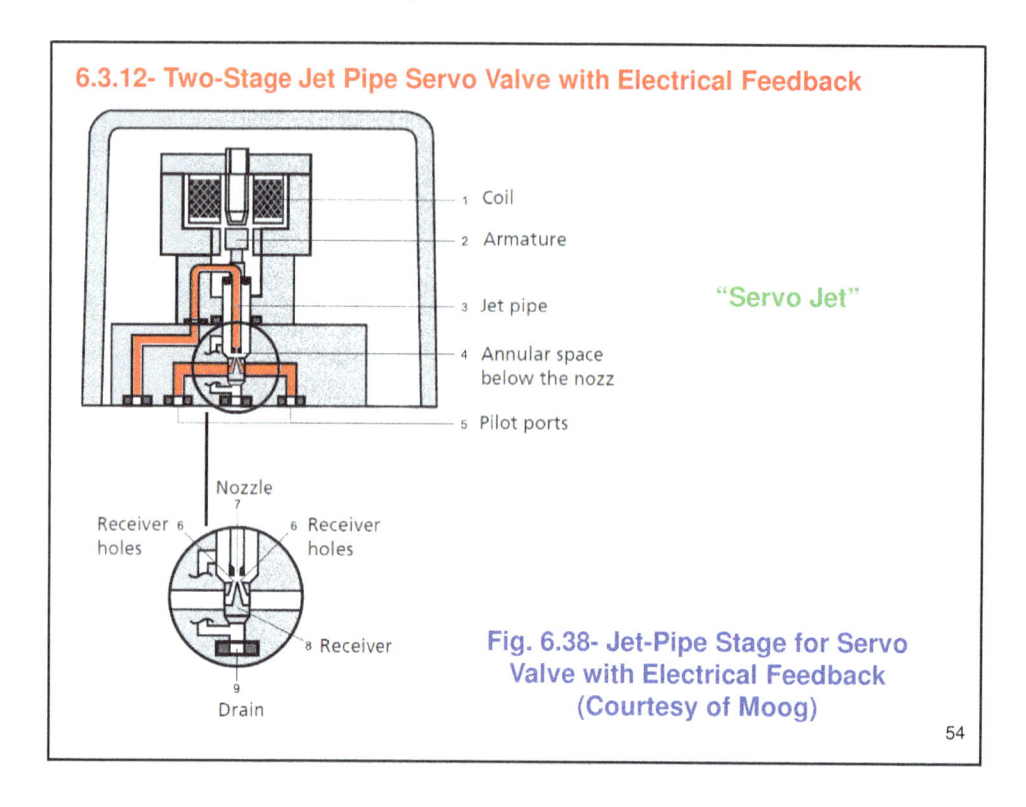

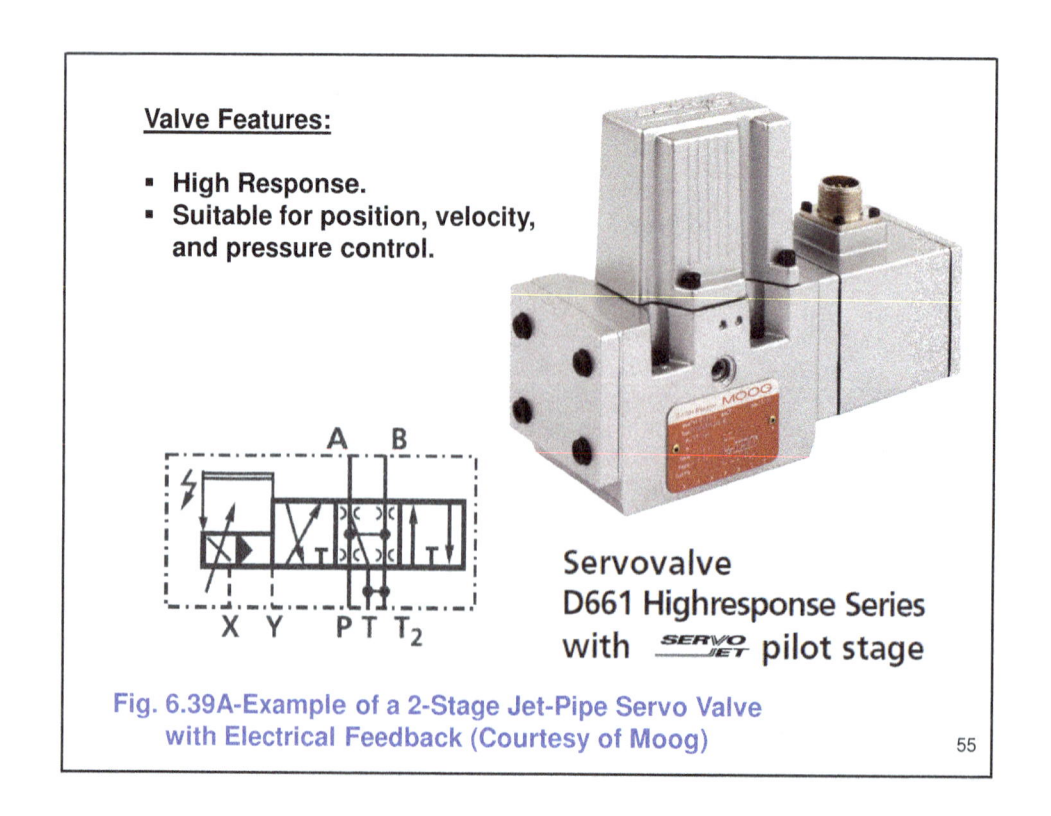

Fig. 6.39B-Valve Sectional View (Courtesy of Moog)

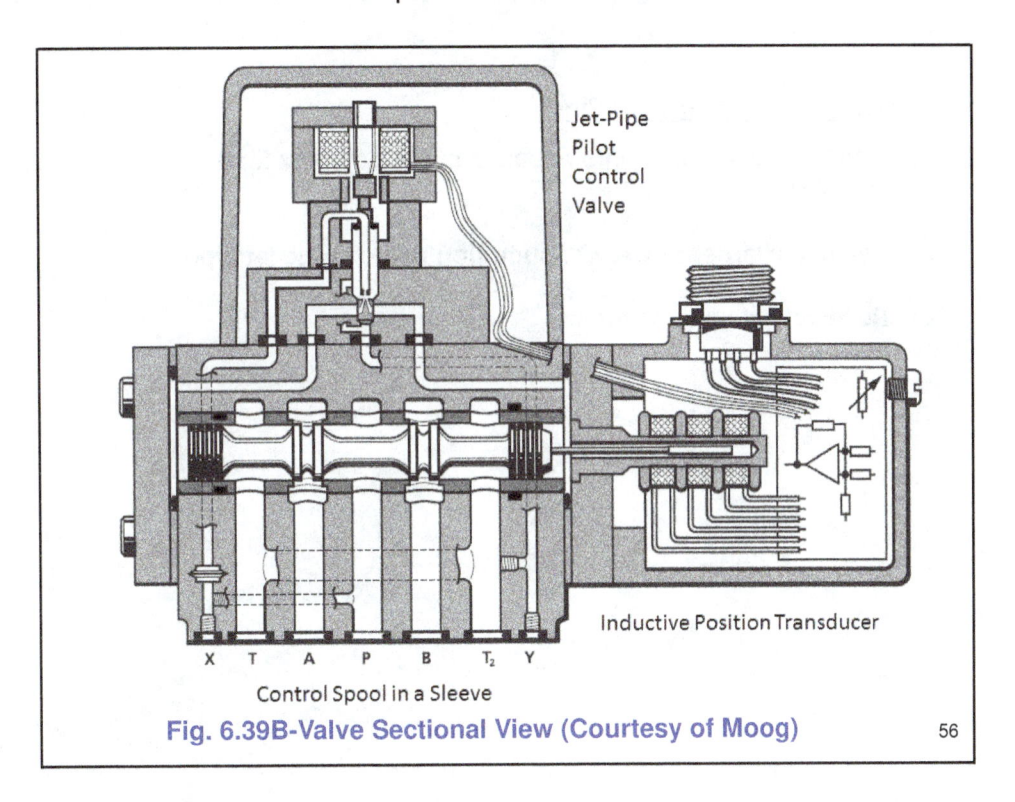

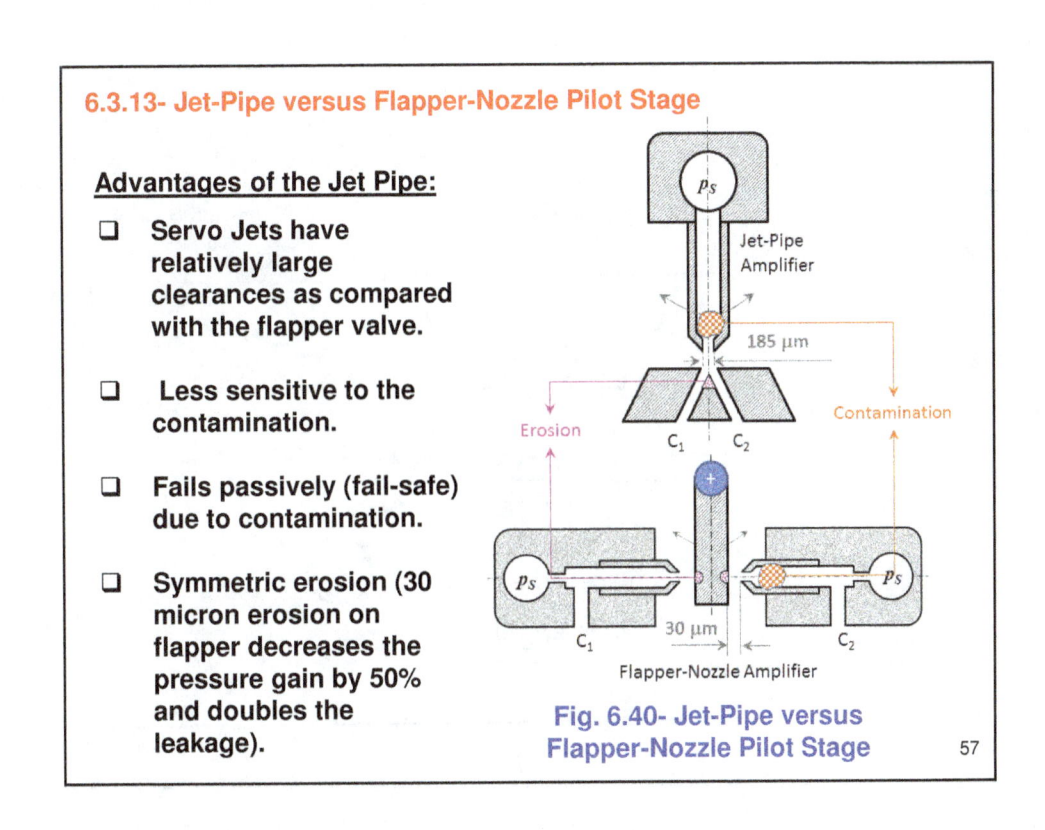

6.3.13- Jet-Pipe versus Flapper-Nozzle Pilot Stage

Advantages of the Jet Pipe:

- Servo Jets have relatively large clearances as compared with the flapper valve.

- Less sensitive to the contamination.

- Fails passively (fail-safe) due to contamination.

- Symmetric erosion (30 micron erosion on flapper decreases the pressure gain by 50% and doubles the leakage).

Fig. 6.40- Jet-Pipe versus Flapper-Nozzle Pilot Stage

Disadvantages of the Jet Pipe:

- Difficult null adjustments because of the complex fluid flow characteristics.

- Relatively larger power consumption to move the jet pipe.

- Relatively slower dynamics.

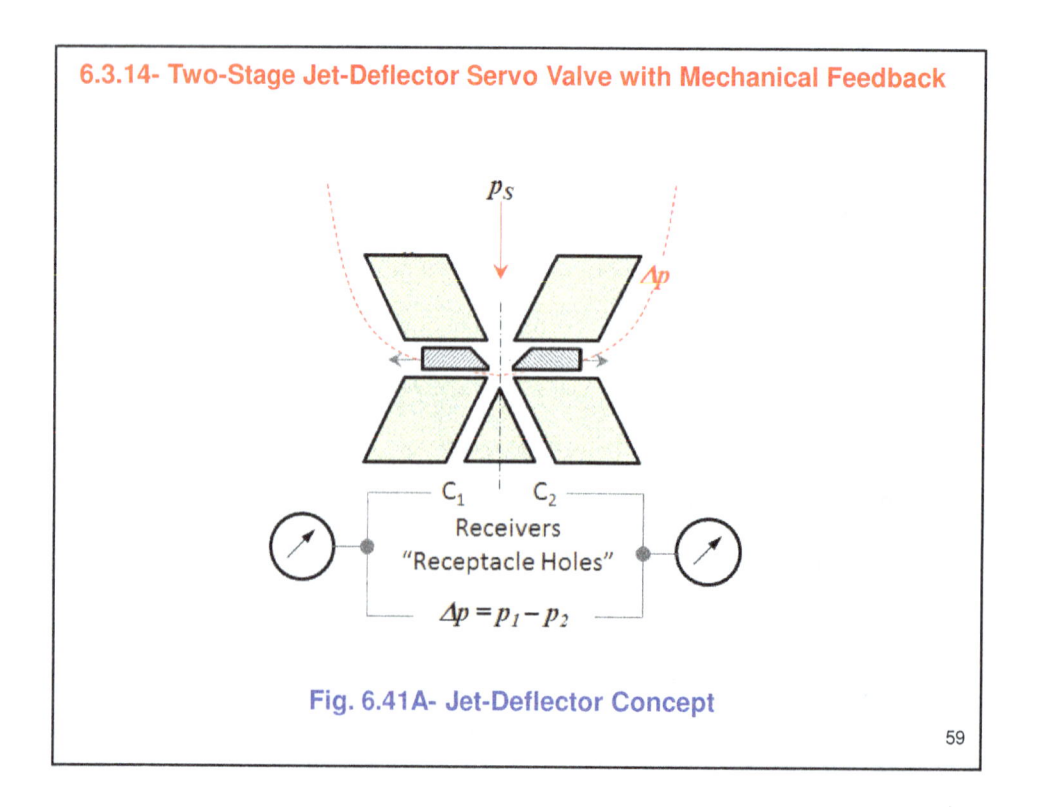

6.3.14- Two-Stage Jet-Deflector Servo Valve with Mechanical Feedback

p_S

Δp

C_1 C_2
Receivers
"Receptacle Holes"

$\Delta p = p_1 - p_2$

Fig. 6.41A- Jet-Deflector Concept

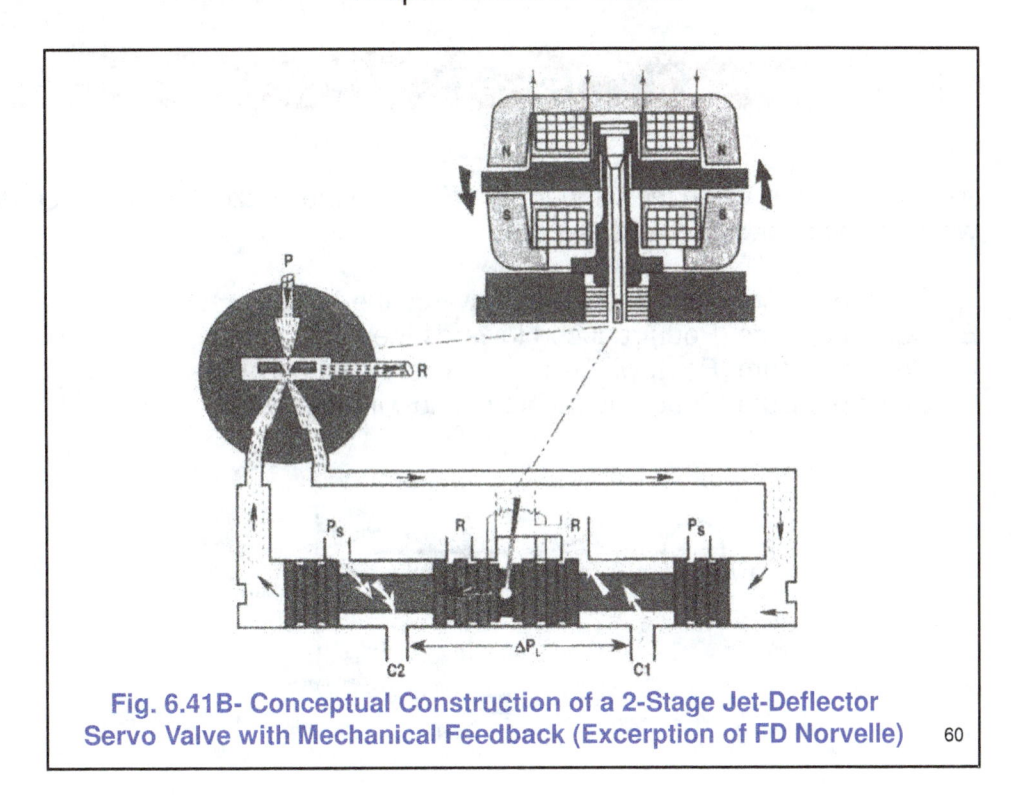

Workbook: Electro-Hydraulic Components and Systems
Chapter 6: Servo Valves

Fig. 6.41B- Conceptual Construction of a 2-Stage Jet-Deflector Servo Valve with Mechanical Feedback (Excerption of FD Norvelle)

Workbook: Electro-Hydraulic Components and Systems
Chapter 6: Reviews and Assignments

Chapter 6 Reviews

1. As shown below, when the flapper gets closer to the nozzle (distance X reduces), which of the following effects is **True**?

    A. Outlet pressure (Pout) increased towards the supply pressure (Ps).
    B. Outlet pressure (Pout) reduced towards the atmospheric pressure.
    C. Outlet pressure (Pout) will became half of the supply pressure (Ps).
    D. Outlet pressure (Pout) has nothing to do with the flapper movement.

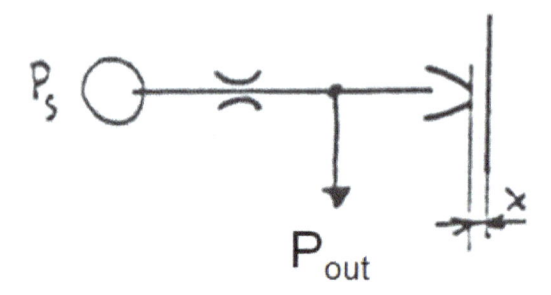

2. Servo valves are faster than proportional valves because?

    A. Torque motors are faster than the proportional solenoid.
    B. Servo valves have a built-in internal feedback mechanism that assures the correct position of the spool relative to the input command.
    C. Pilot pressure builds up inside a servo valve against the main spool faster than in a proportional valve.
    D. All of the above.

3. The figure shown below represents which of the following types of torque motor wiring?

    A. Parallel connection?
    B. Series connection?
    C. Single connection?
    D. None of the above.

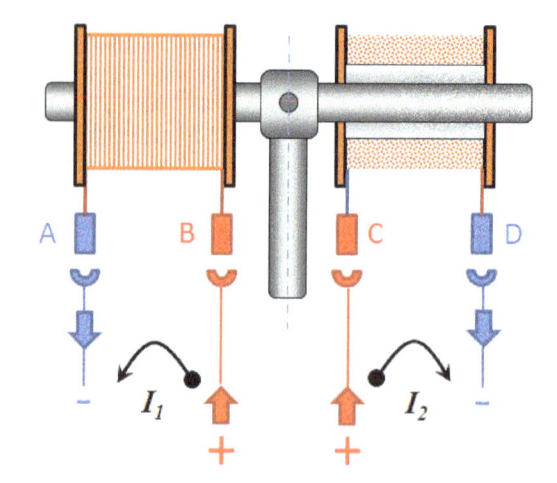

210

Workbook: Electro-Hydraulic Components and Systems
Chapter 6: Reviews and Assignments

4. The parallel connection is preferred over the other types of torque motor wiring because?

   A. It is the simplest type of wiring since only 2 legs will be wired.
   B. It offers safer operation. If one coil is broken, the other coil will work. It also offers the least power consumption.
   C. Coils are powered independently so that the angular position can be held without mechanical feedback.
   D. It drives the torque motor faster.

5. For the shown below servo valve, what torque motor wiring method should be used?

   A. Parallel connection?
   B. Series connection?
   C. Single connection?
   D. None of the above.

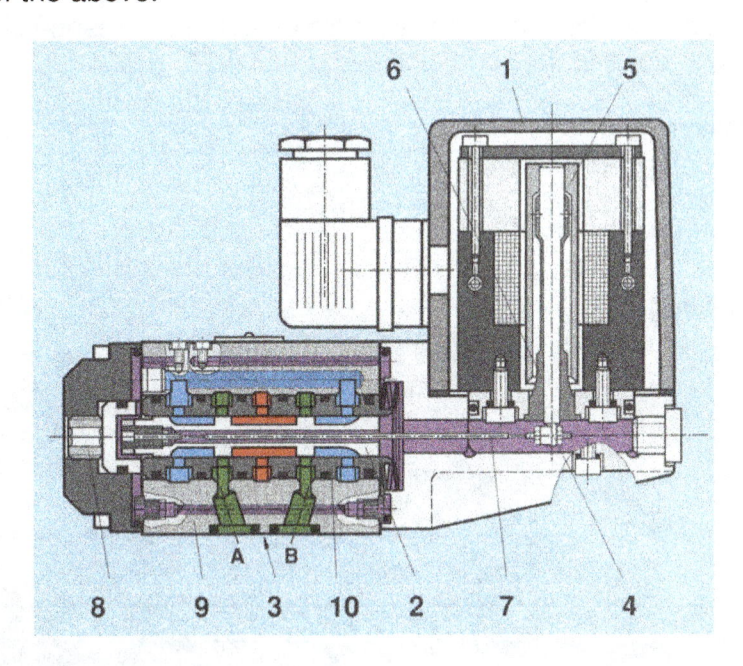

6. What is the power gain of a 2-stage servo valve that has a torque motor of 0.1 Watt power consumption, a flapper-nozzle produces 10 Watt, and the main stage handles hydraulic power of 1000 Watt?

   A. 0.1.
   B. 100.
   C. 1000.
   D. 10,000.

211

Workbook: Electro-Hydraulic Components and Systems
Chapter 6: Reviews and Assignments

7. In the servo valve shown below, what is the type of the pilot stage?

    A. Jet-deflection type.
    B. Jet-pipe type.
    C. Flapper-nozzle type
    D. Torque motor.

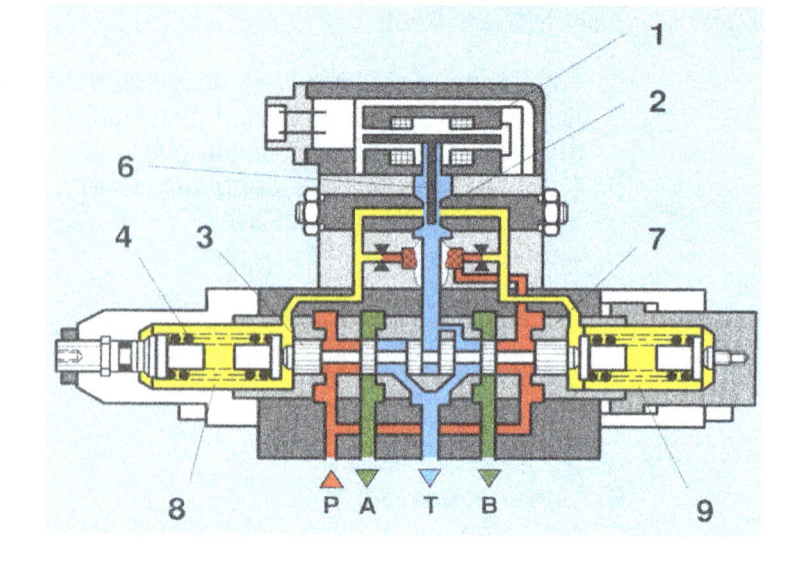

8. In the servo valve shown below, the element that is responsible for balancing the torque motor and the spool position is named?

    A. Main stage.
    B. Counterbalance spring.
    C. Nozzle.
    D. Flapper.

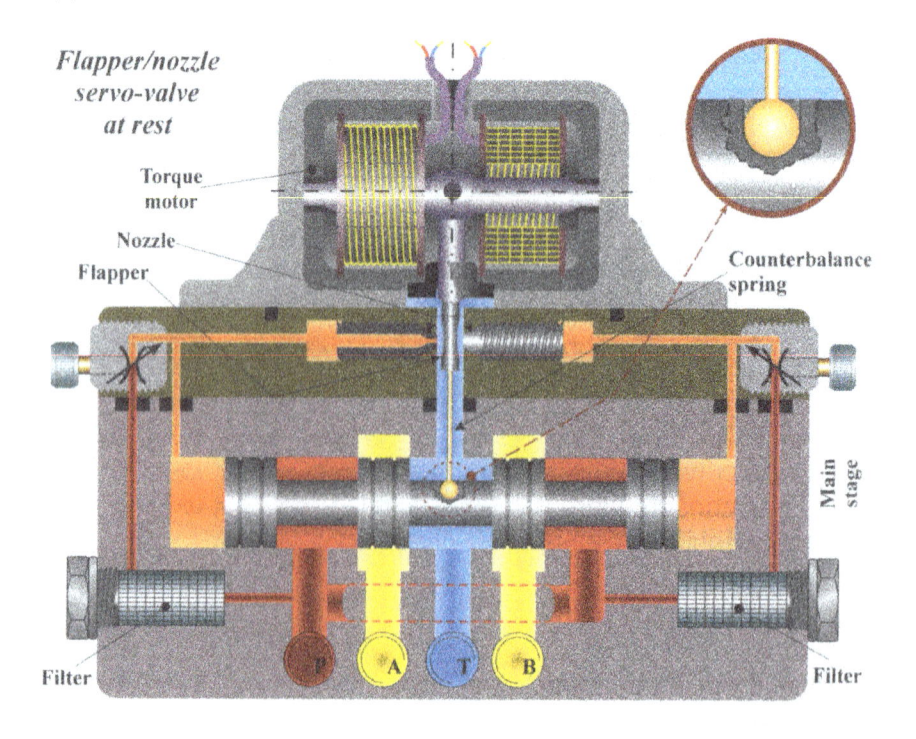

9. What type of feedback is the servo valve shown below?

   A. Mechanical Feedback.
   B. Electrical Feedback.
   C. Mechanical and Electrical Feedback.
   D. No feedback.

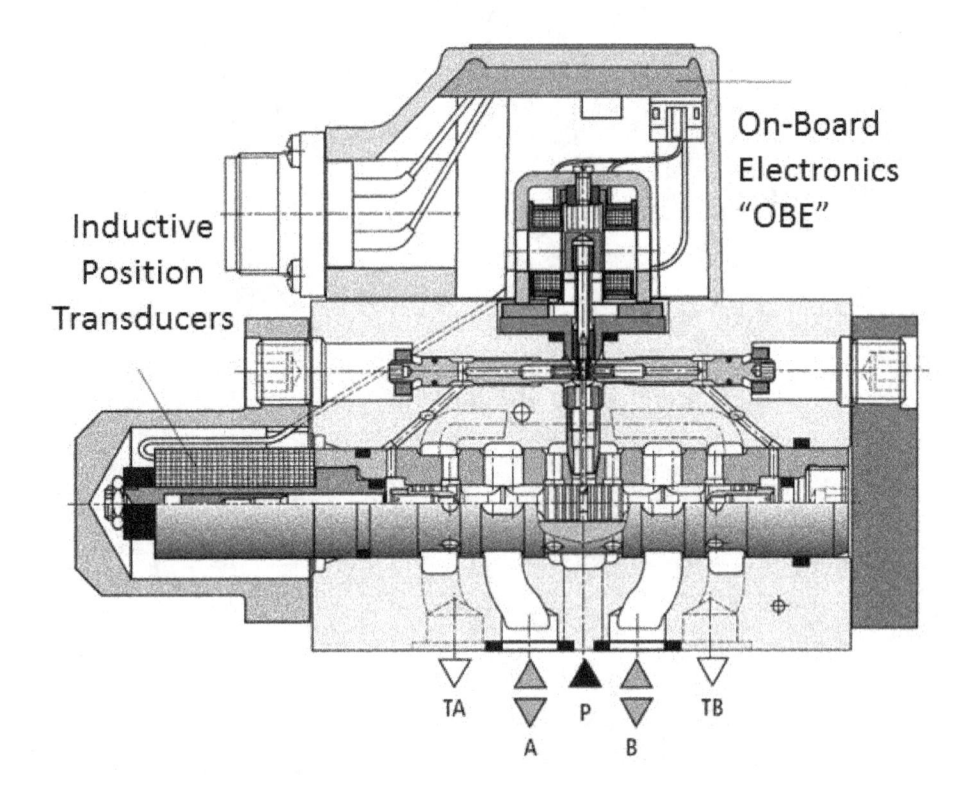

10. From safety point of view, a Jet-Pipe pilot stage is preferred over a flapper-nozzle one because?

    A. Jet-pipe pilot stage responds faster.
    B. Jet-pipe pilot stage consumes less power.
    C. Jet-pipe pilot stage has larger clearances.
    D. Jet pipe pilot stage fails passively terminating the pilot pressure from the main stage when the jet is contaminated.

Workbook: Electro-Hydraulic Components and Systems
Chapter 6: Reviews and Assignments

Chapter 6 Assignment

Student Name: -- Student ID: ------------------

Date: -- Score: ------------------------

A- Draw a simple diagram for a 2-stage servo valve that has flapper-nozzle pilot stage with electrical feedback

B- Draw a simple diagram for the suitable wiring method of the torque motor used with the valve

C- Calculate the power gain of the valve if the torque motor consumes 0.1 Watt, the pilot stage produces 10 Watt, main stage pressure is 100 bar, and main stage flow is 6 liter/min.

Chapter 7
Electro-Hydraulic Valve Selection Criteria

Objectives:

This chapter presents the various criteria to select a specific valve for an application. These criteria include, valve type, spool design, operating conditions, static and dynamic characteristics. The chapter also provides examples of the current valves that are produced by existing manufacturers.

Brief Contents:

7.1- Importance of Proper Selection of an EH Valve
7.2- Electro-Hydraulic Valve Selection Criteria
7.3- Valve Type
7.4- Valve Spool Design
7.5- Valve Static Characteristics
7.6- Valve Dynamic Characteristics
7.7- Valve Operating Conditions
7.8- Examples of Published Data

Workbook: Electro-Hydraulic Components and Systems
Chapter 7: Electro-Hydraulic Valve Selection Criteria

7.1- Importance of Proper Selection of an EH Valve

Proper EH valve selection is the most critical and important action item in designing an electro-hydraulic closed loop control system, based on which the entire system performance will be determined.

- Proper valve selection will result in:

  o Optimized closed loop overall gain to maintain stable and responsive system.

  o Better control resolution.

  o Reduced energy wasting and heat generation in the system.

Fig. 7.1- Ammunition Handling and Control Systems

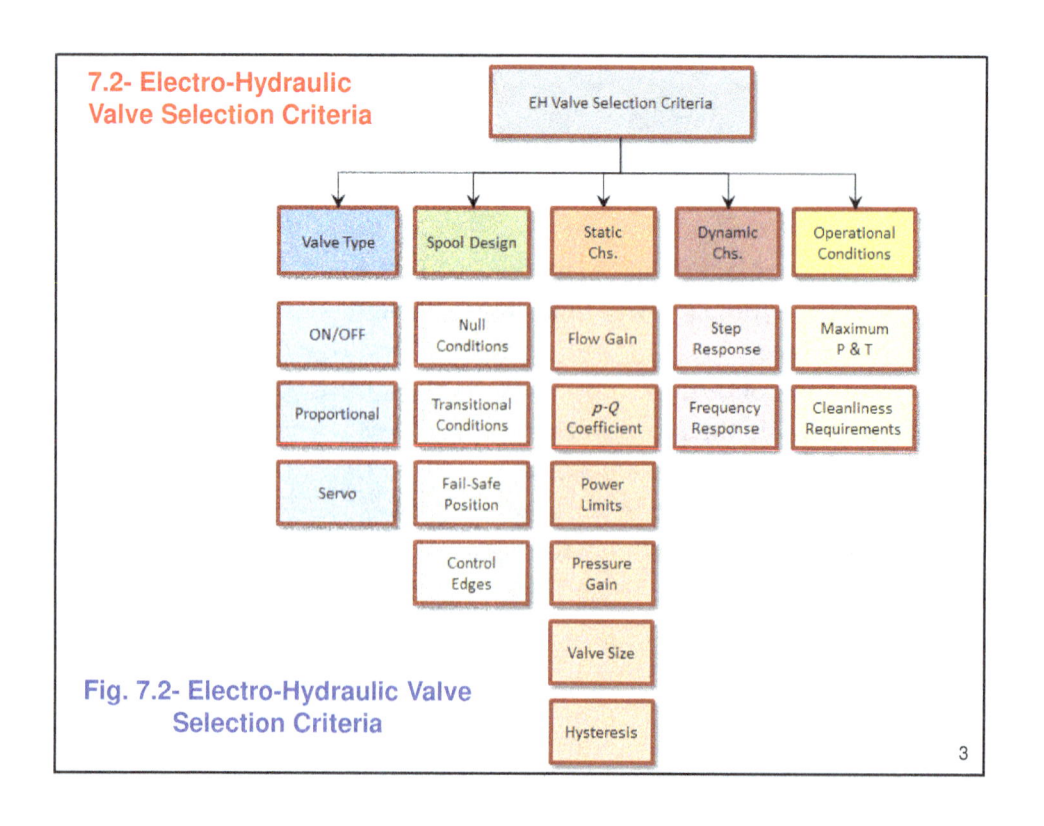

7.2- Electro-Hydraulic Valve Selection Criteria

EH Valve Selection Criteria

Valve Type	Spool Design	Static Chs.	Dynamic Chs.	Operational Conditions
ON/OFF	Null Conditions	Flow Gain	Step Response	Maximum P & T
Proportional	Transitional Conditions	p-Q Coefficient	Frequency Response	Cleanliness Requirements
Servo	Fail-Safe Position	Power Limits		
	Control Edges	Pressure Gain		
		Valve Size		
		Hysteresis		

Fig. 7.2- Electro-Hydraulic Valve Selection Criteria

Workbook: Electro-Hydraulic Components and Systems
Chapter 7: Electro-Hydraulic Valve Selection Criteria

7.3- Valve Type

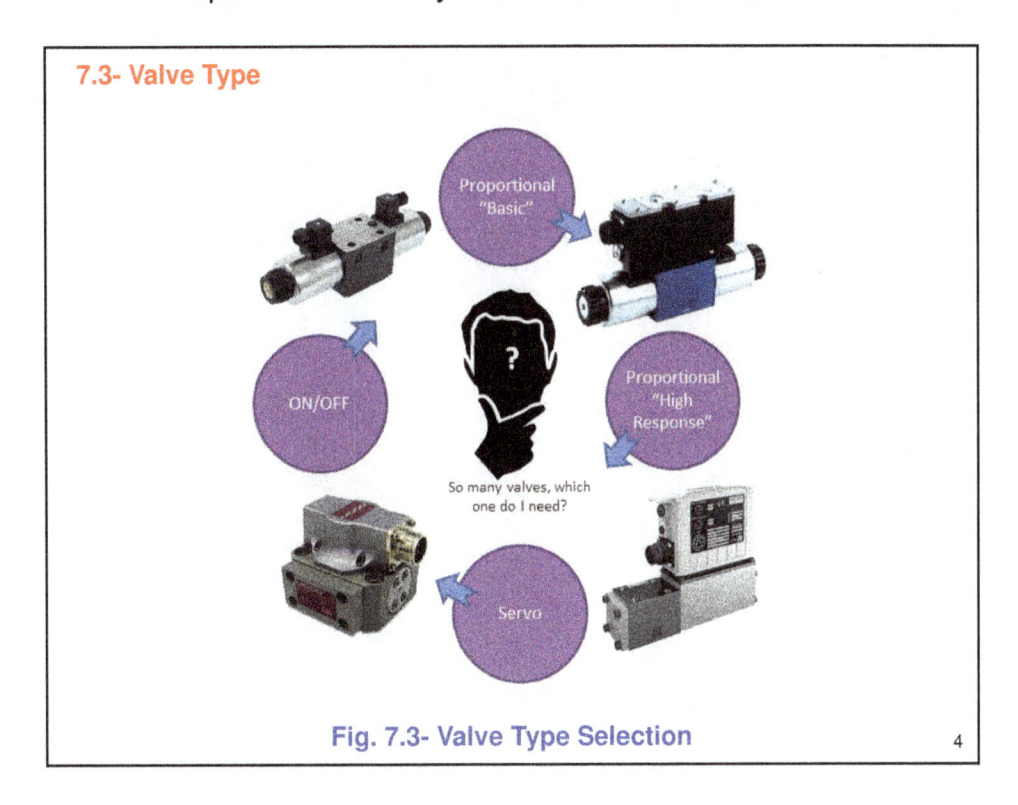

Fig. 7.3- Valve Type Selection

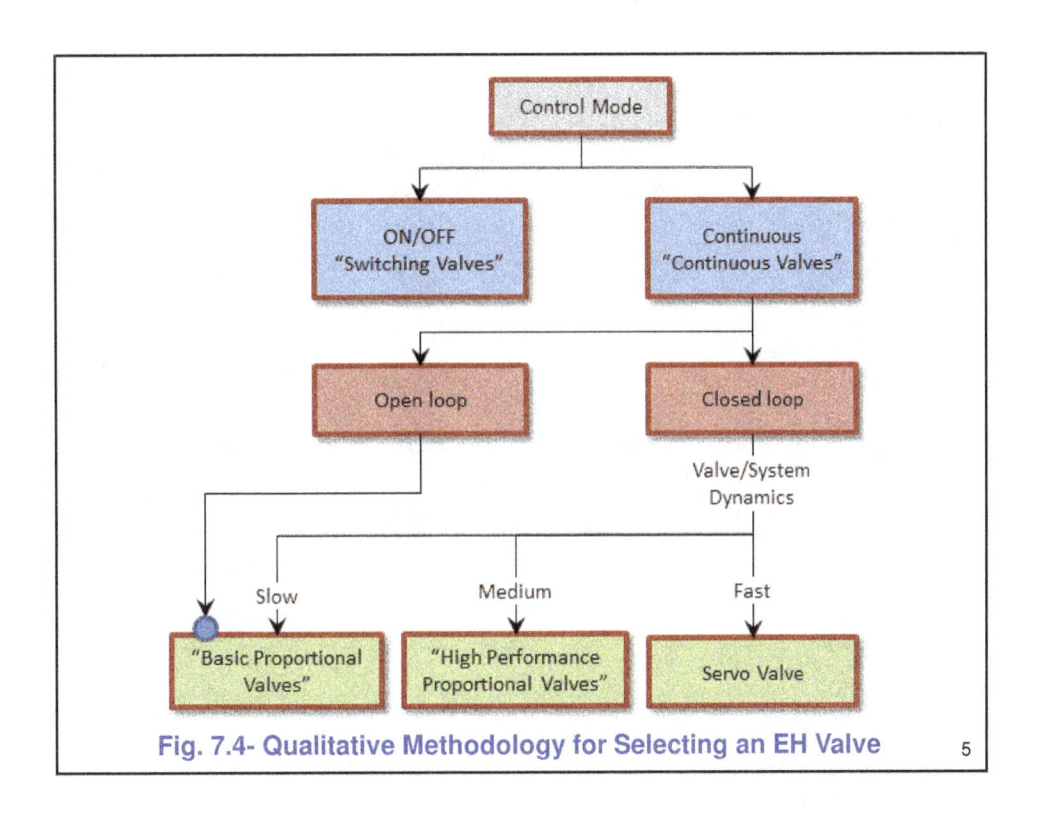

Fig. 7.4- Qualitative Methodology for Selecting an EH Valve

Fig. 7.5- ON/OFF Cooling Fan Speed Control

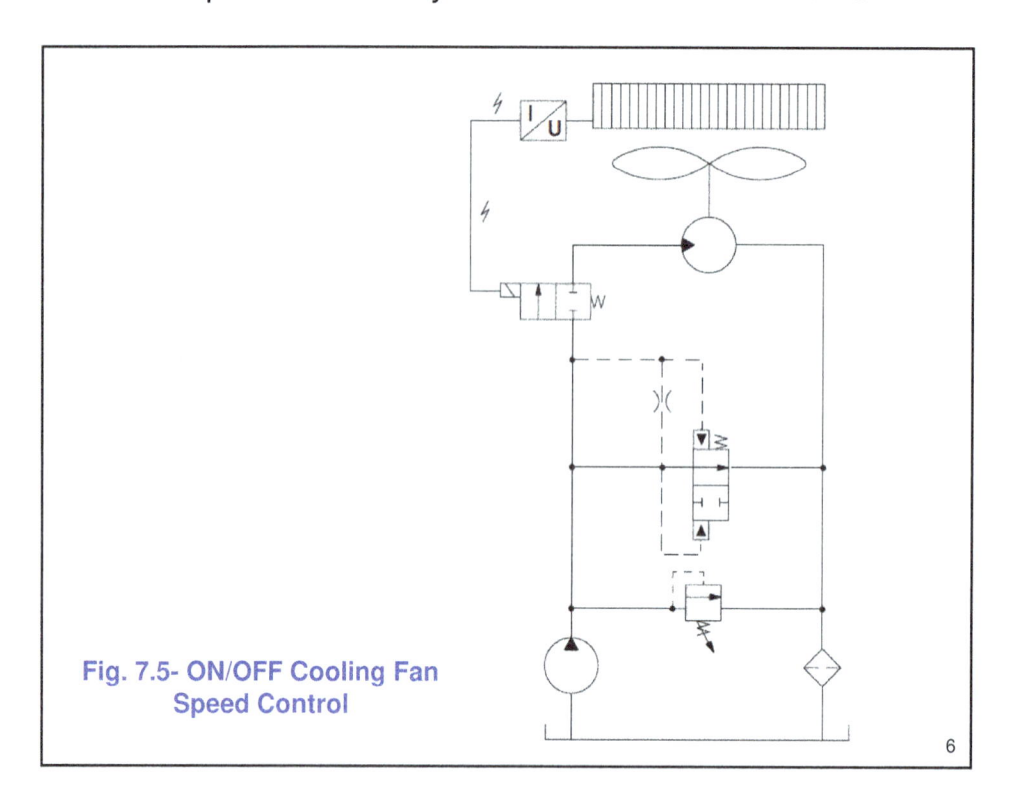

	Basic Proportional Valves	High Response Proportional Valves	Servo Valves
Position	1 🟡	🟢	🟢
Flow/Speed (one direction)	🟢	🟢	2 🟡
Flow/Speed (two directions)	🟢	🟢	🟢
Pressure/Load (one direction)	🟢	🟢	3 🟡
Pressure/Load (two directions)	4 🟡	🟢	🟢

Table 7.1- Valve Selection Based on the Controlled Parameter in a Closed Loop

Workbook: Electro-Hydraulic Components and Systems
Chapter 7: Electro-Hydraulic Valve Selection Criteria

Matrix of proportional directional valves

	Valve model	Nominal flow (l/min)	Nominal Δp (bar)	Data sheet RE..	Overlap compensation (with E, W spool)	Valve dynamics (natural frequency) [1]	Typical application [2]			
							Open control loop	Closed-loop position control		Closed-loop pressure control
								With low precision	With high precision	
Direct operated	4WRA(E)	Size 6: 7, 15, 30 Size 10: 30, 60	10	29055	Yes	Very low	✓	▽	STOP	STOP
	4WRP(E)	Size 6: 8, 18, 32 Size 10: 50, 80	10	29022 29025	Yes	Low	✓	✓	STOP	STOP
	4WRE(E)	Size 6: 4, 8, 16, 32 Size 10: 25, 50, 75	10	29061	No	Medium	✓	✓	✓	✓
	4WRSE	Size 6: 4, 10, 20, 35 Size 10: 25, 50, 80	10	29067	No	High	STOP	▽	✓	✓
	4WRPE(H)	Size 6: 2, 4, 12, 24, 40 Size 10: 50, 100	70	29035 29037 29028 29032	No	High	STOP	▽	✓	✓
	4WRREH	Size 6: 4, 8, 12, 24, 40	70	29041	No	Very high	STOP	STOP	✓	✓
	4WS(E)2E	Size 6: 2, 5, 10, 15, 20 Size 10: 10, 20, 30, 45, 60, 75, 90	70	29564 29583	No	Very high	STOP	STOP	✓	✓
	4WRZ(E)	Size 10: 25, 50, 65 Size 16: 100, 150 Size 25: 220, 325 Size 32: 360, 520 Size 52: 1000	10	29115	No	Very low	✓	▽	STOP	STOP

Table 7.2- Valve Selection Chart (Courtesy of Bosch Rexroth)

7.4- Valve Spool Design

7.4.1- Spool Null Conditions

- **Null Conditions of a Spool:**

  Dimension of a spool land versus the ports on the sleeve at the central position.

- **Null Leakage:**

  Flow at null point
  (i.e. when the valve is de-energized)

- **Flow Gain:**

  p → A or p → B as function of spool displacement or input command.

7.4.1.1- Overlapped (Closed-Center) Null Conditions

For $|x| \leq x_O, Q_N = 0$

For $|x| > x_O, Q_N = q_A$ or q_B

Fig. 7.6- Overlapped (Closed-Center) Null Conditions

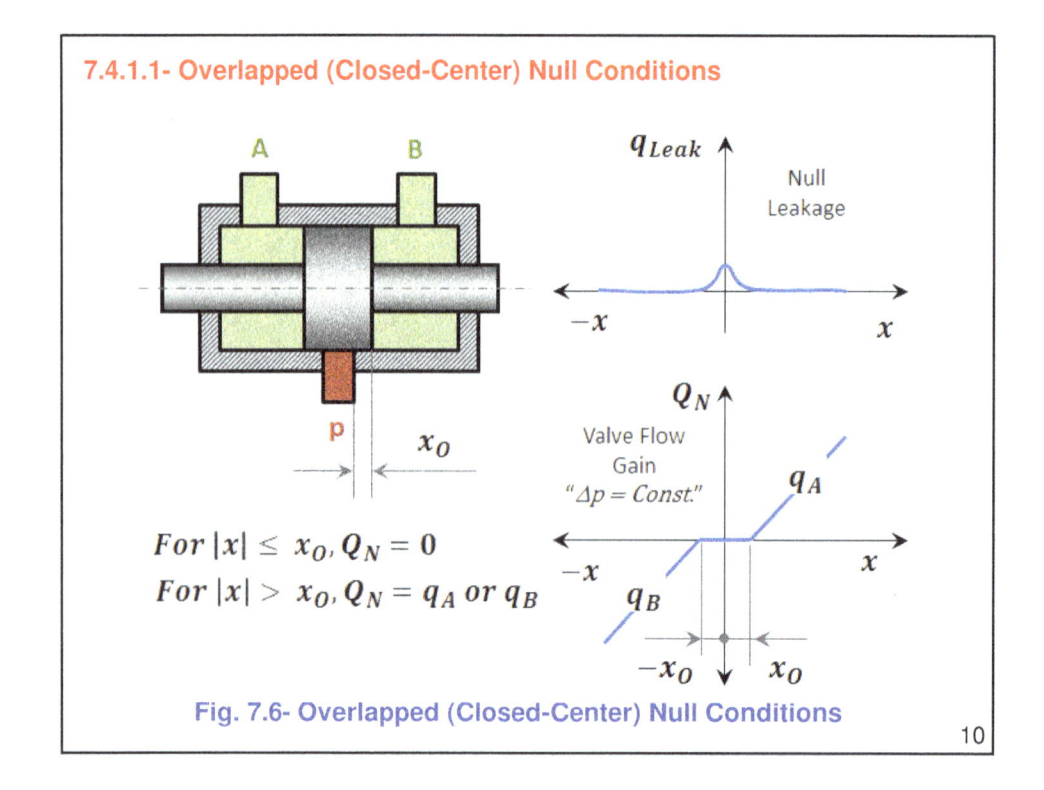

Features:

- **Null Leakage:** Reduced null leakage.
- **Flow Gain:** Nonlinear with dead zone.
- **Cost:** Inexpensive to produce.

Applications:

- **Position Control (Not Recommended): Results in steady state error.**

- **Flow/Speed Control (Recommended):**
  - the valve is open most of the time. Therefore it does not really matter what the valve null condition is.
  - A valve with overlap null conditions stops the actuator completely before changing its direction of motion. This reduces pressure spikes and mechanical stresses due to the inertia of the actuator.

- **Pressure/Load Control (N/A): because pressure control requires underlapped valve.**

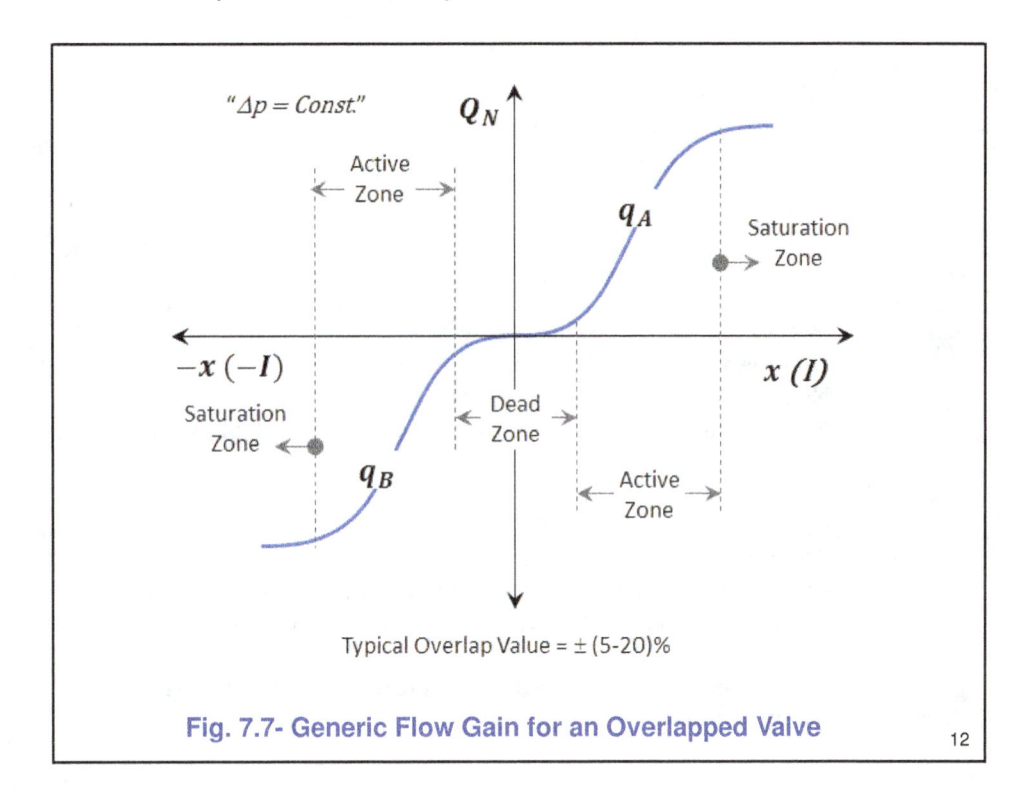

Workbook: Electro-Hydraulic Components and Systems
Chapter 7: Electro-Hydraulic Valve Selection Criteria

"$\Delta p = Const.$"

Typical Overlap Value = $\pm (5-20)\%$

Fig. 7.7- Generic Flow Gain for an Overlapped Valve

7.4.1.2- Zero-Lapped (Critical-Center) Null Conditions

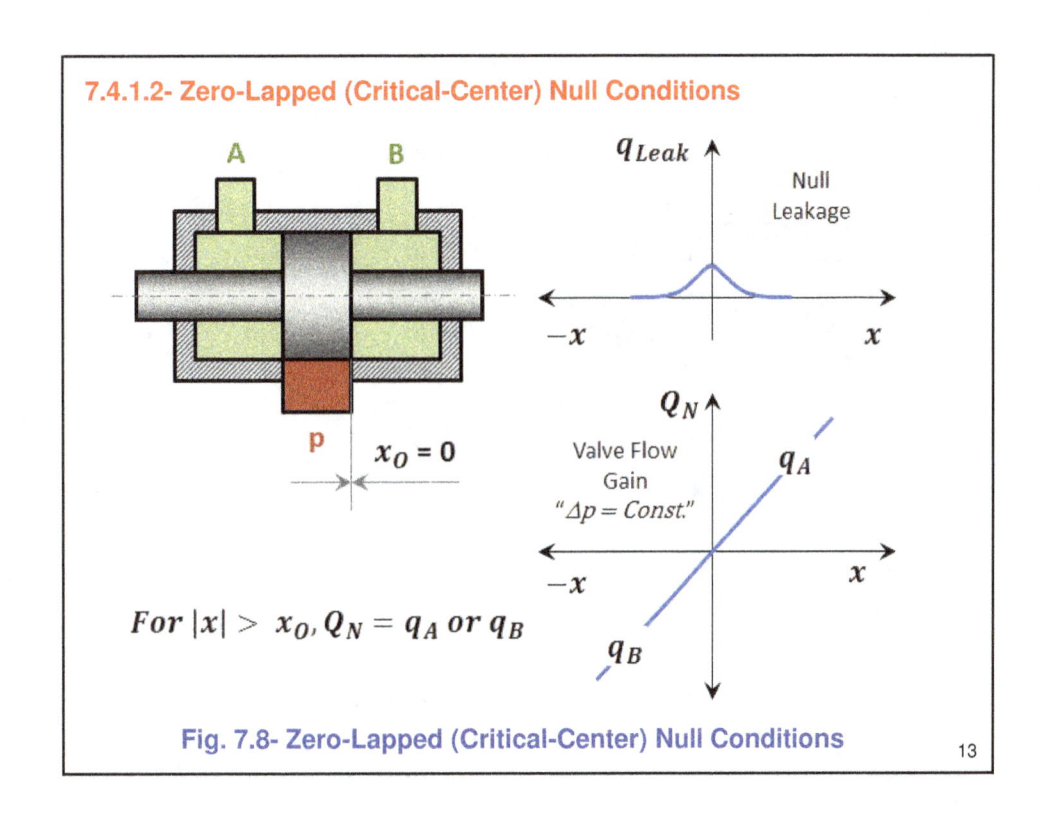

$x_O = 0$

For $|x| > x_O$, $Q_N = q_A$ or q_B

"$\Delta p = Const.$"

Fig. 7.8- Zero-Lapped (Critical-Center) Null Conditions

Features:

- **Null Leakage:** Medium null leakage.
- **Flow Gain:** Linear flow gain.
- **Cost:** Expensive to produce

Applications:

🟢 - **Position Control (Recommended):**
  - No steady state error.
  - No need to enable/tune dead band eliminator.
  - Easy to design a control loop using linear control theory.

🟡 - **Speed Control (Not Recommended):**
  - Expensive for no reason.
  - Possible load creep.
  - Pressure spikes during motion reversal because no pause period to stop the actuator before reversing its motion.

🔴 - **Pressure Control (N/A):** because pressure control requires underlapped valve.

7.4.1.3- Underlapped (Open-Center) Null Conditions

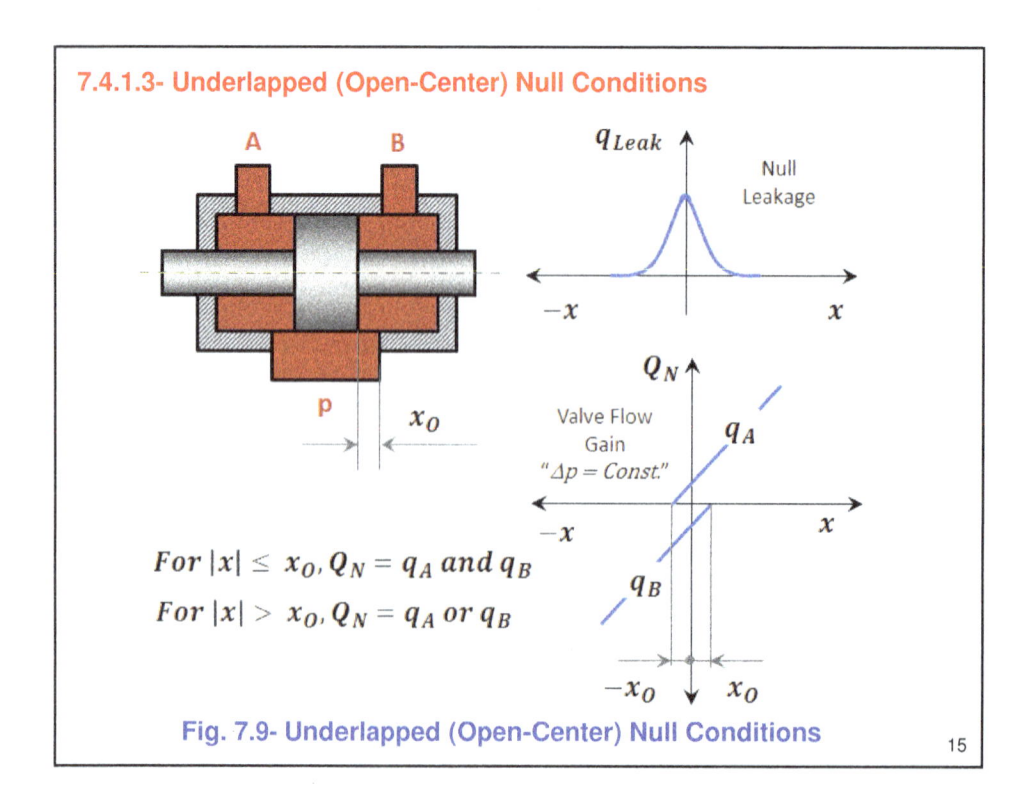

For $|x| \leq x_O$, $Q_N = q_A$ and q_B
For $|x| > x_O$, $Q_N = q_A$ or q_B

Fig. 7.9- Underlapped (Open-Center) Null Conditions

Features:

- **Null Leakage:** Large leakage.
- **Flow Gain:** nonlinear with discontinuity around the null position. Continuous flow around null leakage increases energy wasting and heat generation.
- **Cost:** Inexpensive to produce
- **Applications:**
- **Position Control (N/A):** because the load is float at null position.
- **Speed Control (Conditionally Recommended):**
  - for some cases where:
    - Actuator should be float at null position, e.g. cooling fan control.
    - Cycling slow actuator or small loads that can be reversed without pressure spikes.
- **Pressure Control (Only usable valve for this application):**

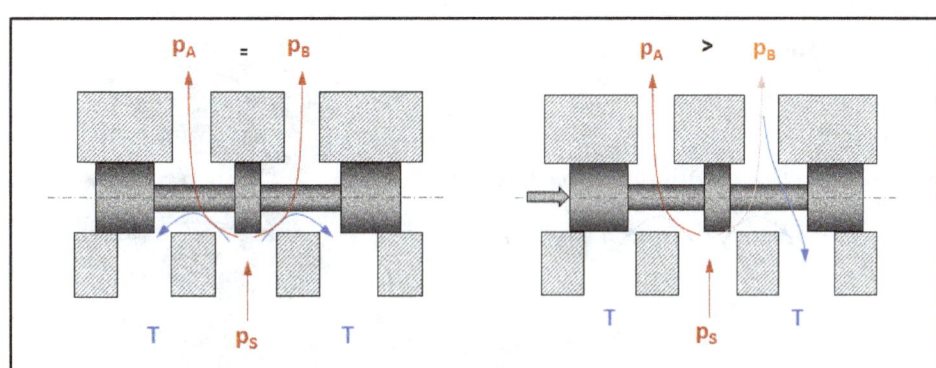

Fig. 7.10- Underlapped Valve in Pressure Control

The Spool that is only Applicable for Pressure/Load Control?

- **To create a differential pressure across the hydraulic actuator, the valve must be underlapped.**
- **Control pressure is performed around the null position.**
- **Once one land is closed, the valve is saturated and the differential pressure across the actuator will equal the supply pressure.**
- **If an overlapped or zero lapped valve is used, differential pressure won't be created because one land is always closed.**

Workbook: Electro-Hydraulic Components and Systems
Chapter 7: Electro-Hydraulic Valve Selection Criteria

	Overlapped "Closed-Center"	Zero Lapped "Critical-Center"	Underlapped "Open-Center"
Position control	🟡 Steady state error	🟢 Recommended	🔴 Float actuator around valve null position
Speed/Flow Control	🟢 Recommended	🟡 Expensive valve is used for no reason	🟡 Float actuator around valve null position. Good for slow actuators and small loads
Load/Pressure Control	🔴 One land is always closed	🔴 One land is always closed	🟢 Recommended

Table 7.3 – Selection of Null Conditions Based on Control Solutions

---

7.4.2- Spool Transitional Conditions

Fig. 7.11- Transitional Conditions of Selected Spools (Courtesy of Bosch Rexroth)

Type 4WRPE

= E
= E1

= W6
= W8

= V
= V1

= EA

With symbol E1-, V1- and W8-:
P → A: $q_{v\,max}$ B → T: $q_v/2$
P → B: $q_v/2$ A → T: $q_{v\,max}$

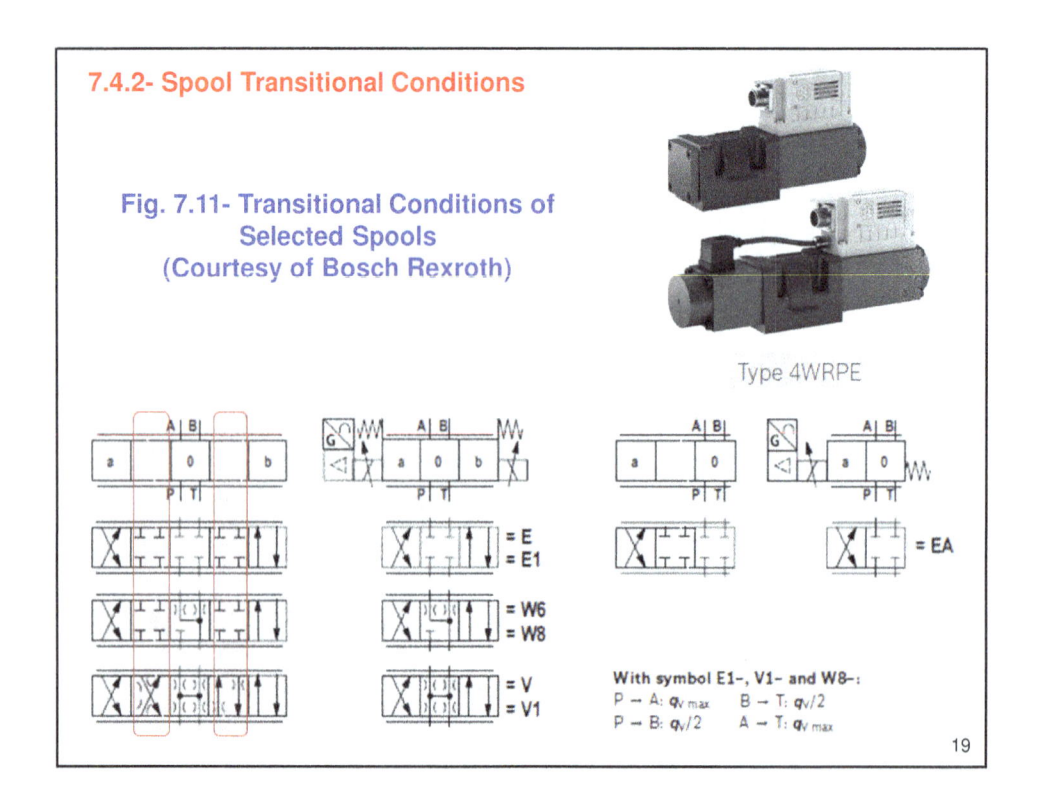

Workbook: Electro-Hydraulic Components and Systems
Chapter 7: Electro-Hydraulic Valve Selection Criteria

7.4.2- Spool Transitional Conditions

Fig. 7.11- Transitional Conditions of Selected Spools (Courtesy of Bosch Rexroth)

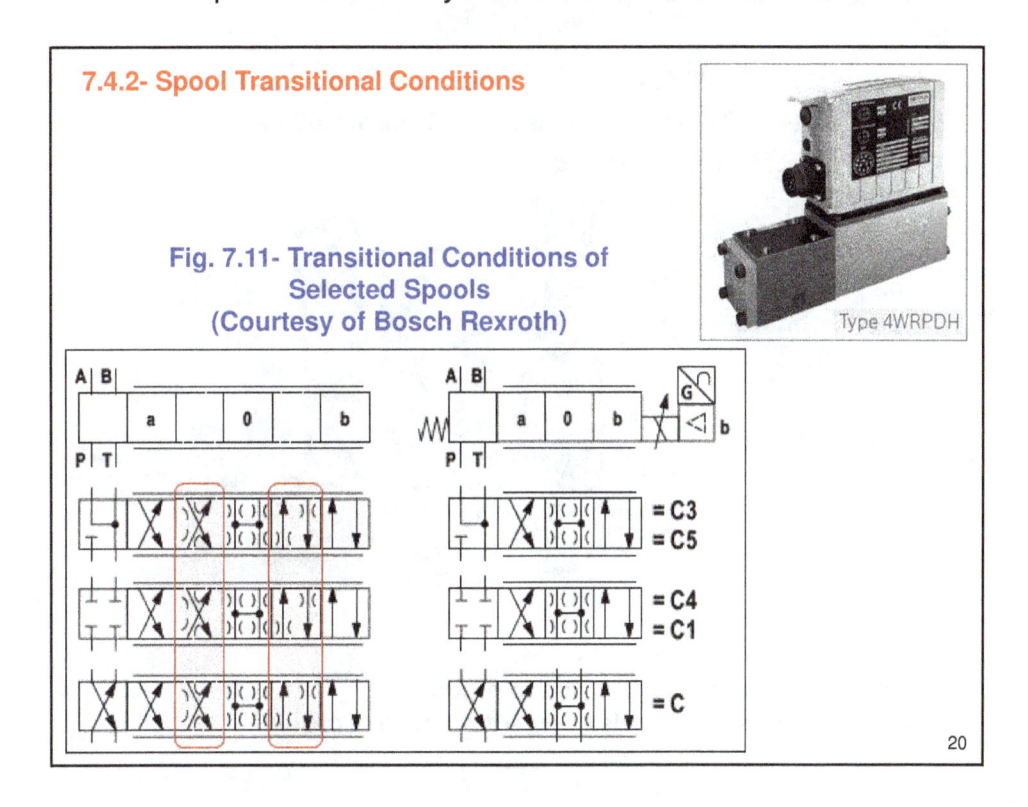

Type 4WRPDH

7.4.3- Spool Fail-Safe Position

Fig. 7.12- Fail-Safe Position of Selected Spools (Courtesy of Bosch Rexroth)

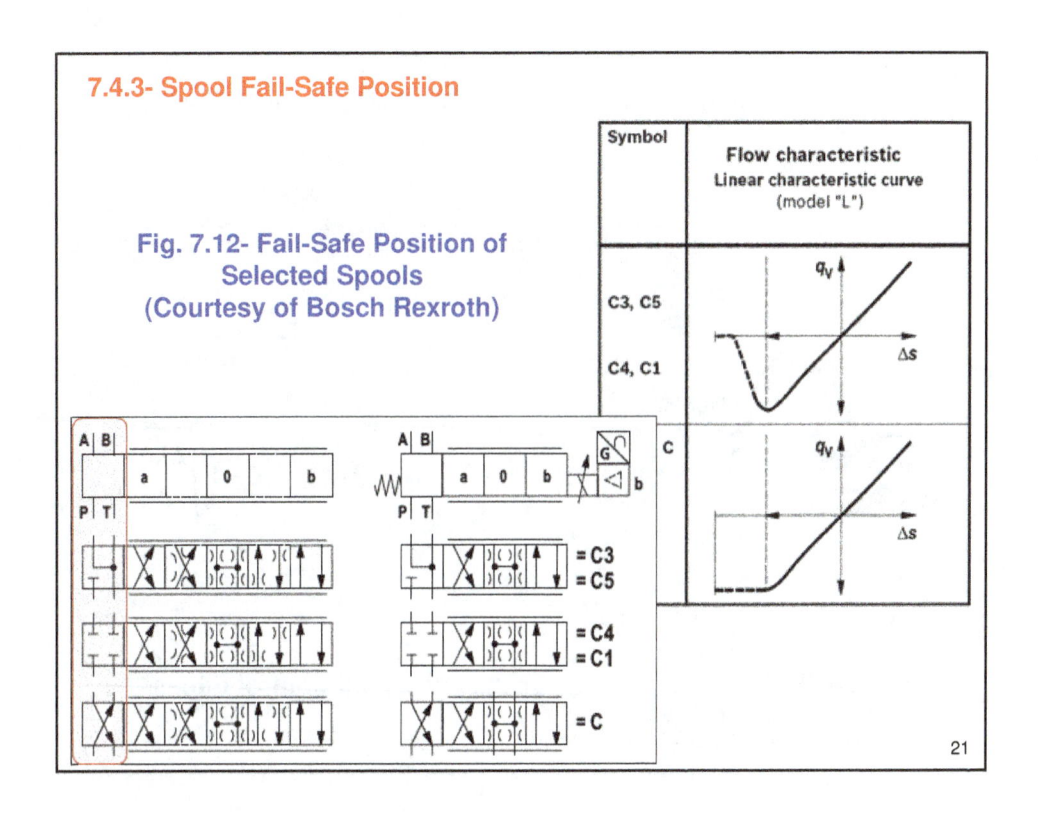

Symbol	Flow characteristic Linear characteristic curve (model "L")
C3, C5 C4, C1	q_V vs Δs
C	q_V vs Δs

Workbook: Electro-Hydraulic Components and Systems
Chapter 7: Electro-Hydraulic Valve Selection Criteria

7.4.4- Spool Control Edges
❑ Referred to as "Metering Orifices" or "Control Notches"

7.4.4.1- Holes versus Control Edges

Fig. 7.13- Holes versus Control Edges

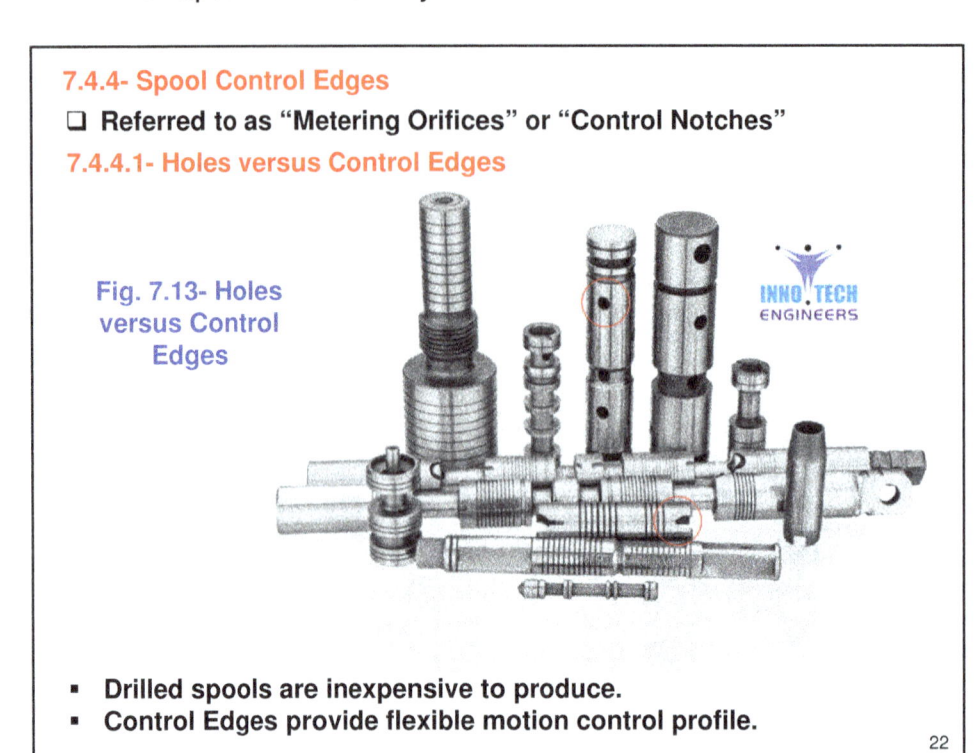

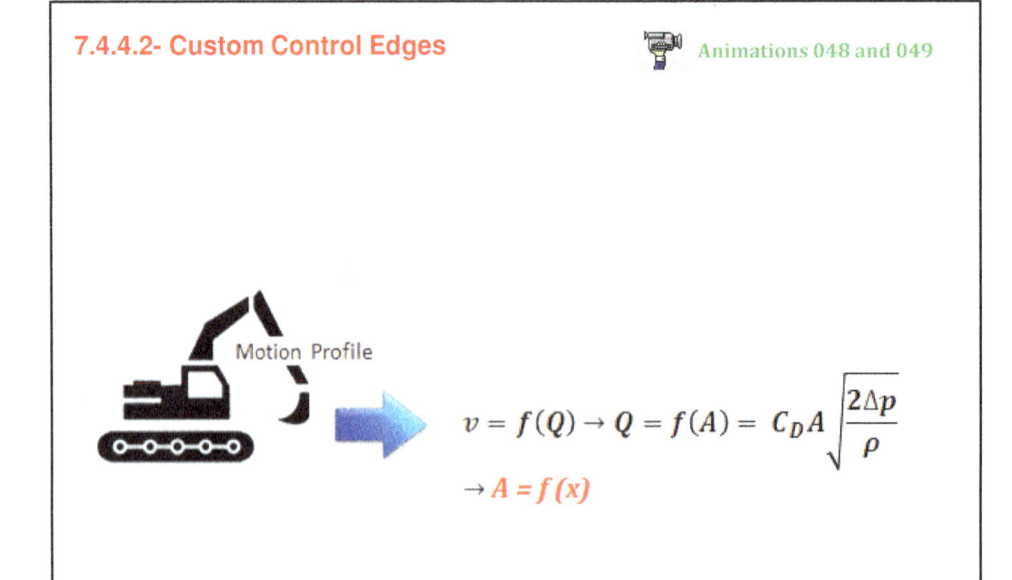

- Drilled spools are inexpensive to produce.
- Control Edges provide flexible motion control profile.

7.4.4.2- Custom Control Edges

Animations 048 and 049

Motion Profile

$$v = f(Q) \rightarrow Q = f(A) = C_D A \sqrt{\frac{2\Delta p}{\rho}}$$

$$\rightarrow A = f(x)$$

Fig. 7.14- Design Methodology for Custom Spool Control Edges

7.4.4.3- Standard Control Edges
❑ Square Notches

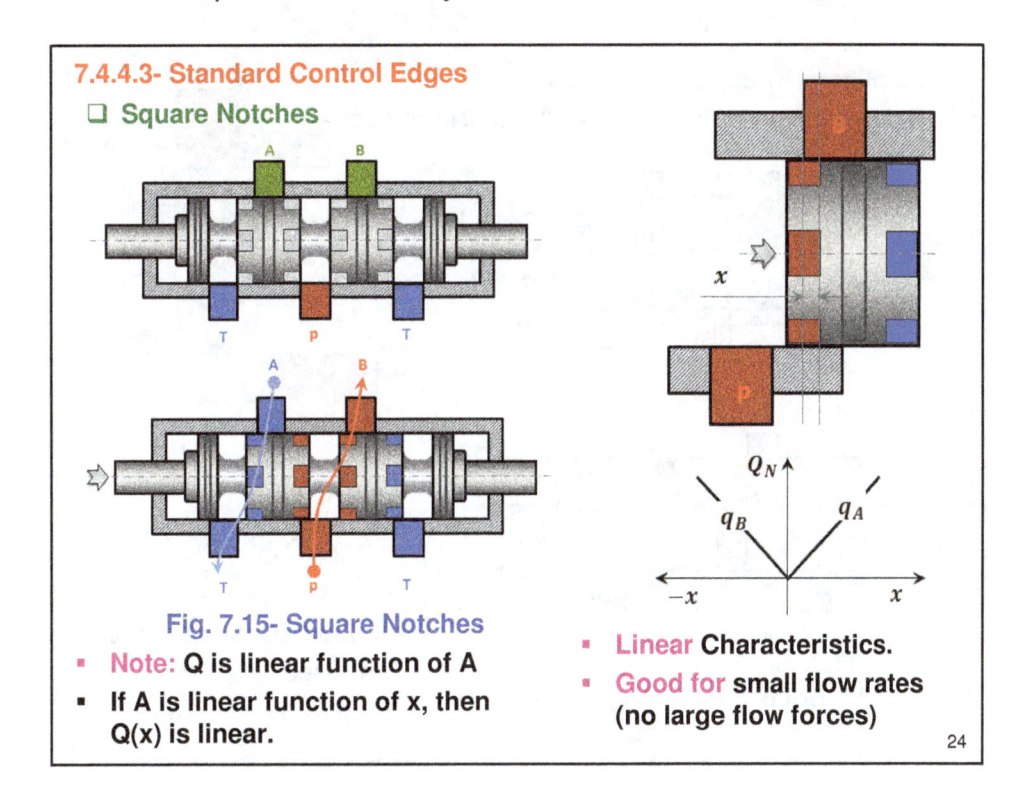

Fig. 7.15- Square Notches

- **Note: Q is linear function of A**
- If A is linear function of x, then Q(x) is linear.
- **Linear** Characteristics.
- **Good for** small flow rates (no large flow forces)

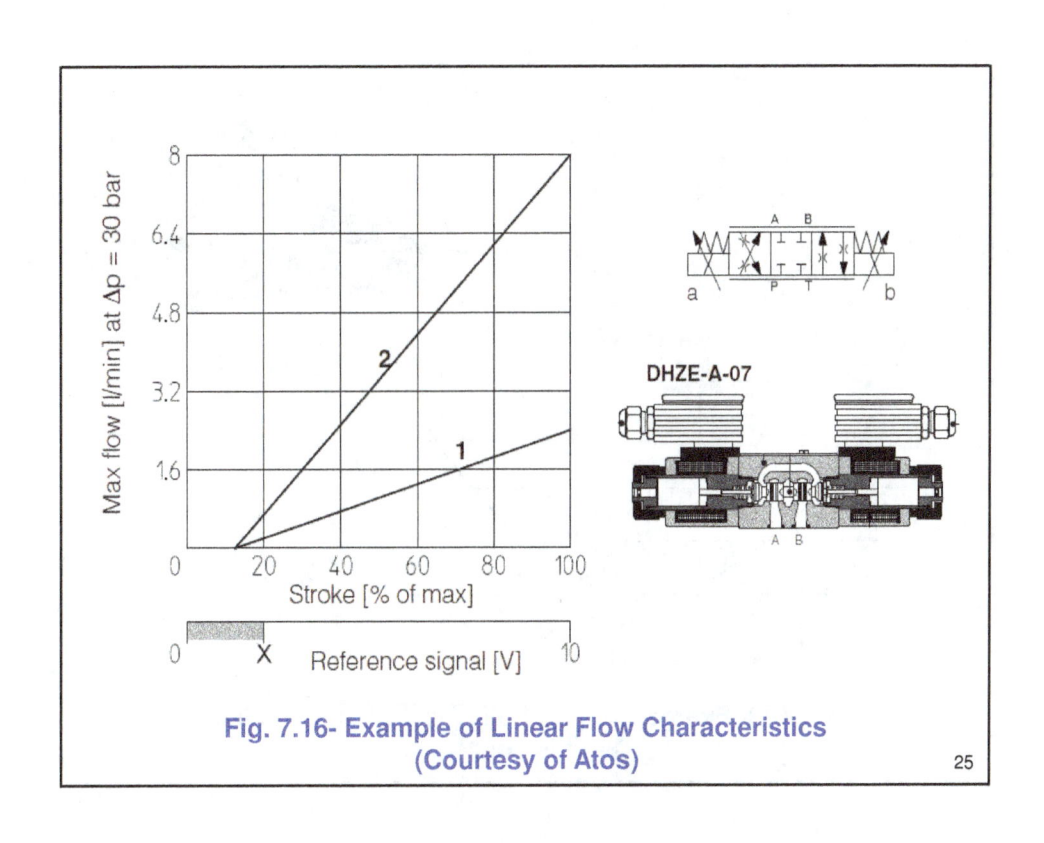

Fig. 7.16- Example of Linear Flow Characteristics (Courtesy of Atos)

DHZE-A-07

❑ **Equilateral Triangular Notches**
❑ **Referred to as "V Notches" or "Delta Notches"**

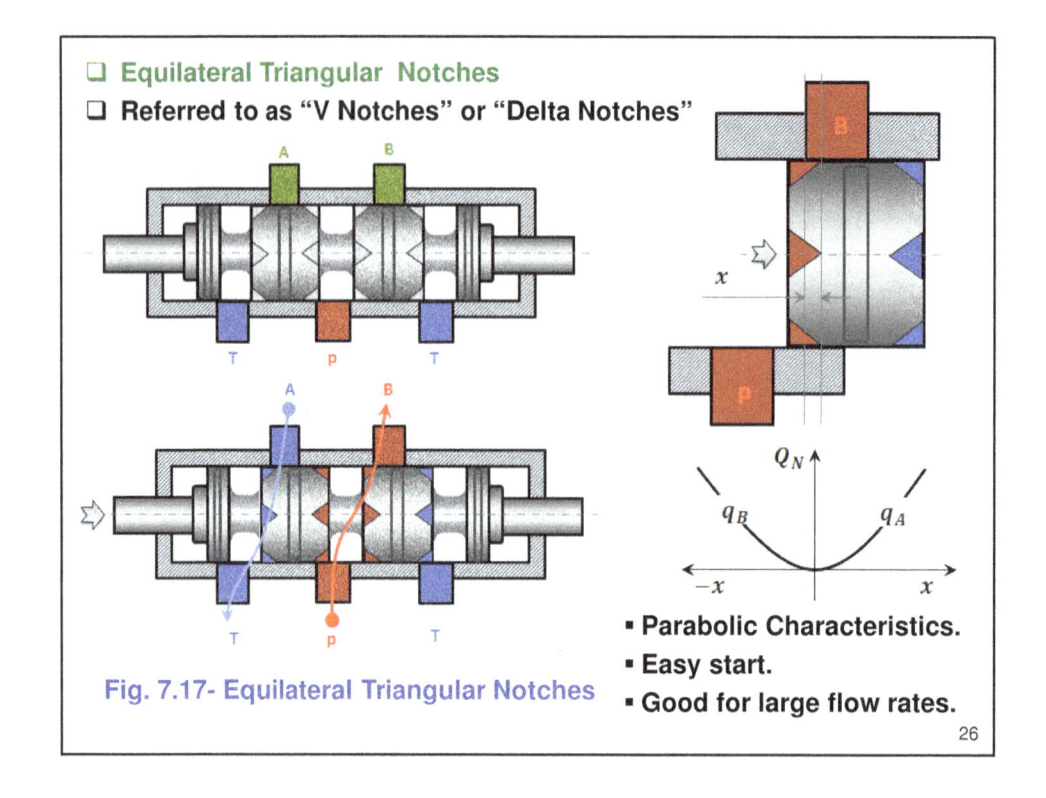

- Parabolic Characteristics.
- Easy start.
- Good for large flow rates.

Fig. 7.17- Equilateral Triangular Notches

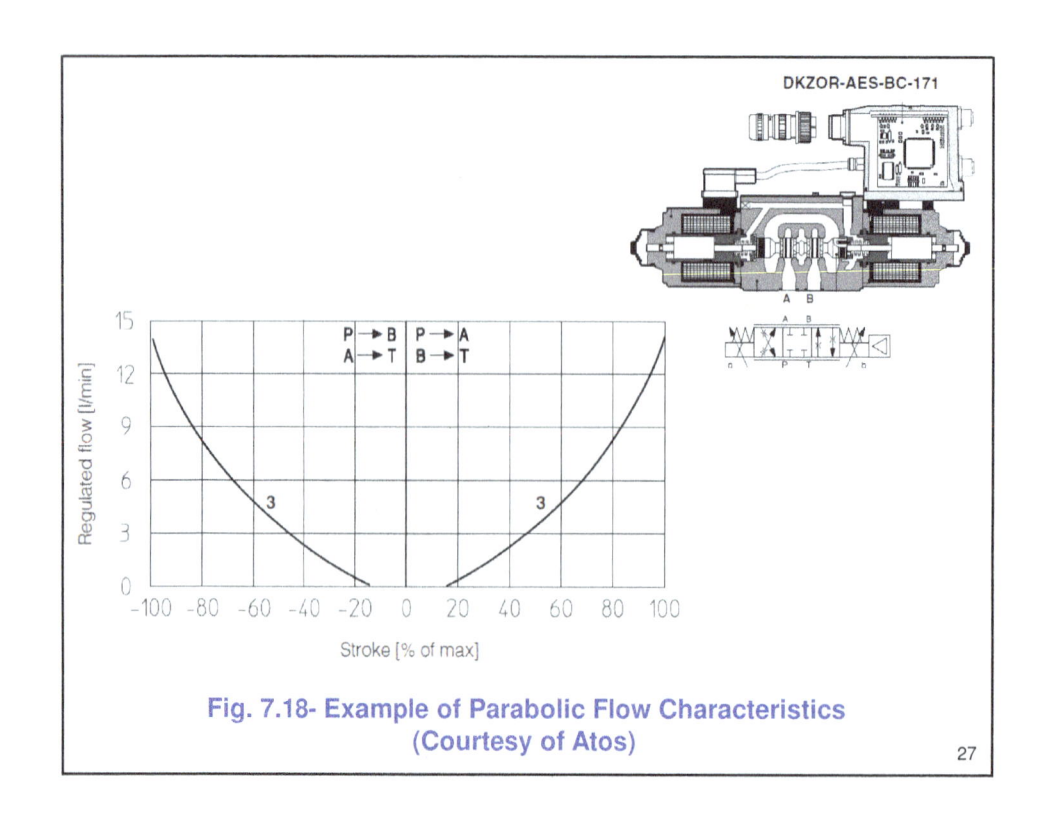

Fig. 7.18- Example of Parabolic Flow Characteristics
(Courtesy of Atos)

Semicircular Notches
- Referred to as "D Notches"

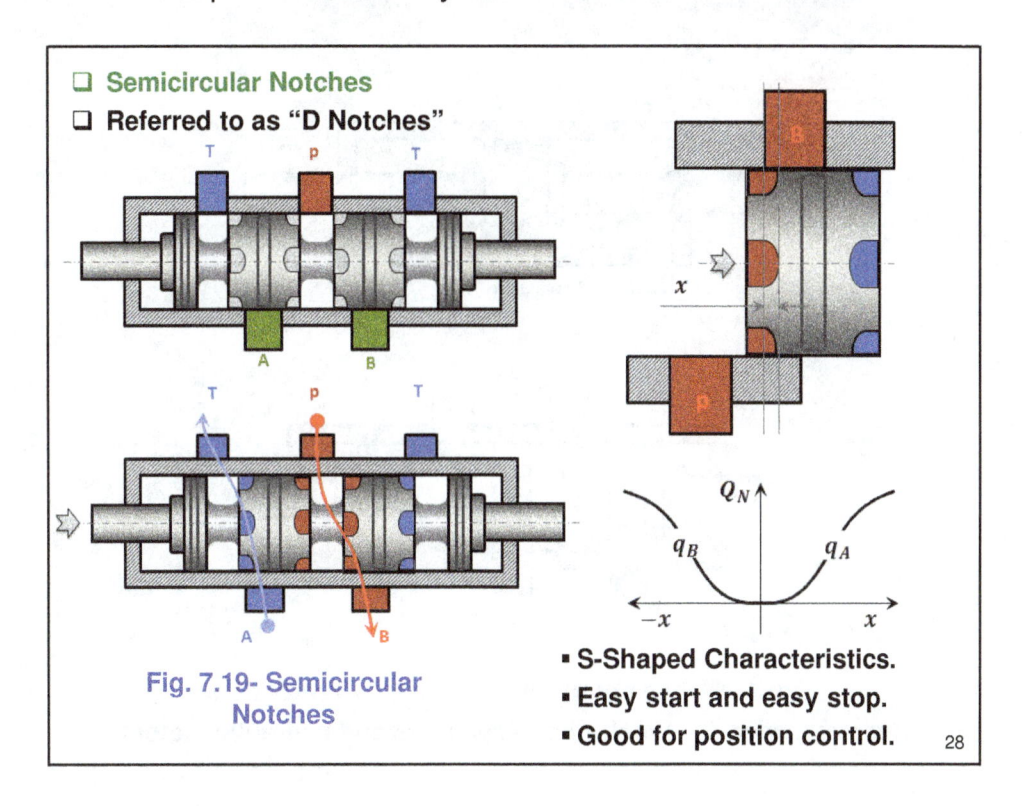

Fig. 7.19- Semicircular Notches

- S-Shaped Characteristics.
- Easy start and easy stop.
- Good for position control.

Volume flow-signal-characteristics (Δp = 10 bar)
[Type: S4D41]

NG4-Mini®

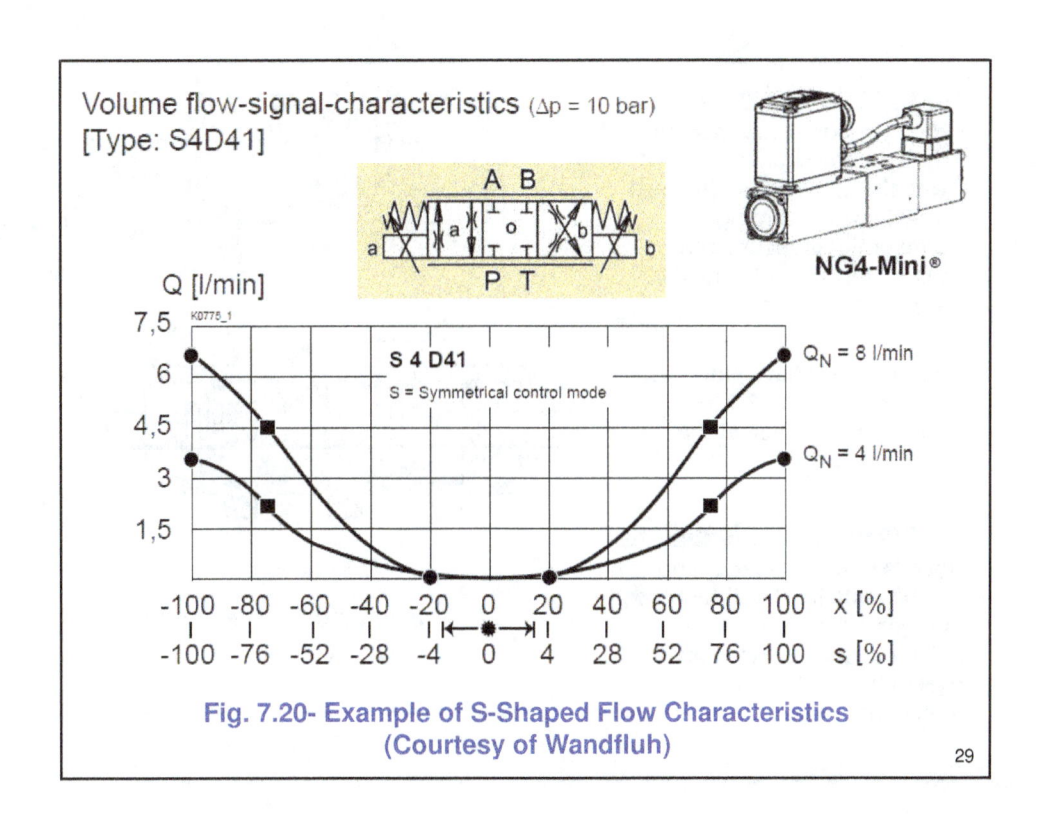

S 4 D41
S = Symmetrical control mode

Q_N = 8 l/min
Q_N = 4 l/min

Fig. 7.20- Example of S-Shaped Flow Characteristics
(Courtesy of Wandfluh)

7.4.4.4- Distribution of Control Edges

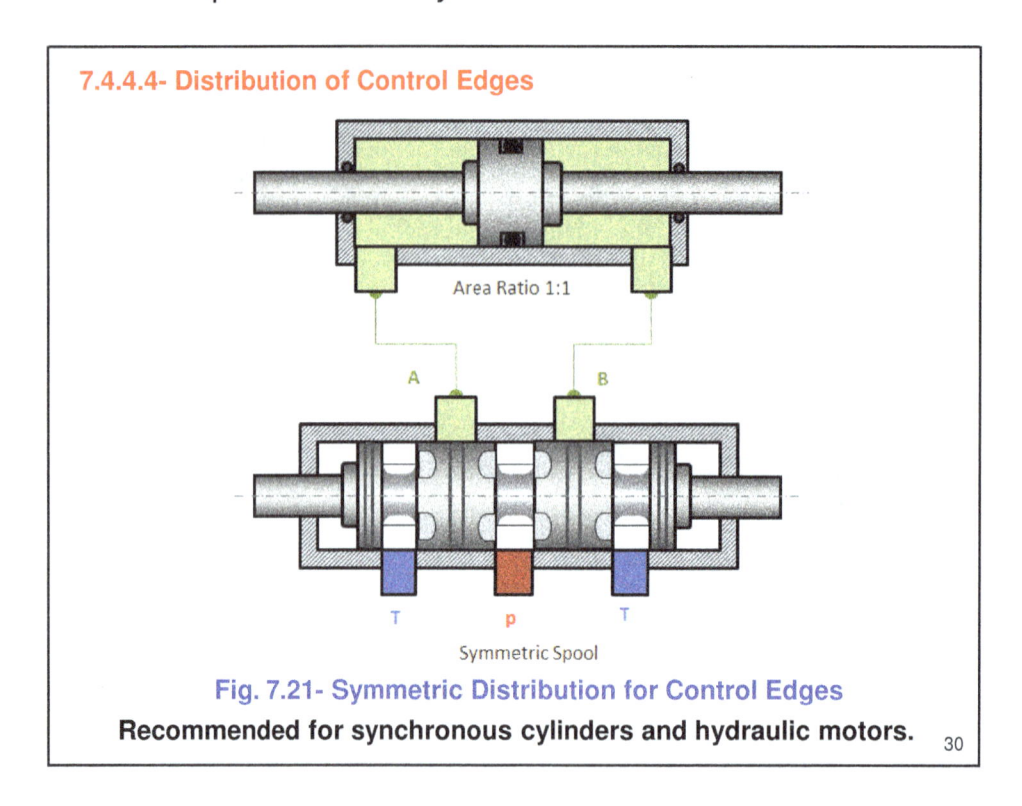

Fig. 7.21- Symmetric Distribution for Control Edges

Recommended for synchronous cylinders and hydraulic motors.

For Open-Loop Operation:
- Highly recommended so that the cylinder could be cycled with the same speed in both directions (assuming same differential pressure across the valve) without a need to apply flow control.

- For a differential cylinder that has an area ratio 2:1, control edges are doubled at the piston side.

For Closed-Loop Operation:
Despite that the closed loop can compensate for the difference in the cylinder effective area, it is still recommended to unify the overall loop gain in both directions.

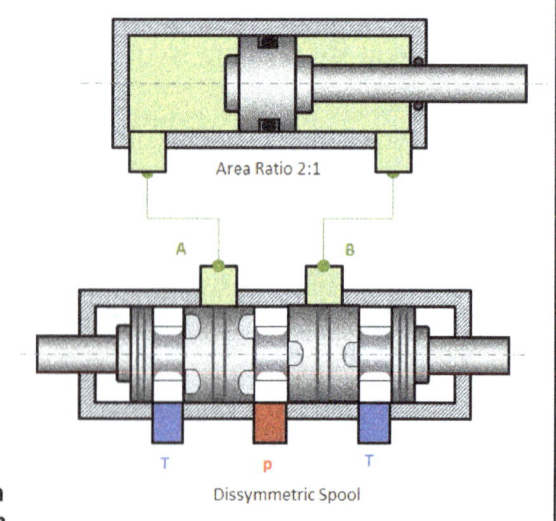

Fig. 7.22- Dissymmetric Distribution for Control Edges

7.5- Valve Static Characteristics

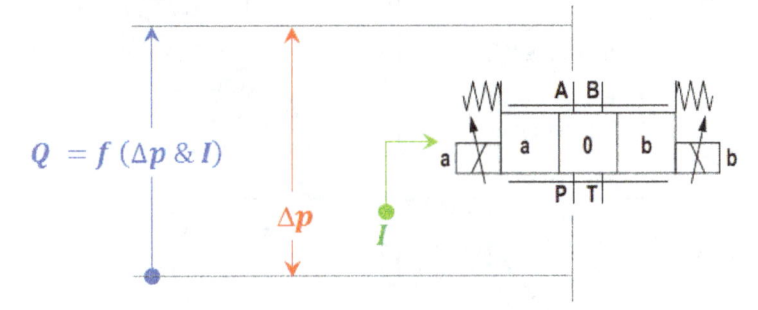

$$Q = f(\Delta p \,\&\, I)$$

Fig. 7.23- Flow Dependency through a Continuous Valve

❑ **Valve Flow Gain:**
- Required to size the valve for position, flow or speed control.

❑ **Valve Flow-Pressure Sensitivity:**
- Similar to flow gain + quantify the power losses.

❑ **Valve Pressure Gain:**
- Required to size the valve for pressure or load control.

7.5.1- Valve Flow Gain

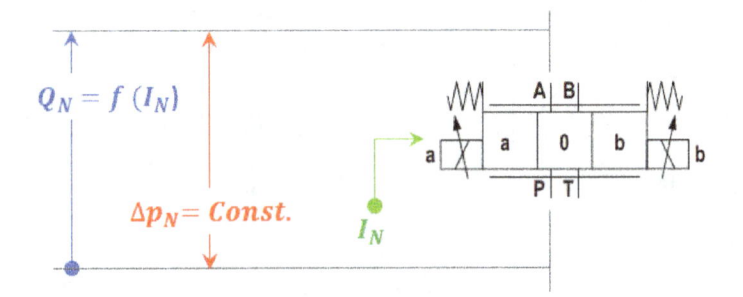

$$Q_N = f(I_N)$$
$$\Delta p_N = Const.$$

Fig. 7.24- Interpretation of Valve Flow Gain

$$Flow\ Gain = k_{QI} = \left[\frac{\partial Q_N}{\partial I_N}\right]_{\Delta p_N = Constant} \qquad 7.1$$

❑ Q_N = Output Nominal "Rated" Flow (developed by testing the valve).
❑ Δp_N = Constant Differential Pressure (used to test the valve).
❑ I_N = Variable Nominal Input Command to the valve.

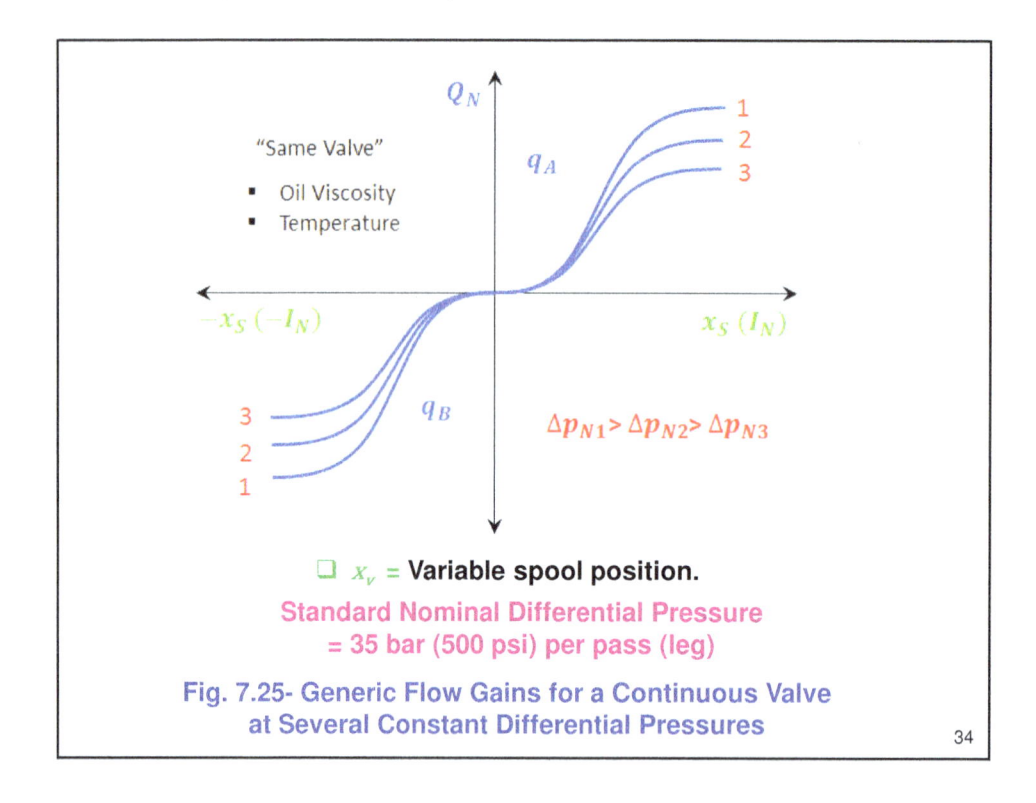

Workbook: Electro-Hydraulic Components and Systems
Chapter 7: Electro-Hydraulic Valve Selection Criteria

"Same Valve"
- Oil Viscosity
- Temperature

$\Delta p_{N1} > \Delta p_{N2} > \Delta p_{N3}$

☐ x_v = **Variable spool position.**

**Standard Nominal Differential Pressure
= 35 bar (500 psi) per pass (leg)**

**Fig. 7.25- Generic Flow Gains for a Continuous Valve
at Several Constant Differential Pressures**

Flow Gain Linearization:

Δp_N = Constant

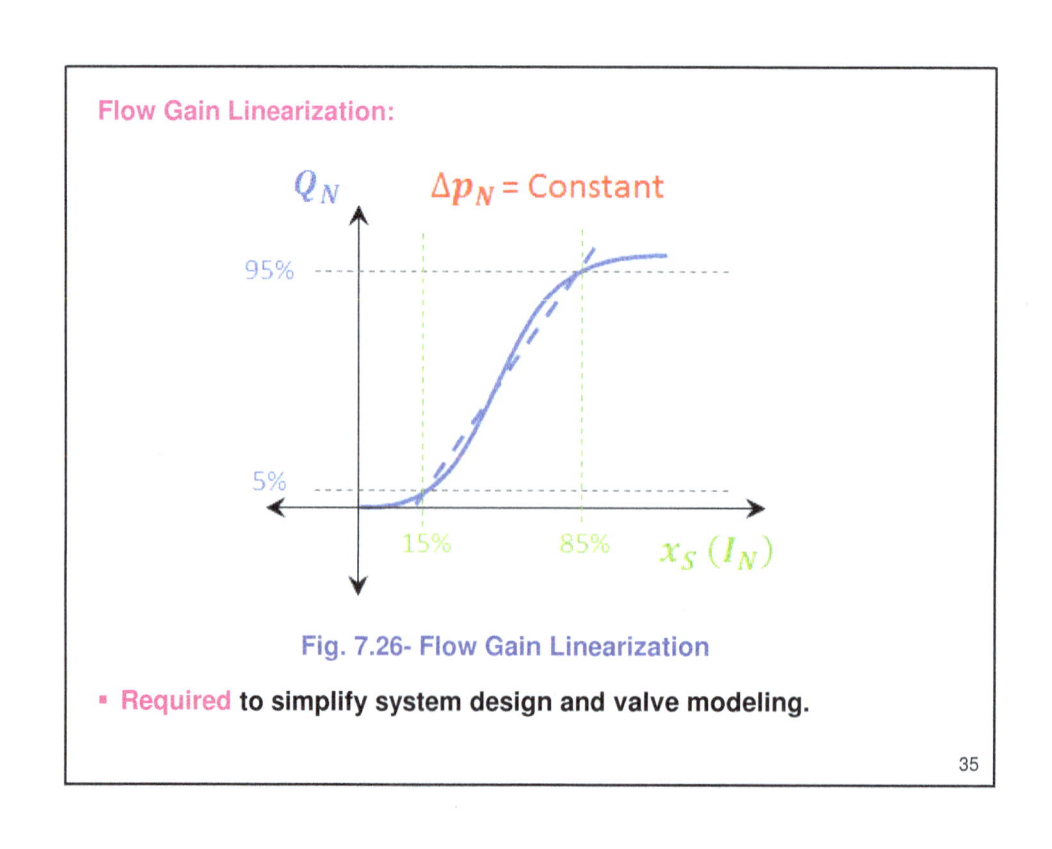

Fig. 7.26- Flow Gain Linearization

- **Required** to simplify system design and valve modeling.

Characteristic curves: Size 6 – flow characteristic (measured with HLP46, ϑ_{oil} = 40 ± 5 °C)

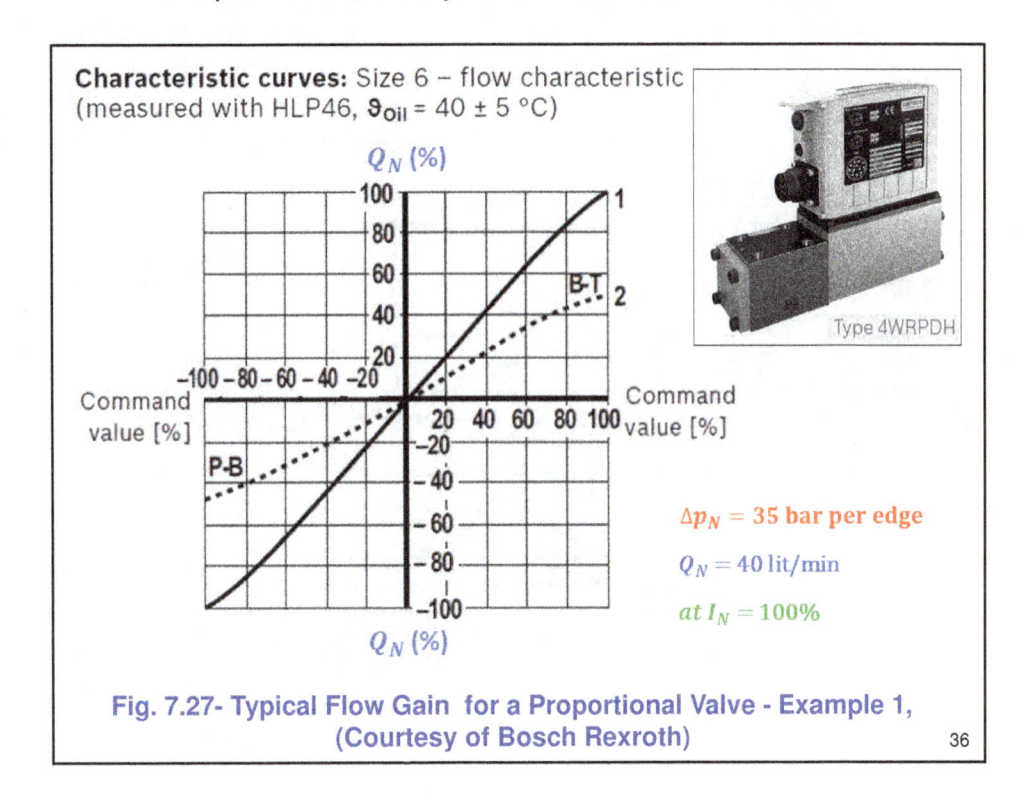

$\Delta p_N = 35 \text{ bar per edge}$

$Q_N = 40 \text{ lit/min}$

$at\ I_N = 100\%$

Fig. 7.27- Typical Flow Gain for a Proportional Valve - Example 1, (Courtesy of Bosch Rexroth)

4WRE

Characteristic curves: Type 4WREE (measured with HLP46, ϑ_{Oil} = 40 °C ± 5 °C and p = 100 bar) Size 6

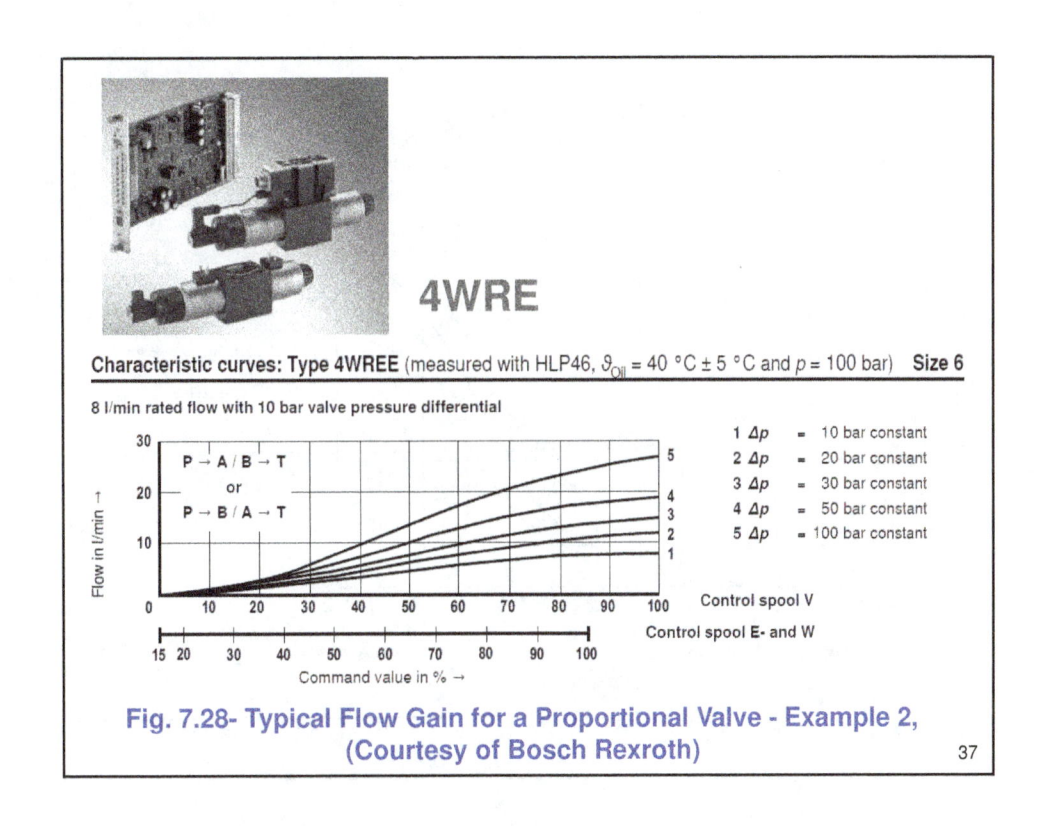

8 l/min rated flow with 10 bar valve pressure differential

1 Δp	= 10 bar constant
2 Δp	= 20 bar constant
3 Δp	= 30 bar constant
4 Δp	= 50 bar constant
5 Δp	= 100 bar constant

Control spool V
Control spool E- and W

Fig. 7.28- Typical Flow Gain for a Proportional Valve - Example 2, (Courtesy of Bosch Rexroth)

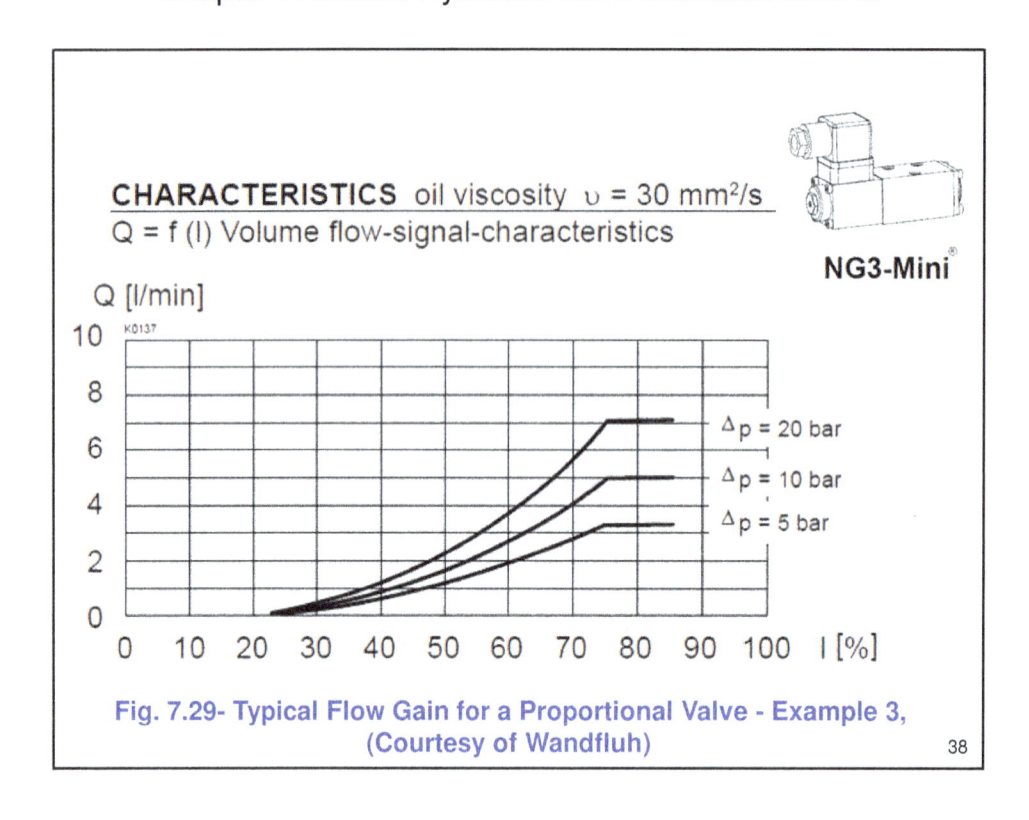

Workbook: Electro-Hydraulic Components and Systems
Chapter 7: Electro-Hydraulic Valve Selection Criteria

CHARACTERISTICS oil viscosity $\upsilon = 30$ mm²/s
$Q = f(I)$ Volume flow-signal-characteristics

NG3-Mini

Q [l/min]

$\Delta p = 20$ bar
$\Delta p = 10$ bar
$\Delta p = 5$ bar

I [%]

Fig. 7.29- Typical Flow Gain for a Proportional Valve - Example 3, (Courtesy of Wandfluh)

FLOW vs. CURRENT
103 bar/1500 psi - - -; 206 bar/3000 psi ———

SPCL10-30

FLOW lpm/gpm

Ranges Shaded

Amps: 0.1 0.2 0.3 0.4 0.5 0.6 0.7 0.8 0.9 1.0 1.1 1.2
% of Max. Control Current: 30 40 50 60 70 80 90 100

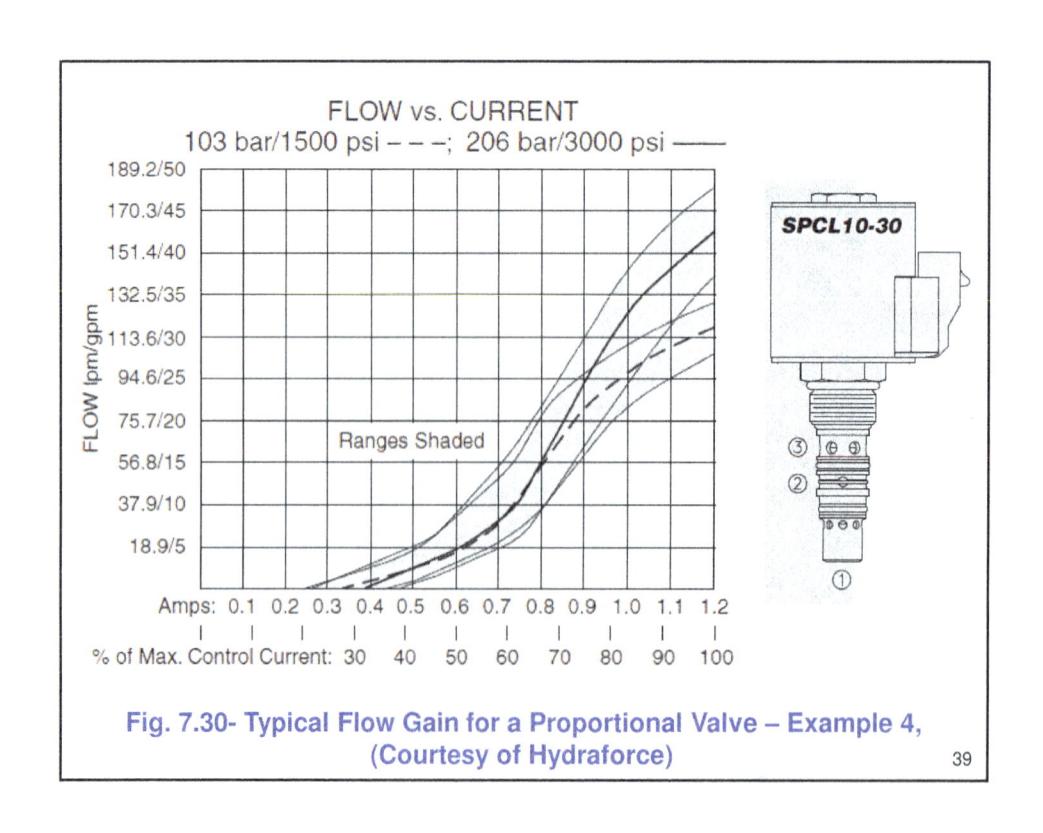

Fig. 7.30- Typical Flow Gain for a Proportional Valve – Example 4, (Courtesy of Hydraforce)

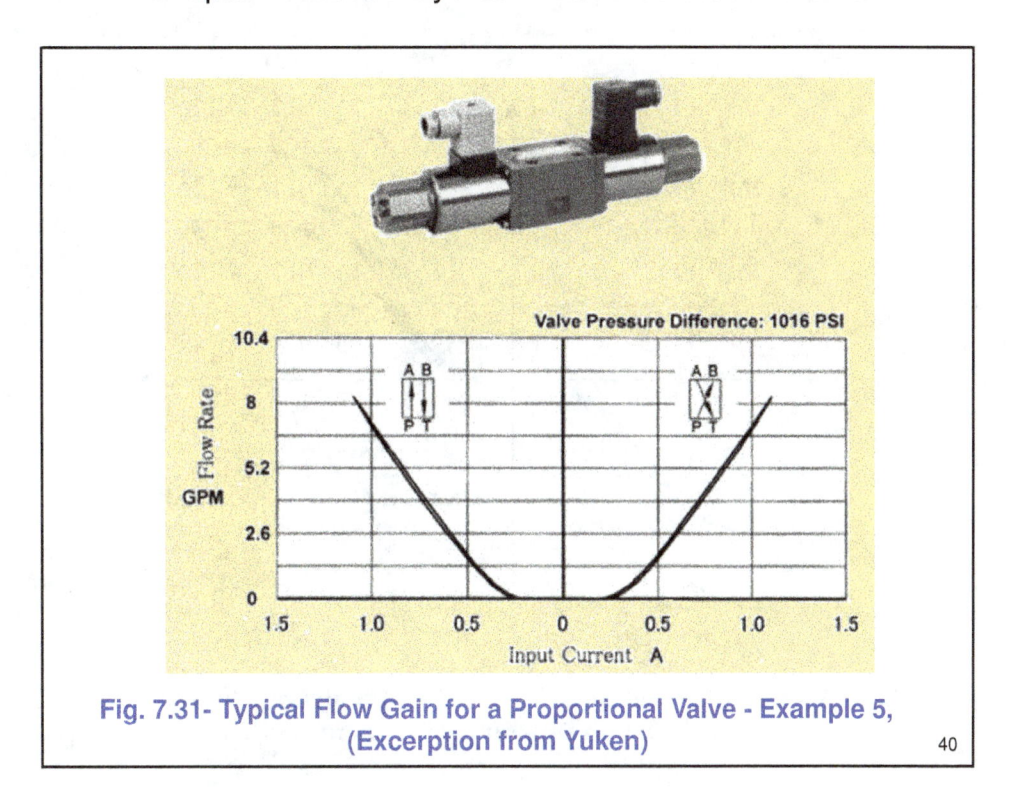

Fig. 7.31- Typical Flow Gain for a Proportional Valve - Example 5, (Excerption from Yuken)

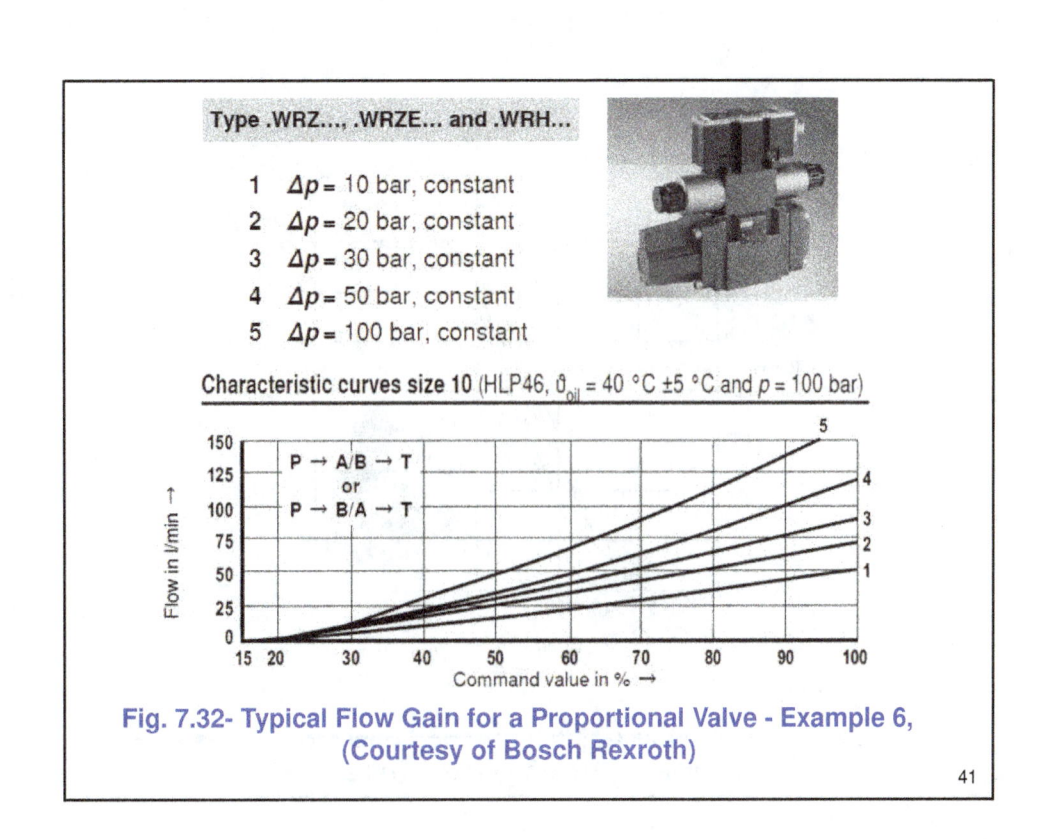

Fig. 7.32- Typical Flow Gain for a Proportional Valve - Example 6, (Courtesy of Bosch Rexroth)

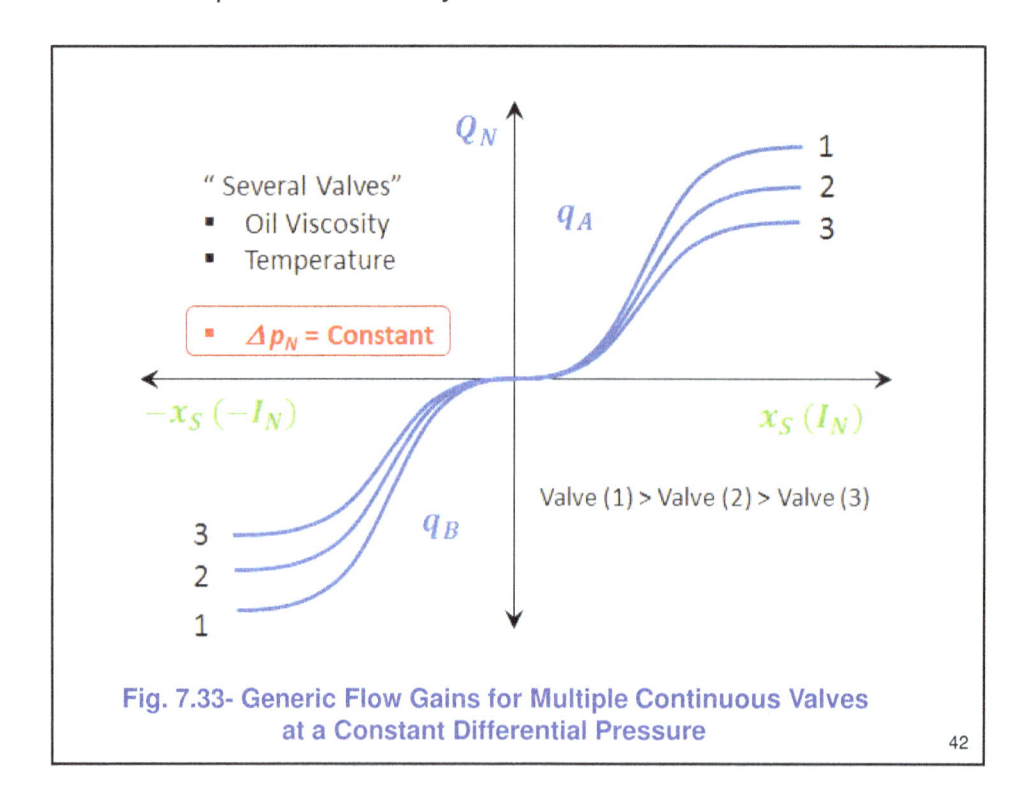

Fig. 7.33- Generic Flow Gains for Multiple Continuous Valves at a Constant Differential Pressure

"Several Valves"
- Oil Viscosity
- Temperature
- Δp_N = Constant

Valve (1) > Valve (2) > Valve (3)

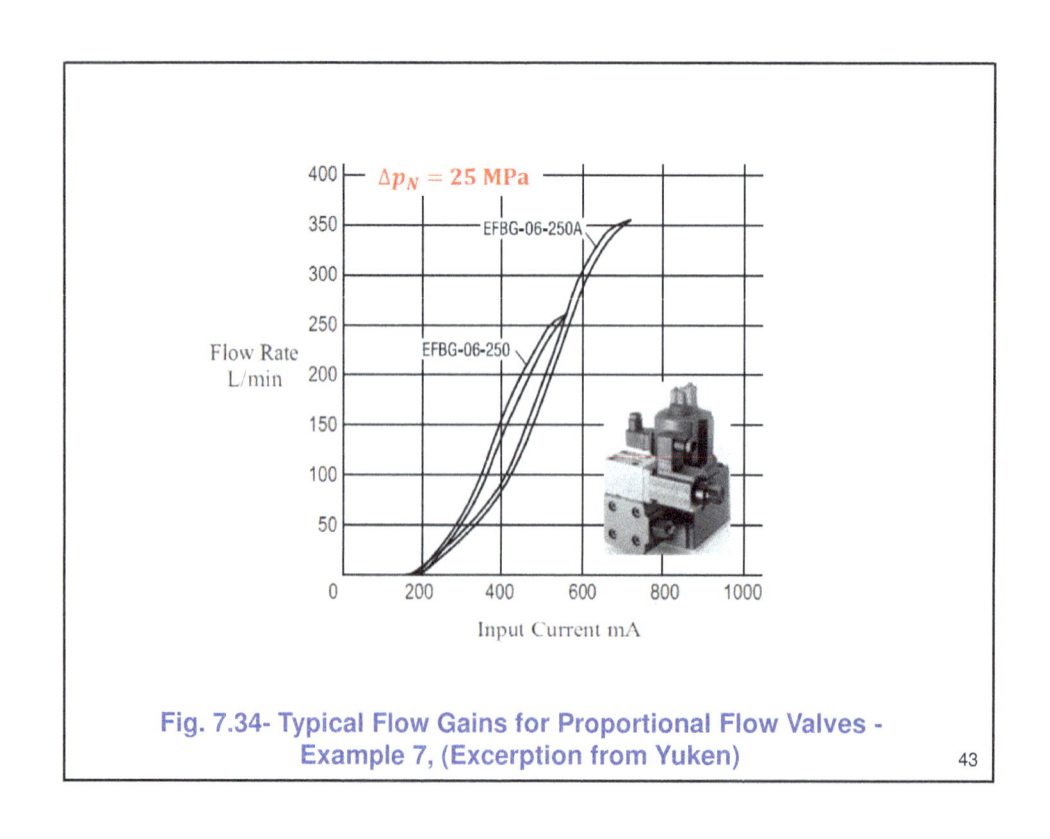

Fig. 7.34- Typical Flow Gains for Proportional Flow Valves - Example 7, (Excerption from Yuken)

$\Delta p_N = 25$ MPa

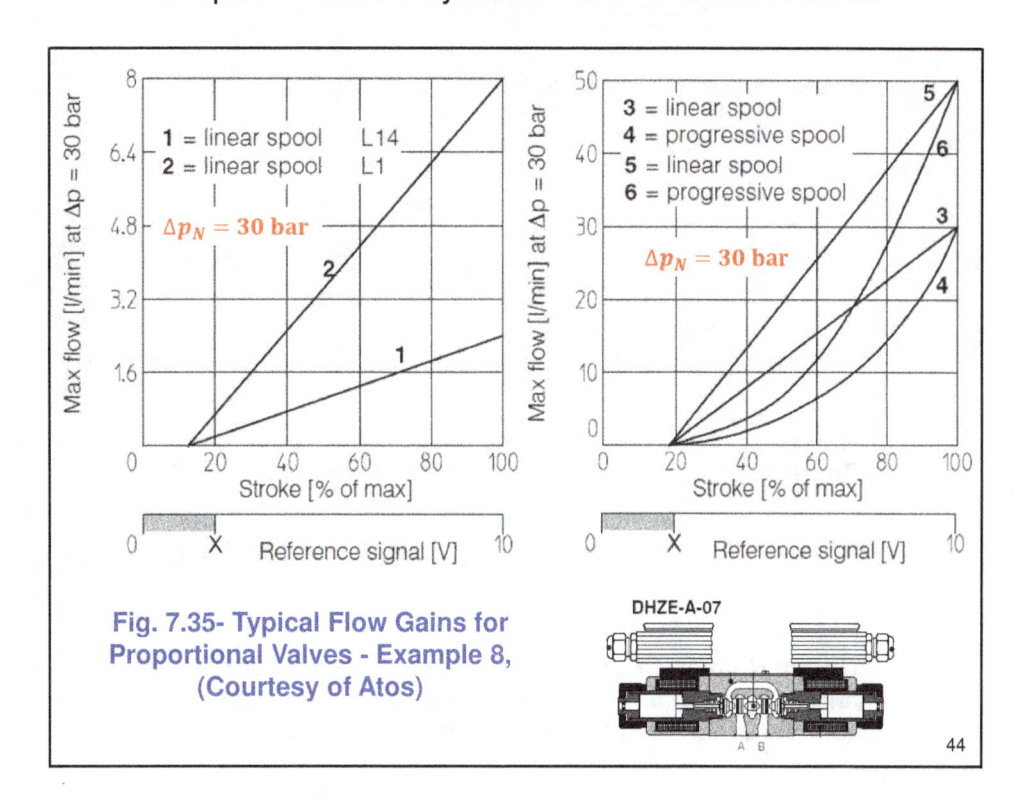

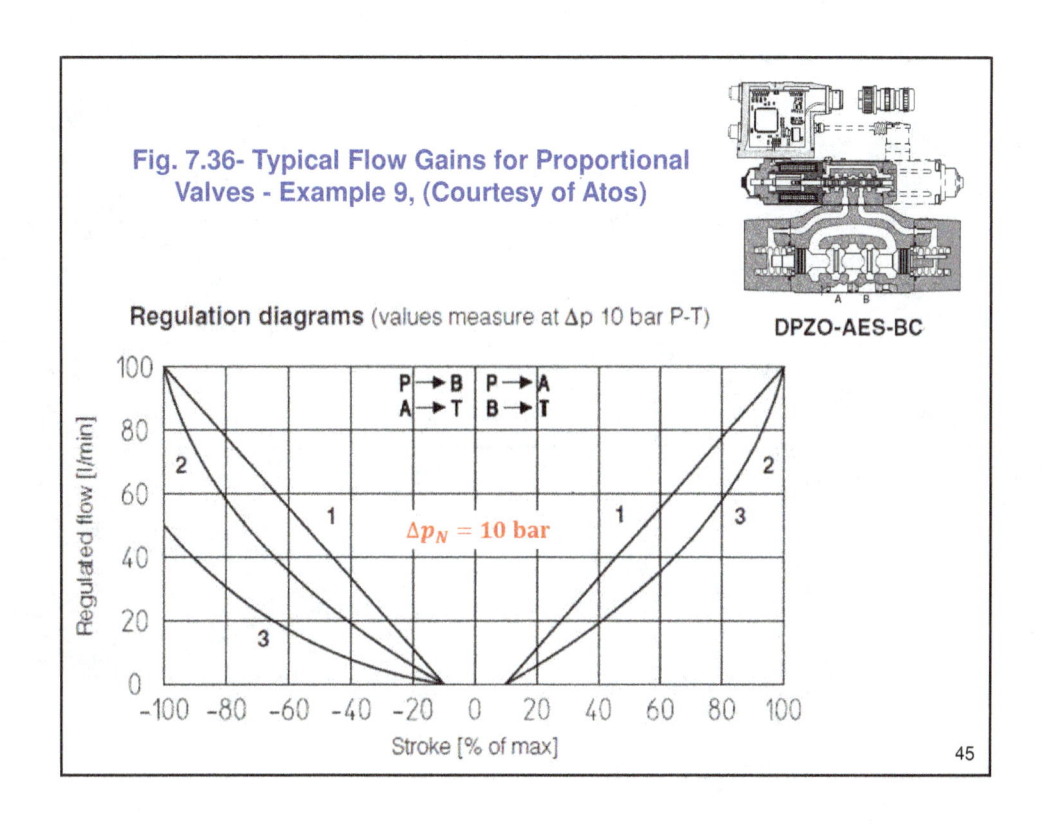

Fig. 7.35- Typical Flow Gains for Proportional Valves - Example 8, (Courtesy of Atos)

Fig. 7.36- Typical Flow Gains for Proportional Valves - Example 9, (Courtesy of Atos)

7.5.2- Valve Flow-Pressure Coefficient

$Q_N = f(\Delta p_N)$

Δp_N

$I_N = Const.$

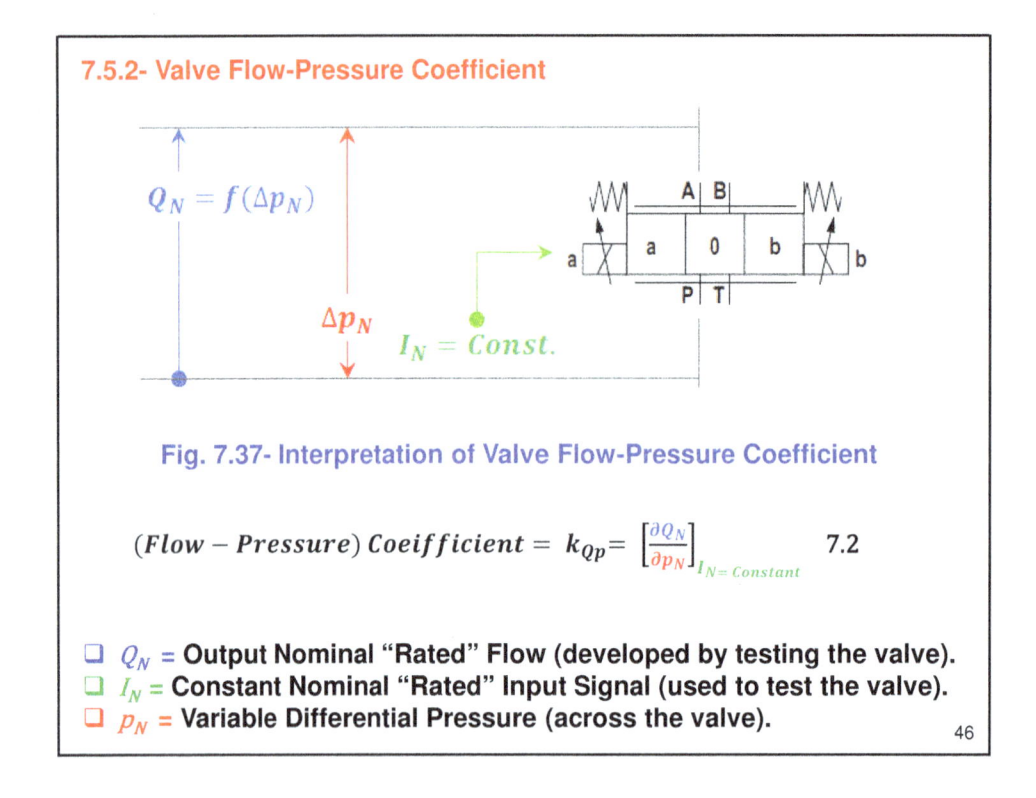

Fig. 7.37- Interpretation of Valve Flow-Pressure Coefficient

$$(Flow - Pressure)\ Coefficient = k_{Qp} = \left[\frac{\partial Q_N}{\partial p_N}\right]_{I_N = Constant} \qquad 7.2$$

- ❏ Q_N = Output Nominal "Rated" Flow (developed by testing the valve).
- ❏ I_N = Constant Nominal "Rated" Input Signal (used to test the valve).
- ❏ p_N = Variable Differential Pressure (across the valve).

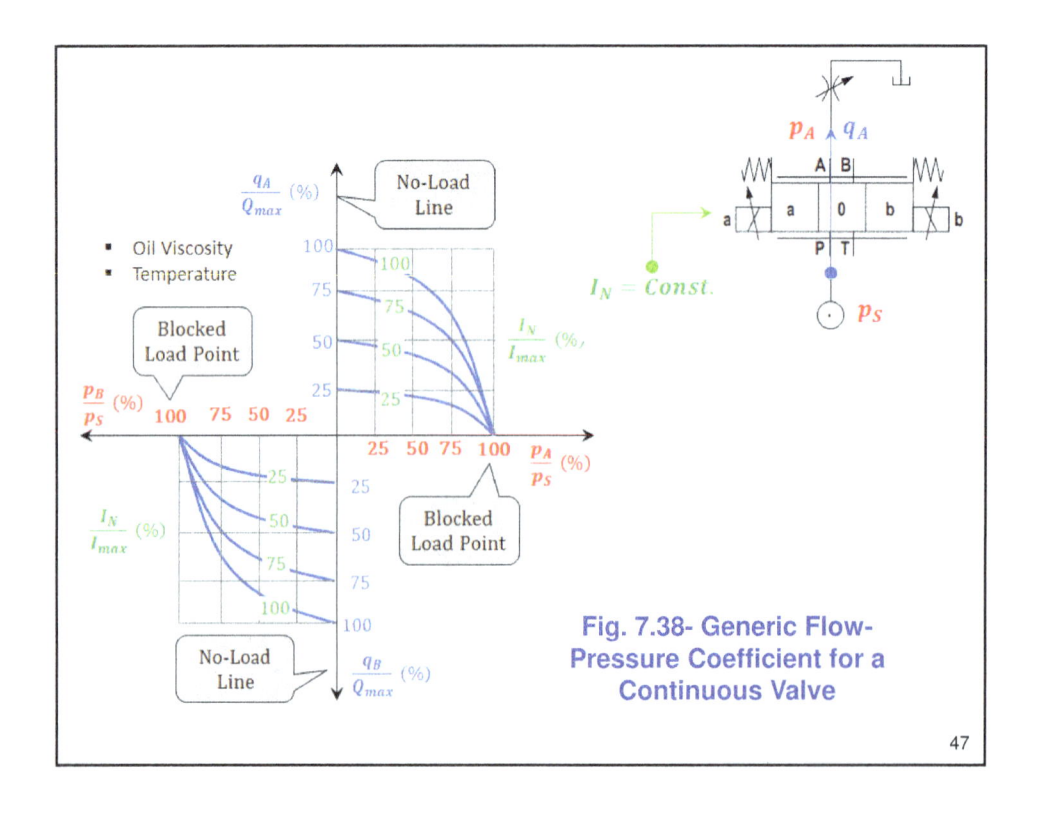

Fig. 7.38- Generic Flow-Pressure Coefficient for a Continuous Valve

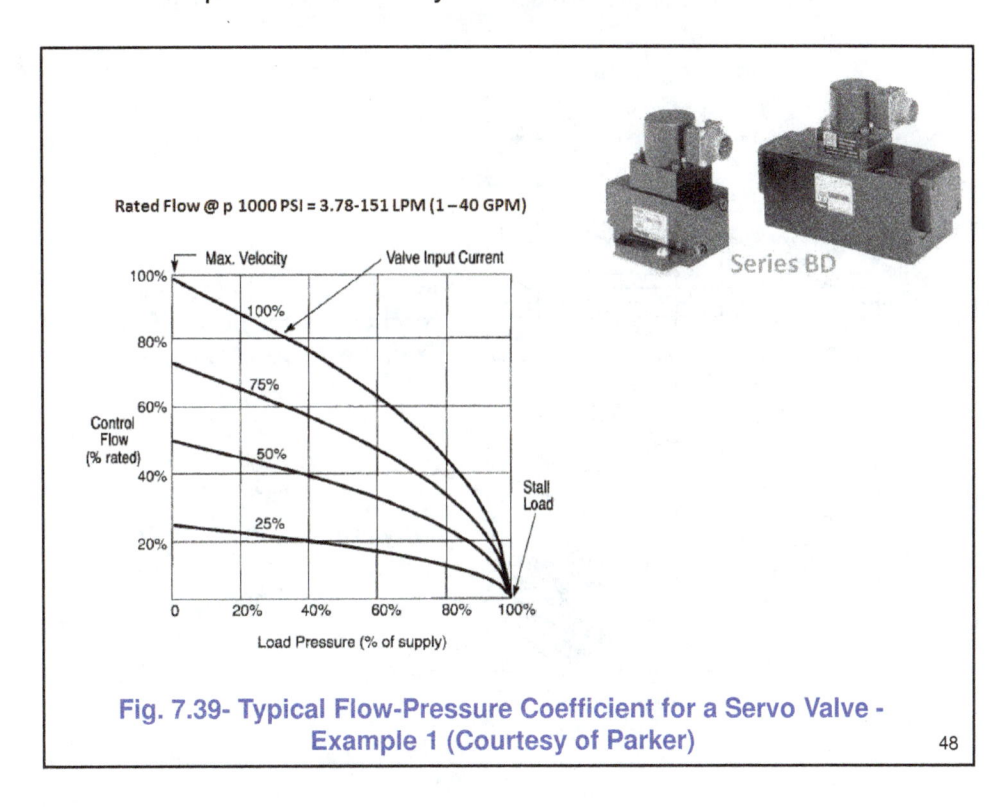

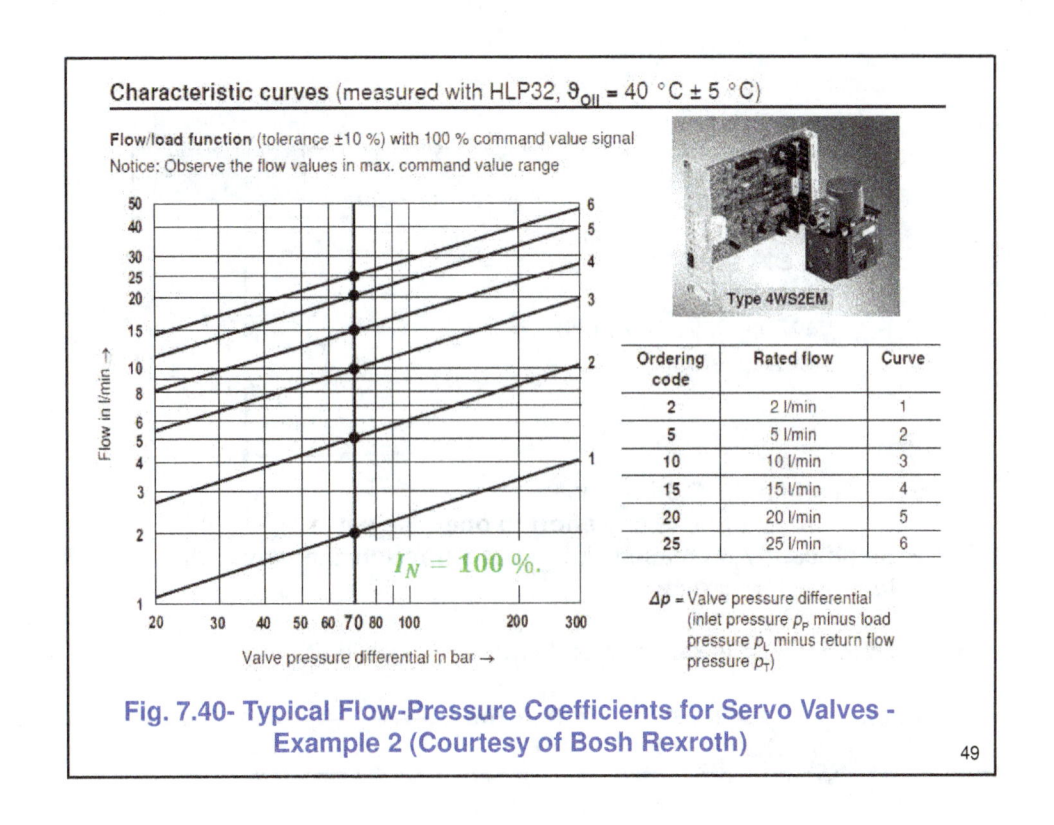

Workbook: Electro-Hydraulic Components and Systems
Chapter 7: Electro-Hydraulic Valve Selection Criteria

Fig. 7.39- Typical Flow-Pressure Coefficient for a Servo Valve - Example 1 (Courtesy of Parker)

$I_N = 100\%$.

Fig. 7.40- Typical Flow-Pressure Coefficients for Servo Valves - Example 2 (Courtesy of Bosh Rexroth)

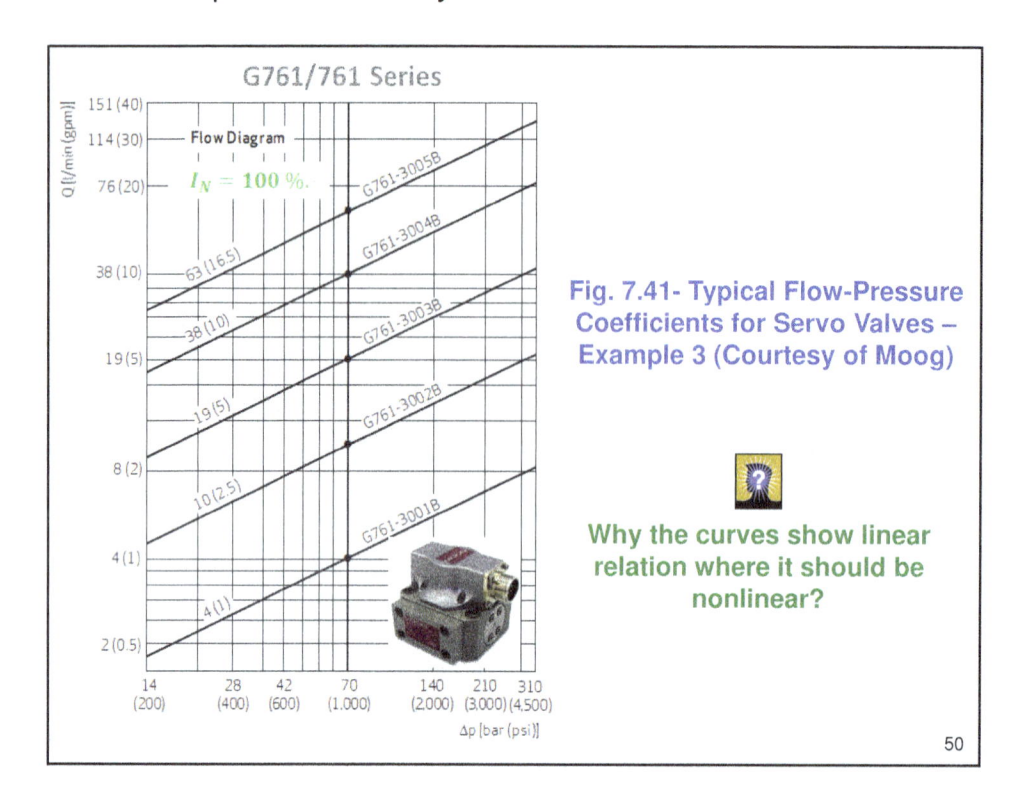

Fig. 7.41- Typical Flow-Pressure Coefficients for Servo Valves – Example 3 (Courtesy of Moog)

Why the curves show linear relation where it should be nonlinear?

7.5.3- Power Limits

$$Q \propto \sqrt{\Delta p}$$

$$Power = Q \times \Delta p$$

Fig. 7.42- Valve Power Limit Interpretation

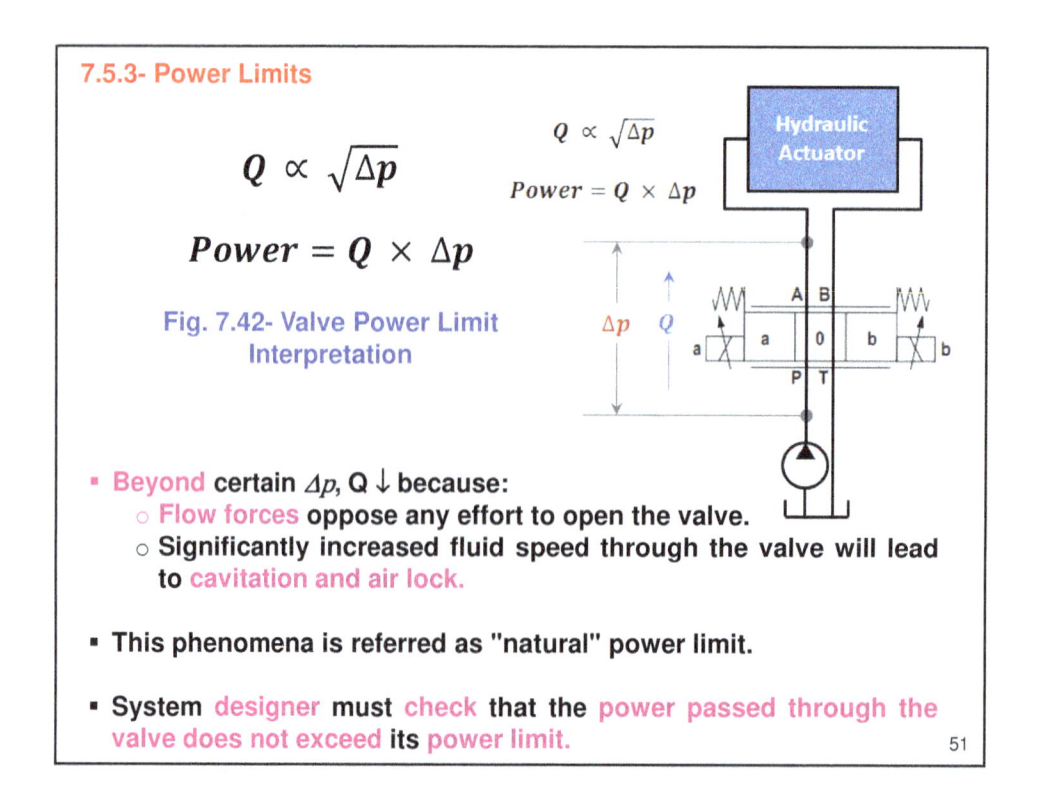

- **Beyond** certain Δp, Q ↓ because:
  - **Flow forces** oppose any effort to open the valve.
  - Significantly increased fluid speed through the valve will lead to **cavitation and air lock**.

- This phenomena is referred as "natural" power limit.

- System **designer** must **check** that the **power passed through the valve does not exceed** its **power limit**.

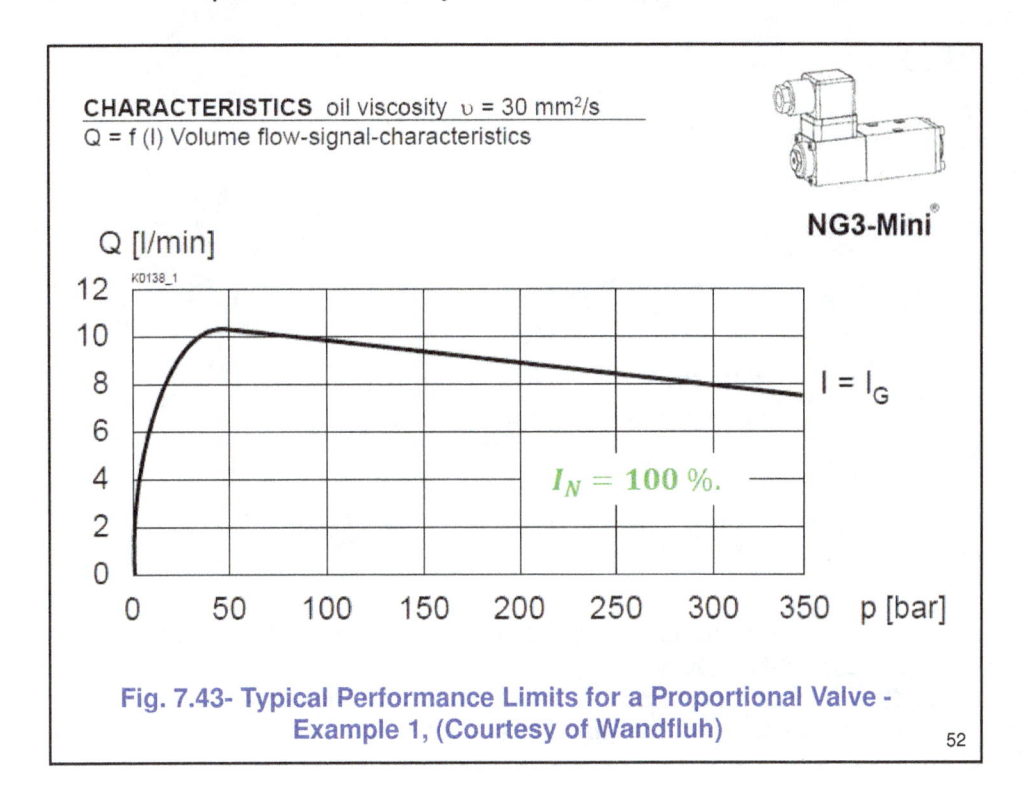

Workbook: Electro-Hydraulic Components and Systems
Chapter 7: Electro-Hydraulic Valve Selection Criteria

CHARACTERISTICS oil viscosity $\upsilon = 30$ mm²/s
$Q = f(I)$ Volume flow-signal-characteristics

NG3-Mini®

$I = I_G$

$I_N = 100\ \%.$

Fig. 7.43- Typical Performance Limits for a Proportional Valve - Example 1, (Courtesy of Wandfluh)

DHZE-A-07

1 = spool L3, S3, D3
2 = spool L5, S5, D5

$I_N = 100\ \%$

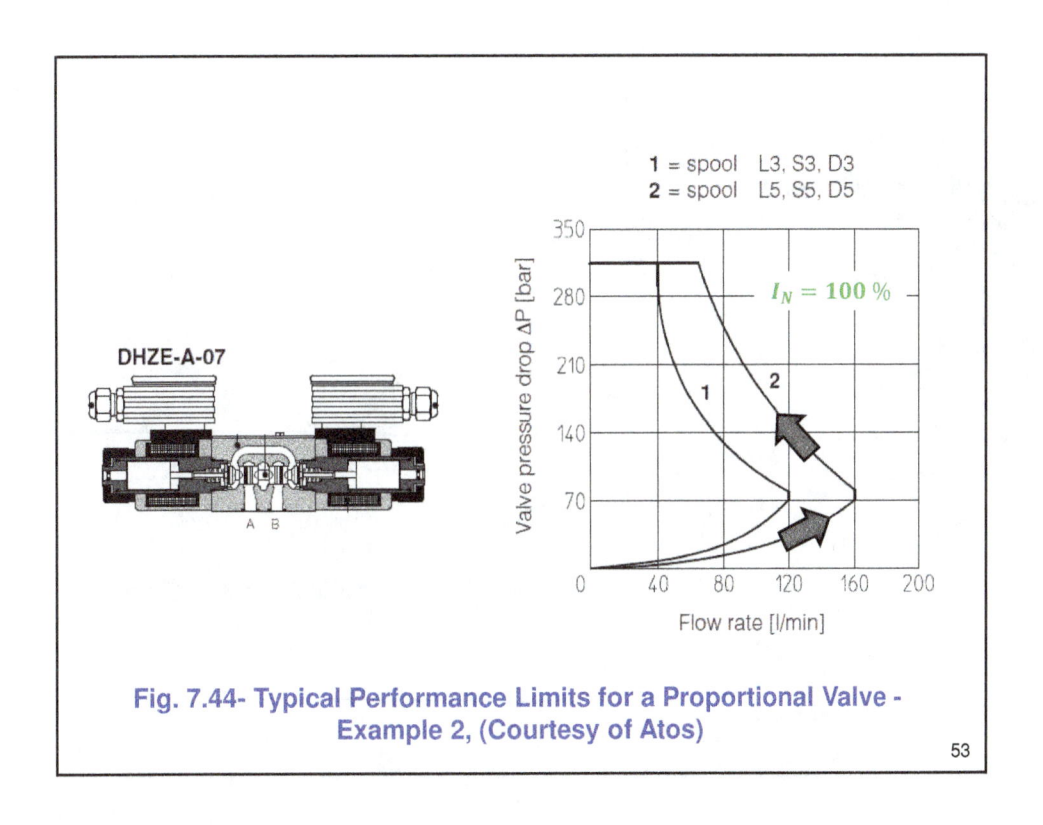

Fig. 7.44- Typical Performance Limits for a Proportional Valve - Example 2, (Courtesy of Atos)

7.5.4- Pressure Gain

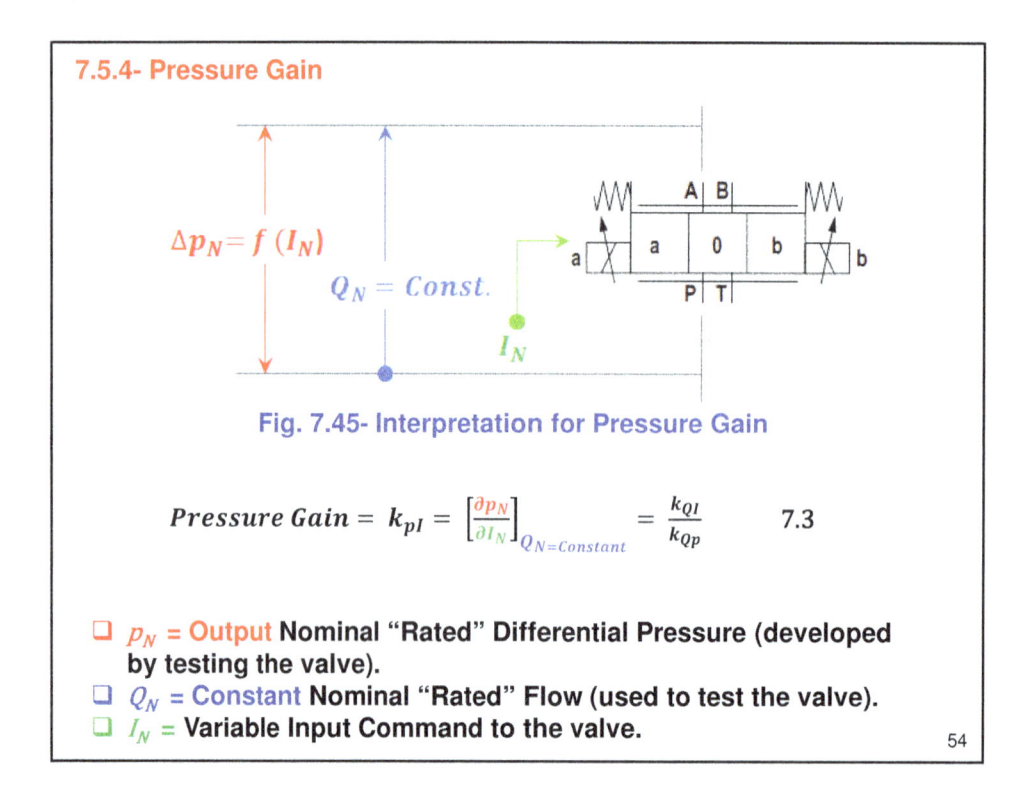

Fig. 7.45- Interpretation for Pressure Gain

$$Pressure\ Gain = k_{pI} = \left[\frac{\partial p_N}{\partial I_N}\right]_{Q_N=Constant} = \frac{k_{QI}}{k_{Qp}} \qquad 7.3$$

- p_N = **Output** Nominal "Rated" Differential Pressure (developed by testing the valve).
- Q_N = **Constant** Nominal "Rated" Flow (used to test the valve).
- I_N = **Variable** Input Command to the valve.

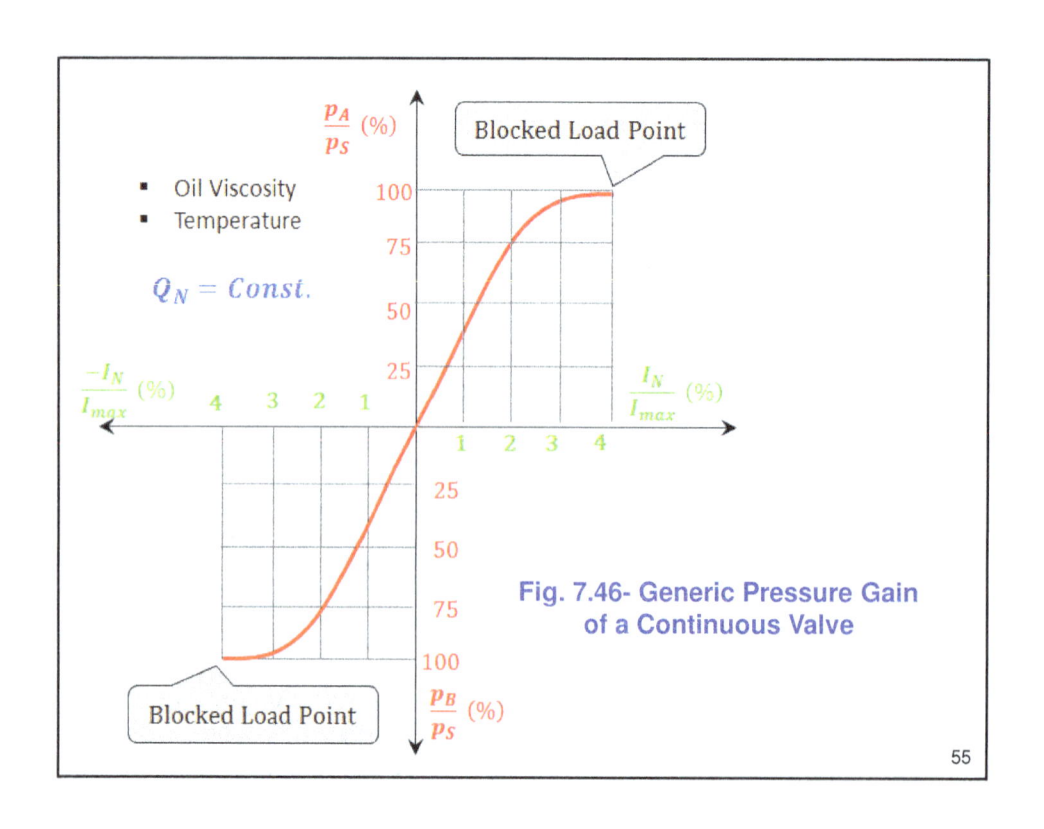

Fig. 7.46- Generic Pressure Gain of a Continuous Valve

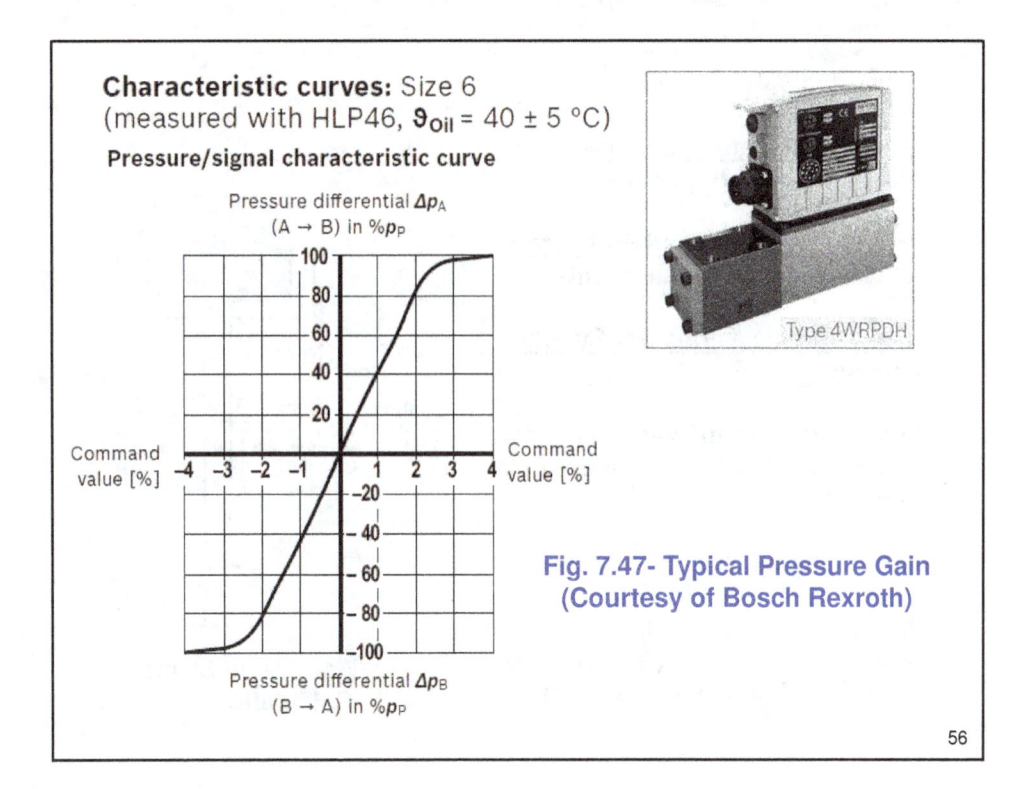

Characteristic curves: Size 6
(measured with HLP46, ϑ_{oil} = 40 ± 5 °C)
Pressure/signal characteristic curve

Fig. 7.47- Typical Pressure Gain (Courtesy of Bosch Rexroth)

7.5.5- Valve Sizing

Valve # of Stages:

- Valve sizing is the most crucial step in designing an EH system.

- Valves, and all other hydraulic components, are <u>primarily sized based on flow</u> (NOT pressure).

- Consequences of under sizing and oversizing a valve?

- Depending on the flow, an EH valve may be:
  - ❑ Direct-Operated (single-stage)
  - ❑ Pilot-Operated (two-stage or three-stage)

- Valves that handle below (80 – 100 liter/min) are direct-operated.

- Above that range, valves are pilot-operated,

What is the consequences of under-sizing or over-sizing the valve ?

Step-by-Step Valve Sizing:

- **Step 1: Calculate Load Flow** Q_L to meet the speed requirements.

- **Step 2: Calculate Load Pressure** p_L to meet the load requirements.

- **Step 3: Define a constant Supply Pressure** p_S.

  - In order to ease the valve selection, it is recommended to choose p_s so that:

  $$\Delta p_L = \Delta p_N$$

  - Where Δp_N is the **Nominal Differential Pressure** (the pressure under which the valve has been tested)

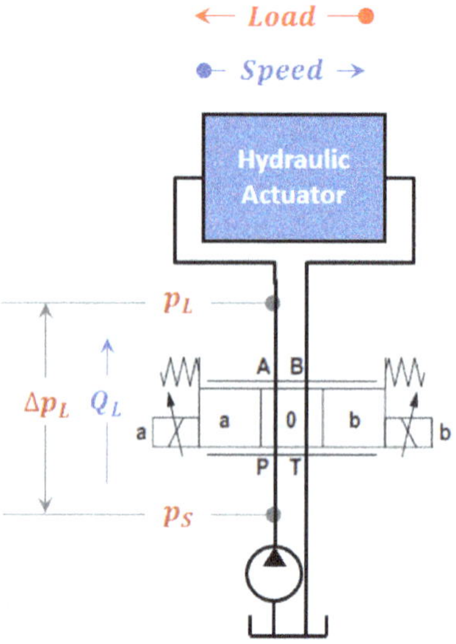

Fig. 7.48- Valve Sizing Calculations

- **Step 4: Selecting a valve from a catalogue.**

  - **Choice A: based on (Flow-Pressure) Coefficient :**

  $$k_{Qp} = \left[\frac{\partial Q_N}{\partial p_v}\right]_{I_N = Constant}$$

  - **By default**, valve is sized based on **maximum input signal** where the valve is fully opened.

  - As **best practices**, valve could be sized based on (80-90)% of input signal to **avoid** the nonlinear portion of the characteristics when the **valve is saturated.**

Workbook: Electro-Hydraulic Components and Systems
Chapter 7: Electro-Hydraulic Valve Selection Criteria

- **Examples:** Assuming: $\Delta p_L = 100$ bar and $Q_L = 30$ lit/min

- This valve is OK and met the application requirement at 95% command signal

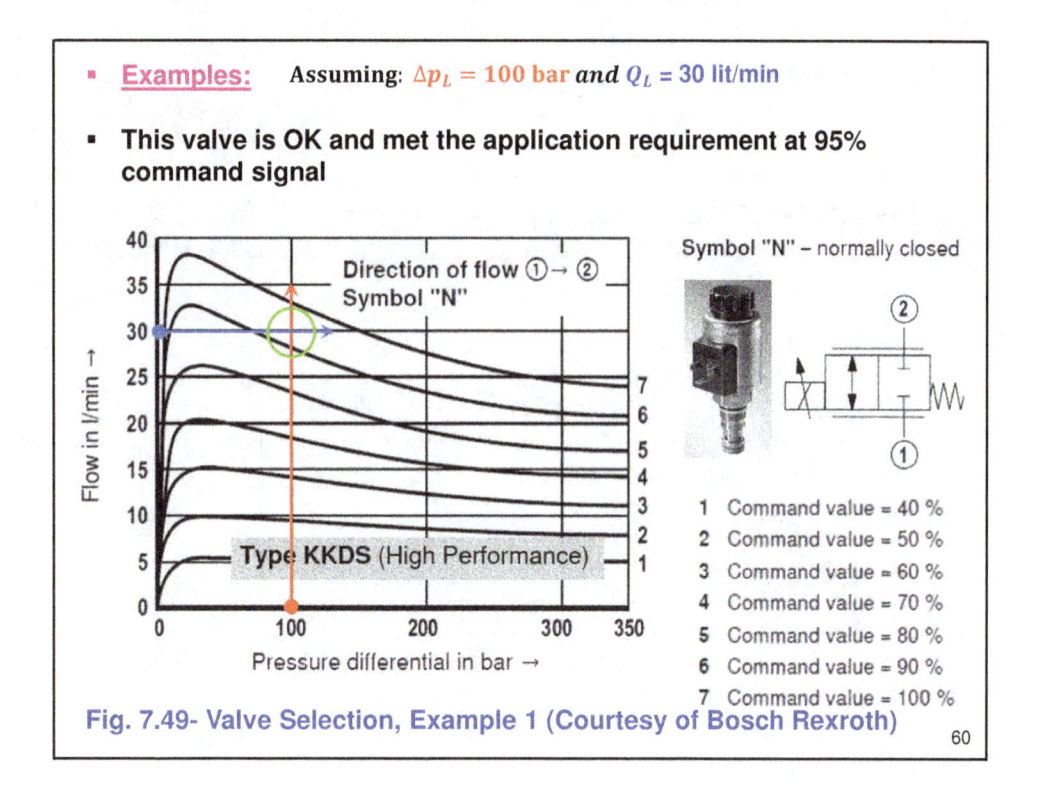

Symbol "N" – normally closed

Type KKDS (High Performance)

Direction of flow ① → ②
Symbol "N"

1 Command value = 40 %
2 Command value = 50 %
3 Command value = 60 %
4 Command value = 70 %
5 Command value = 80 %
6 Command value = 90 %
7 Command value = 100 %

Fig. 7.49- Valve Selection, Example 1 (Courtesy of Bosch Rexroth)

- The correct valve selection here is: Ordering Code 25 (Curve 6)
- The valve meets the operating point requirements at 100% opening.

Characteristic curves (measured with HLP32, $\vartheta_{Oil} = 40\ °C \pm 5\ °C$)

Flow/load function (tolerance ±10 %) with 100 % command value signal
Notice: Observe the flow values in max. command value range

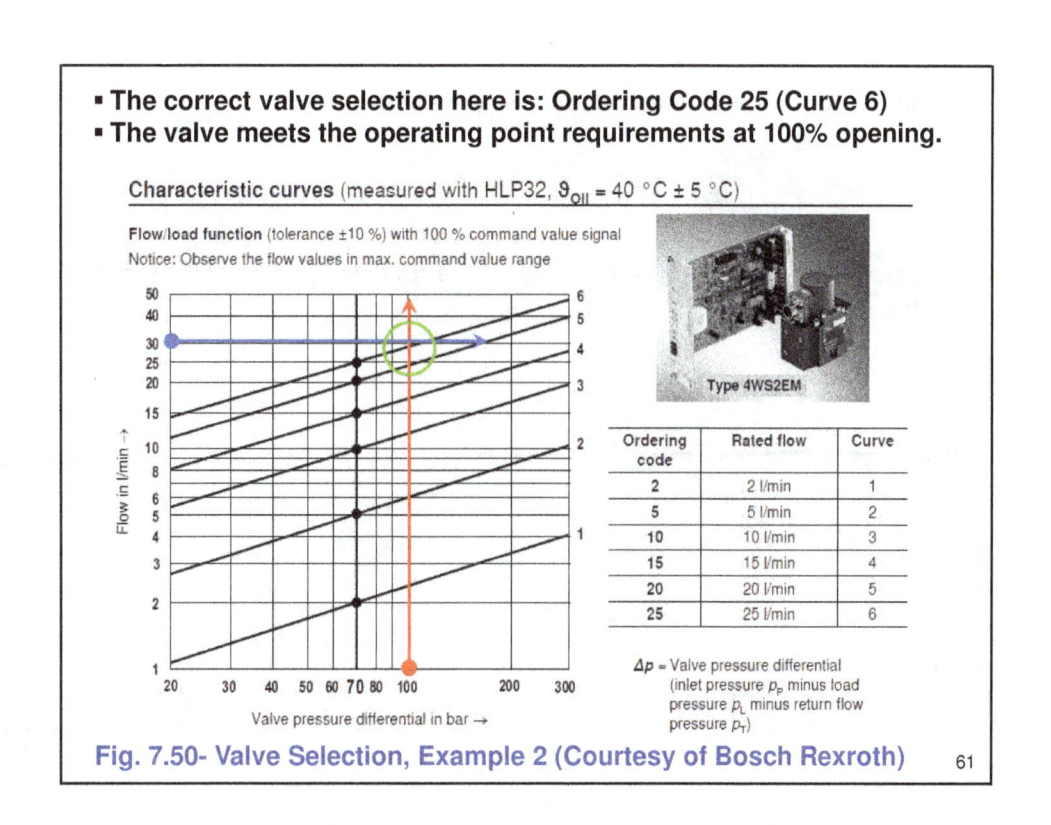

Type 4WS2EM

Ordering code	Rated flow	Curve
2	2 l/min	1
5	5 l/min	2
10	10 l/min	3
15	15 l/min	4
20	20 l/min	5
25	25 l/min	6

Δp = Valve pressure differential
(inlet pressure p_P minus load pressure p_L minus return flow pressure p_T)

Fig. 7.50- Valve Selection, Example 2 (Courtesy of Bosch Rexroth)

Workbook: Electro-Hydraulic Components and Systems
Chapter 7: Electro-Hydraulic Valve Selection Criteria

- **This valve is larger than needed (oversize).**

  Why?

  CHARACTERISTICS oil viscosity $\upsilon = 30$ mm²/s   **M33 x 2**
  $Q = f(\Delta p)$ Pressure loss/flow-characteristics   Wandfluh standard
  over 1 metering edge

  [Graph: Q [l/min] vs p [bar], curves labeled A-T, B-T, and P-A, P-B; $I_N = 100\%$]

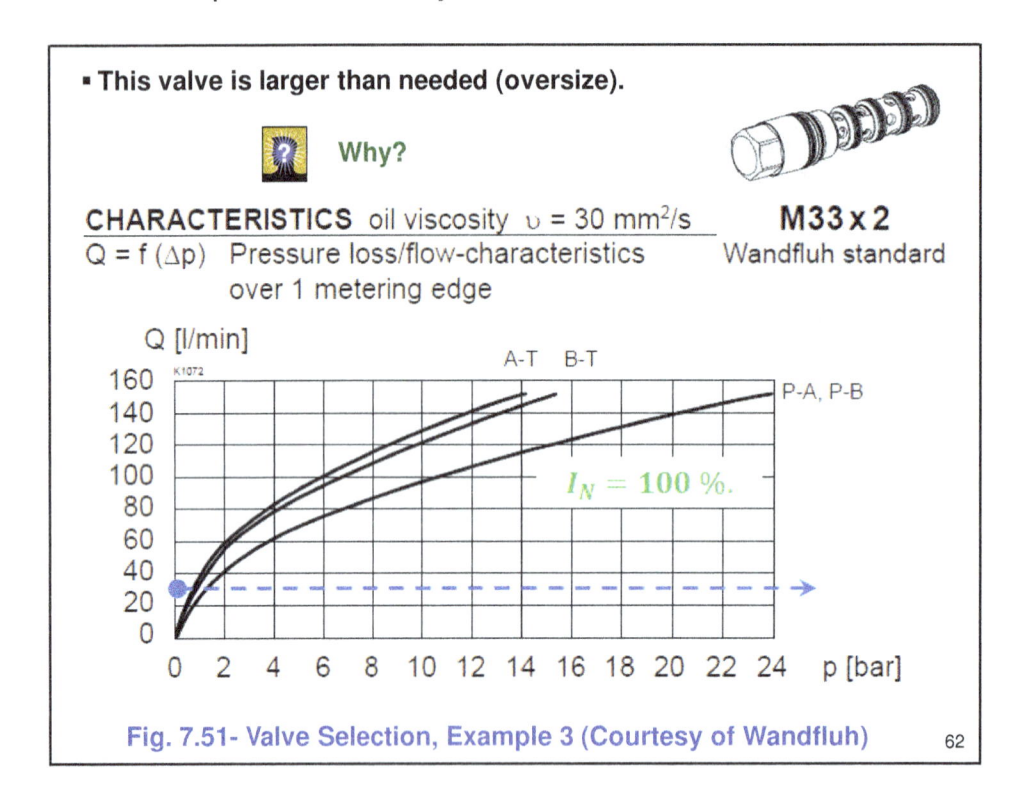

  Fig. 7.51- Valve Selection, Example 3 (Courtesy of Wandfluh)

- **This valve is smaller than needed (undersize).**

  Why?

  CHARACTERISTICS oil viscosity $\upsilon = 30$ mm²/s   **NG3-Mini**
  $Q = f(I)$ Volume flow-signal-characteristics

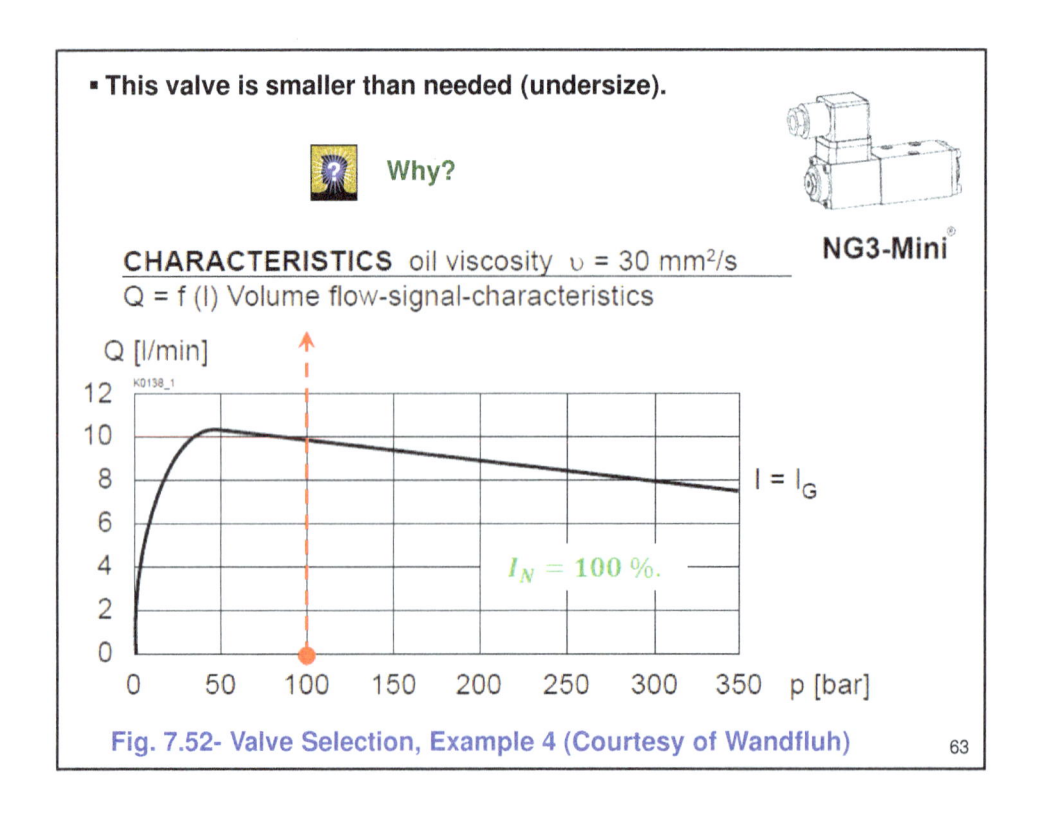

  [Graph: Q [l/min] vs p [bar], curve labeled $I = I_G$; $I_N = 100\%$]

  Fig. 7.52- Valve Selection, Example 4 (Courtesy of Wandfluh)

- **Choice B: Selecting a valve based on Flow Gain:**

$$k_{QI} = \left[\frac{\partial Q_L}{\partial I_v}\right]_{p_N = Constant}$$

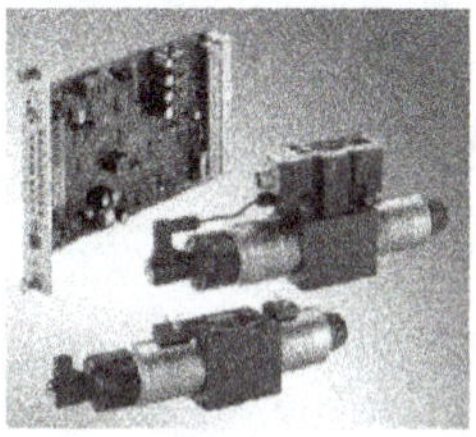

➤ **Case 1 (Direct Selection) if** $\Delta p_L = \Delta p_N$

- **Example:**
- This valve is little smaller but still can be used with a little increase in the supply pressure.

4WRE

Characteristic curves: Type 4WREE (measured with HLP46, ϑ_{Oil} = 40 °C ± 5 °C and p = 100 bar)   Size 6

8 l/min rated flow with 10 bar valve pressure differential

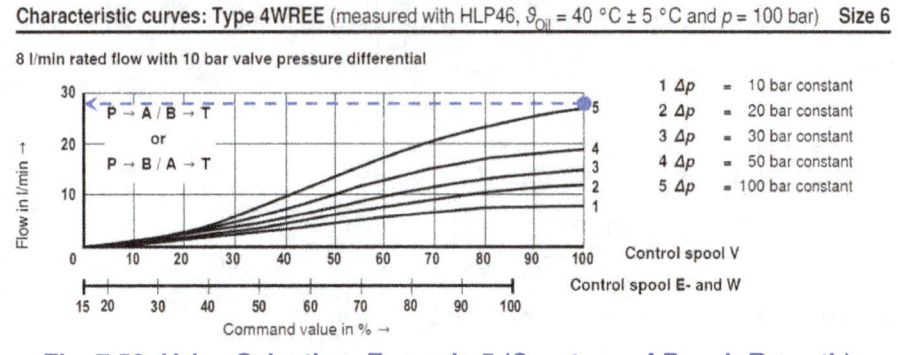

1 Δp	= 10 bar constant
2 Δp	= 20 bar constant
3 Δp	= 30 bar constant
4 Δp	= 50 bar constant
5 Δp	= 100 bar constant

Fig. 7.53- Valve Selection, Example 5 (Courtesy of Bosch Rexroth)

- **This valve is larger than needed (oversize).**

Type .WRZ..., .WRZE... and .WRH...

1. Δp = 10 bar, constant
2. Δp = 20 bar, constant
3. Δp = 30 bar, constant
4. Δp = 50 bar, constant
5. Δp = 100 bar, constant

Characteristic curves size 10 (HLP46, ϑ_{oil} = 40 °C ±5 °C and p = 100 bar)

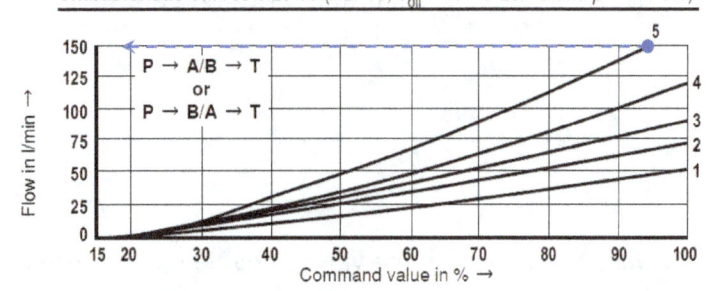

Fig. 7.54- Valve Selection, Example 6 (Courtesy of Bosch Rexroth)

- This valve is still considered oversize unless being used within 60% of the maximum input signal (spool stroke).

Fig. 7.55- Valve Selection, Example 7 (Courtesy of Hydraforce)

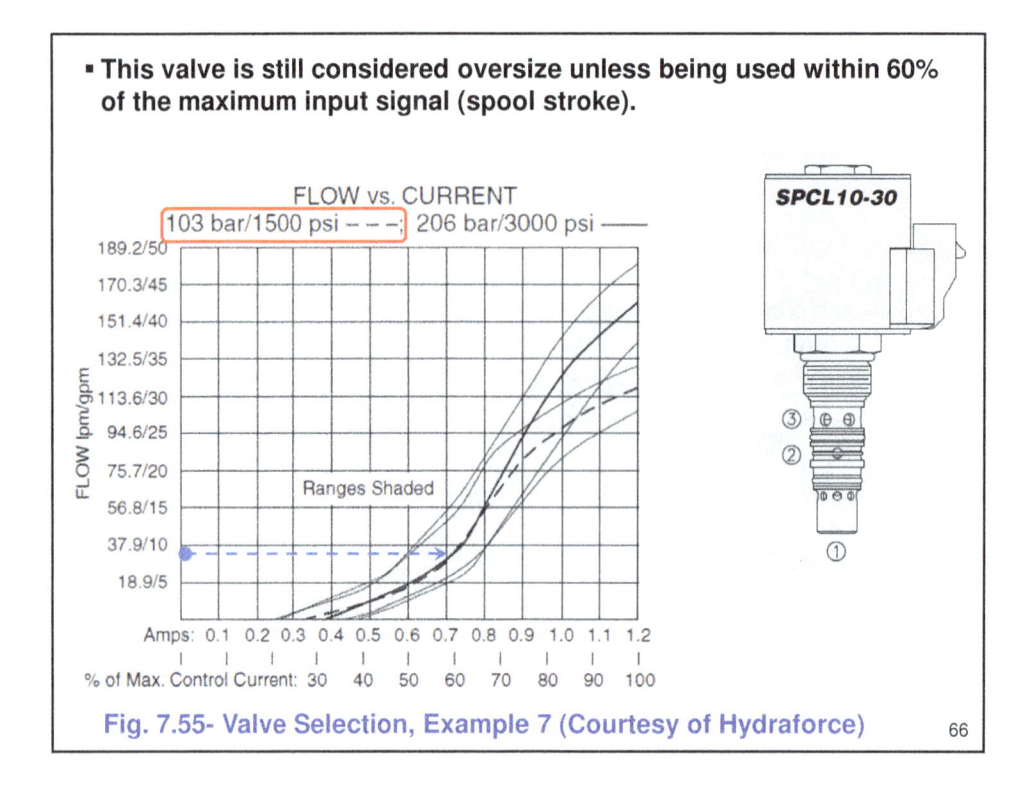

> Case 2 (Indirect Selection) if $\quad \Delta p_L \neq \Delta p_N$

The valve manufacturer can't test a valve under infinite number of differential pressure to cover all possible applications.

How to select a valve if it is tested at a differential pressure different from the differential pressure in my application?

$$Q \propto \sqrt{\Delta p} \rightarrow \frac{Q_N}{Q_L} = \sqrt{\frac{\Delta p_N}{\Delta p_L}} \; at\; a\; constnat\; input\; signal$$

$$\rightarrow Q_N = Q_L \sqrt{\frac{\Delta p_N}{\Delta p_L}} \qquad 7.4$$

❑ **Note:** Review power limit of the valve Variable spool position.

Workbook: Electro-Hydraulic Components and Systems
Chapter 7: Electro-Hydraulic Valve Selection Criteria

$$Q_N\left(\frac{lit}{min}\right) = 30\left(\frac{lit}{min}\right)\sqrt{\frac{30\ bar}{100\ bar}} = 16.4$$

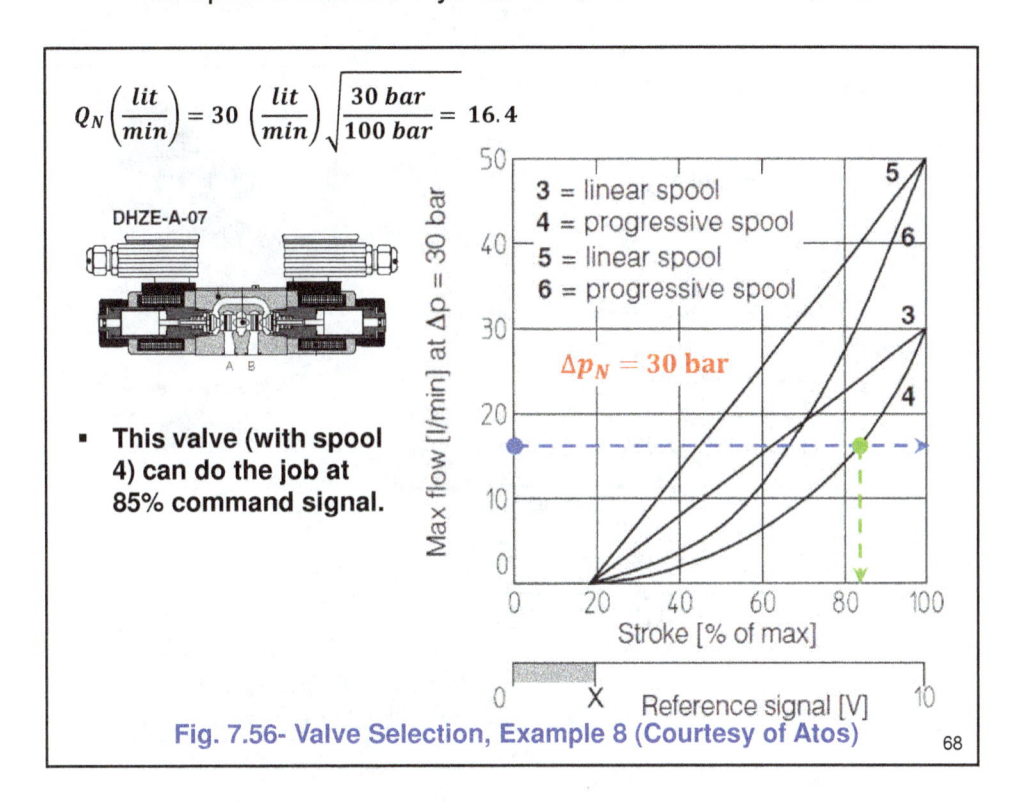

DHZE-A-07

- This valve (with spool 4) can do the job at 85% command signal.

Graph: Max flow [l/min] at $\Delta p = 30$ bar vs Stroke [% of max]
- 3 = linear spool
- 4 = progressive spool
- 5 = linear spool
- 6 = progressive spool

$\Delta p_N = 30$ bar

Reference signal [V]

Fig. 7.56- Valve Selection, Example 8 (Courtesy of Atos)

DKZOR-AES-BC-171

- Since the nominal pressure is the same, the previous calculation is valid
- This valve is a little smaller but still can be used with a little increase in the supply pressure.

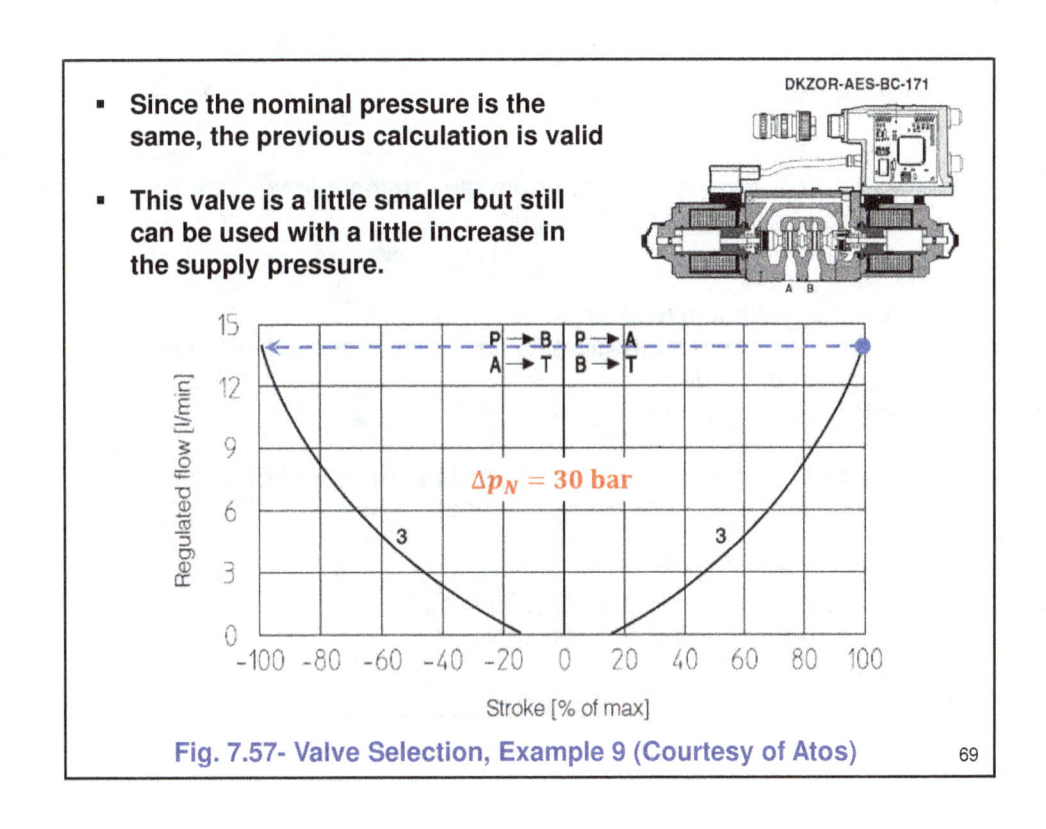

Graph: Regulated flow [l/min] vs Stroke [% of max]

P→B | P→A
A→T | B→T

$\Delta p_N = 30$ bar

Fig. 7.57- Valve Selection, Example 9 (Courtesy of Atos)

251

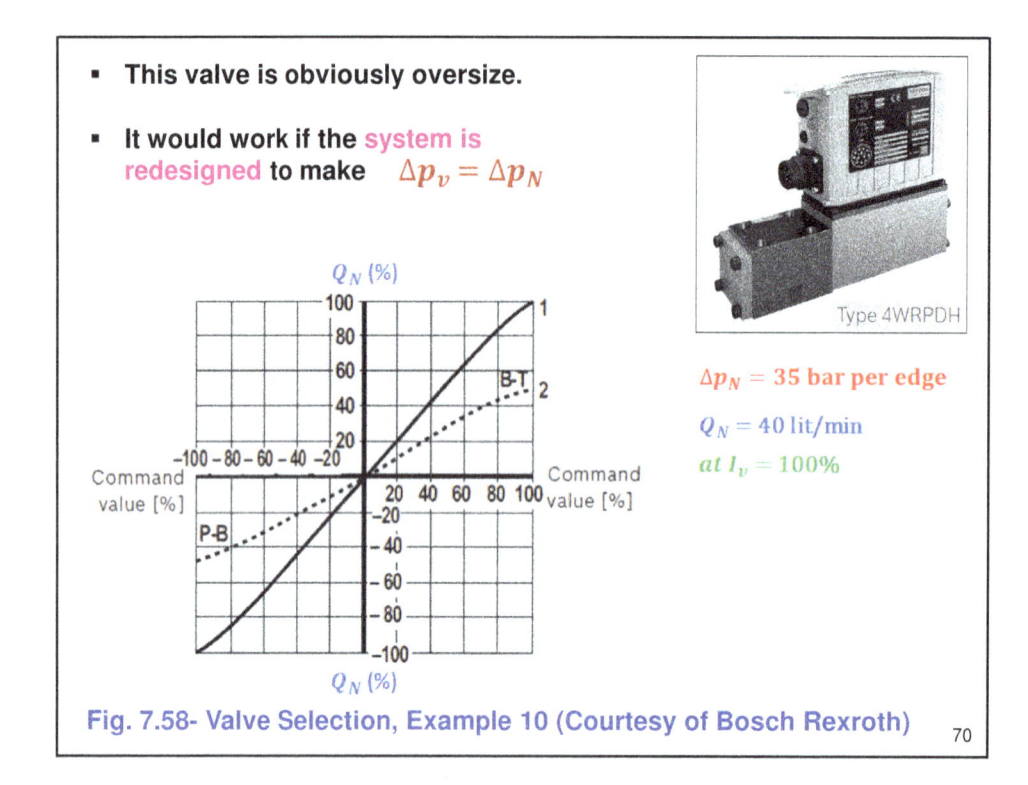

- This valve is obviously oversize.

- It would work if the system is redesigned to make $\Delta p_v = \Delta p_N$

$\Delta p_N = 35$ bar per edge

$Q_N = 40$ lit/min

$at\ I_v = 100\%$

Type 4WRPDH

Fig. 7.58- Valve Selection, Example 10 (Courtesy of Bosch Rexroth)

7.5.6- Valve Hysteresis

❑ **Definition:**
- Deviation of a valve response from the original (design) values.
- Provided as curves or just a value.

❑ **Hysteresis Measures:**
- Absolute maximum hysteresis.
- Hysteresis around the Null (almost maximum because of high friction around Null).
- Hysteresis at a specific signal.

❑ **Best practice** is to check the hysteresis at the spool position around which the valve works the most.

❑ **Order of magnitude:**
- Force-controlled Proportional Valves ≤ 5%.
- Stroke-controlled Proportional Valves ≤ 1%
- Servo Valves ≤ 0.5%

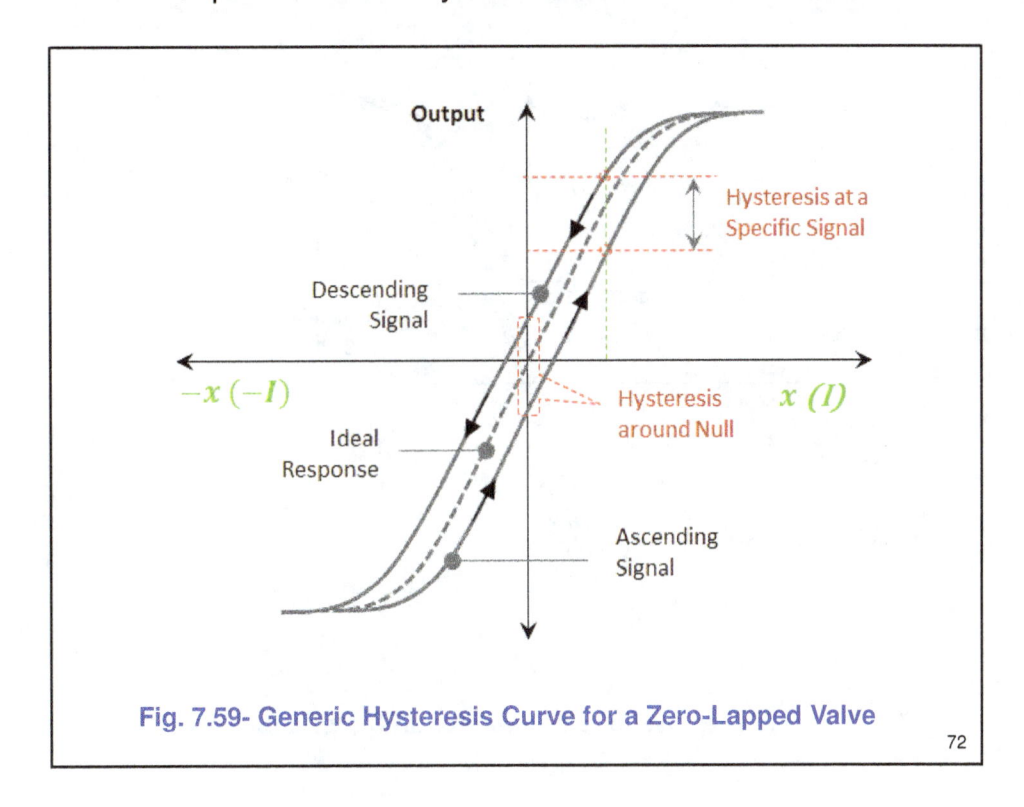

Fig. 7.59- Generic Hysteresis Curve for a Zero-Lapped Valve

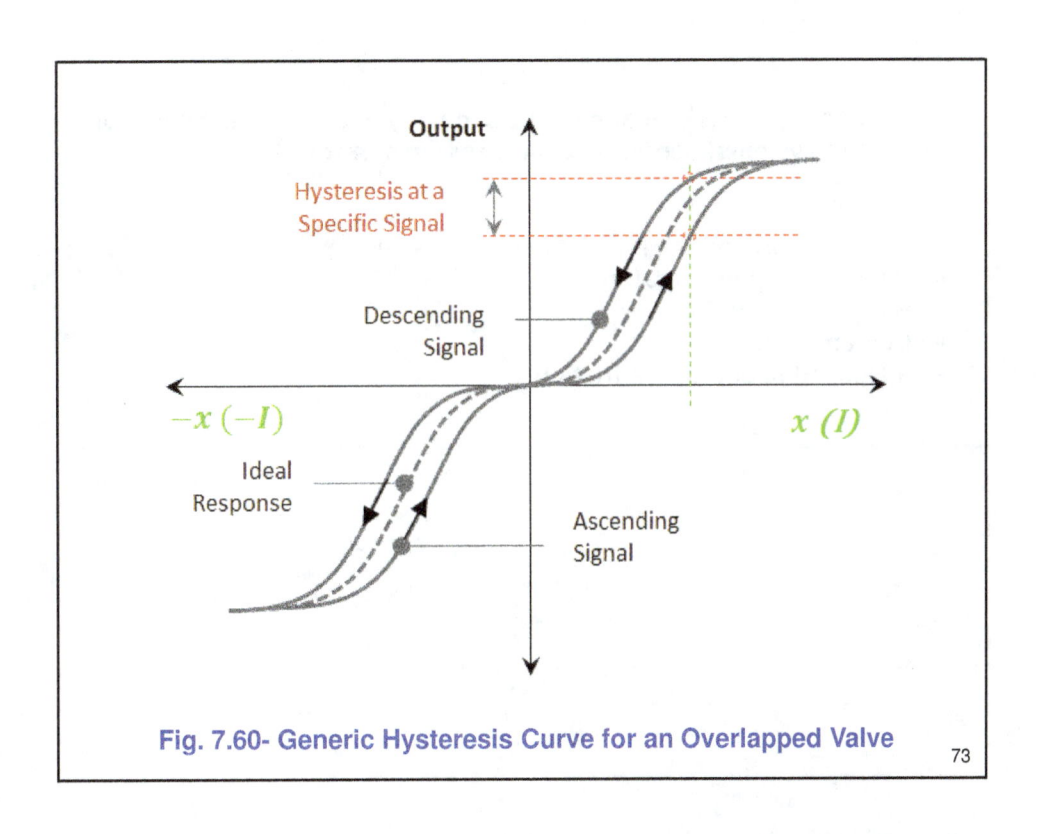

Fig. 7.60- Generic Hysteresis Curve for an Overlapped Valve

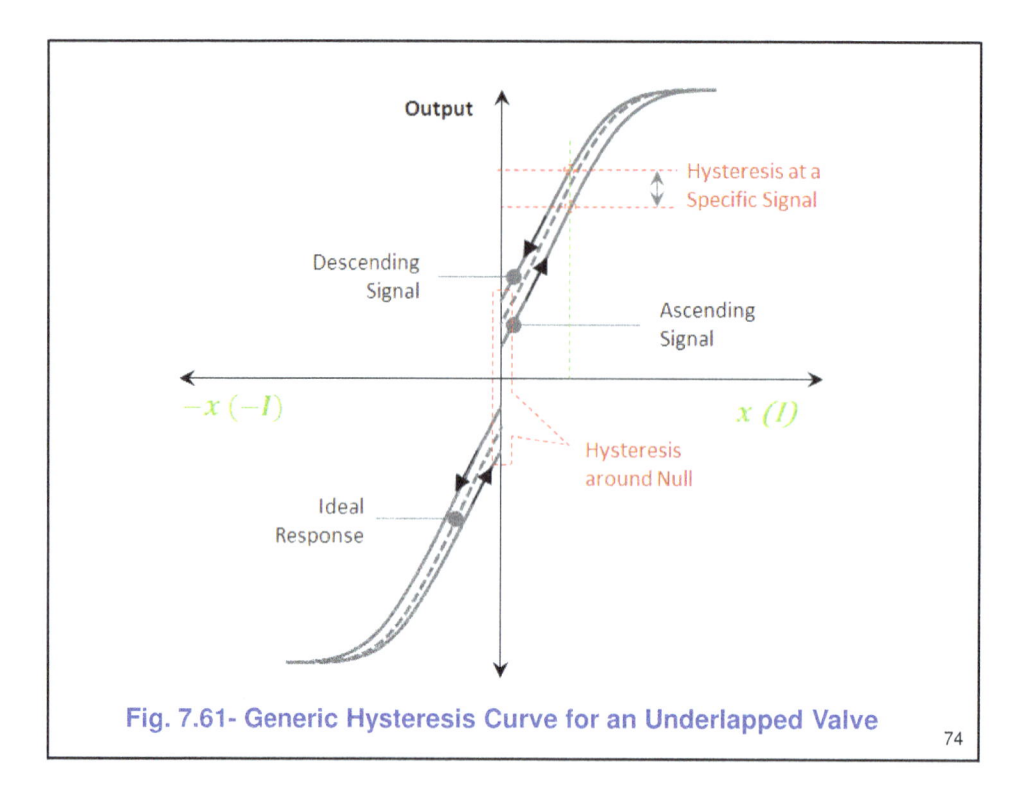

Fig. 7.61- Generic Hysteresis Curve for an Underlapped Valve

- **Effect of hysteresis:** it affects the accuracy and the repeatability of control systems, particularly the position control.

- **Working conditions that increase hysteresis:**
  - High working temperature.
  - High viscous fluid.
  - Contamination.
  - Increased spool-sleeve friction.

Threshold (Reponse Sensitivity)

- **Is the minimum current required to move forward the spool from a stationary position (other than the null).**

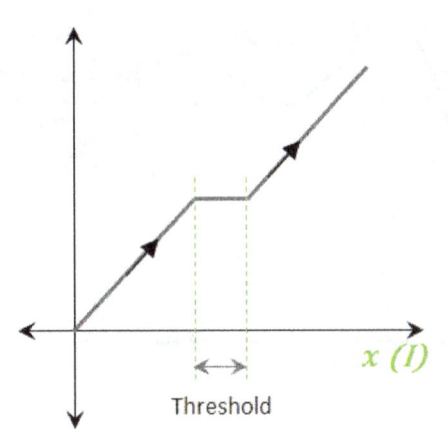

Fig. 7.62- Threshold for a Continuous Valve

Inversion Range (Reversal Error)

- **Is the minimum current required to move backward the spool from a stationary position (other than the null).**

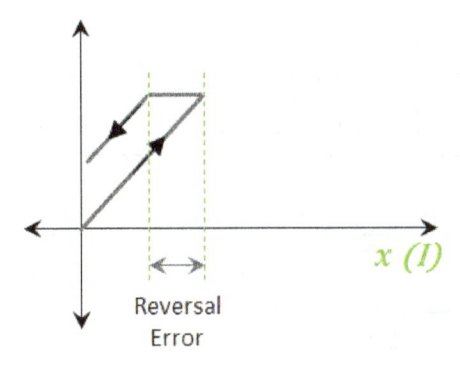

Fig. 7.63- Reversal Error for a Continuous Valve

Workbook: Electro-Hydraulic Components and Systems
Chapter 7: Electro-Hydraulic Valve Selection Criteria

7.6- Valve Dynamic Characteristics
7.6.1- Step Response

Fig. 7.64- Generic Step Response for a Continuous Valve

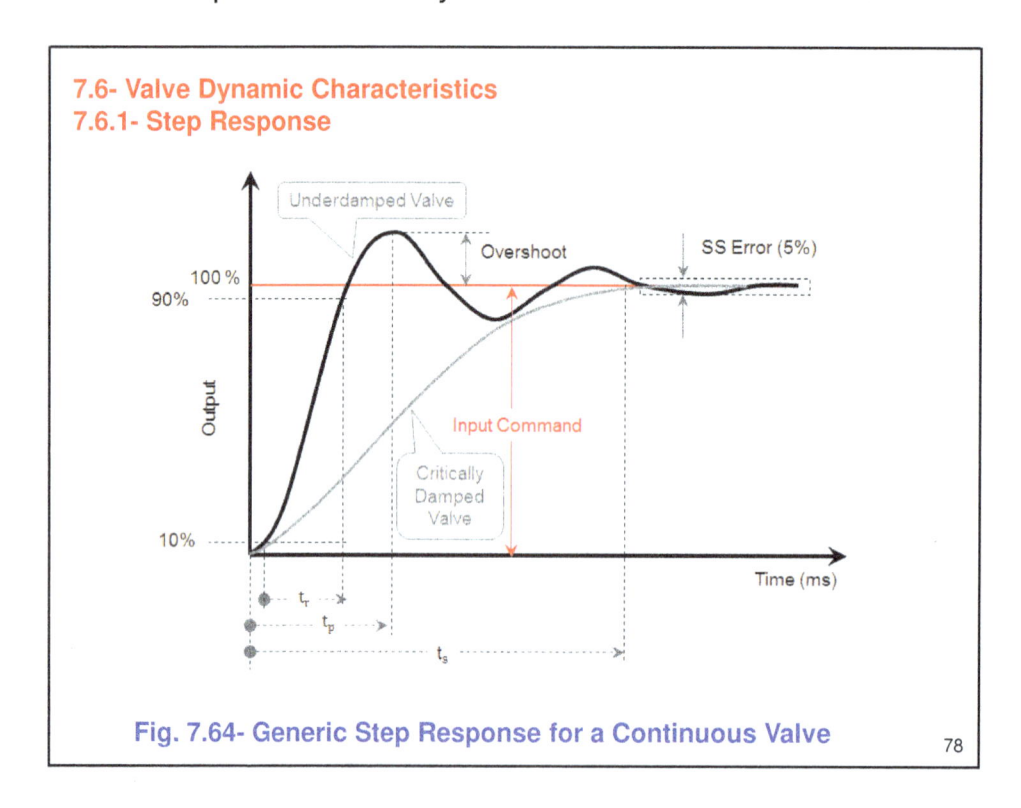

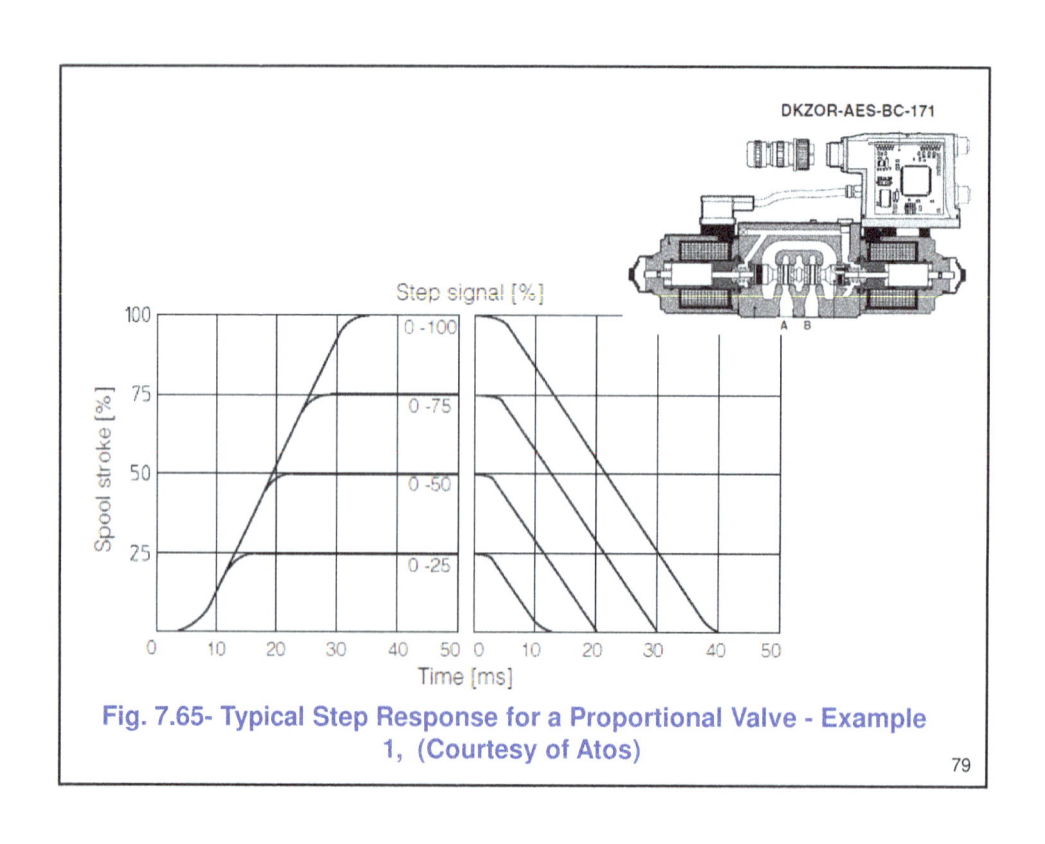

Fig. 7.65- Typical Step Response for a Proportional Valve - Example 1, (Courtesy of Atos)

Characteristic curves: Size 16
(measured with HLP46, ϑ_{Oil} = 40 ± 5 °C)

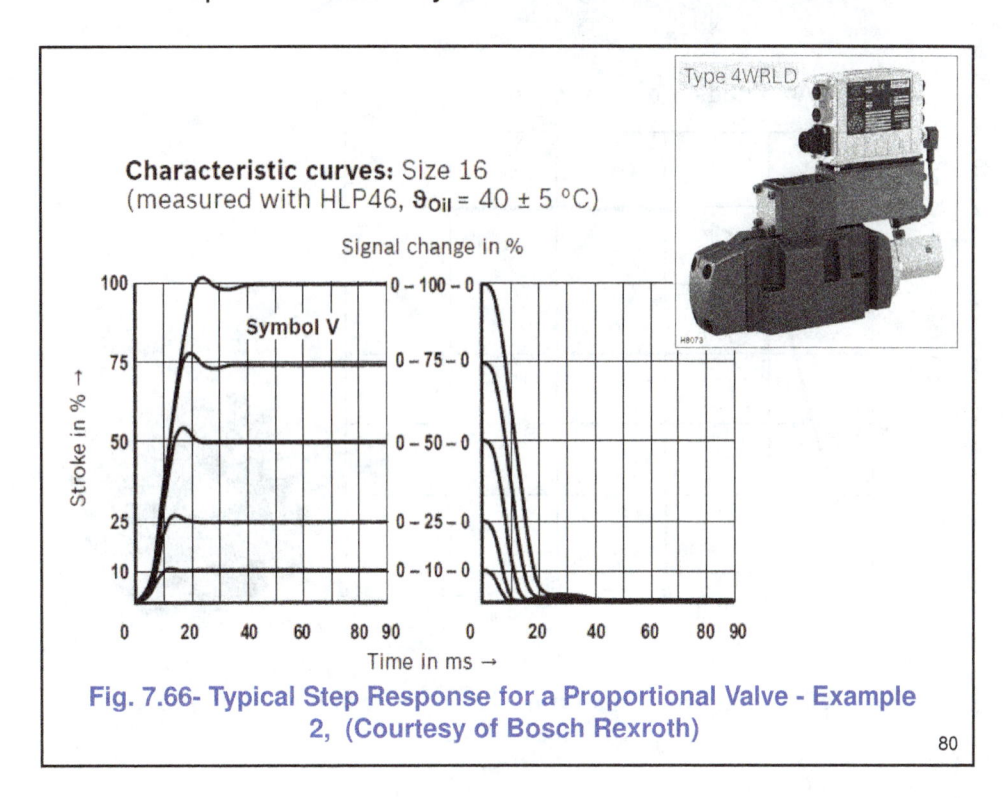

Fig. 7.66- Typical Step Response for a Proportional Valve - Example 2, (Courtesy of Bosch Rexroth)

Characteristic curves (measured with HLP32, ϑ_{Oil} = 40 °C ± 5 °C)

Transition function with pressure rating 315 bar, step response without flow

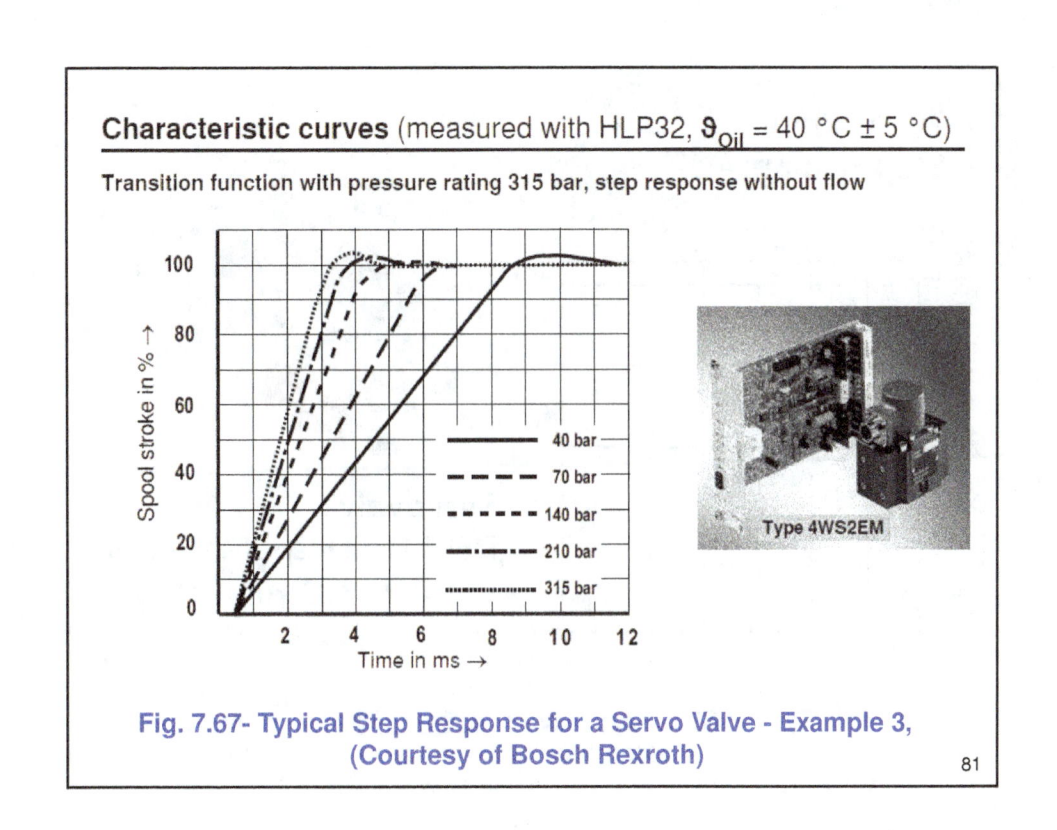

Fig. 7.67- Typical Step Response for a Servo Valve - Example 3, (Courtesy of Bosch Rexroth)

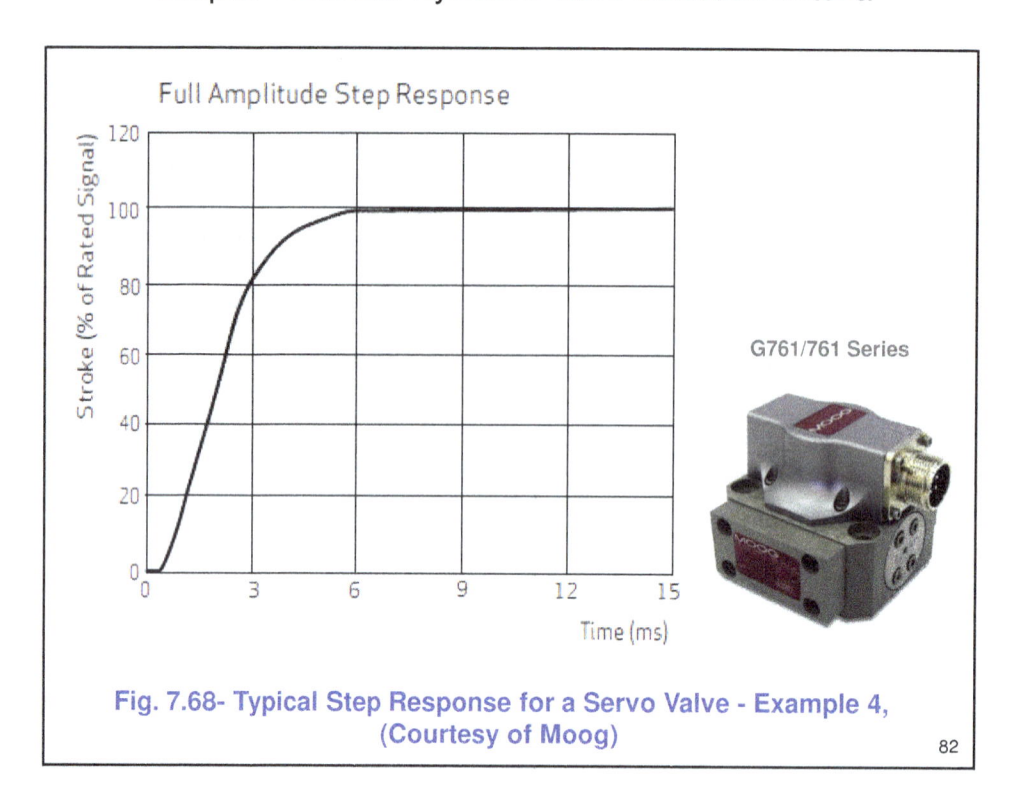

Fig. 7.68- Typical Step Response for a Servo Valve - Example 4, (Courtesy of Moog)

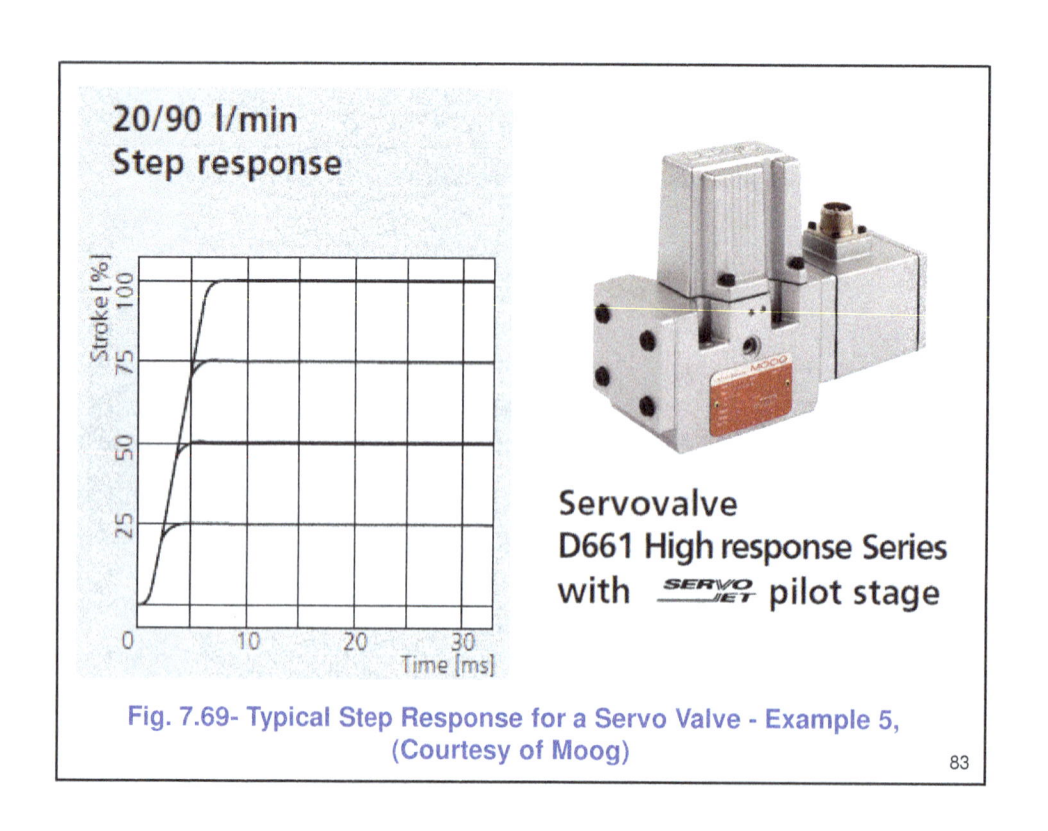

Fig. 7.69- Typical Step Response for a Servo Valve - Example 5, (Courtesy of Moog)

Workbook: Electro-Hydraulic Components and Systems
Chapter 7: Electro-Hydraulic Valve Selection Criteria

7.6.2- Frequency Response

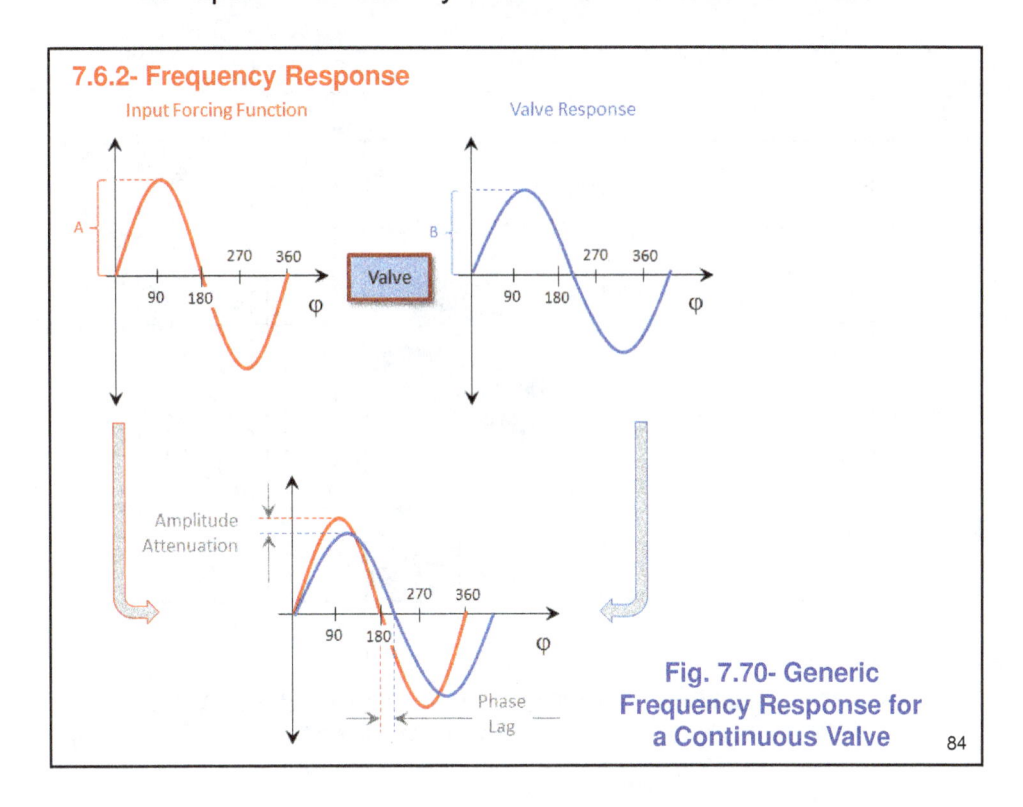

Fig. 7.70- Generic Frequency Response for a Continuous Valve

$$AR[dB] = 20 \, log \left[\frac{B}{A}\right] \quad 7.5$$

- Log (AR>1) = +ve
- Log (AR=1) = 0
- Log (AR<1) = -ve

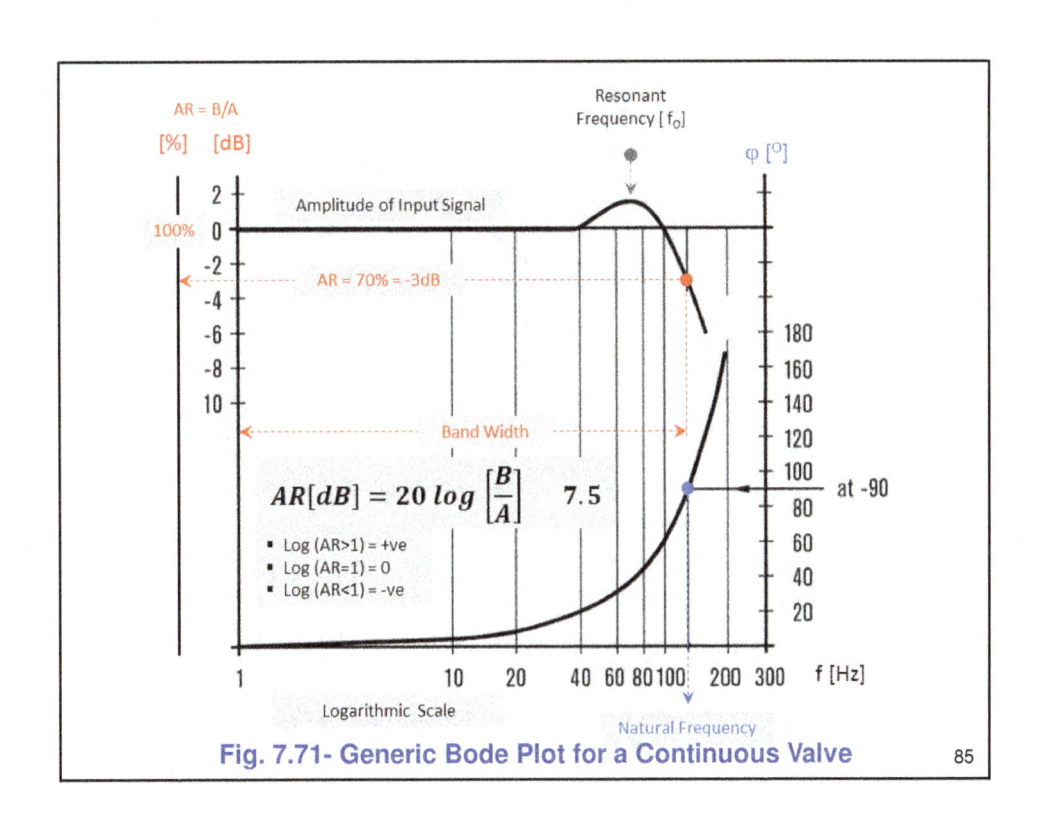

Fig. 7.71- Generic Bode Plot for a Continuous Valve

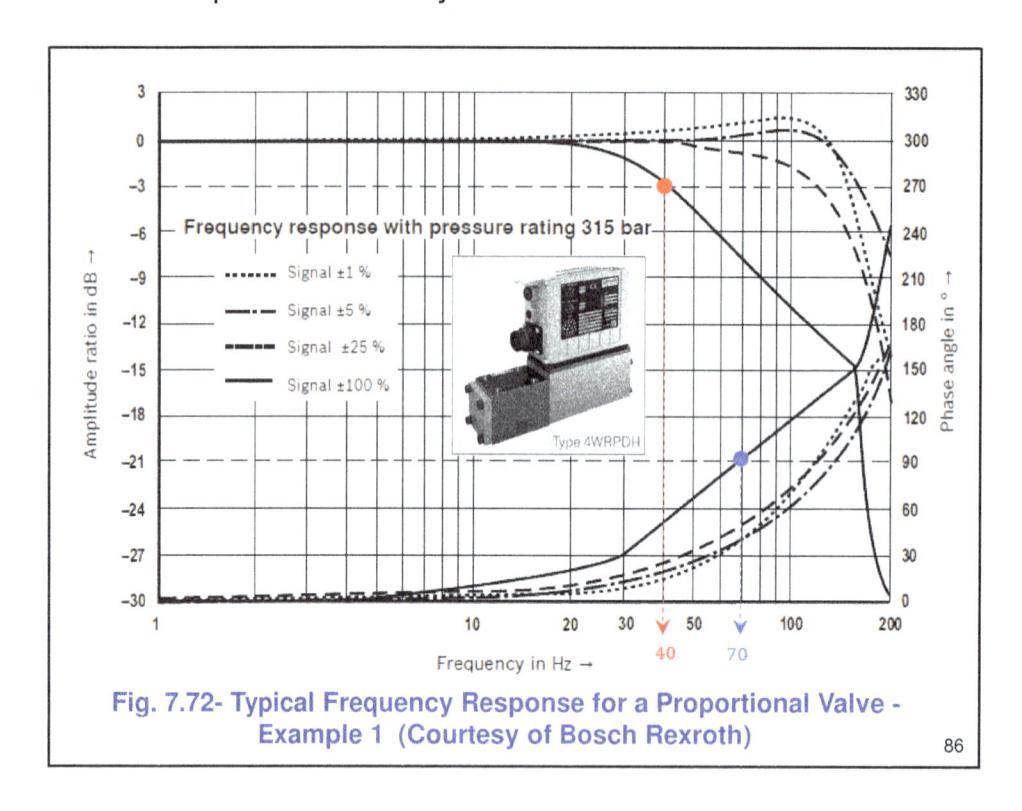

Workbook: Electro-Hydraulic Components and Systems
Chapter 7: Electro-Hydraulic Valve Selection Criteria

Fig. 7.72- Typical Frequency Response for a Proportional Valve - Example 1 (Courtesy of Bosch Rexroth)

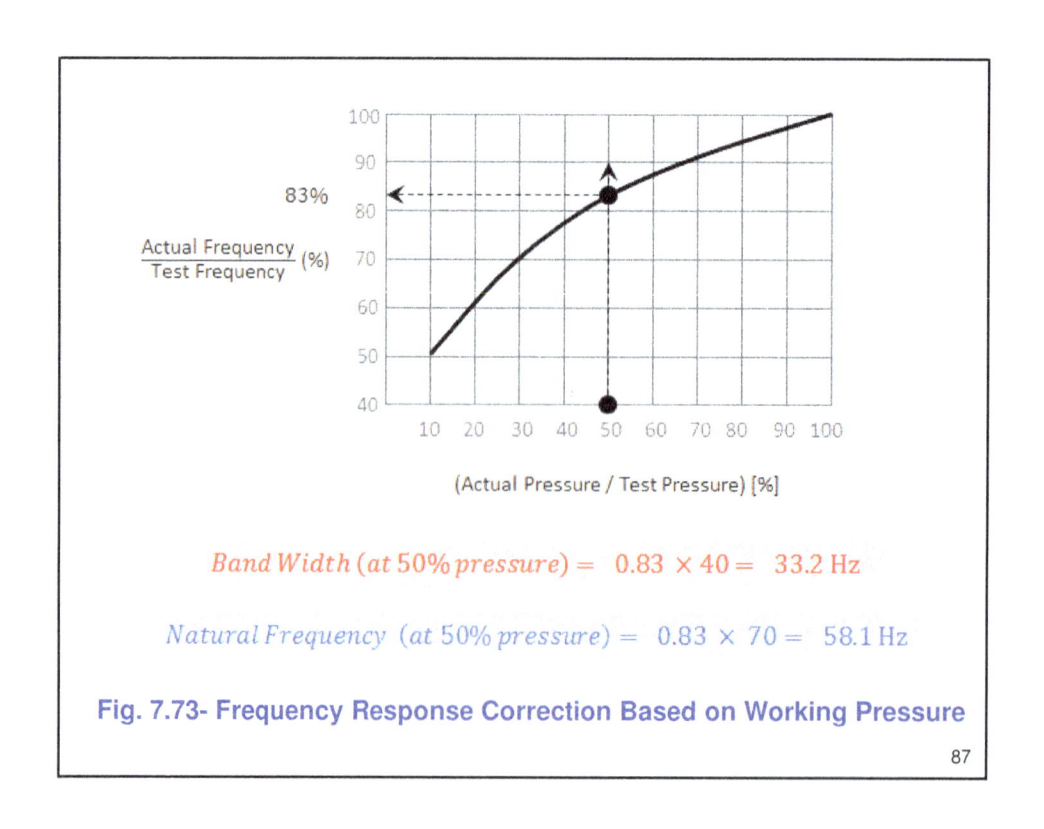

$Band\ Width\ (at\ 50\%\ pressure) = 0.83 \times 40 = 33.2\ Hz$

$Natural\ Frequency\ (at\ 50\%\ pressure) = 0.83 \times 70 = 58.1\ Hz$

Fig. 7.73- Frequency Response Correction Based on Working Pressure

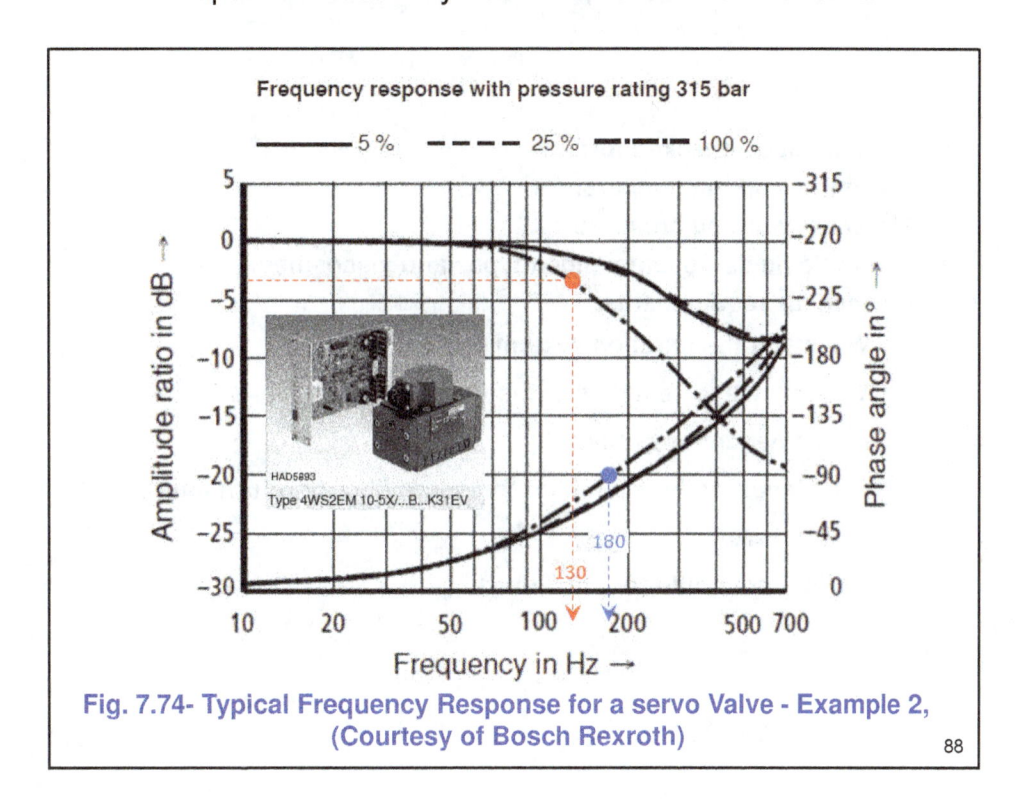

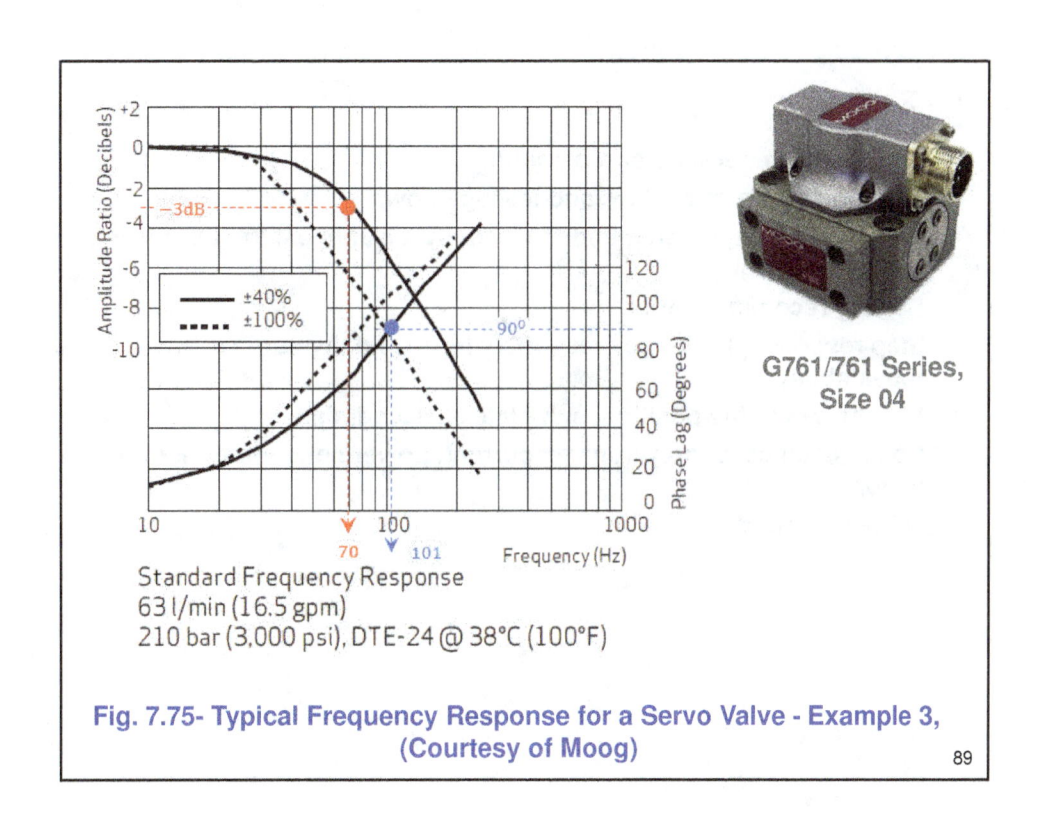

Fig. 7.74- Typical Frequency Response for a servo Valve - Example 2, (Courtesy of Bosch Rexroth)

Fig. 7.75- Typical Frequency Response for a Servo Valve - Example 3, (Courtesy of Moog)

7.7- Valve Operating Conditions

❑ Valve must be checked for:
- Maximum working temperature.
- Maximum working pressure.
- Hydraulic fluids: recommended types and viscosities.
- Cleanliness requirements:
    - Will affect the filtration system design.
    - You must also be aware of the valve manufacturer's recommendations about:
        - Location of the main filter (pressure filter or return filter).
        - Mesh size.
        - Filter beta ratio (or efficiency).

7.8- Examples of Published Data

- Maximum temperature and pressure.
- Rated flow, tolerance flow, and leakage flow.
- Hydraulic fluids, recommended viscosity, cleanliness requirements, and filter rating.
- Sealing recommendation.
- Step response, frequency response, threshold, hysteresis, and repeatability.
- Weight, assembly position, sub plate surface finish.
- Coil resistance, supply voltage, current, power consumption, duty factor.
- Other custom data.

Workbook: Electro-Hydraulic Components and Systems
Chapter 7: Electro-Hydraulic Valve Selection Criteria

HYDRAULIC CHARACTERISTICS (based on mineral oil ISO VG 46 at 50 °C)

DHZE-A-07

Hydraulic symbols	*71	*73	*51	*53	*51/B (1)	*53/B (1)
Valve model	DHZE				DKZE	
Spool overlapping	1, 3	1	1, 3	1, 3	1, 3	1, 3
Spool type and size (2)	L14	L1	S3, L3, D3	S5, L5, D5	S3, L3, D3	S5, L5, D5
Pressure limits [bar]	ports P, A, B = 350; T = 210				ports P, A, B = 315; T = 210	
Max flow (3) [l/min]						
at Δp = 10 bar (P-T)	1	4,5	17	28	45	60
at Δp = 30 bar (P-T)	2	8	30	50	80	105
at Δp = 70 bar (P-T)	3	12	45	70	120	160
Response time (4) [ms]	< 30				< 40	
Hysteresis [%]	≤ 5%				≤ 5%	
Repeatability	± 1%				± 1%	

Fig. 7.76- Published Data for a Proportional Valve - Example 1 – Part 1, (Courtesy of Atos)

MAIN CHARACTERISTICS, SEALS AND HYDRAULIC FLUID

Assembly position / location	Any position					
Subplate surface finishing	Roughness index Ra 0,4 - flatness ratio 0,01/100 (ISO 1101)					
Ambient temperature	Standard execution = -30°C ÷ +70°C; /PE option = -20°C ÷ +70°C; /BT option = -40°C ÷ +70°C					
Seals, recommended fluid temperature	NBR seals (standard) = -20°C ÷ +60°C, with HFC hydraulic fluids = -20°C ÷ +50°C FKM seals (/PE option)= -20°C ÷ +80°C HNBR seals (/BT option)= -40°C ÷ +60°C, with HFC hydraulic fluids = -40°C ÷ +50°C					
Recommended viscosity	15÷100 mm²/s - max allowed range 2,8 ÷ 500 mm²/s					
Fluid contamination class	ISO 4406 class 21/19/16 NAS 1638 class 10, in line filters of 25 μm (β10 ≥75 recommended)					
Hydraulic fluid	**Suitable seals type**		**Classification**		**Ref. Standard**	
Mineral oils	NBR, FKM, HNBR		HL, HLP, HLPD, HVLP, HVLPD		DIN 51524	
Flame resistant without water	FKM		HFDU, HFDR		ISO 12922	
Flame resistant with water	NBR, HNBR		HFC			
Flow direction	As shown in the symbols of table 3					
Coil code	DHZE-A*			DKZE-A*		
	standard	option /6 (1)	option /18 (2)	standard	option /6 (1)	option /18 (2)
Coil resistance R at 20°C	3 ÷ 3,3 Ω	2 ÷ 2,2 Ω	13 ÷ 13,4 Ω	3,8 ÷ 4,1 Ω	2,2 ÷ 2,4 Ω	12 ÷ 12,5 Ω
Max. solenoid current	2,2 A	2,75 A	1 A	2.6 A	3.25 A	1,2 A
Max. power	30 Watt			35 Watt		
Protection degree (CEI EN-60529)	IP65					
Duty factor	Continuous rating (ED=100%)					

Fig. 7.77- Published Data for a Proportional Valve - Example 1 – Part 2, (Courtesy of Atos)

> i.e. if the deviation is greater than 10%, return the valve or you should be authorized to do the adjustments

G761/761 Series

Hydraulic Data

Maximum operating pressure to ports P, T, A, B, X	315 bar (4.500 psi)
Rated flow at p_N 35 bar/spool land (500 psi/spool land)	4/10/19/38/63 l/min (1/2.5/5/10/16.5 gpm)
Maximum main stage leakage flow rate (zero lap)	2.3 l/min (0.60 gpm)
Null adjust authority	Greater than 10% of rated flow
Hydraulic fluid	Hydraulic oil as per DIN 51524 parts 1 to 3 and ISO 11158. Other fluids on request.
Temperature range	-40 to +60 °C (-40 to +140 °F)
Recommended viscosity range	10 to 97 mm^2/s (cSt)
Maximum permissible viscosity range	5 to 1,250 mm^2/s (cSt)
Recommended cleanliness class as per ISO 4406	
For functional safety	17/14/11
For longer life	15/13/10
Recommended filter rating	
For functional safety	$\beta_{10} \leq 75$ (10µm absolute)
For longer life	$\beta_5 \leq 75$ (5 µm absolute)

Fig. 7.78- Published Data for a Servo Valve - Example 2 – Part 1, (Courtesy of Moog)

Static and Dynamic Data

Rated Flow	4 l/min (1 gpm)	10 l/min (2.5 gpm)	19 l/min (5 gpm)	38 l/min (10 gpm)	63 l/min (16.5 gpm)
Tolerance of rated flow	± 10% of rated flow				
Step response time for 0 to 100% stroke	5 ms	5 ms	5 ms	7 ms	16 ms
Threshold	$\leq 0.5\%$ of rated signal				
Hysteresis	$\leq 3.0\%$ of rated signal				
Null shift at $\Delta T = 38°C$ (100°C)	$\leq 2.0\%$ of rated signal				

Fig. 7.79- Published Data for a Servo Valve - Example 2 – Part 2, (Courtesy of Moog)

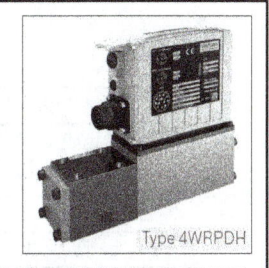

general			
Size	Size	6	10
Design		Directional spool valve, direct operated, with steel sleeve	
Actuation		Proportional solenoid with position control, OBE	
Type of connection		Plate connection, porting pattern according to ISO 4401	
Installation position		Any	
Ambient temperature range	°C	−20 ... +60	
Storage temperature range	°C	+5 ... +40	
Sine test according to DIN EN 60068-2-6		10 ... 2000 Hz / maximum of 10 g / 10 cycles / 3 axes	
Noise test according to DIN EN 60068-2-64		20 ... 2000 Hz / 10 g$_{RMS}$ / 30 g peak / 30 min / 3 axes	
Transport shock according to DIN EN 60068-2-27		15g / 11 ms / 3 axes	
Weight	kg	3.2	7.2
Maximum relative humidity (no condensation)	%	95	
Maximum surface temperature	°C	150	

Fig. 7.80- Published Data for a Proportional Valve - Example 3 – Part 1, (Courtesy of Bosch Rexroth)

hydraulic										
Maximum operating pressure	▶ Port A, B, P	bar	350 (size 6); 315 (size 10)							
	▶ Port T	bar	250							
Rated flow (Δp = 35 bar per edge [1])		l/min	2	4	12	15	24/25	40	50	100
Limitation of use (transition in fail safe position)	▶ Symbol C3, C5	bar	350	350	350	350	350	160	315	160
	▶ Symbol C4, C1	bar	350	350	350	280	250	100	250	100
Leakage flow (at 100 bar)	▶ Linear characteristic curve "L"	cm³/min	< 150	< 180	< 300	–	< 500	< 900	< 1200	< 1500
	▶ Inflected characteristic curve "P"	cm³/min	–	–	–	< 180	< 300	< 450	< 600 (1:1) < 500 (2:1)	< 600
Hydraulic fluid			See table page 7							
Viscosity range	▶ Recommended	mm²/s	20 ... 100							
	▶ Maximum admissible	mm²/s	10 ... 800							
Hydraulic fluid temperature range (flown-through)		°C	−20 ... +60							
Maximum admissible degree of contamination of the hydraulic fluid, cleanliness class according to ISO 4406 (c)			Class 18/16/13 [2]							

Fig. 7.81- Published Data for a Proportional Valve - Example 3 – Part 2, (Courtesy of Bosch Rexroth)

static / dynamic		
Hysteresis	%	≤ 0.2
Manufacturing tolerance q_{Vmax}	%	≤ 10
Temperature drift	%/10 K	Zero shift < 0.25
Pressure drift	%/100 bar	Zero shift < 0.15
Zero compensation		Ex plant ±1 %

Technical data
(For applications outside these values, please consult us!)

Hydraulic fluid		Classification	Suitable sealing materials
Mineral oils		HL, HLP, HLPD, HVLP, HVLPD	NBR, FKM
Bio-degradable	► Insoluble in water	HETG	NBR, FKM
		HEES	FKM
	► Soluble in water	HEPG	FKM

Fig. 7.82- Published Data for a Proportional Valve - Example 3 – Part 3, (Courtesy of Bosch Rexroth)

electrical, integrated electronics (OBE)			
Relative duty cycle		%	100 (continuous operation)
Protection class according to EN 60529			IP 65 with mounted and locked plug-in connectors
Supply voltage 3), 4)	► Nominal voltage	VDC	24
	► Lower limit value	VDC	18
	► Upper limit value	VDC	36
	► Maximum admissible residual ripple	Vpp	2.5 (Comply with absolute supply voltage limit value)
Current consumption	► Maximum 5)	A	2.5
	► Impulse current	A	4
Maximum power consumption	► Size 6	W	40
	► Size 10	W	60
AD/DA resolution	► Analog inputs		12 bit
	► Analog output		10 bit

Fig. 7.83- Published Data for a Proportional Valve - Example 3 – Part 4, (Courtesy of Bosch Rexroth)

Chapter 7 Reviews

1. Which of the following statements is True?

    A. Proportional valves are recommended over servo valves for a closed loop operation.
    B. Proportional valves have smaller hysteresis range than servo valves.
    C. Proportional valves have faster response than servo valves.
    D. All the above statements are false.

2. In an application, a cylinder is to perform reciprocating motion at high frequency. Cylinder positioning is to be controlled by a closed loop feedback control system to achieve high accuracy. Which of the following valves are recommended for such application?

    A. ON/OFF solenoid-operated valve.
    B. Basic proportional valve.
    C. High performance proportional valve or servo valve.
    D. None of the above.

3. The term "bandwidth" indicates?

    A. Range of frequency to which a continuous valve can respond with a magnitude of at least 70% of the amplitude of the exciting frequency.
    B. Range of position accuracy.
    C. Steady state error in speed control.
    D. Deadband zone for the valve's spool.

4. The term "Settling Time" indicates?

    A. Range of frequency to which a system can respond with magnitude of at least 70% of the exciting frequency.
    B. Value of the steady state error is in the system.
    C. How much time it takes for a valve to respond to a step input before it settles at the steady state condition.
    D. None of the above.

5. If a valve is claimed to have a positive dead band, this means that the valve is of?

    A. Under lapped (open) center.
    B. Zero lapped (critical) center.
    C. Overlapped (closed) center.
    D. None of the above.

6. Linear flow gain can be achieved by?

    A. A spool with D notches.
    B. A spool with square notches.
    C. A spool with V notches.
    D. None of the above.

7. Natural frequency of a valve is defined at?

    A. Amplitude ratio equal – 3 dB.
    B. Amplitude ratio equal 70%.
    C. Phase lead equal 90 degrees.
    D. Phase lag equal 90 degrees.

8. Which of the following valves are the most suitable for a position control application?

    A. ON/OFF valve.
    B. Basic proportional valve with overlapped spool.
    C. High performance proportional valve with underlapped spool.
    D. Servo valve with zero-lapped spool.

9. Which of the following valves are the most suitable for a pressure control application?

    A. ON/OFF valve.
    B. Basic proportional valve with overlapped spool.
    C. High performance proportional valve with underlapped spool.
    D. Servo valve with zero-lapped spool.

10. Which of the following valves are the most suitable for a temperature control application?

    A. ON/OFF valve.
    B. Basic proportional valve with overlapped spool.
    C. High performance proportional valve with underlapped spool.
    D. Servo valve with zero-lapped spool.

Workbook: Electro-Hydraulic Components and Systems
Chapter 7: Reviews and Assignments

Chapter 7 Assignment

Student Name: -- Student ID: -------------------

Date: -- Score: ------------------------

Find out the nominal flow of a valve that has been tested under 200 bar. The valve will be used to supply a maximum of 60 liters/min to a hydraulic motor. The inlet pressure of the motor is 100 bar.

Chapter 8
Open-Loop versus Closed-Loop EH Applications

Objectives:

This chapter explores the differences between electro-hydraulic open-loop and closed-loop systems. Several examples have been presented to discuss various ways to control load, speed and position of a hydraulic actuator. The chapter concludes by introducing examples from industrial and mobile applications.

Brief Contents:

8.1- Electro-Hydraulic Open-Loop Control Systems
8.2- Electro-Hydraulic Closed-Loop Control Systems
8.3- Closed-Loop Performance Analysis
8.4- Electro-Hydraulic Control Systems Application Example

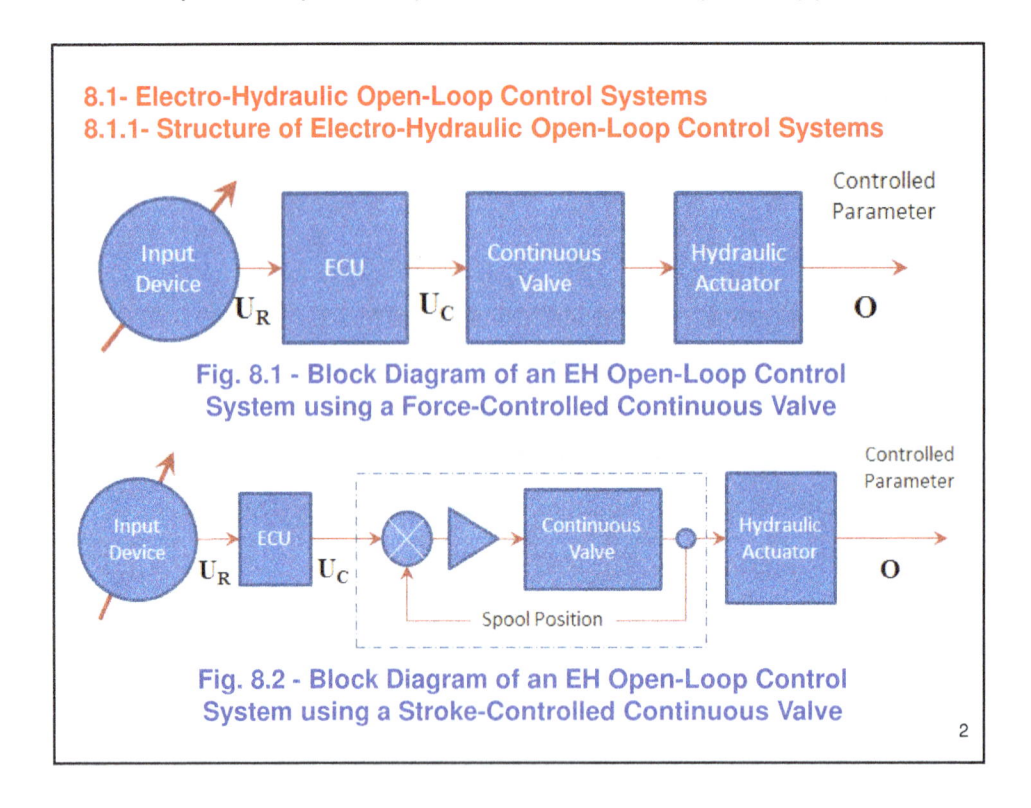

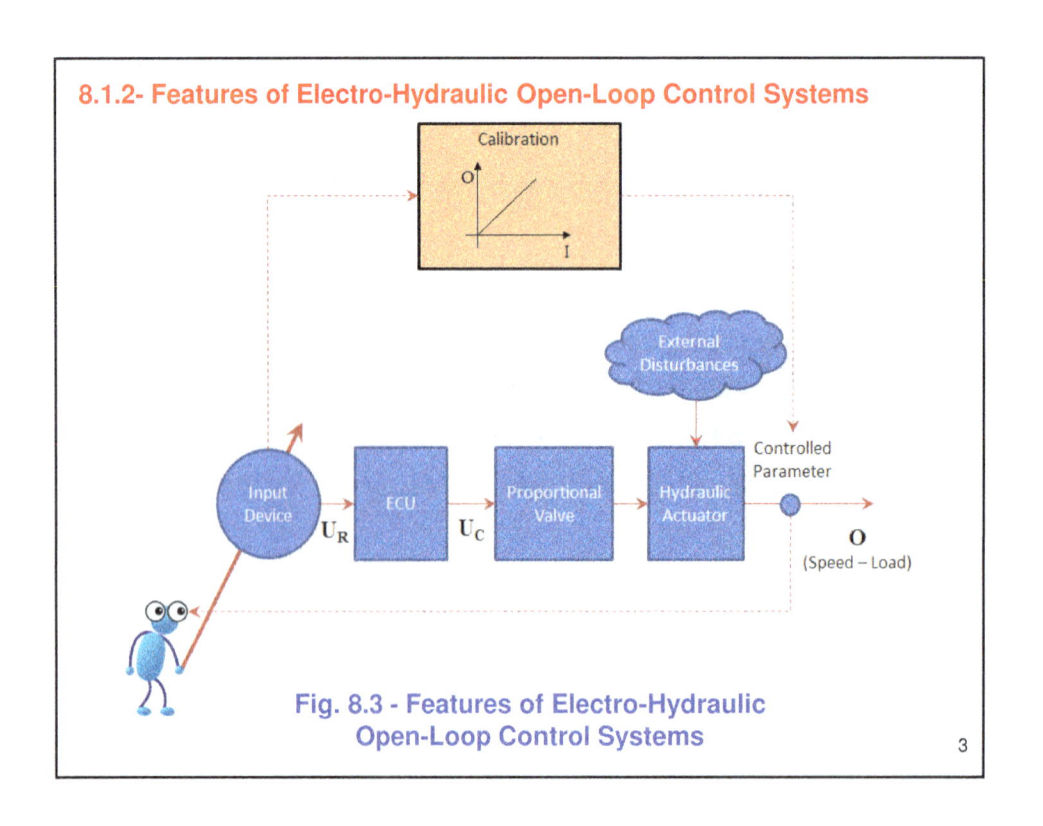

Workbook: Electro-Hydraulic Components and Systems
Chapter 8: Open-Loop versus Closed-Loop EH Applications

8.1- Electro-Hydraulic Open-Loop Control Systems
8.1.1- Structure of Electro-Hydraulic Open-Loop Control Systems

Fig. 8.1 - Block Diagram of an EH Open-Loop Control System using a Force-Controlled Continuous Valve

Fig. 8.2 - Block Diagram of an EH Open-Loop Control System using a Stroke-Controlled Continuous Valve

8.1.2- Features of Electro-Hydraulic Open-Loop Control Systems

Fig. 8.3 - Features of Electro-Hydraulic Open-Loop Control Systems

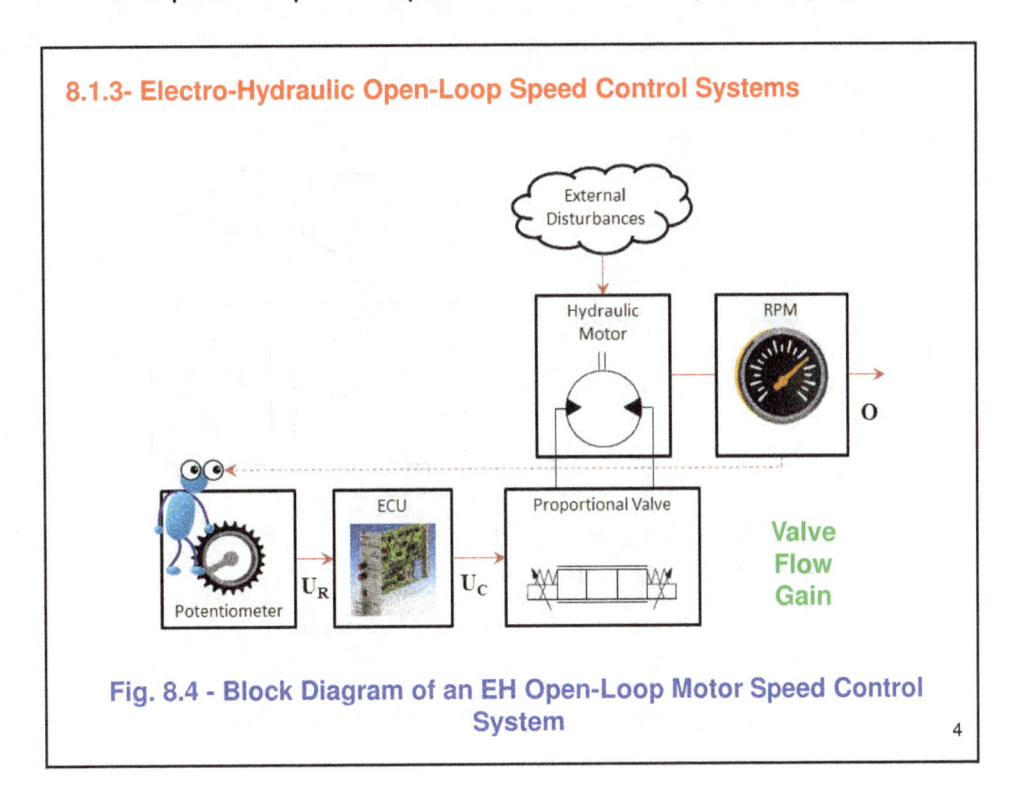

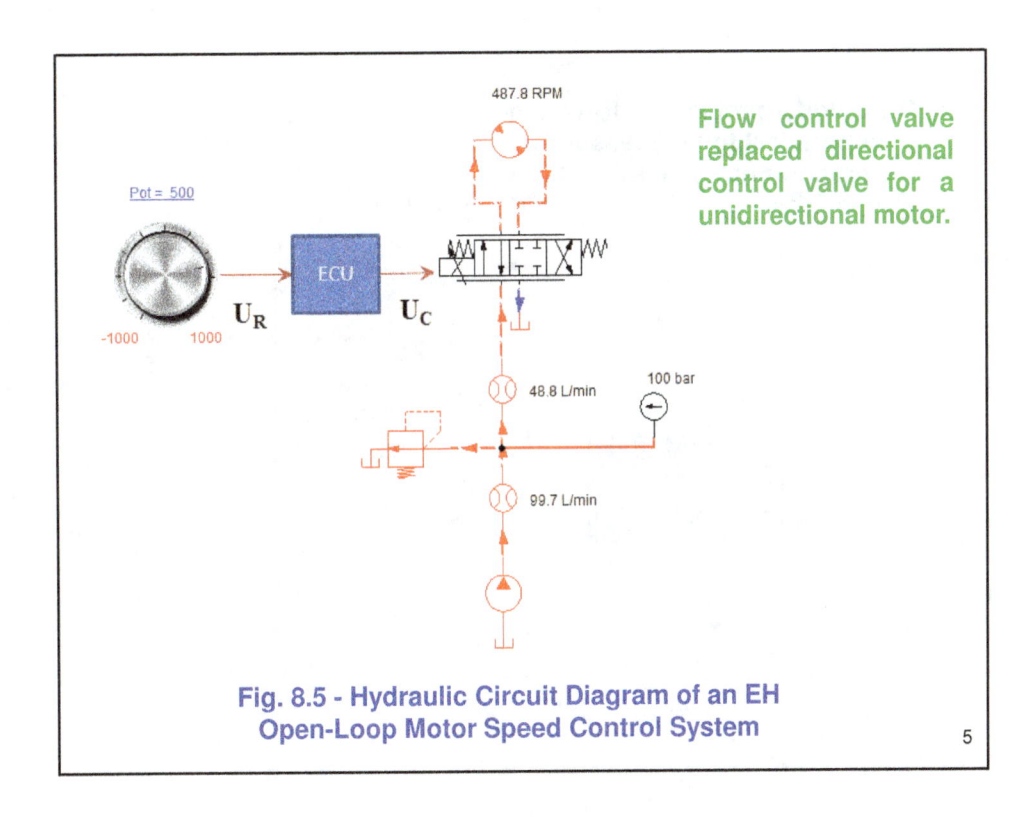

Workbook: Electro-Hydraulic Components and Systems
Chapter 8: Open-Loop versus Closed-Loop EH Applications

8.1.3- Electro-Hydraulic Open-Loop Speed Control Systems

Valve Flow Gain

Fig. 8.4 - Block Diagram of an EH Open-Loop Motor Speed Control System

Flow control valve replaced directional control valve for a unidirectional motor.

Fig. 8.5 - Hydraulic Circuit Diagram of an EH Open-Loop Motor Speed Control System

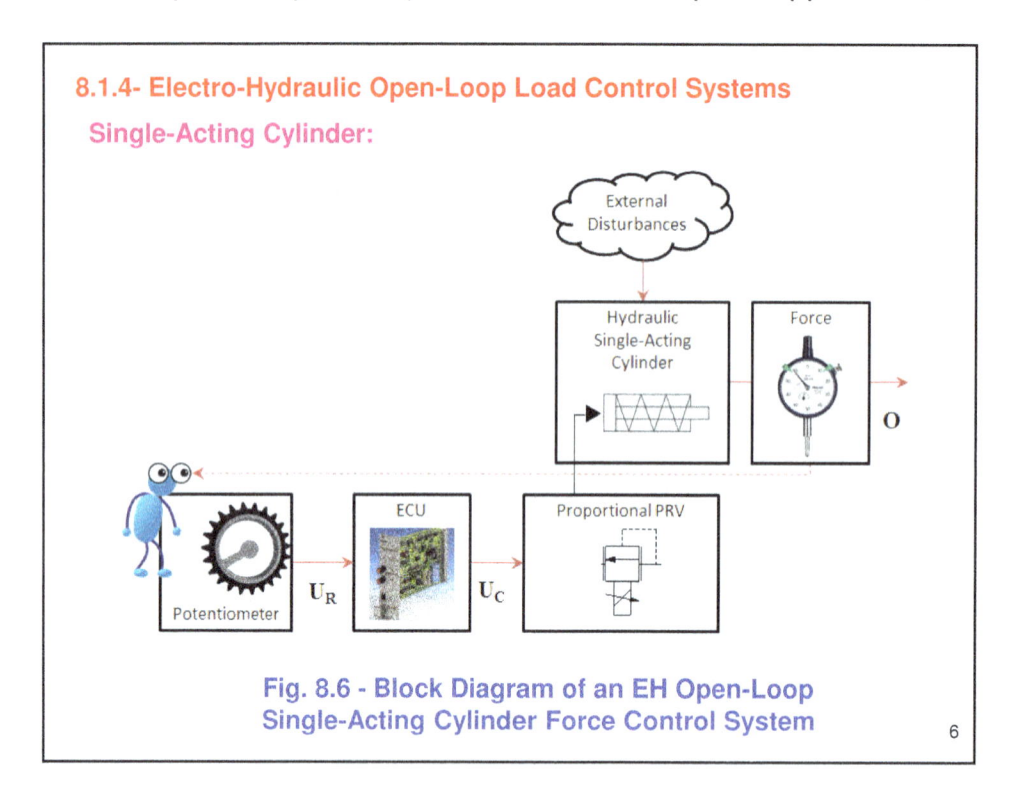

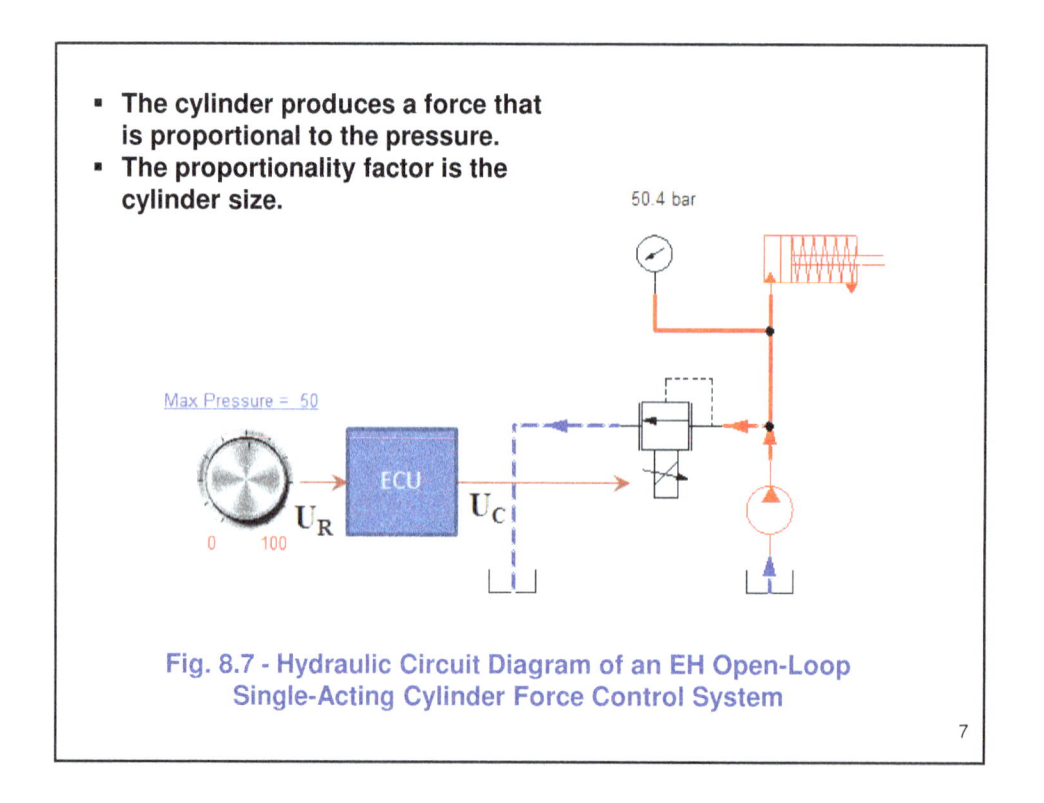

Workbook: Electro-Hydraulic Components and Systems
Chapter 8: Open-Loop versus Closed-Loop EH Applications

8.1.4- Electro-Hydraulic Open-Loop Load Control Systems

Single-Acting Cylinder:

Fig. 8.6 - Block Diagram of an EH Open-Loop Single-Acting Cylinder Force Control System

- The cylinder produces a force that is proportional to the pressure.
- The proportionality factor is the cylinder size.

Fig. 8.7 - Hydraulic Circuit Diagram of an EH Open-Loop Single-Acting Cylinder Force Control System

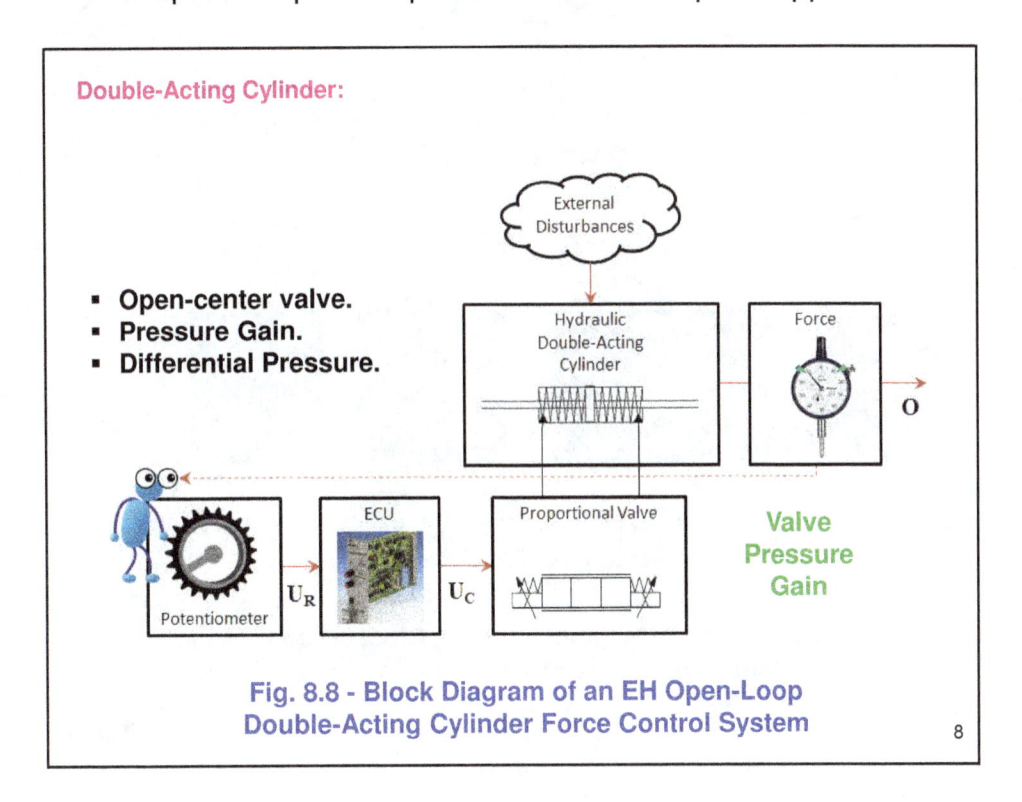

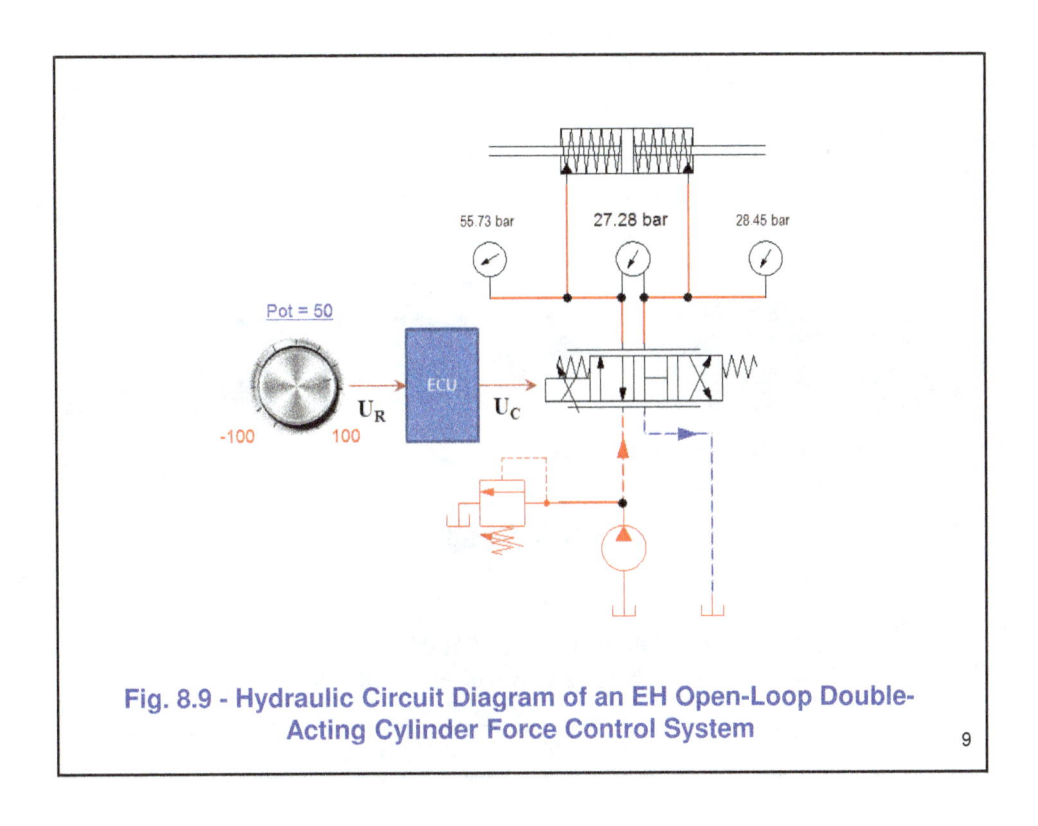

Workbook: Electro-Hydraulic Components and Systems
Chapter 8: Open-Loop versus Closed-Loop EH Applications

Double-Acting Cylinder:

- Open-center valve.
- Pressure Gain.
- Differential Pressure.

Fig. 8.8 - Block Diagram of an EH Open-Loop Double-Acting Cylinder Force Control System

Fig. 8.9 - Hydraulic Circuit Diagram of an EH Open-Loop Double-Acting Cylinder Force Control System

8.2- Electro-Hydraulic Closed-Loop Control Systems
8.2.1- Structure of Electro-Hydraulic Closed-Loop Control Systems

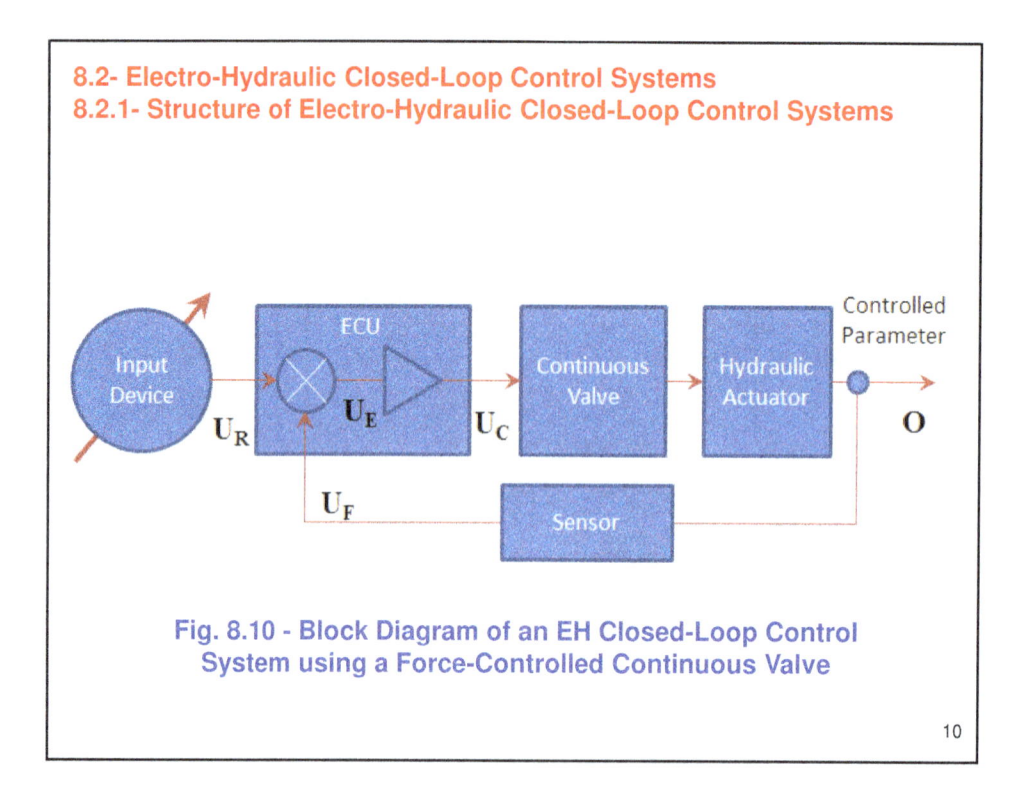

Fig. 8.10 - Block Diagram of an EH Closed-Loop Control System using a Force-Controlled Continuous Valve

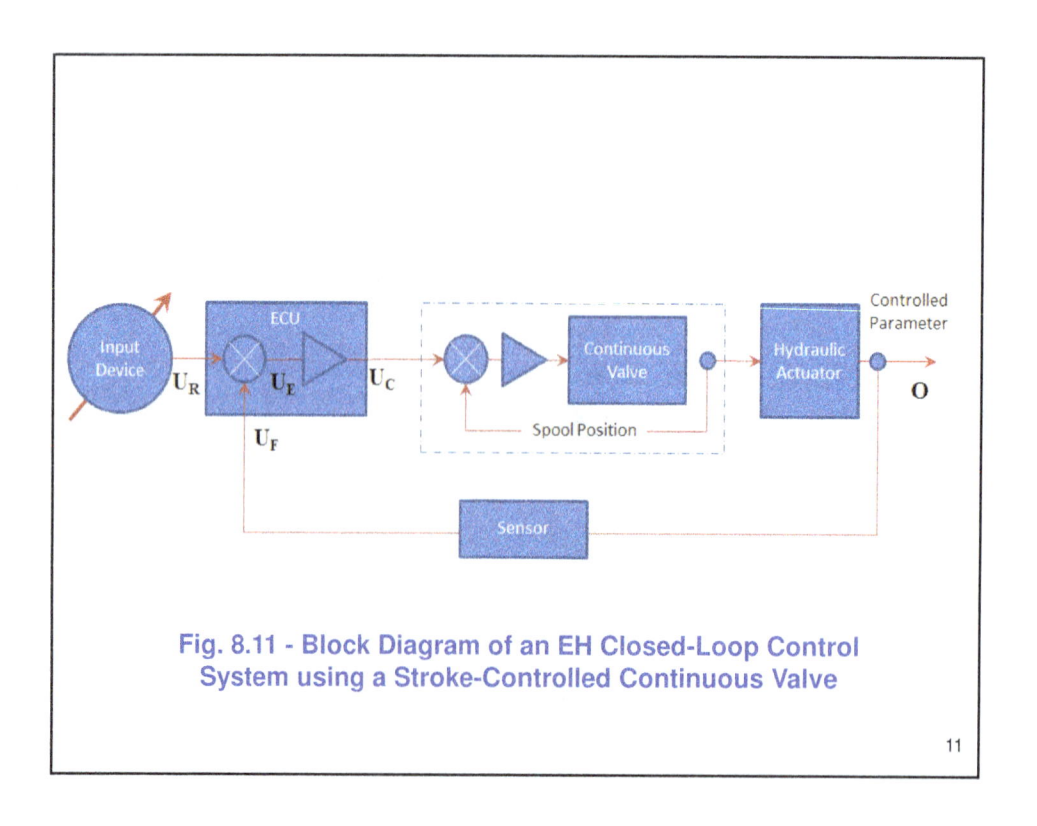

Fig. 8.11 - Block Diagram of an EH Closed-Loop Control System using a Stroke-Controlled Continuous Valve

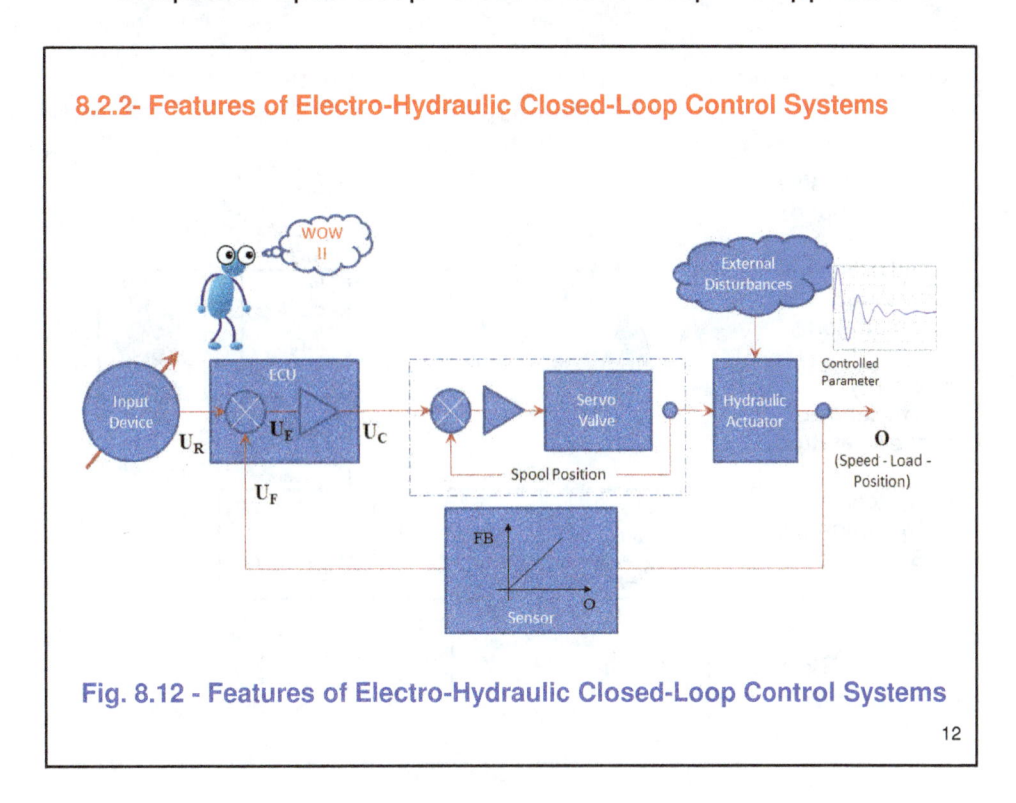

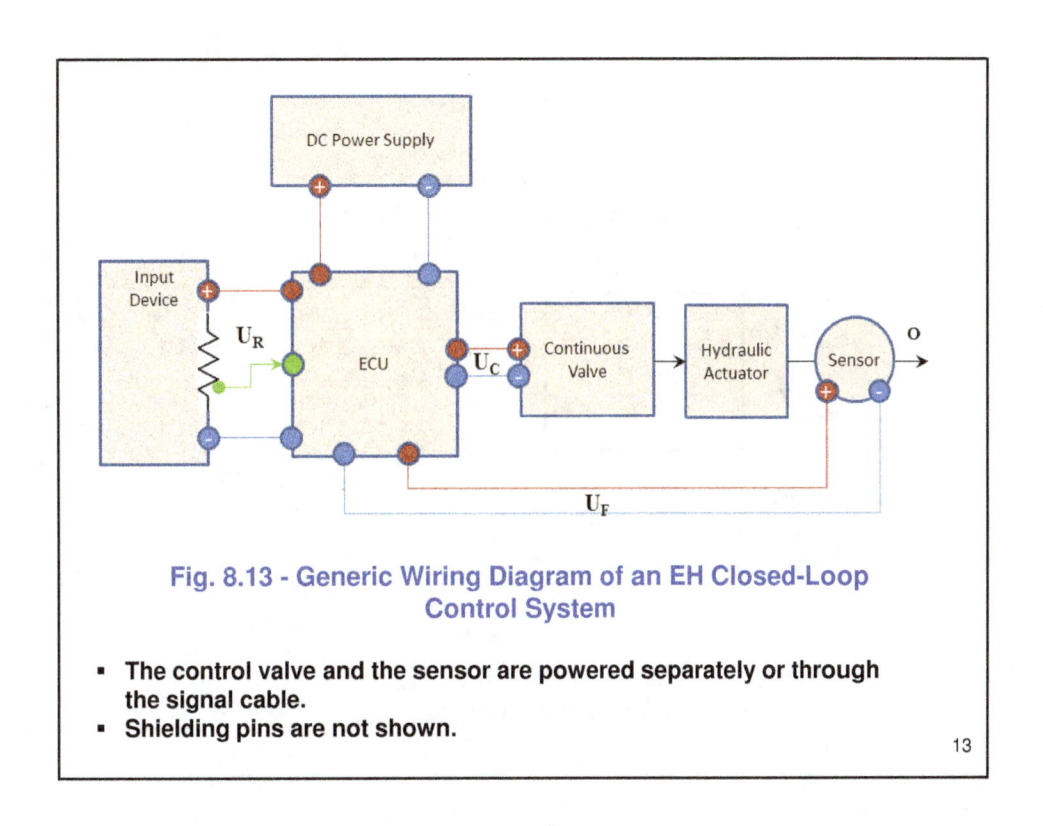

Workbook: Electro-Hydraulic Components and Systems
Chapter 8: Open-Loop versus Closed-Loop EH Applications

8.2.2- Features of Electro-Hydraulic Closed-Loop Control Systems

Fig. 8.12 - Features of Electro-Hydraulic Closed-Loop Control Systems

Fig. 8.13 - Generic Wiring Diagram of an EH Closed-Loop Control System

- The control valve and the sensor are powered separately or through the signal cable.
- Shielding pins are not shown.

Workbook: Electro-Hydraulic Components and Systems
Chapter 8: Open-Loop versus Closed-Loop EH Applications

8.2.3- Electro-Hydraulic Closed-Loop Speed Control Systems
8.2.3.1- Electro-Hydraulic Closed-Loop Flow Control System

- Control Valve: (Overlapped).

- Valve Flow Gain for calibration.

- Energy Efficiency of the Operation.

- System Operation and Sources of Error.

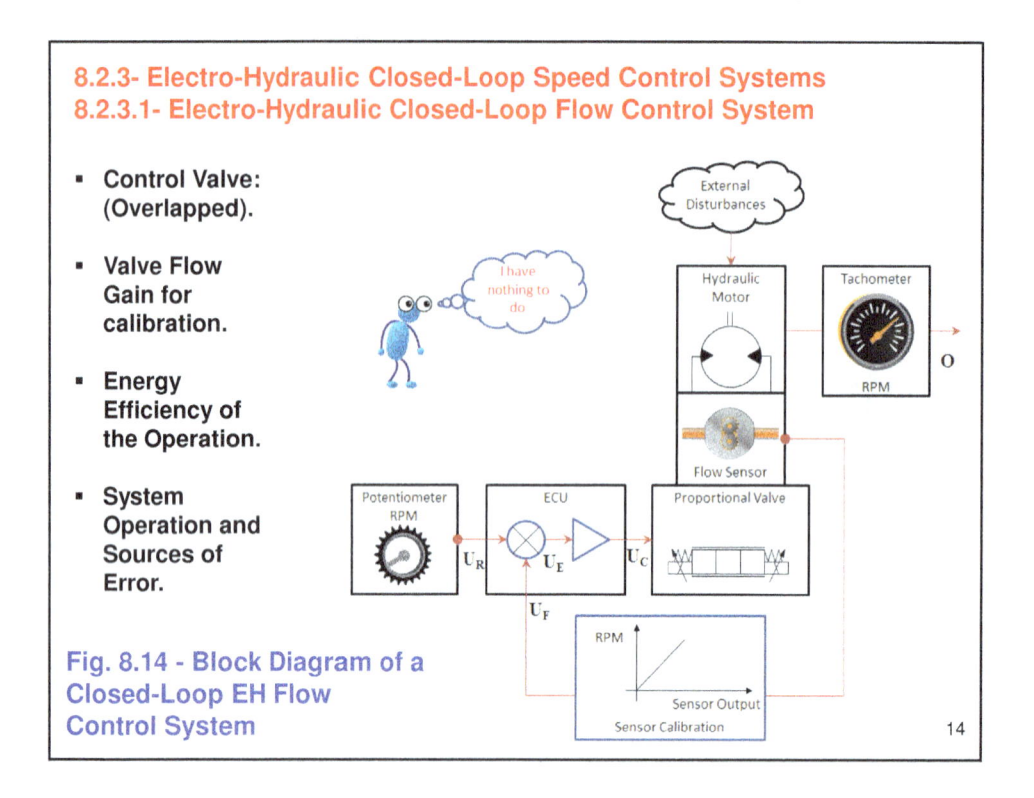

Fig. 8.14 - Block Diagram of a Closed-Loop EH Flow Control System

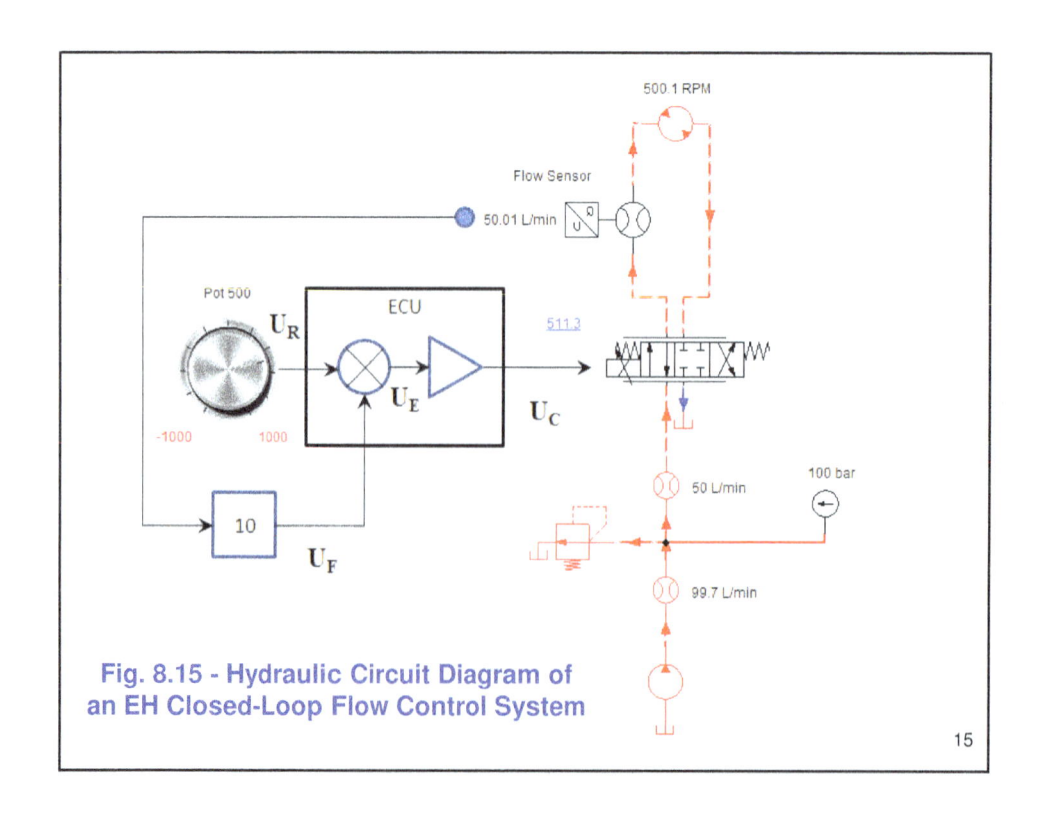

Fig. 8.15 - Hydraulic Circuit Diagram of an EH Closed-Loop Flow Control System

Workbook: Electro-Hydraulic Components and Systems
Chapter 8: Open-Loop versus Closed-Loop EH Applications

8.2.3.2- Electro-Hydraulic Closed-Loop Motor Speed Control System

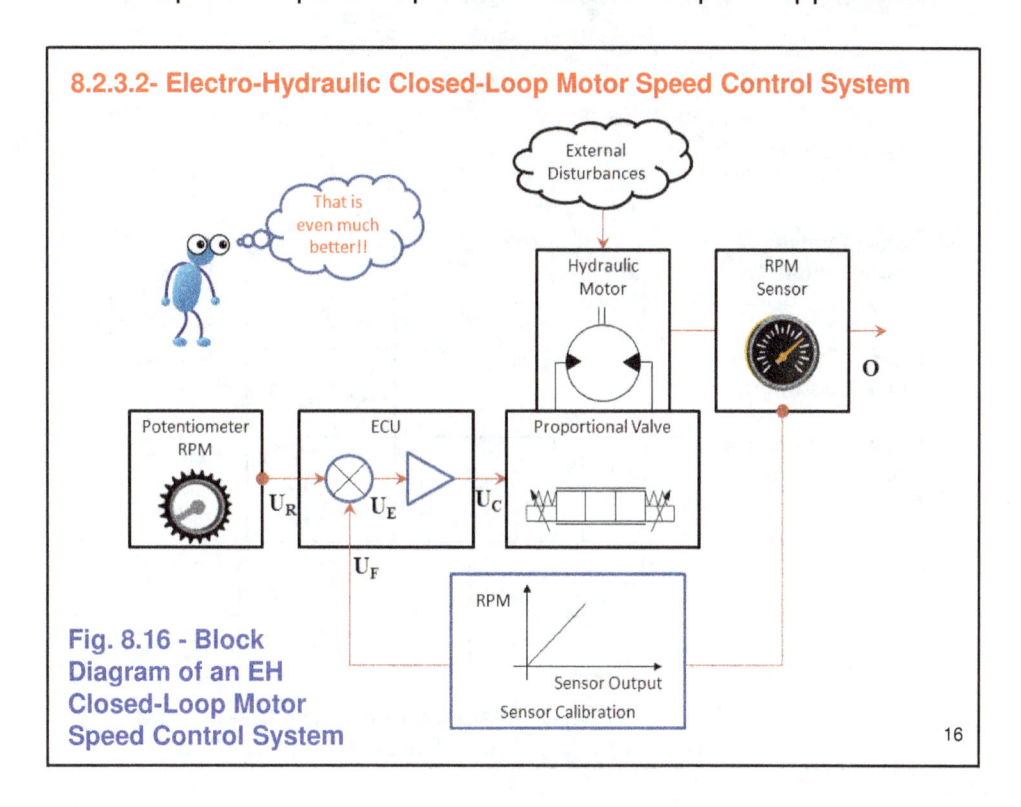

Fig. 8.16 - Block Diagram of an EH Closed-Loop Motor Speed Control System

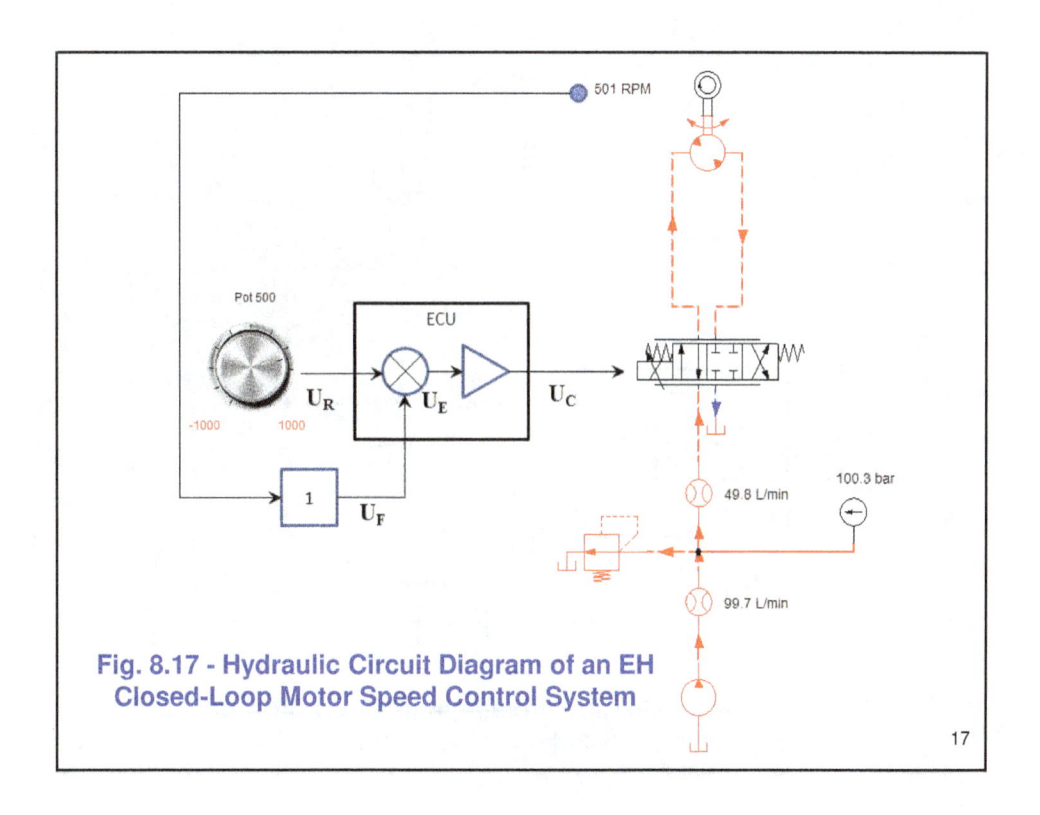

Fig. 8.17 - Hydraulic Circuit Diagram of an EH Closed-Loop Motor Speed Control System

Workbook: Electro-Hydraulic Components and Systems
Chapter 8: Open-Loop versus Closed-Loop EH Applications

8.2.4- Electro-Hydraulic Closed-Loop Load Control Systems
8.2.4.1- Electro-Hydraulic Closed-Loop Single-Acting Cylinder Pressure Control System

- Control Valve: (Proportional PRV).

- Energy Efficiency of the Operation.

- System Operation and Sources of Error.

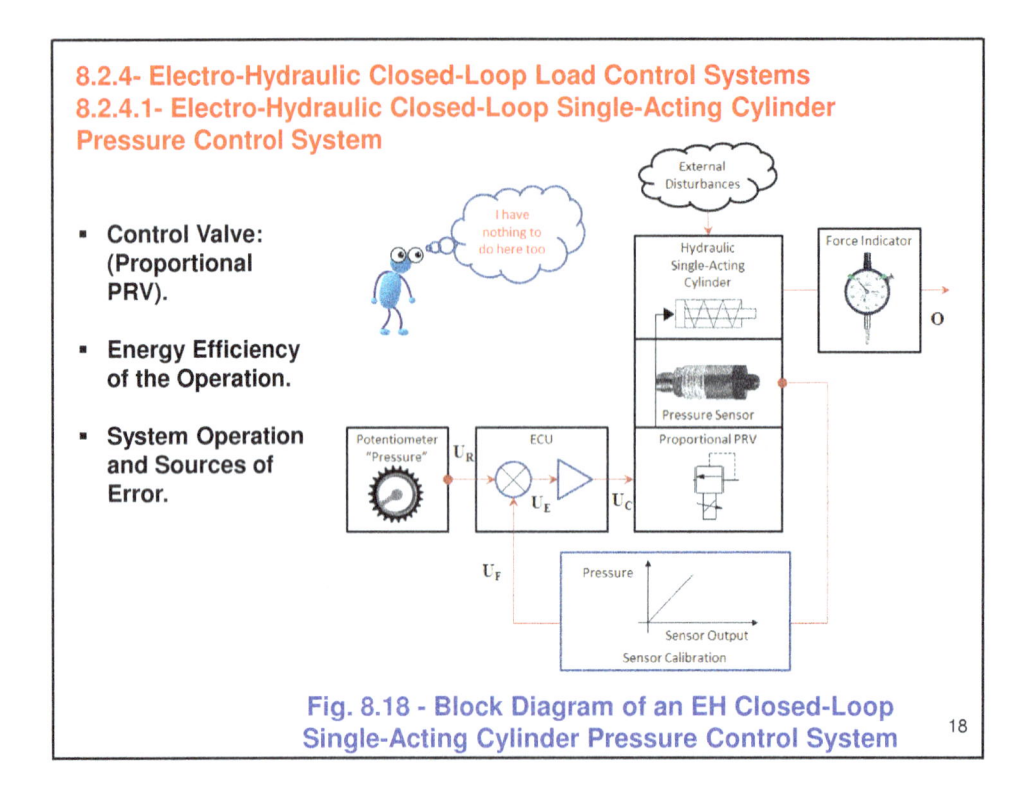

Fig. 8.18 - Block Diagram of an EH Closed-Loop Single-Acting Cylinder Pressure Control System

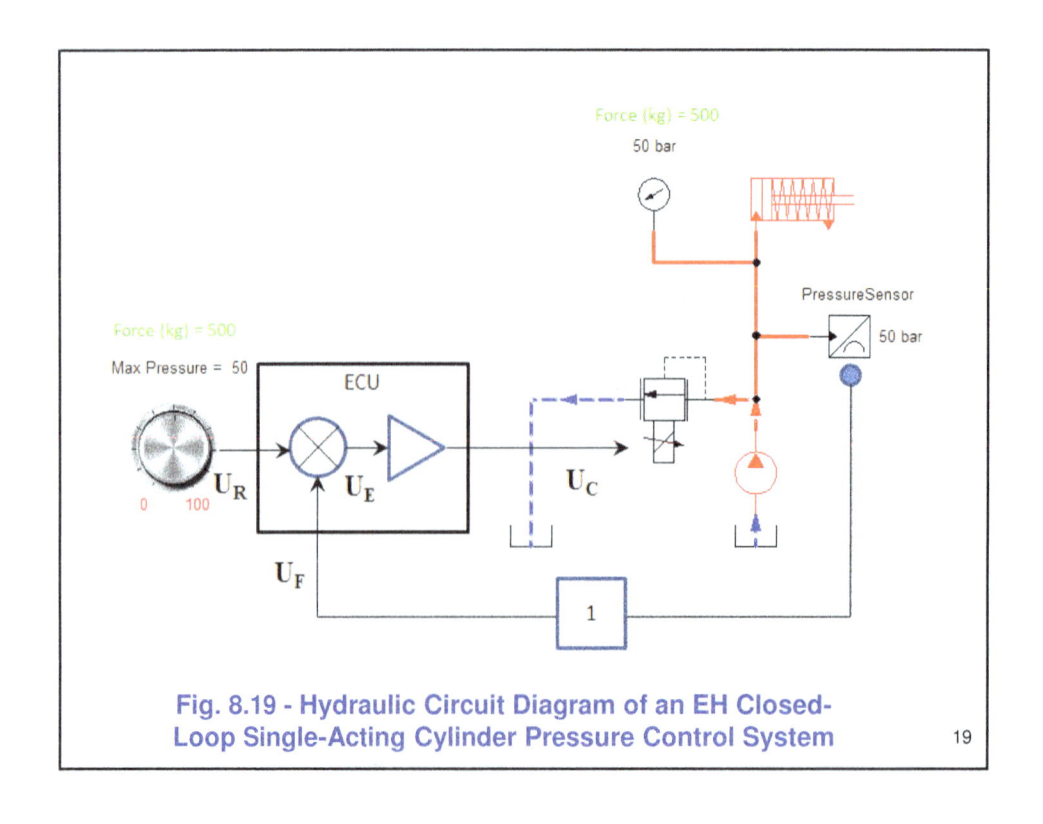

Fig. 8.19 - Hydraulic Circuit Diagram of an EH Closed-Loop Single-Acting Cylinder Pressure Control System

Workbook: Electro-Hydraulic Components and Systems
Chapter 8: Open-Loop versus Closed-Loop EH Applications

8.2.4.2- Electro-Hydraulic Closed-Loop Single-Acting Cylinder Force Control System

That is even much better

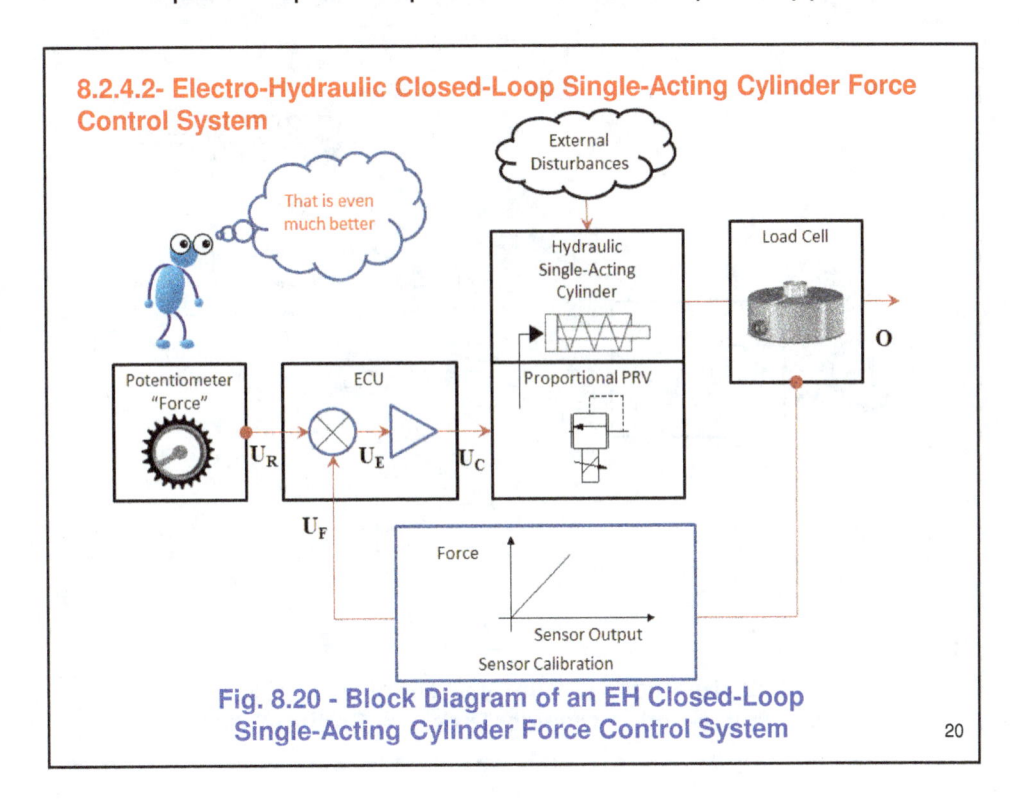

Fig. 8.20 - Block Diagram of an EH Closed-Loop Single-Acting Cylinder Force Control System

---

8.2.4.3- Electro-Hydraulic Closed-Loop Double-Acting Cylinder Pressure Control System

- **Control Valve:** (Underlapped).

- **Valve Pressure Gain for calibration.**

- **Energy Efficiency of the Operation.**

- **System Operation and Sources of Error.**

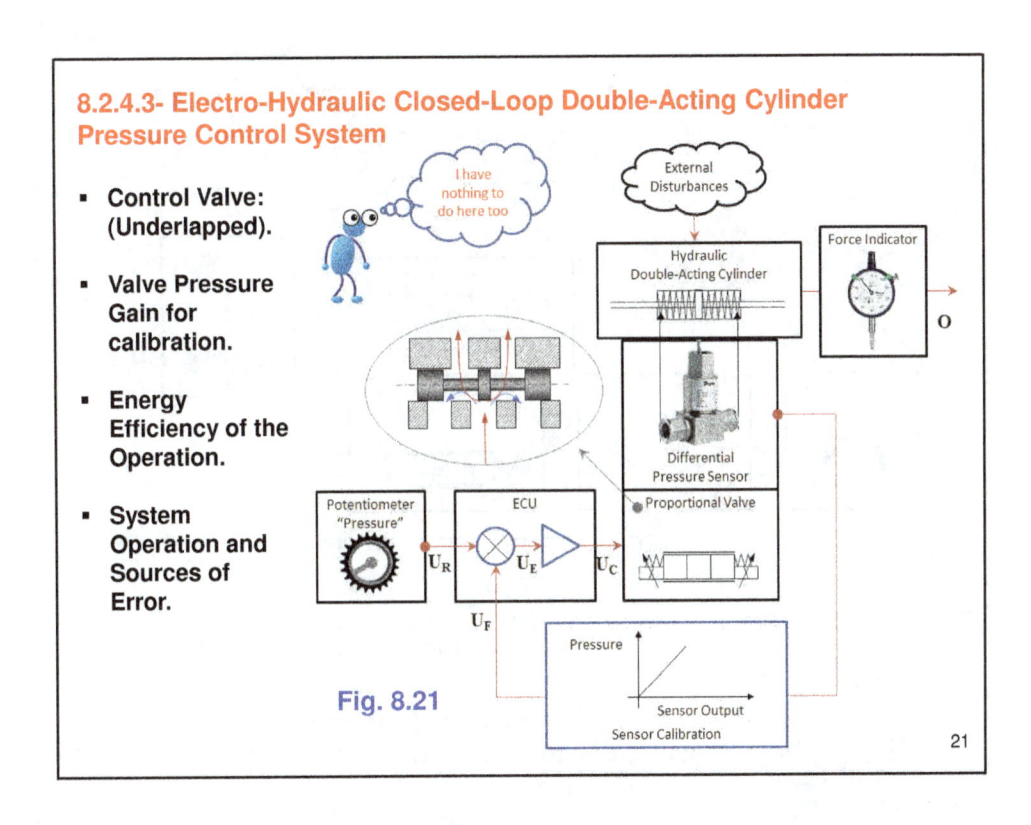

I have nothing to do here too

Fig. 8.21

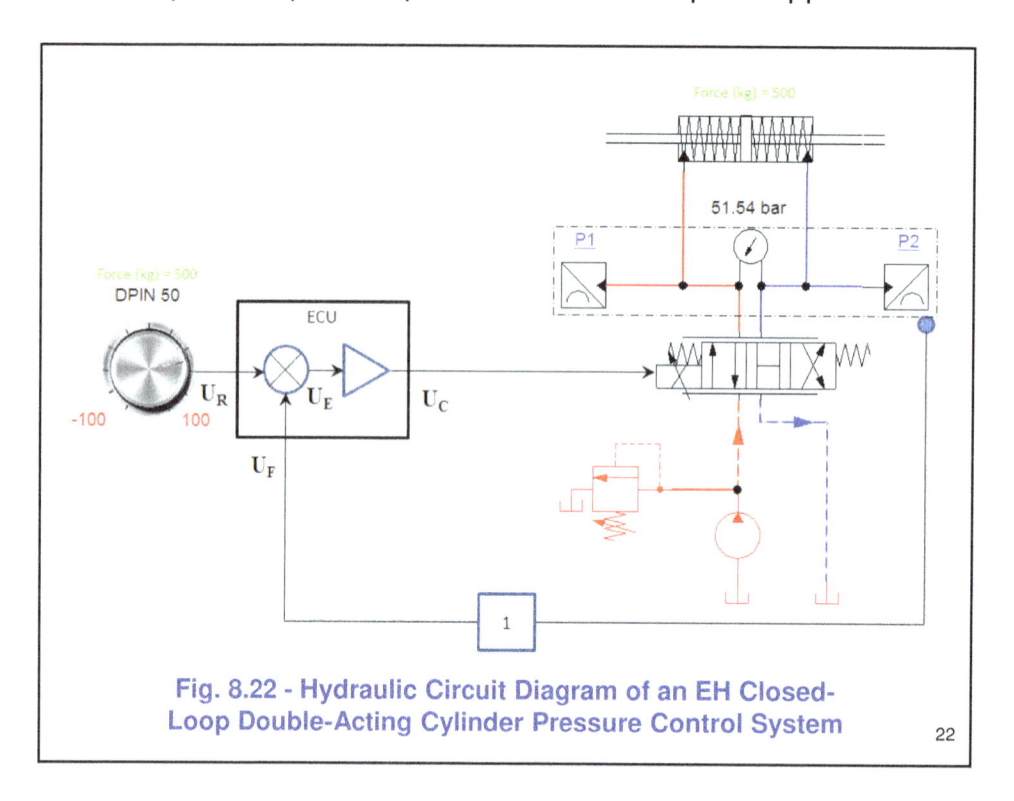

Workbook: Electro-Hydraulic Components and Systems
Chapter 8: Open-Loop versus Closed-Loop EH Applications

Fig. 8.22 - Hydraulic Circuit Diagram of an EH Closed-Loop Double-Acting Cylinder Pressure Control System

8.2.4.4- Electro-Hydraulic Closed-Loop Double-Acting Cylinder Force Control System

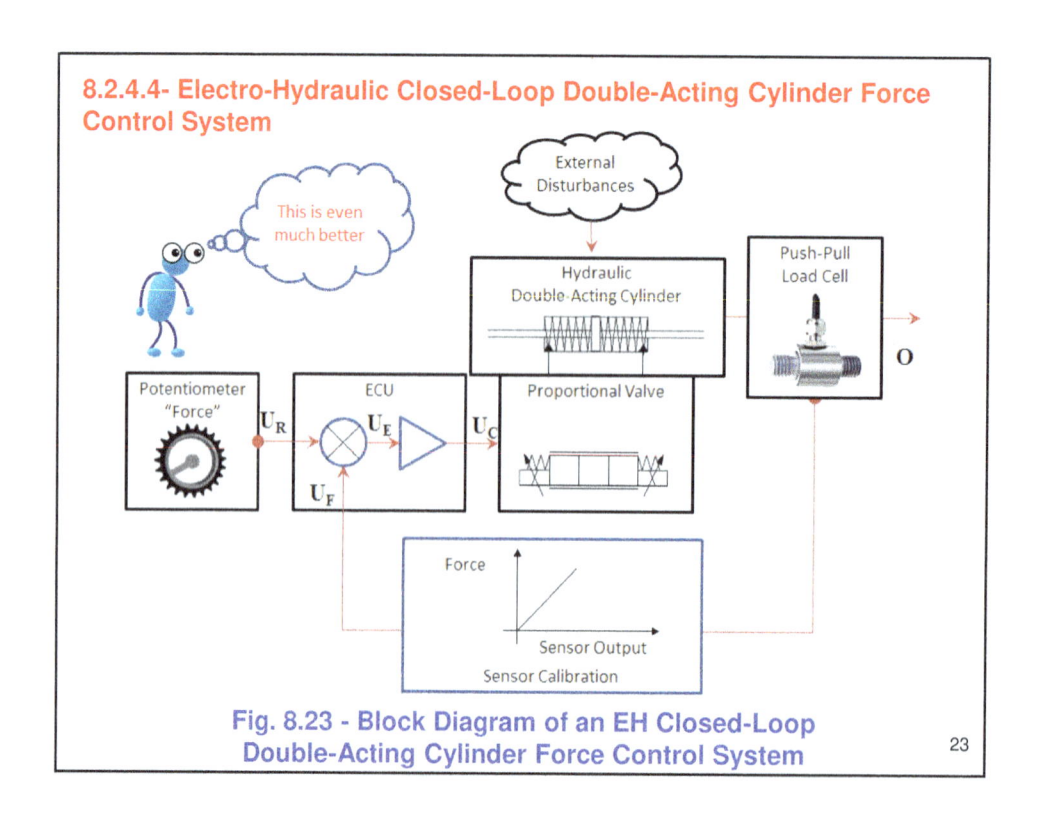

Fig. 8.23 - Block Diagram of an EH Closed-Loop Double-Acting Cylinder Force Control System

Workbook: Electro-Hydraulic Components and Systems
Chapter 8: Open-Loop versus Closed-Loop EH Applications

8.2.5- Electro-Hydraulic Closed-Loop Position Control Systems

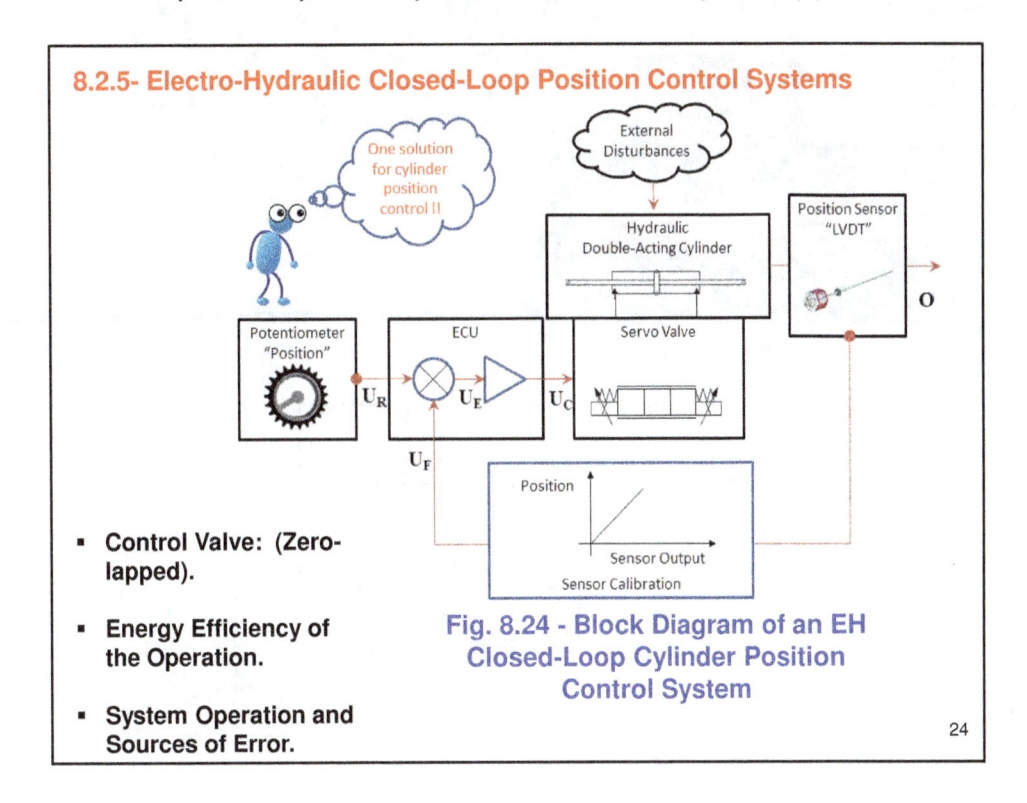

- Control Valve: (Zero-lapped).
- Energy Efficiency of the Operation.
- System Operation and Sources of Error.

Fig. 8.24 - Block Diagram of an EH Closed-Loop Cylinder Position Control System

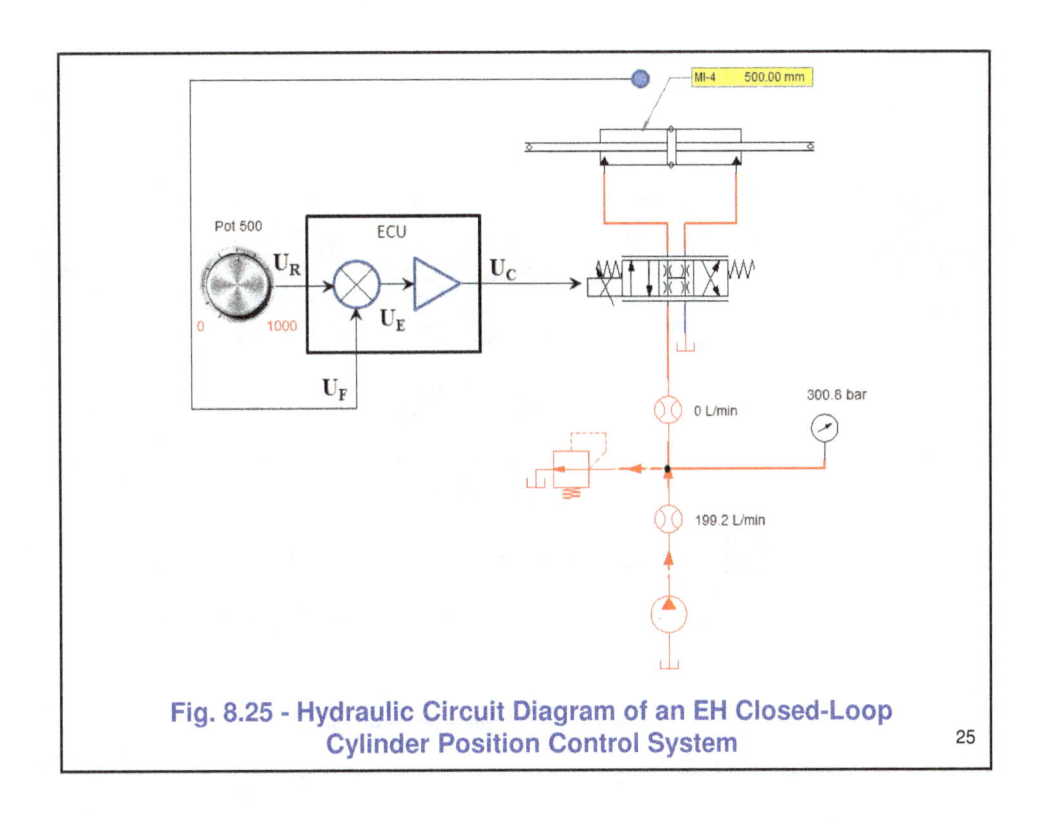

Fig. 8.25 - Hydraulic Circuit Diagram of an EH Closed-Loop Cylinder Position Control System

Workbook: Electro-Hydraulic Components and Systems
Chapter 8: Open-Loop versus Closed-Loop EH Applications

Fig. 7.26 - Servo Cylinder (Courtesy of Atos)

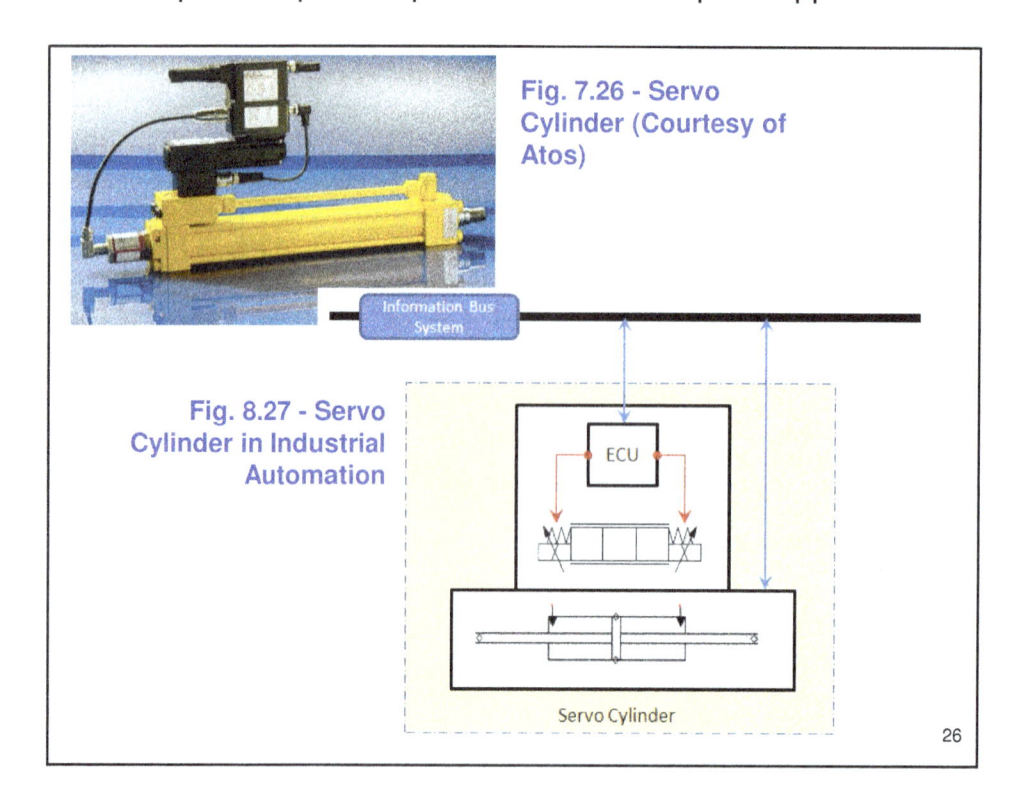

Fig. 8.27 - Servo Cylinder in Industrial Automation

8.3- Closed-Loop Performance Analysis

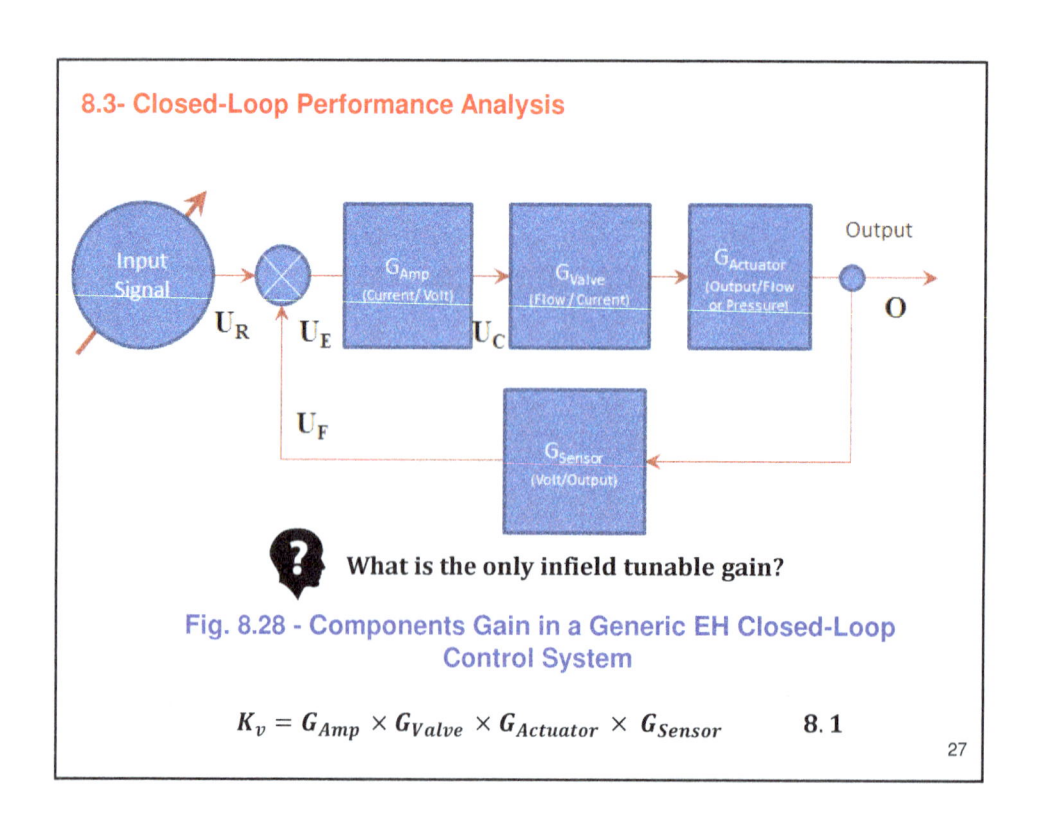

What is the only infield tunable gain?

Fig. 8.28 - Components Gain in a Generic EH Closed-Loop Control System

$$K_v = G_{Amp} \times G_{Valve} \times G_{Actuator} \times G_{Sensor} \qquad 8.1$$

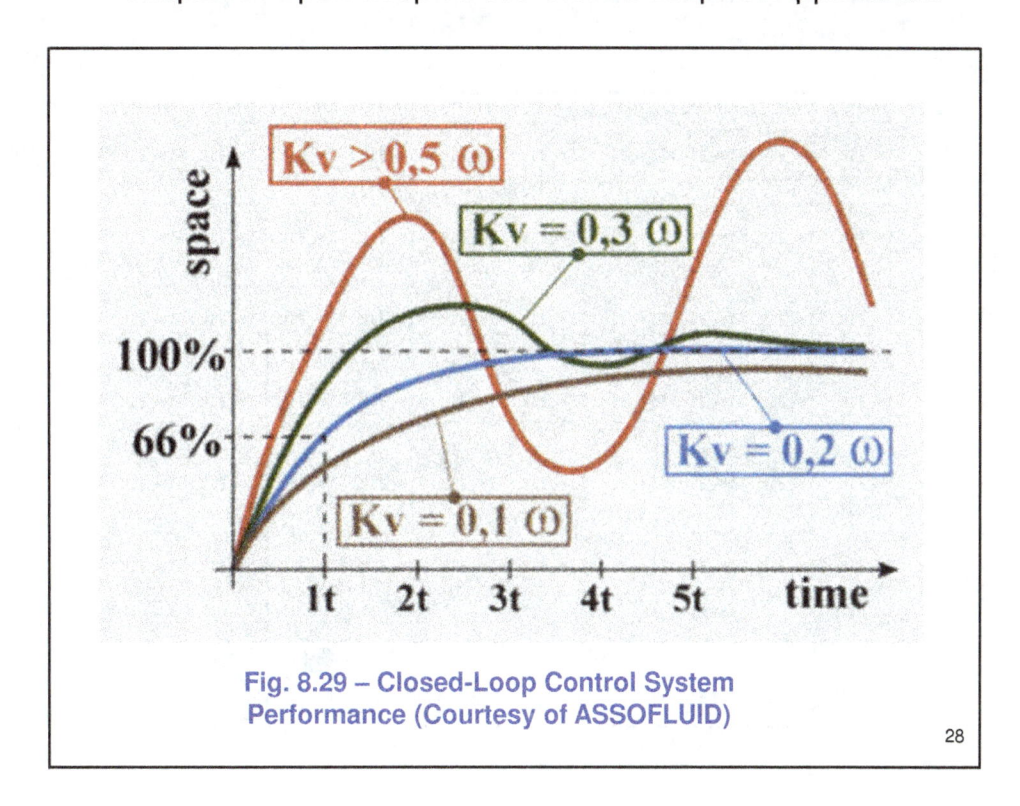

Fig. 8.29 – Closed-Loop Control System Performance (Courtesy of ASSOFLUID)

8.4- Electro-Hydraulic Control Systems Applications Examples
8.4.1- Industrial Applications

Video 260 (1 min)

Video 261 (0.5 min)

Ex.09-Lab 28

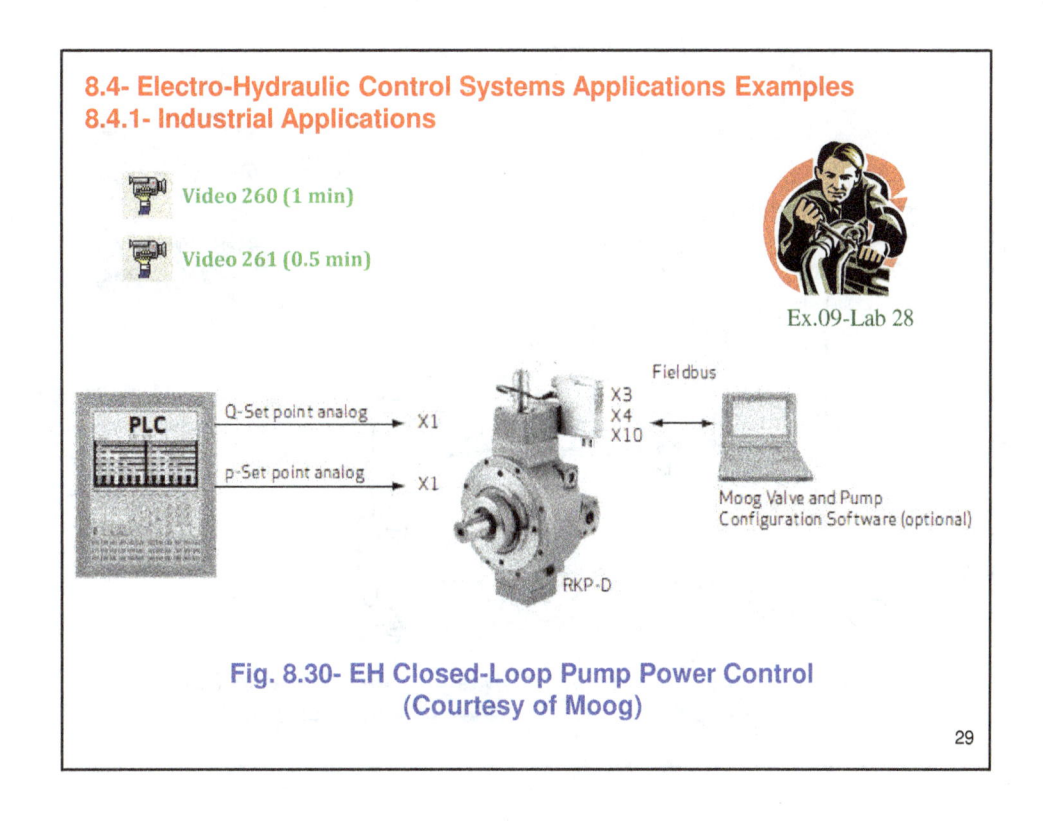

Fig. 8.30- EH Closed-Loop Pump Power Control (Courtesy of Moog)

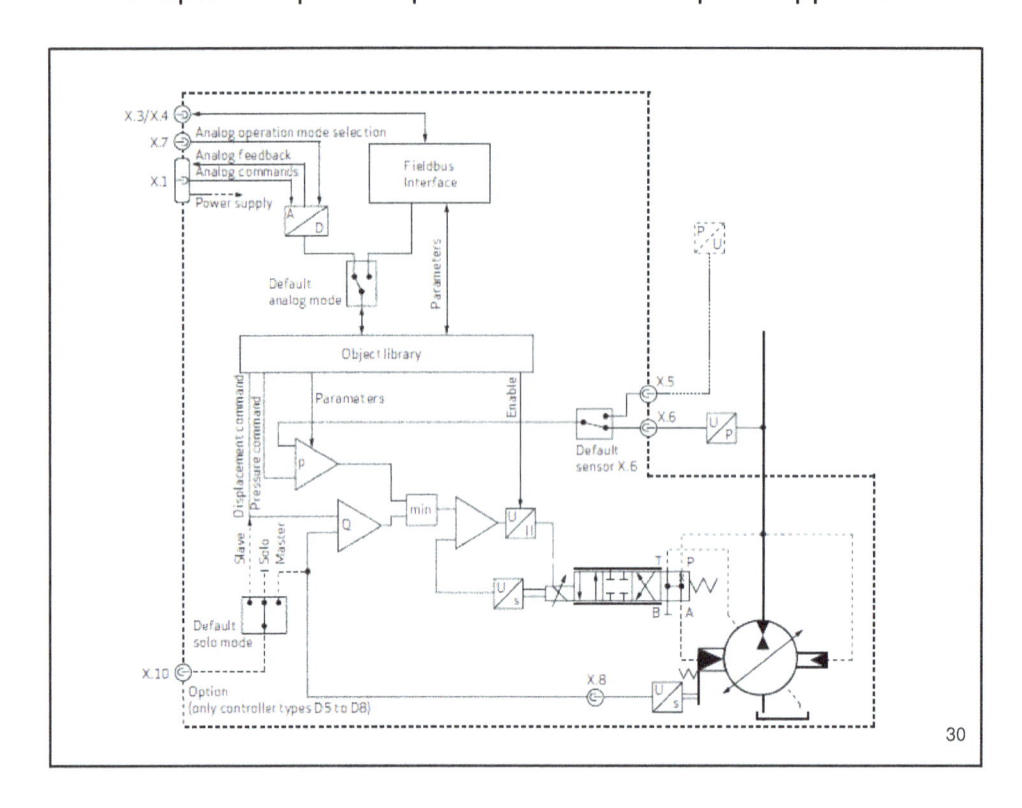

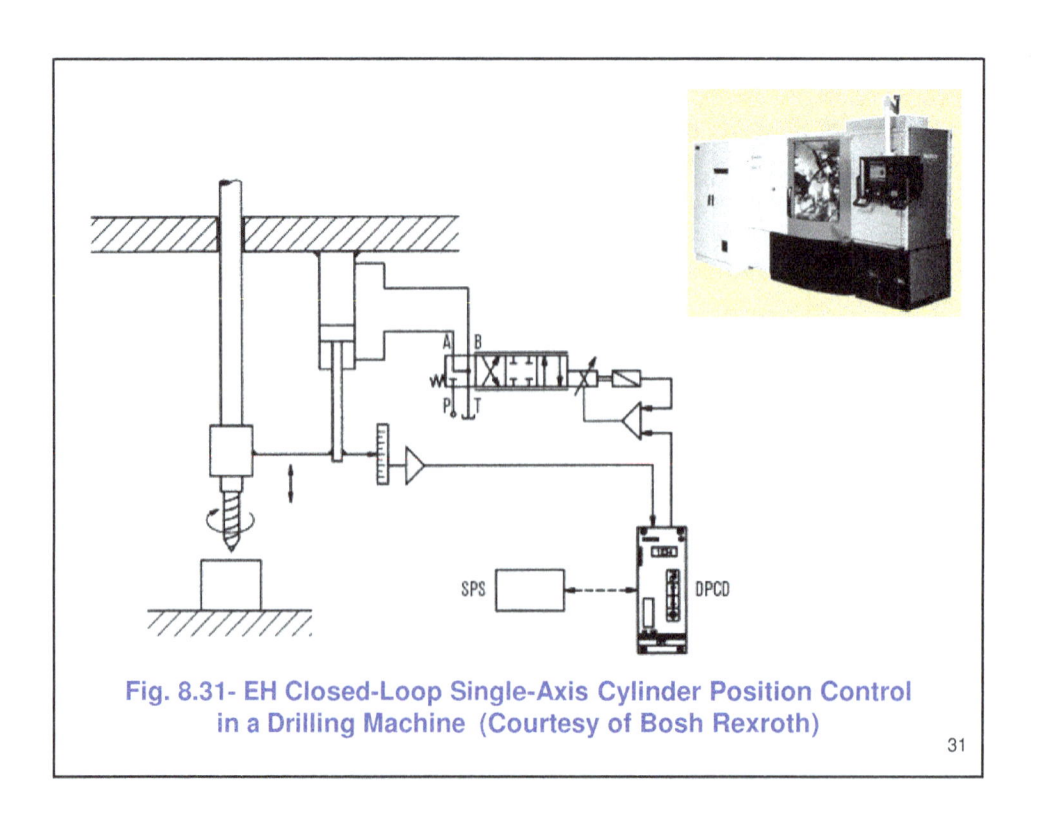

Fig. 8.31- EH Closed-Loop Single-Axis Cylinder Position Control in a Drilling Machine (Courtesy of Bosh Rexroth)

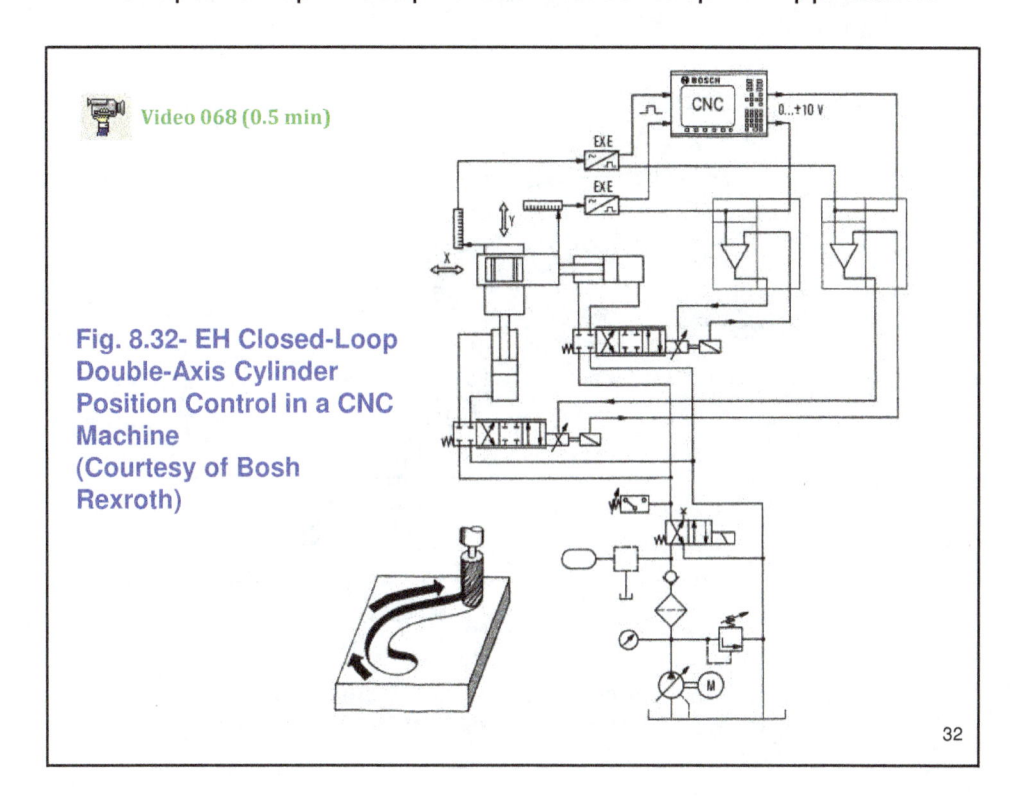

Fig. 8.32- EH Closed-Loop Double-Axis Cylinder Position Control in a CNC Machine (Courtesy of Bosh Rexroth)

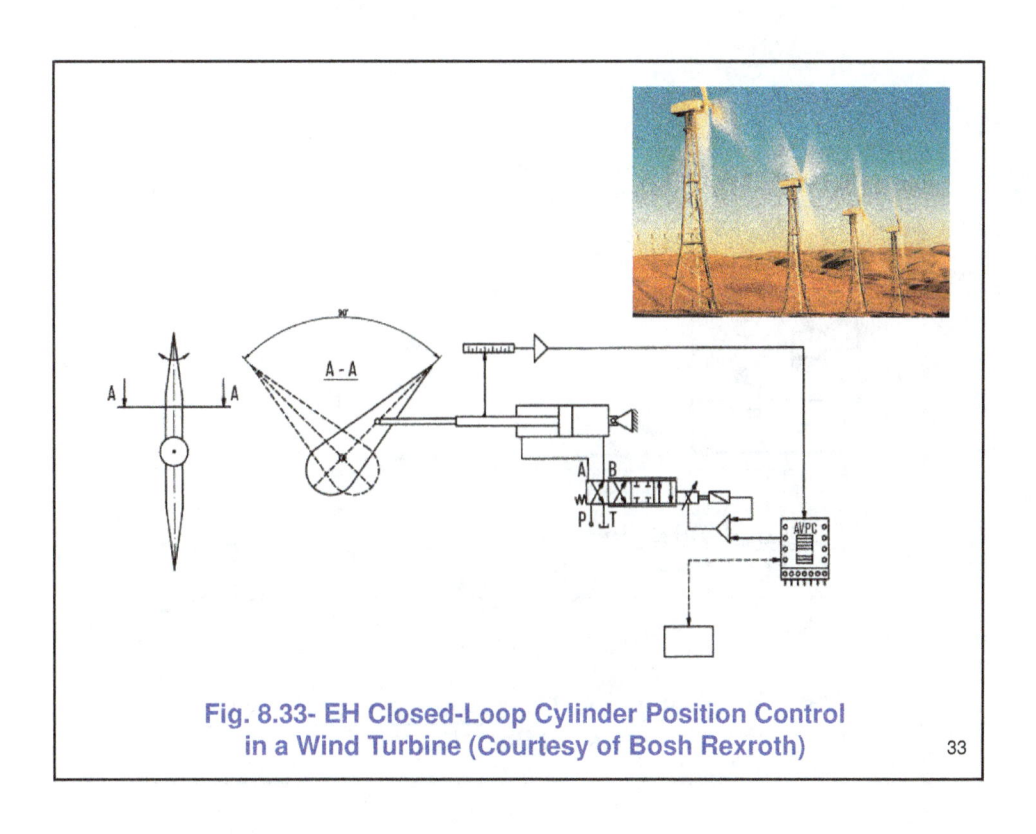

Fig. 8.33- EH Closed-Loop Cylinder Position Control in a Wind Turbine (Courtesy of Bosh Rexroth)

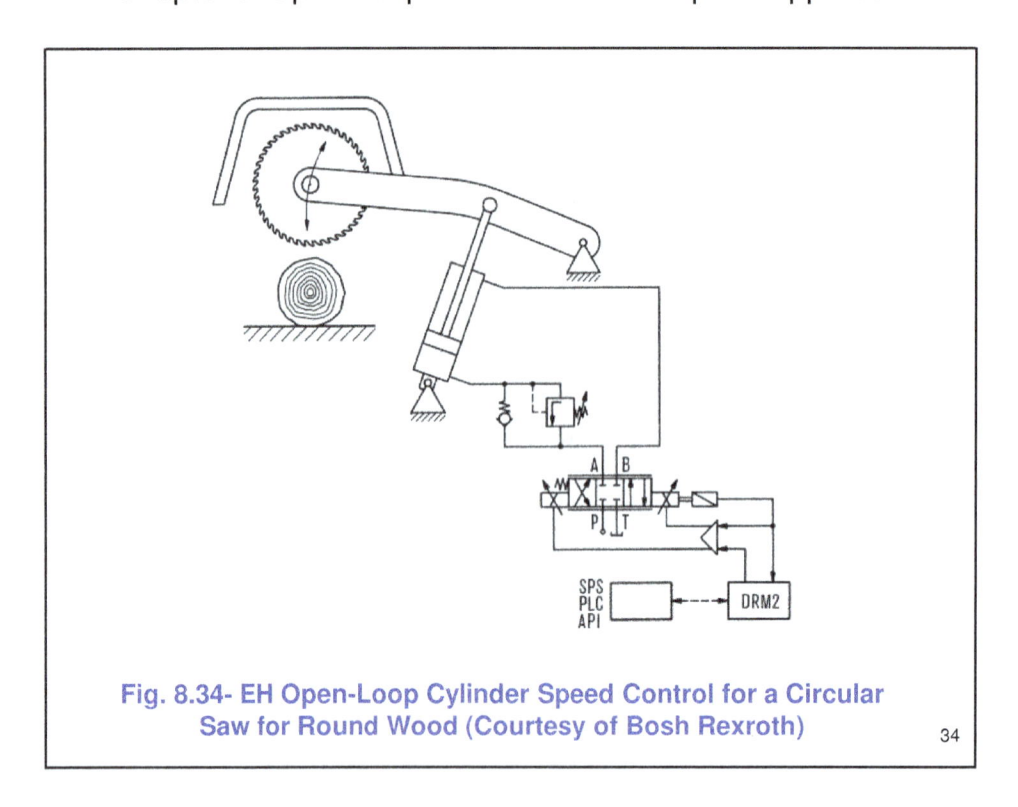

Fig. 8.34- EH Open-Loop Cylinder Speed Control for a Circular Saw for Round Wood (Courtesy of Bosh Rexroth)

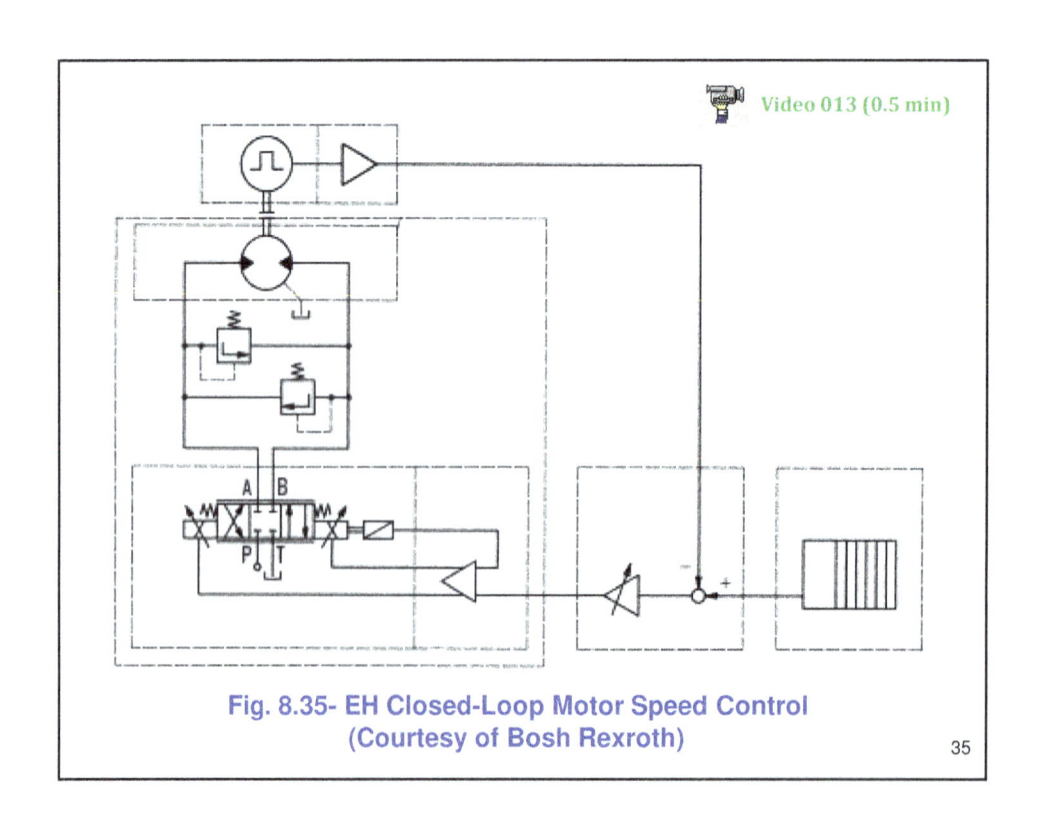

Fig. 8.35- EH Closed-Loop Motor Speed Control (Courtesy of Bosh Rexroth)

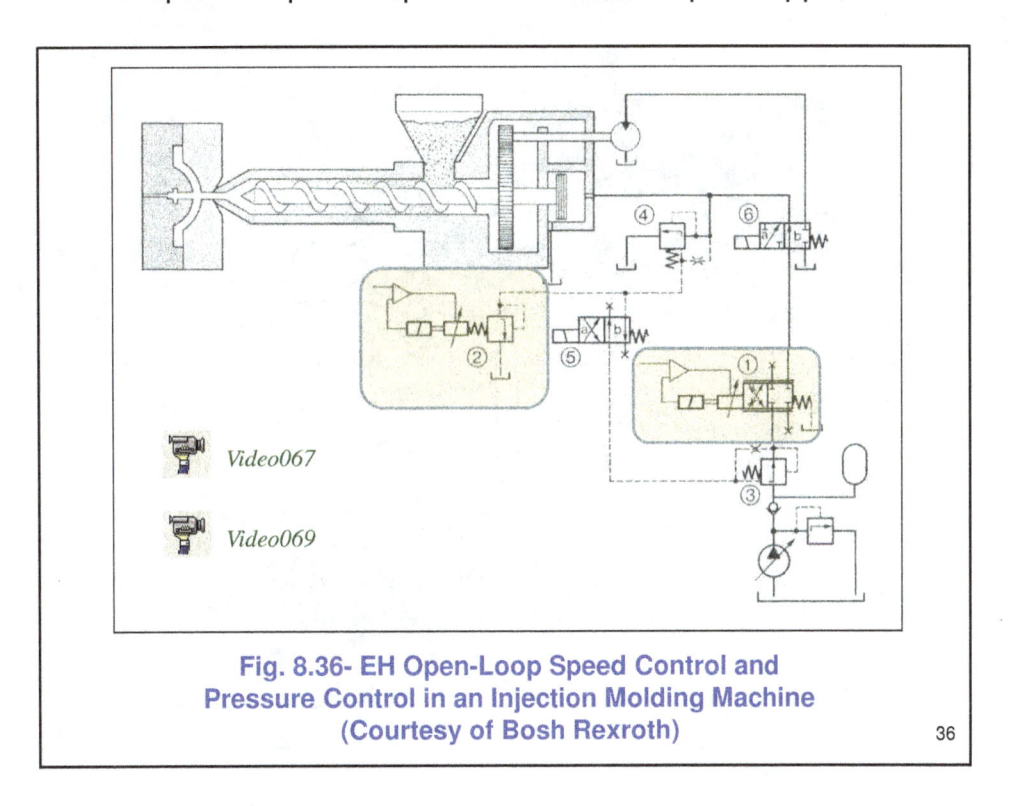

Workbook: Electro-Hydraulic Components and Systems
Chapter 8: Open-Loop versus Closed-Loop EH Applications

Fig. 8.36- EH Open-Loop Speed Control and Pressure Control in an Injection Molding Machine (Courtesy of Bosh Rexroth)

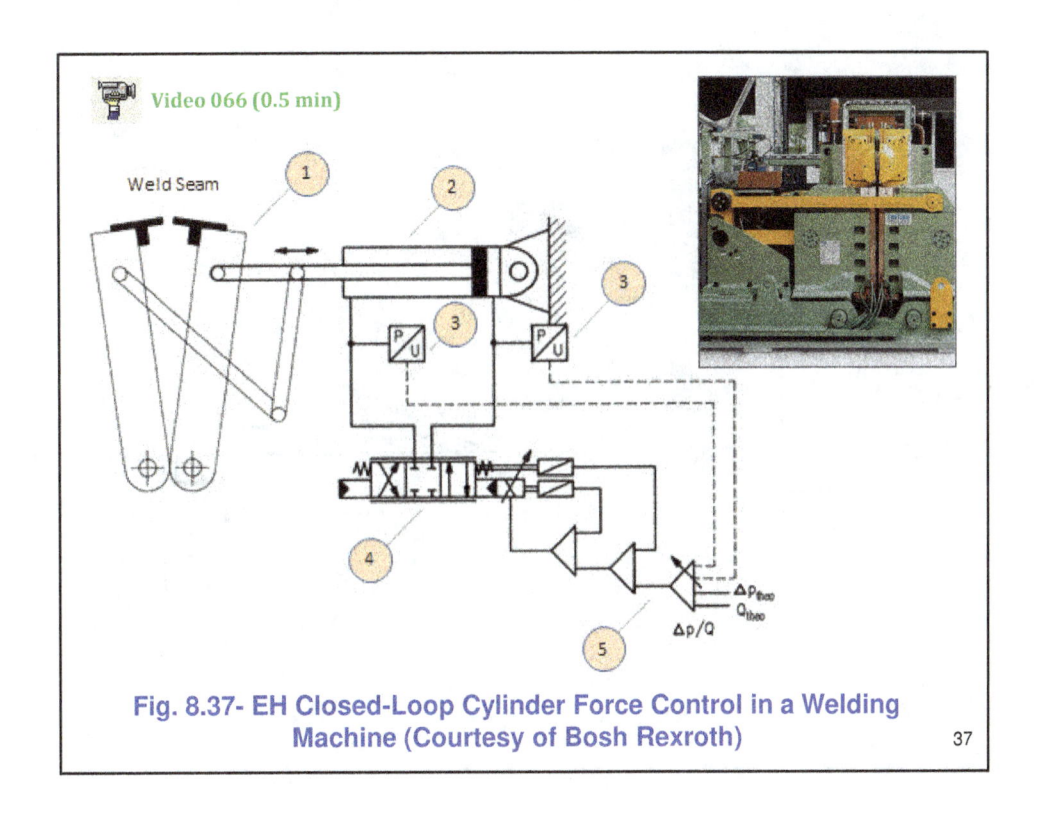

Fig. 8.37- EH Closed-Loop Cylinder Force Control in a Welding Machine (Courtesy of Bosh Rexroth)

Workbook: Electro-Hydraulic Components and Systems
Chapter 8: Open-Loop versus Closed-Loop EH Applications

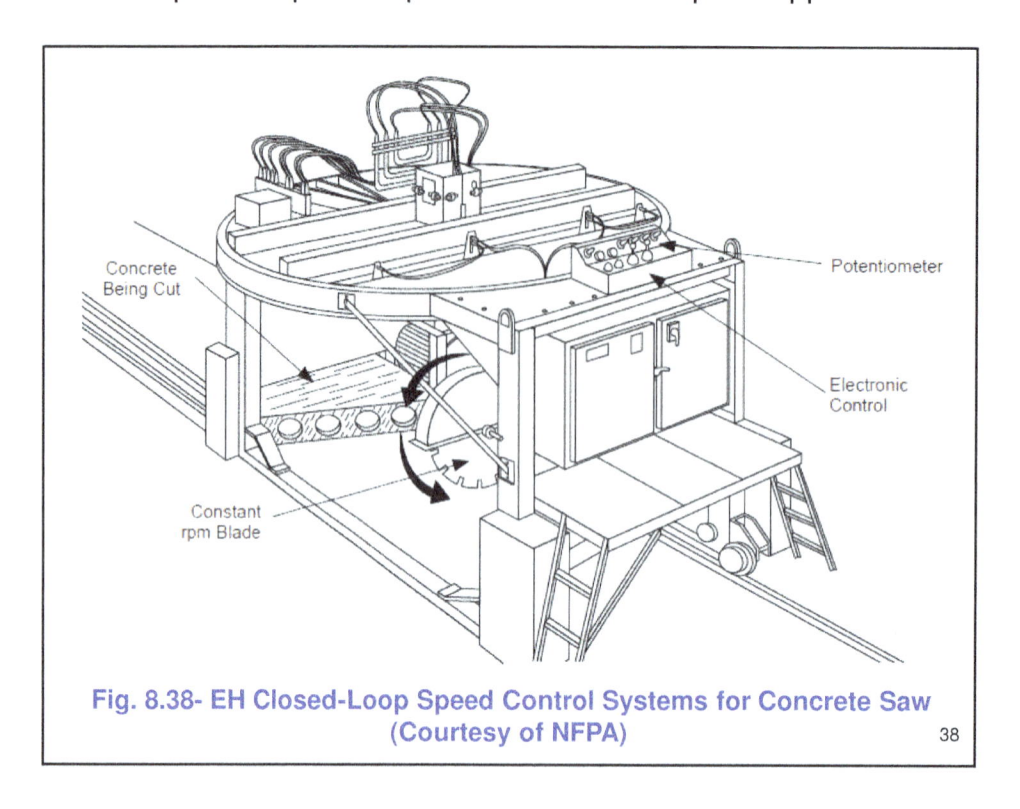

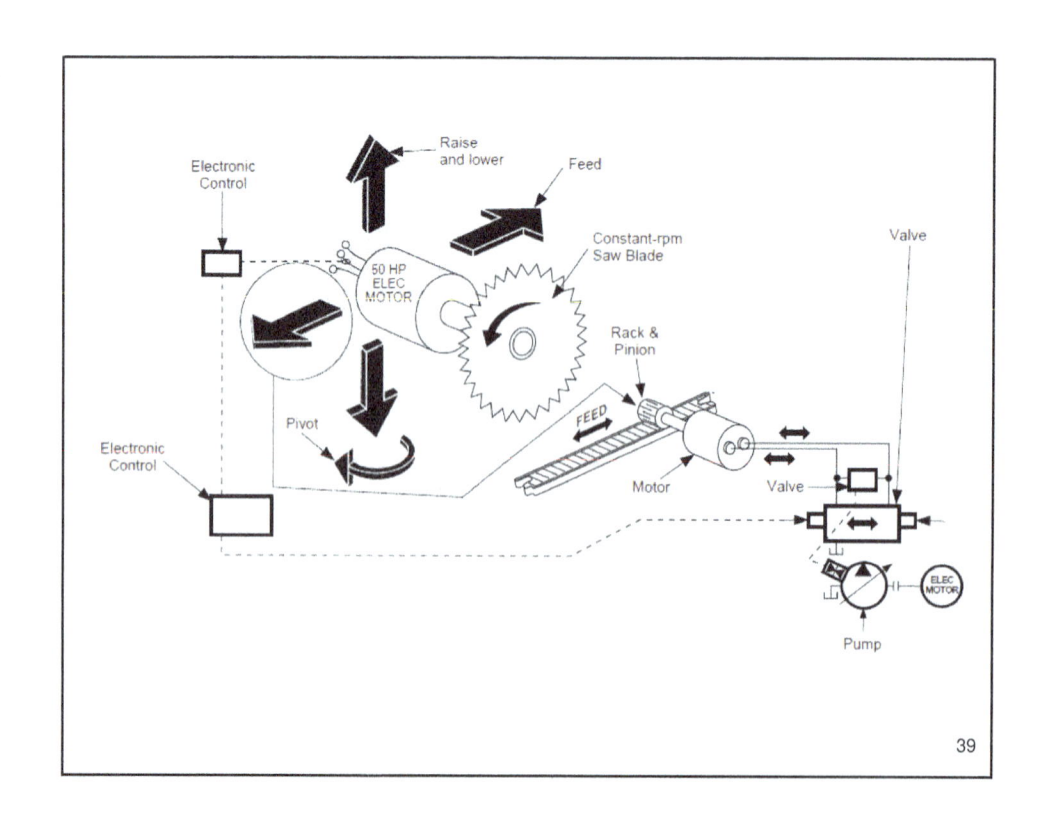

Fig. 8.38- EH Closed-Loop Speed Control Systems for Concrete Saw (Courtesy of NFPA)

Workbook: Electro-Hydraulic Components and Systems
Chapter 8: Open-Loop versus Closed-Loop EH Applications

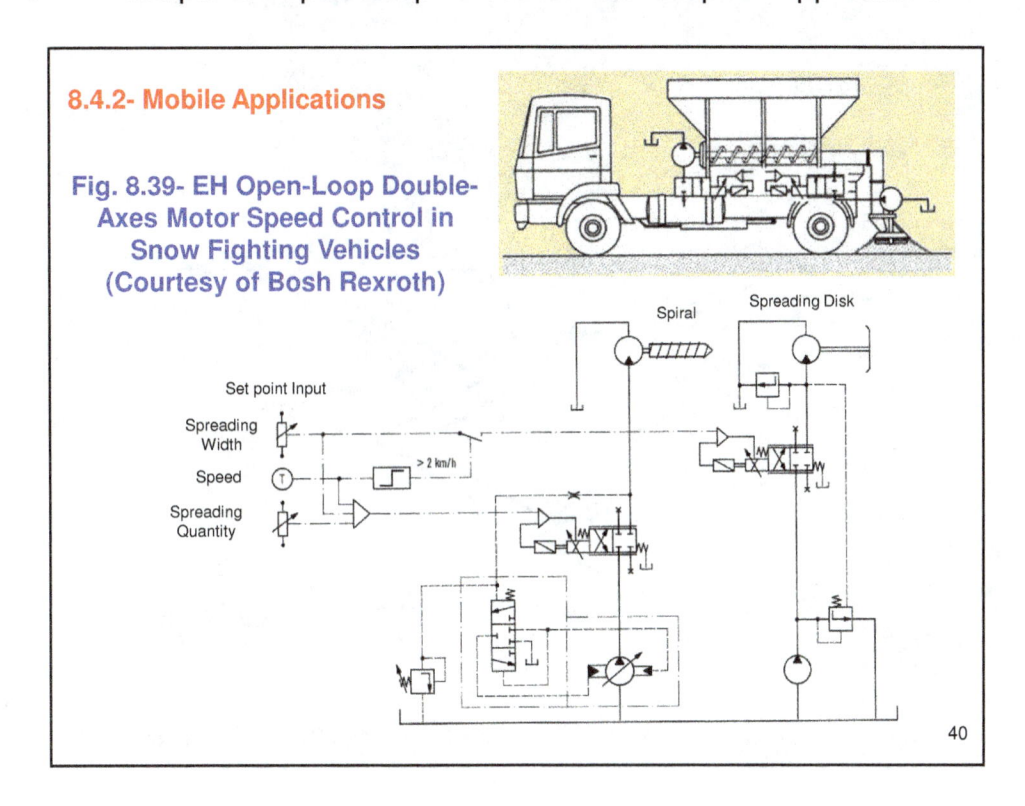

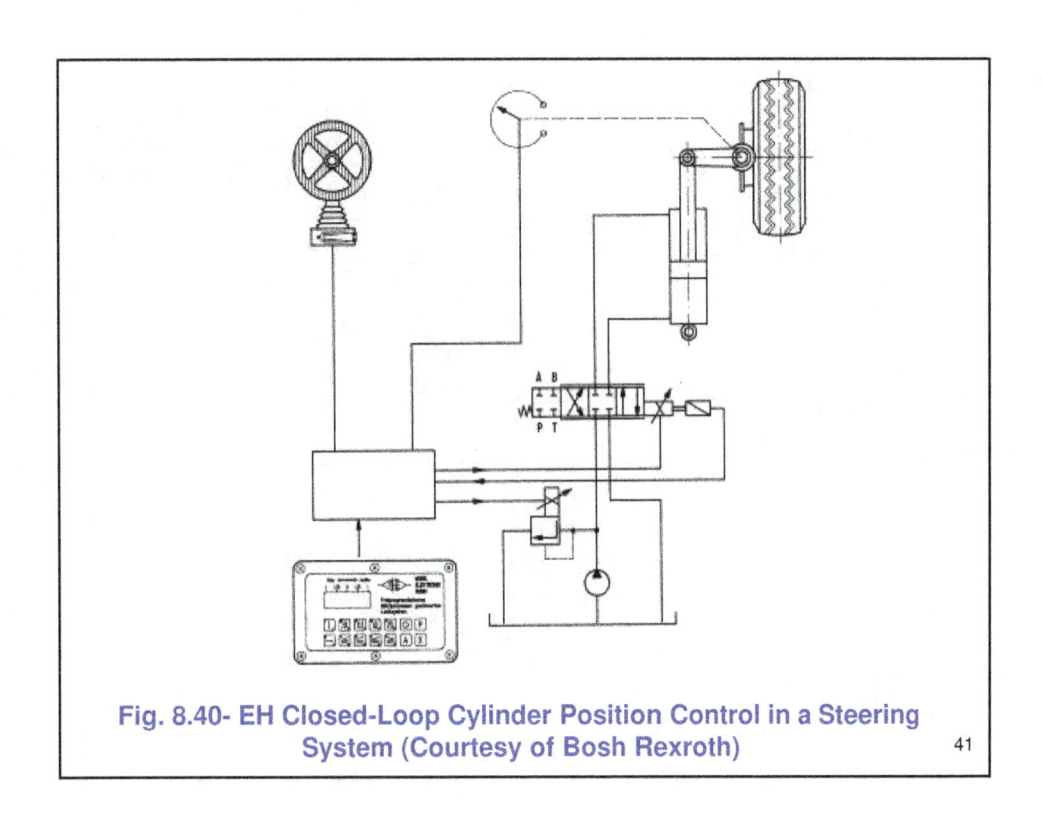

8.4.2- Mobile Applications

Fig. 8.39- EH Open-Loop Double-Axes Motor Speed Control in Snow Fighting Vehicles (Courtesy of Bosh Rexroth)

Fig. 8.40- EH Closed-Loop Cylinder Position Control in a Steering System (Courtesy of Bosh Rexroth)

Workbook: Electro-Hydraulic Components and Systems
Chapter 8: Open-Loop versus Closed-Loop EH Applications

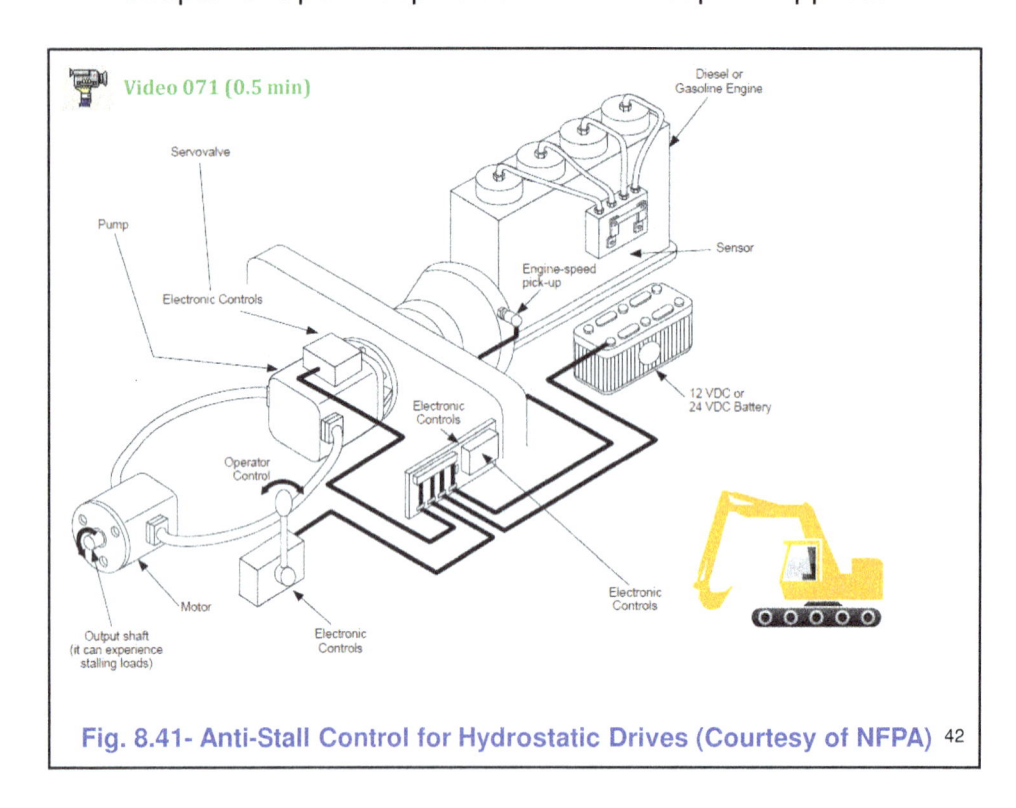

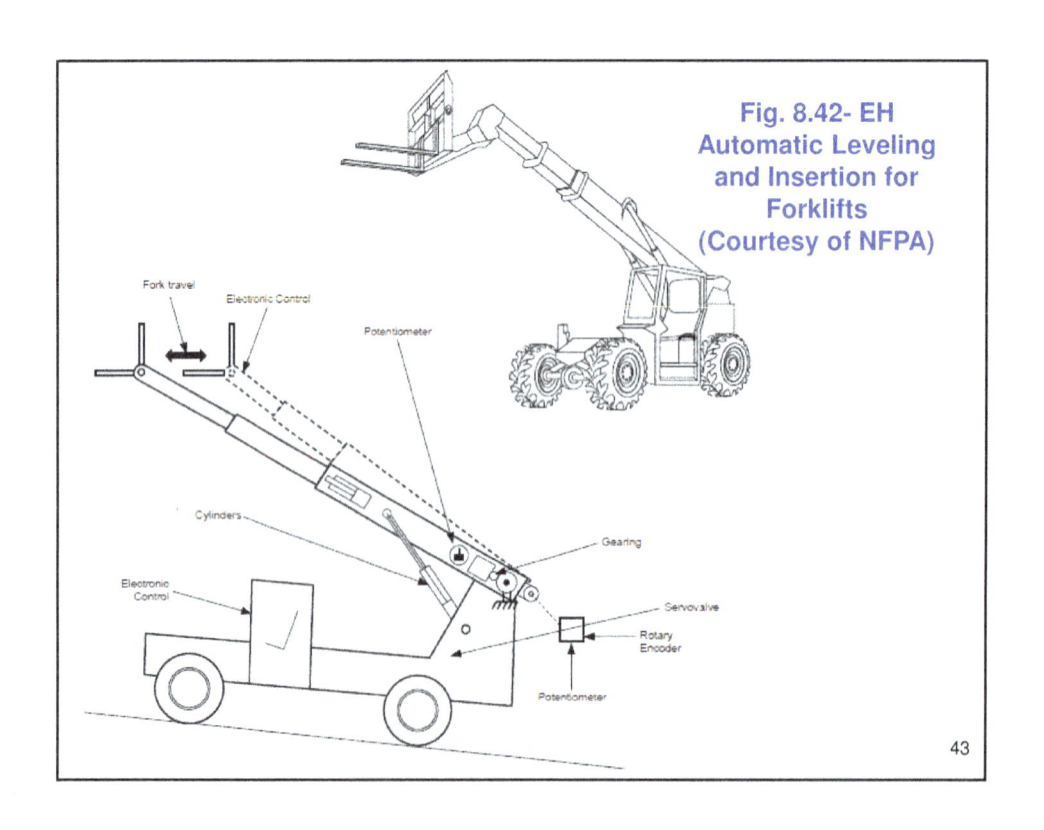

Fig. 8.41- Anti-Stall Control for Hydrostatic Drives (Courtesy of NFPA)

Fig. 8.42- EH Automatic Leveling and Insertion for Forklifts (Courtesy of NFPA)

Workbook: Electro-Hydraulic Components and Systems
Chapter 8: Open-Loop versus Closed-Loop EH Applications

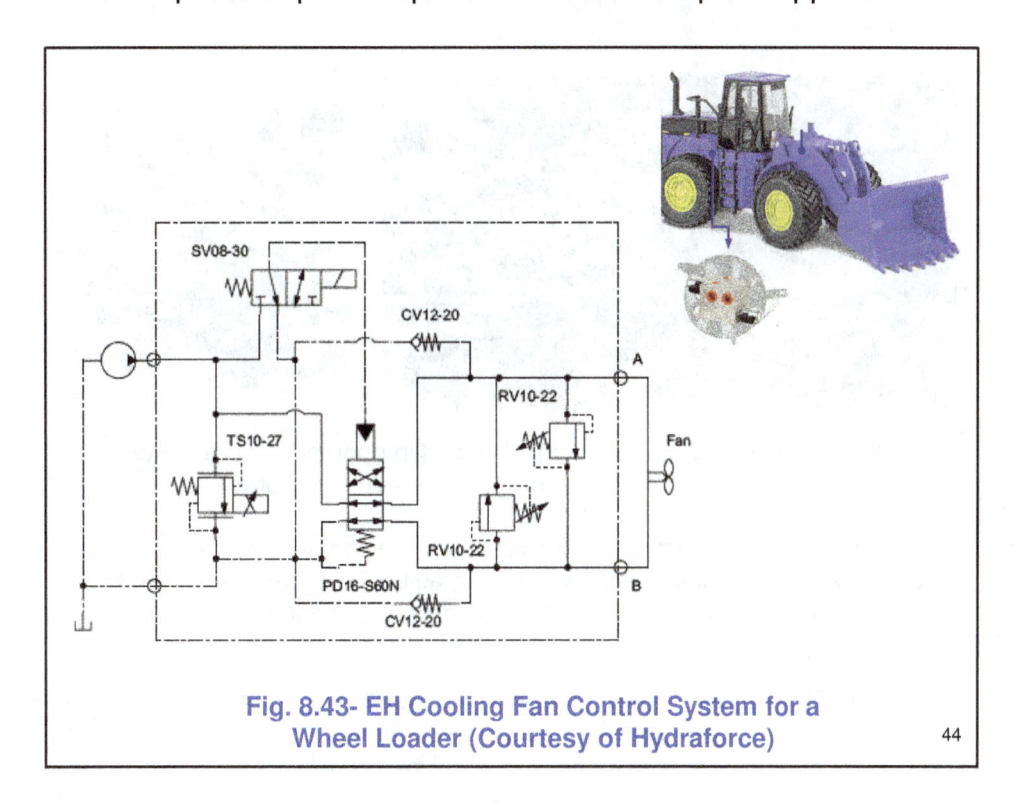

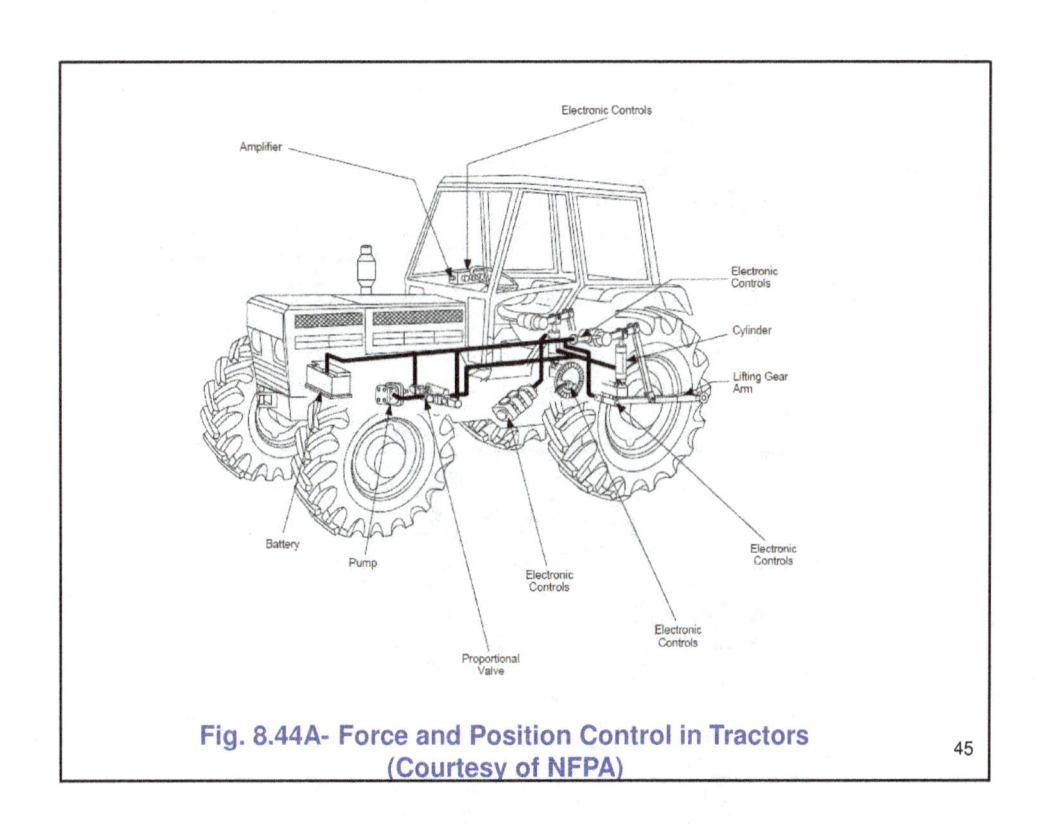

Fig. 8.43- EH Cooling Fan Control System for a Wheel Loader (Courtesy of Hydraforce)

Fig. 8.44A- Force and Position Control in Tractors (Courtesy of NFPA)

Workbook: Electro-Hydraulic Components and Systems
Chapter 8: Open-Loop versus Closed-Loop EH Applications

In the agricultural industry, the digital display proportional valve driver is used to control the height of the booms on crop sprayers.

Working in connection with ultrasonic sensors, which measure the distance, the drivers adjusts the booms when the gradient fluctuates. Resulting in proportionally distribute the nutrients and health crops.

**Fig. 8.44B – Level Control in Agricultural Machines
(Excerption from Lynch)**

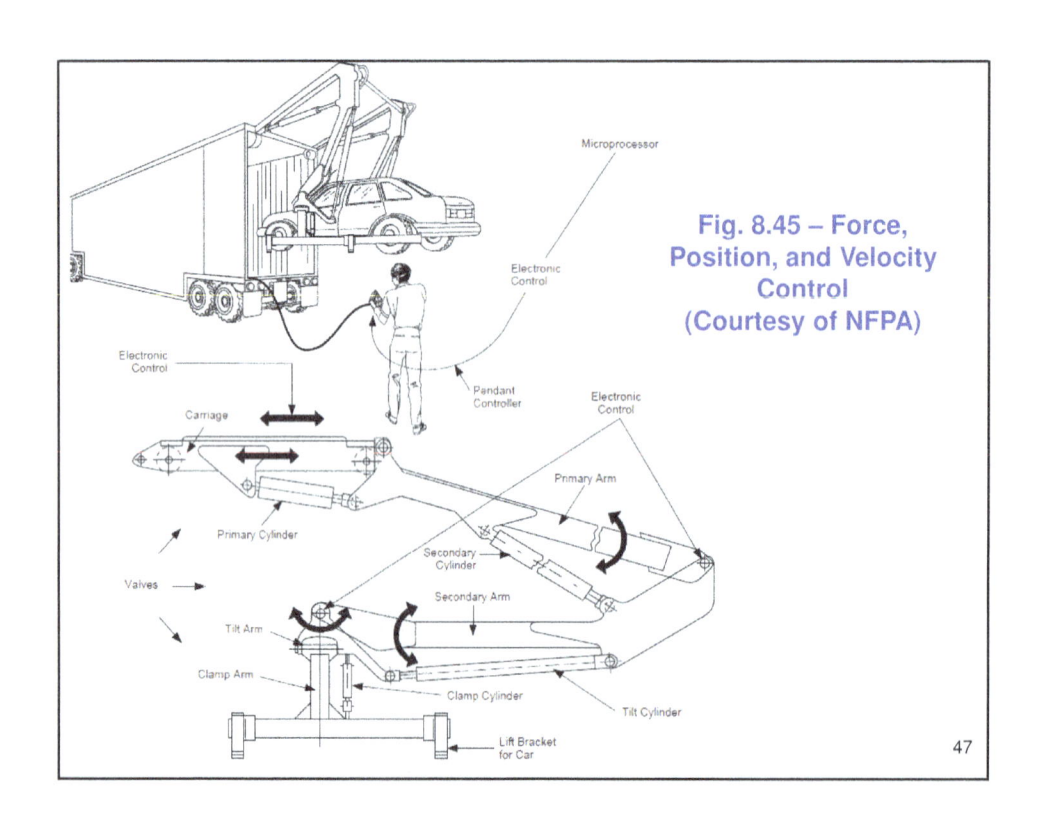

**Fig. 8.45 – Force, Position, and Velocity Control
(Courtesy of NFPA)**

Chapter 8 Reviews

1. Describe the system shown below:
    A. An open-loop system using force-controlled continuous valve.
    B. An open-loop system using stroke-controlled continuous valve.
    C. A closed-loop system using force-controlled continuous valve.
    D. A closed-loop system using stroke-controlled continuous valve.

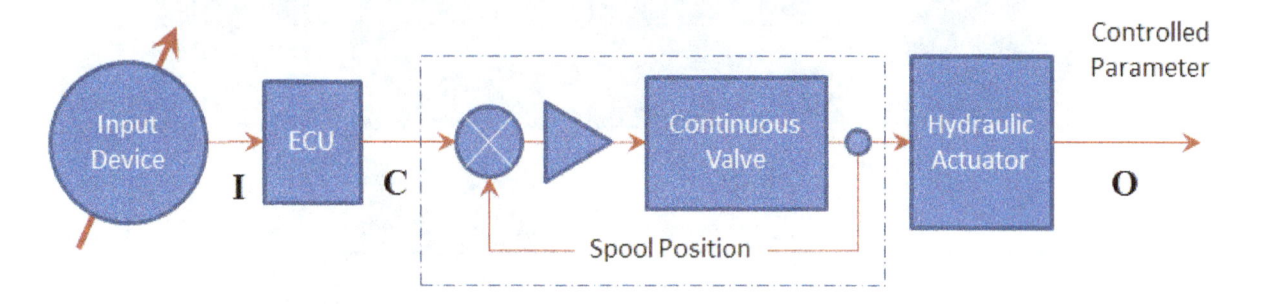

2. Describe the system shown below:
    A. An open-loop system using force-controlled continuous valve.
    B. An open-loop system using stroke-controlled continuous valve.
    C. A closed-loop system using force-controlled continuous valve.
    D. A closed-loop system using stroke-controlled continuous valve.

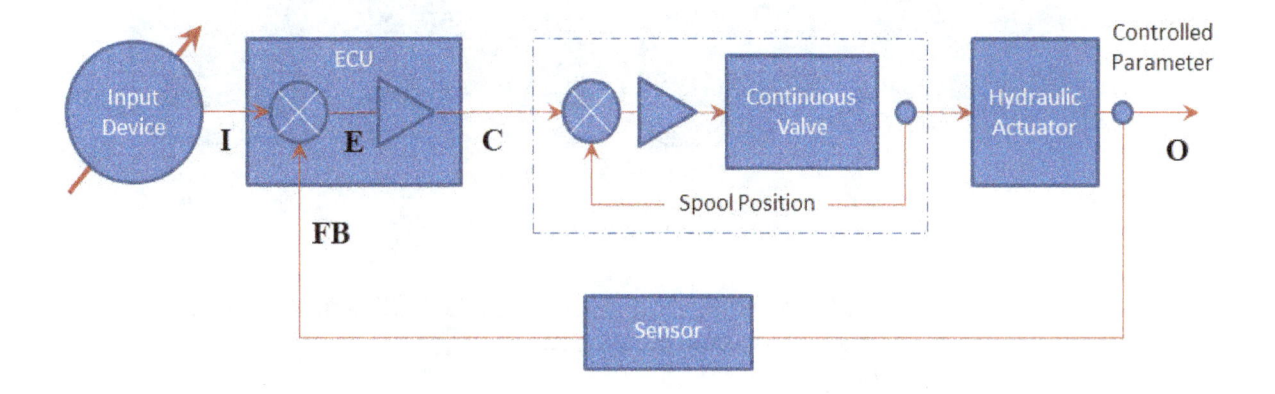

3. Describe the system shown below:
    A. An open-loop system using force-controlled continuous valve.
    B. An open-loop system using stroke-controlled continuous valve.
    C. A closed-loop system using force-controlled continuous valve.
    D. A closed-loop system using stroke-controlled continuous valve.

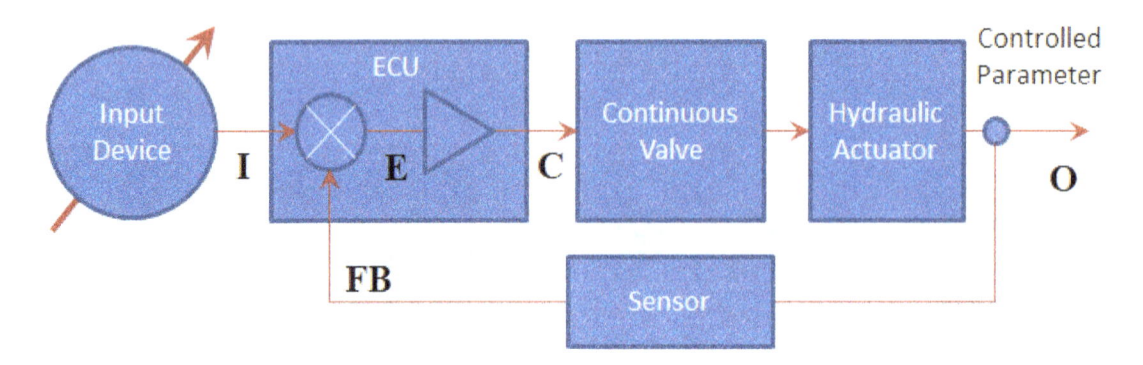

4. Describe the system shown below:
    A. An open-loop system using force-controlled continuous valve.
    B. An open-loop system using stroke-controlled continuous valve.
    C. A closed-loop system using force-controlled continuous valve.
    D. A closed-loop system using stroke-controlled continuous valve.

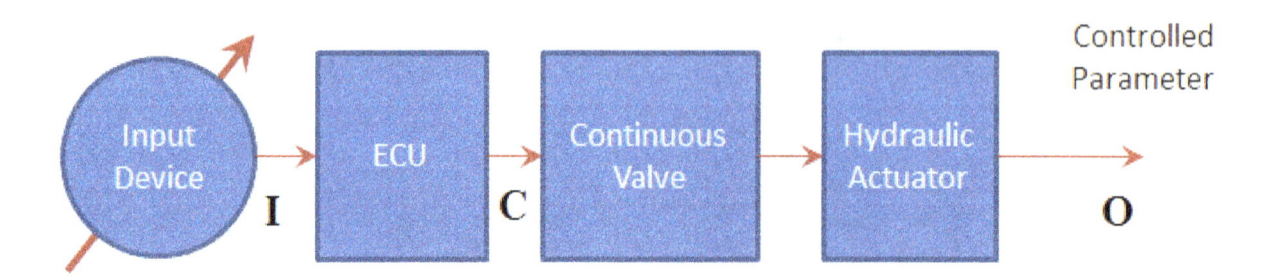

5. Select the most suitable valve for a closed-loop ON/Off temperature control system:
    A. A switching solenoid-operated valve.
    B. A proportional valve.
    C. A servo valve.
    D. None of the above.

6. Select the most suitable valve for an open-loop motor speed control system:
    A. A switching solenoid-operated valve.
    B. A proportional valve.
    C. A servo valve.
    D. None of the above.

Workbook: Electro-Hydraulic Components and Systems
Chapter 8: Reviews and Assignments

7. Select the most suitable valve for a closed-loop high accuracy cylinder position control system:
   A. A switching solenoid-operated valve.
   B. A proportional valve.
   C. A servo valve.
   D. None of the above.

8. Select the most suitable valve for a closed-loop motor speed control system:
   A. A three-position closed-center (overlapped) stroke-controlled proportional valve.
   B. A three-position zero lapped stroke-controlled proportional valve.
   C. A three-position open-center (underlapped) stroke-controlled proportional valve.
   D. None of the above.

9. Select the most suitable valve for a closed-loop double-acting cylinder load control system:
   A. A three-position closed-center (overlapped) stroke-controlled proportional valve.
   B. A three-position zero lapped stroke-controlled proportional valve.
   C. A three-position open-center (underlapped) stroke-controlled proportional valve.
   D. None of the above.

10. Select the most suitable valve for a closed-loop synchronous cylinder position control system:
    A. A three-position closed-center (overlapped) stroke-controlled proportional valve.
    B. A three-position zero lapped stroke-controlled proportional valve.
    C. A three-position open-center (underlapped) stroke-controlled proportional valve.
    D. None of the above.

Chapter 8 Assignment

Student Name: --- Student ID: ------------------

Date: -- Score: ------------------------

Draw a block diagram and the hydraulic circuit for a closed-loop unidirectional motor speed control. Calculate the best feedback calibration factor and amplifier gain where:

- Input Device maximum signal = 10 Volt
- Valve gain = 20 (lit/min)/Amp
- Motor gain = 100 RPM/(lit/min)
- RPM sensor gain = 0.005 Volt/RPM
- Overall system natural frequency = 10 rad/s

Workbook: Electro-Hydraulic Components and Systems
Chapter 9: Control Electronics for Electro-Hydraulic Valves

Chapter 9
Control Electronic Units for Electro-Hydraulic Valves

Objectives:

This chapter presents various configurations and basic functions contained in control electronics for EH valves. The chapter discusses the changes in the system performance when these functions are enabled, including gain adjustor, dead band eliminator, ramp, and dither.

Brief Contents:

9.1- Control Electronics Basic Functions
9.2- DC Power Supply
9.3- Signal Amplifier
9.4- Dither Signal
9.5- Pulse Width Modulation
9.6- Input Signal Generator
9.7- Feedback Sensors
9.8- PID Controller
9.9- Gain Adjustor
9.10- Ramp Adjustor
9.11- Null Adjustor (I_{Bias})
9.12- Dead Band Eliminator (I_{min})
9.13- Saturation Adjustor (I_{max})
9.14- Control Electronics Basic Configurations
9.15- Hardware Configurations
9.16- Typical Electronic Schematics for Continuous Valves

Workbook: Electro-Hydraulic Components and Systems
Chapter 9: Control Electronics for Electro-Hydraulic Valves

9.1 - Control Electronics Basic Functions

- **Power:**
  - Supply the required voltage and DC current.
  - <u>Amplify</u> the input signal into an output current signal proportional to the input signal and <u>powerful enough</u> to drive a proportional solenoid or a torque motor.

- **Regulation:**
  - <u>Regulate</u> the signal received by the proportional solenoid or the torque motor to assure <u>proper and safe operation</u>.
  - Overcome <u>undesirable valve characteristics</u>.

- **Control:**
  - Apply the desired <u>control algorithm</u> within the entire system operation.

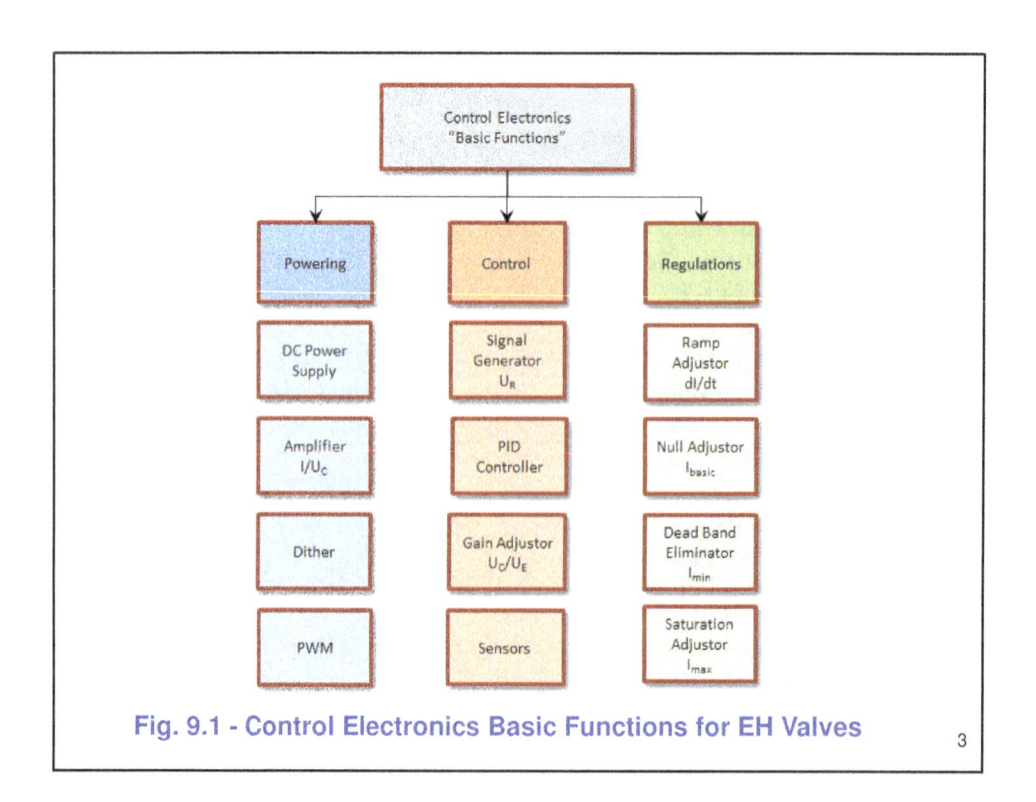

Control Electronics "Basic Functions"

- **Powering**
  - DC Power Supply
  - Amplifier I/U_C
  - Dither
  - PWM

- **Control**
  - Signal Generator U_R
  - PID Controller
  - Gain Adjustor U_C/U_E
  - Sensors

- **Regulations**
  - Ramp Adjustor dI/dt
  - Null Adjustor I_{basic}
  - Dead Band Eliminator I_{min}
  - Saturation Adjustor I_{max}

Fig. 9.1 - Control Electronics Basic Functions for EH Valves

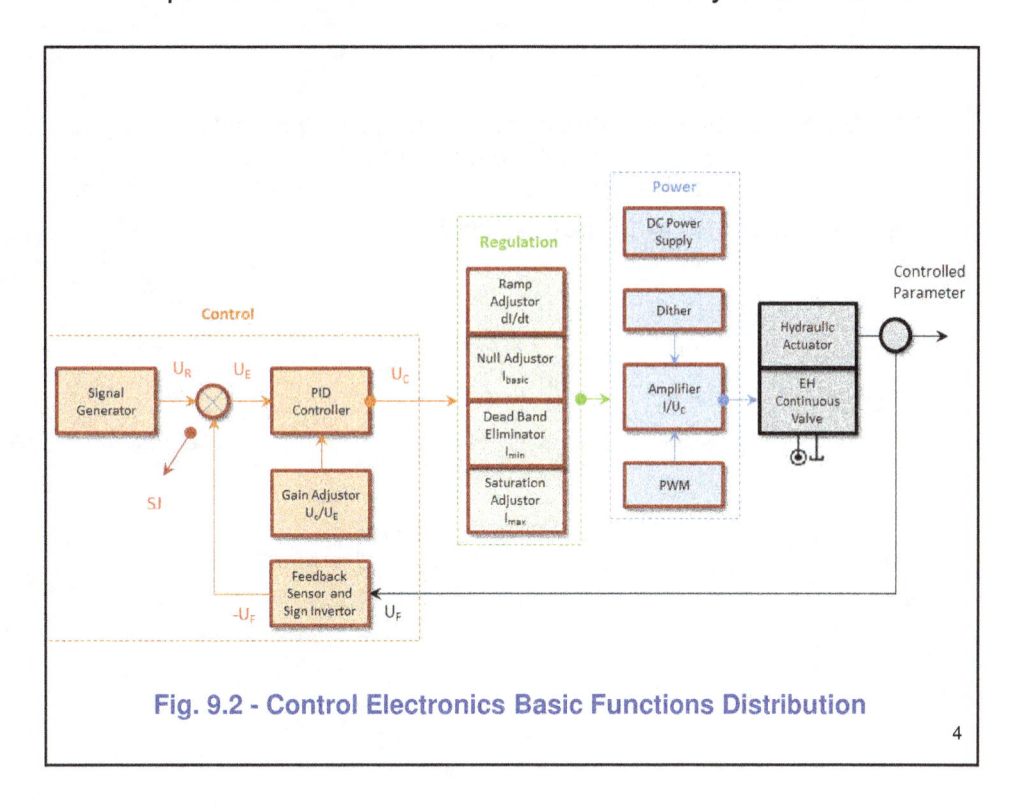

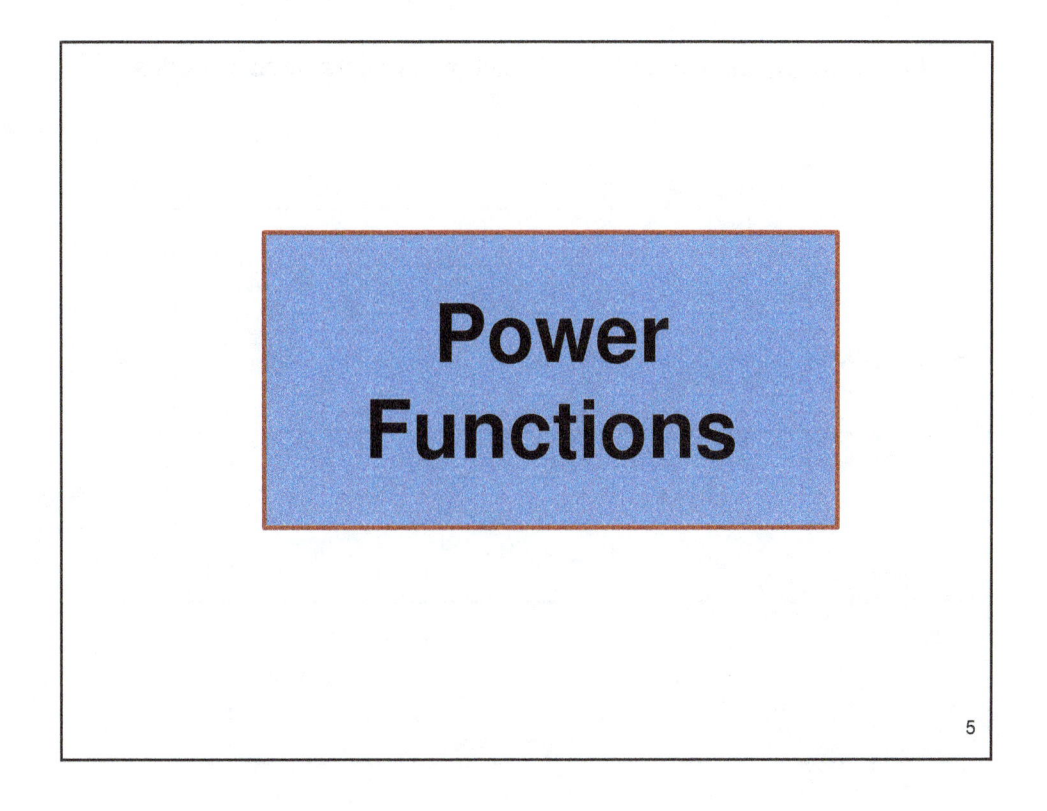

Workbook: Electro-Hydraulic Components and Systems
Chapter 9: Control Electronics for Electro-Hydraulic Valves

Fig. 9.2 - Control Electronics Basic Functions Distribution

Power Functions

9.2 - DC Power Supply

- **Why DC power for Proportional solenoids and torque motors?**
  - DC power is easier to handle and control.
  - Avoid the undesirable phenomena associated with the AC power.
- **Optional built in AC to DC power supply.**

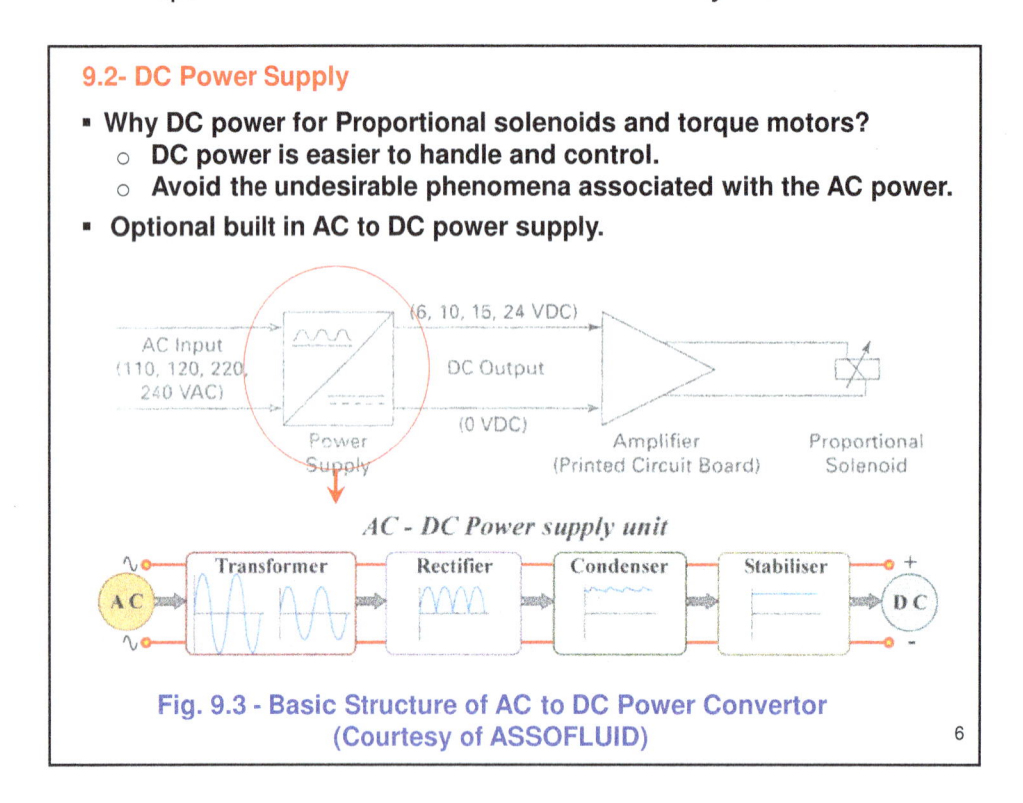

Fig. 9.3 - Basic Structure of AC to DC Power Convertor (Courtesy of ASSOFLUID)

- **Power supply pins must be assigned by numbers or color codes.**

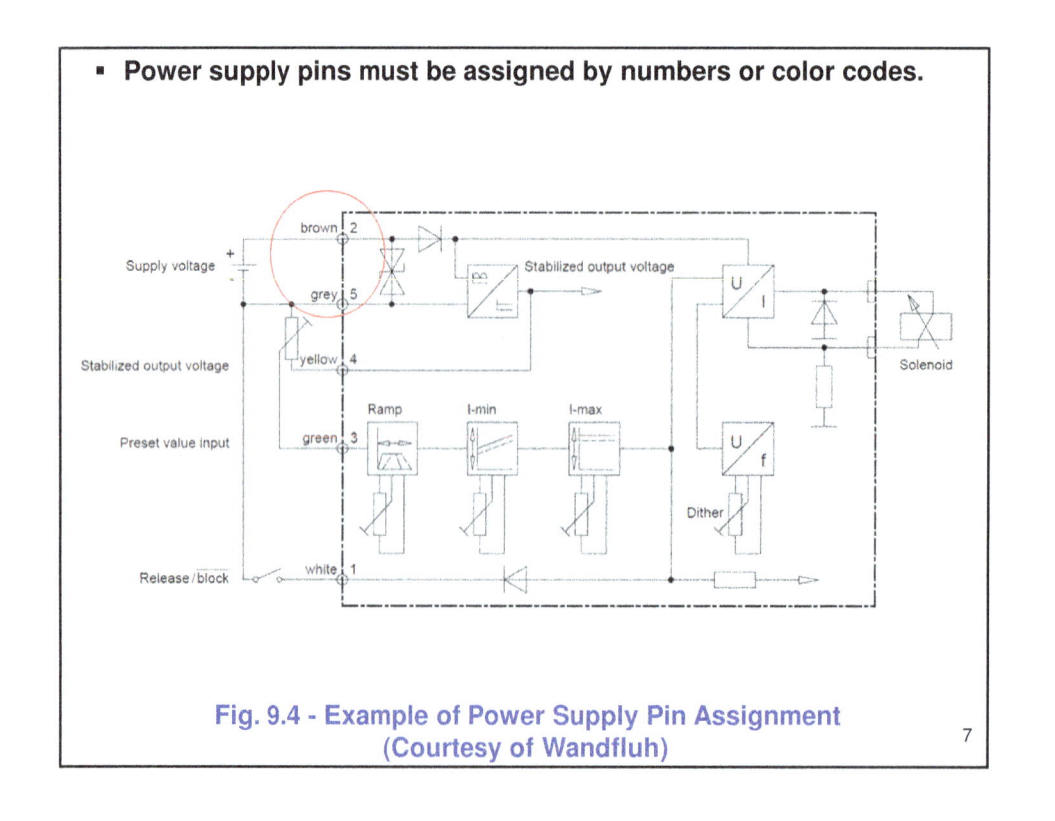

Fig. 9.4 - Example of Power Supply Pin Assignment (Courtesy of Wandfluh)

9.3- Signal Amplifier

- **The key element of ECUs is the amplifier (shortened to 'amp').**

- **Conventionally, the whole ECU is refereed as the "amp".**

- **Its job is to amplify an input voltage low power signal to an output current signal that is powerful enough to drive the solenoid.**

- **It consists of a series of transistor stages.**

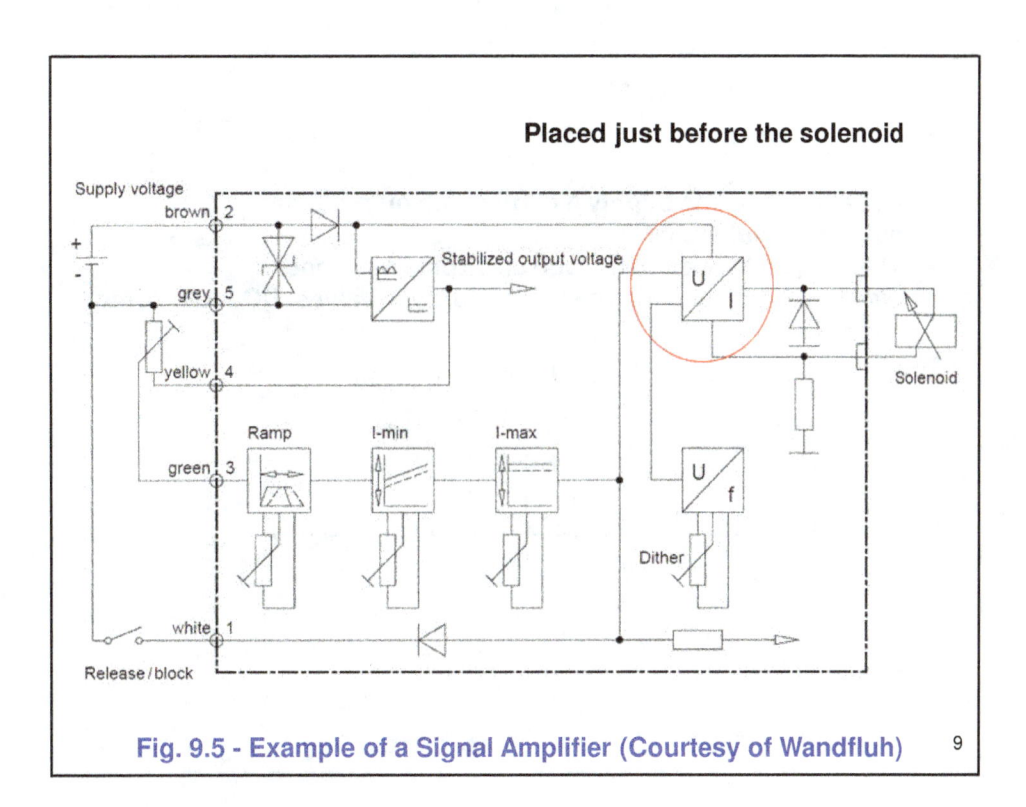

Placed just before the solenoid

Fig. 9.5 - Example of a Signal Amplifier (Courtesy of Wandfluh)

9.4- Dither Signal

- **What is Dither?**
  - A micro-vibration signal that vibrates the spool when it moves and at rest.
- **Why Dither?**
  - Valve spool frictional forces causes hysteresis and nonlinearity.
  - Applying dither signal considerably reduces spool adhesion and thereby improves linearity and hysteresis performance of the valve.
- **How it Works:**
  - Dither signal is superimposed on the valve solenoid drive signal.
- **Dither Amplitude:** micro amplitude and is adjustable on the amplifier card.
- **Dither Frequency:** should be within the valve bandwidth otherwise the valve won't respond.

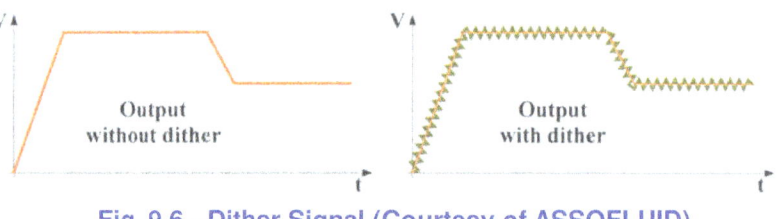

Fig. 9.6 - Dither Signal (Courtesy of ASSOFLUID)

9.5- Pulse Width Modulation

- **What is PWM?**

- It is a technique to supply a continuous output signal via a pulsated input signal.
- Was originally used to encode electronic messages.
- Was found useful to drive electrohydraulic valves, DC motors, etc.

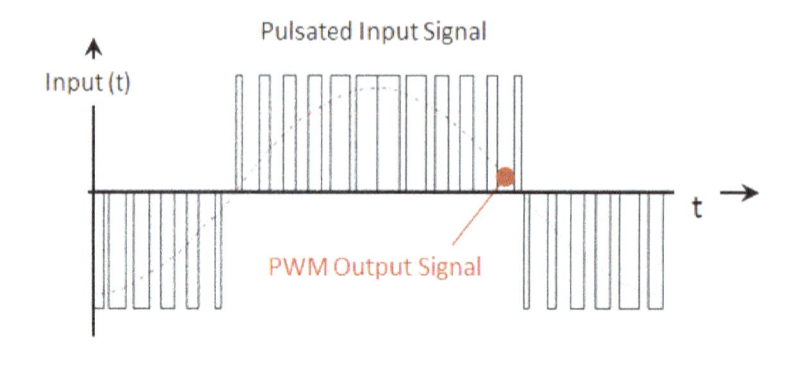

Fig. 9.7 - Pulse Width Modulated Signal

Workbook: Electro-Hydraulic Components and Systems
Chapter 9: Control Electronics for Electro-Hydraulic Valves

- **Why PWM?**

- Applying DC power to a proportional solenoids for long time (100% duty cycle) causes heat generation. A large heat sink is therefore required to dissipate the heat.

- PWM technique maintains reduced power losses at the output stage.

- PWM became a standard for all valve amplifiers in order to reduce amplifier size.

- PWM helps reducing hysteresis effects.

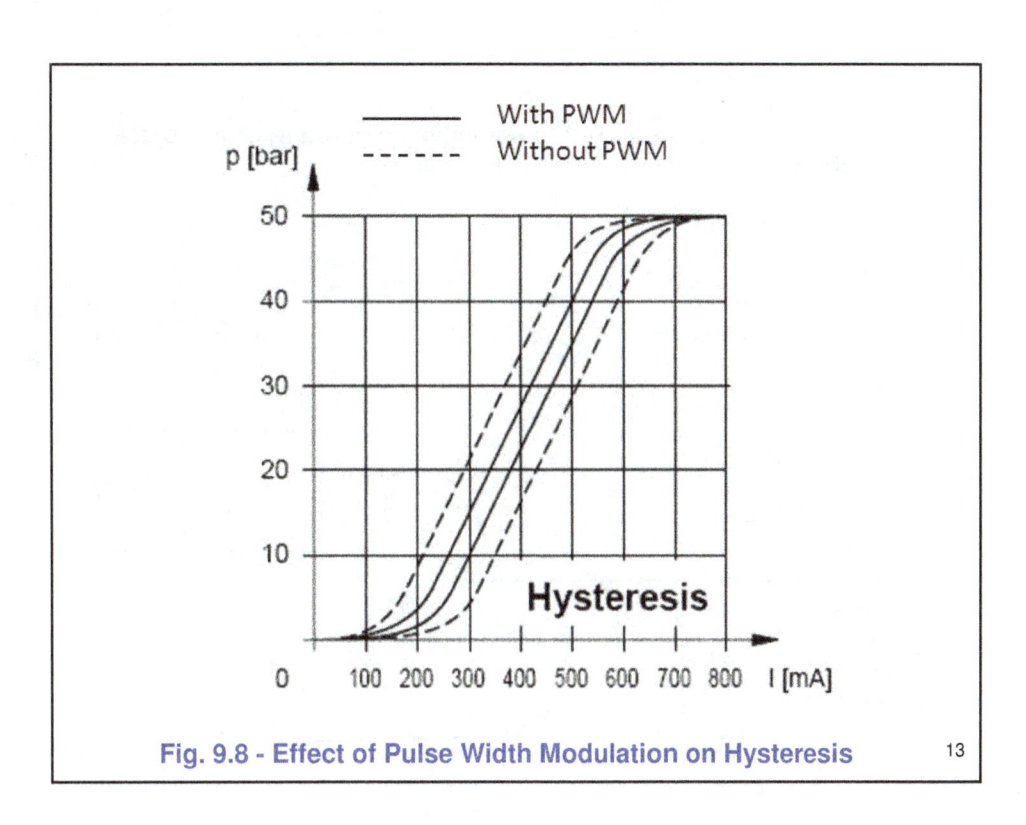

Fig. 9.8 - Effect of Pulse Width Modulation on Hysteresis

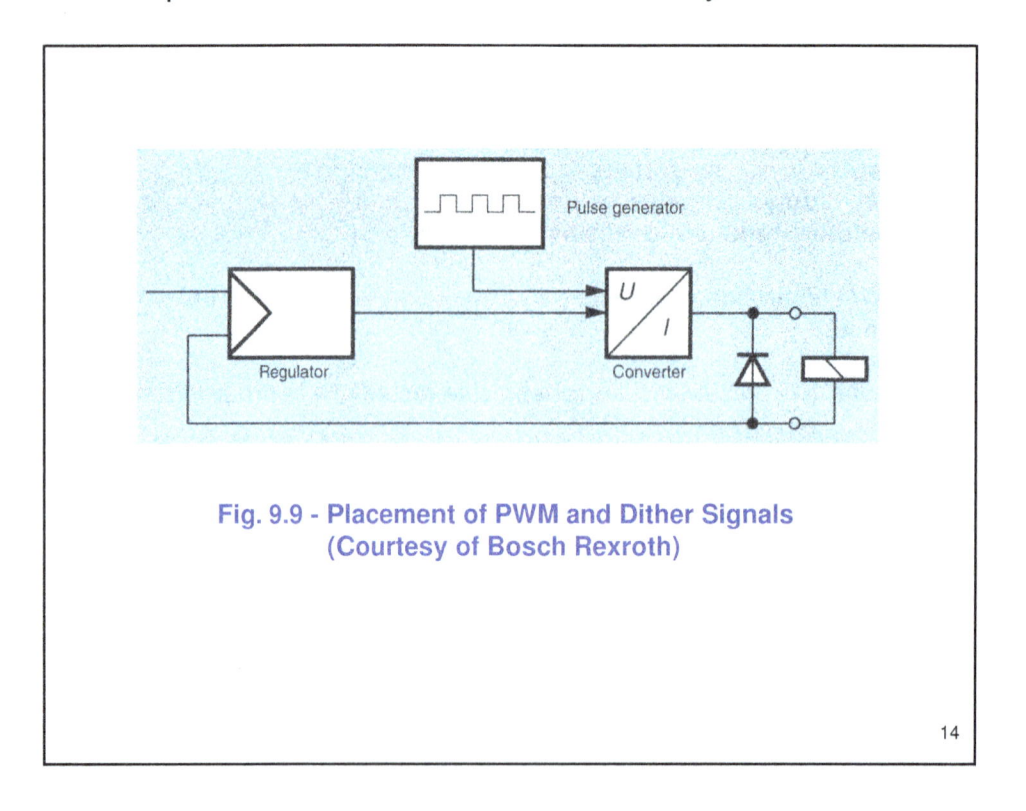

Fig. 9.9 - Placement of PWM and Dither Signals
(Courtesy of Bosch Rexroth)

- **How PWM Works?**
- **No modifications** are required to the valve solenoid in order to use this technique.

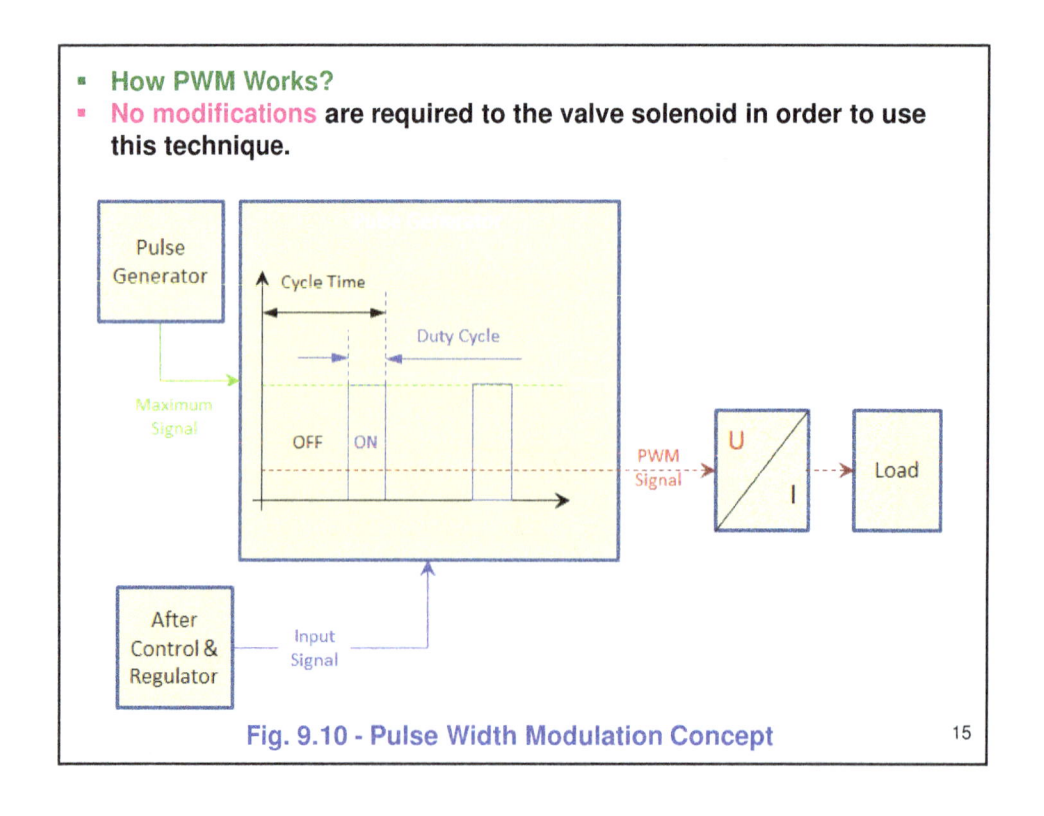

Fig. 9.10 - Pulse Width Modulation Concept

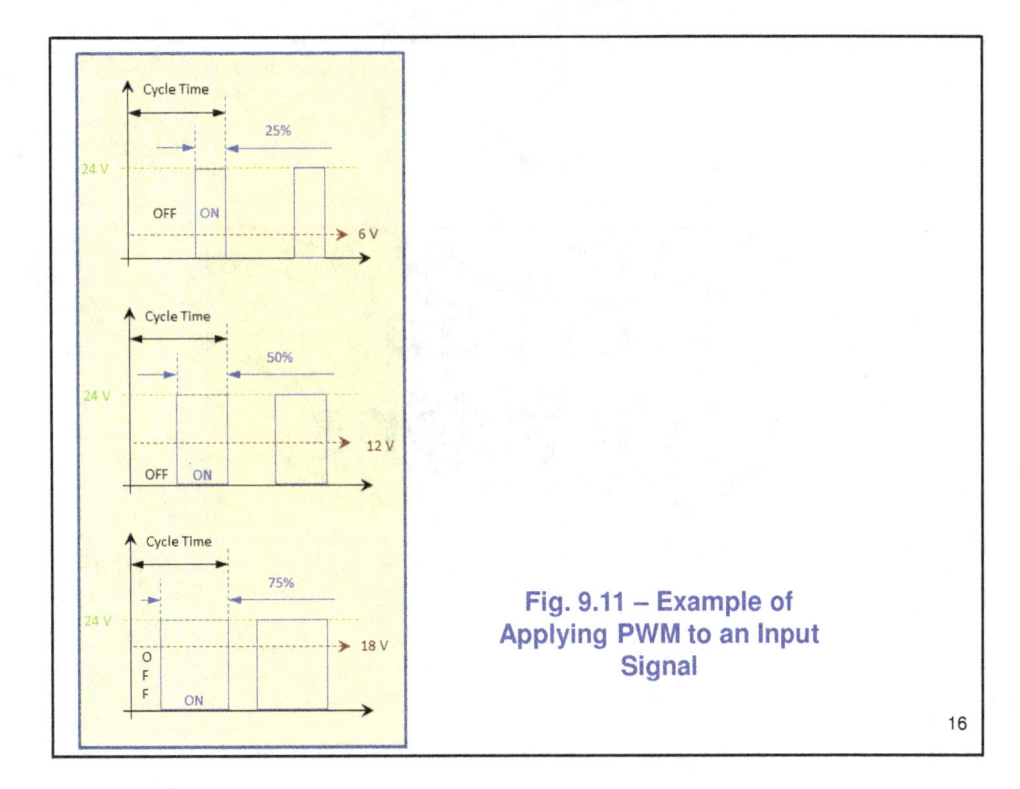

Fig. 9.11 – Example of Applying PWM to an Input Signal

- **PWM Frequency?**

  o It should be much higher than what would affect the load, which is the valve solenoid and the spool.

  o Much higher than the valve bandwidth. Typically 1kHz.

  o Higher frequency improves the continuity of the average output signal.

Control Functions

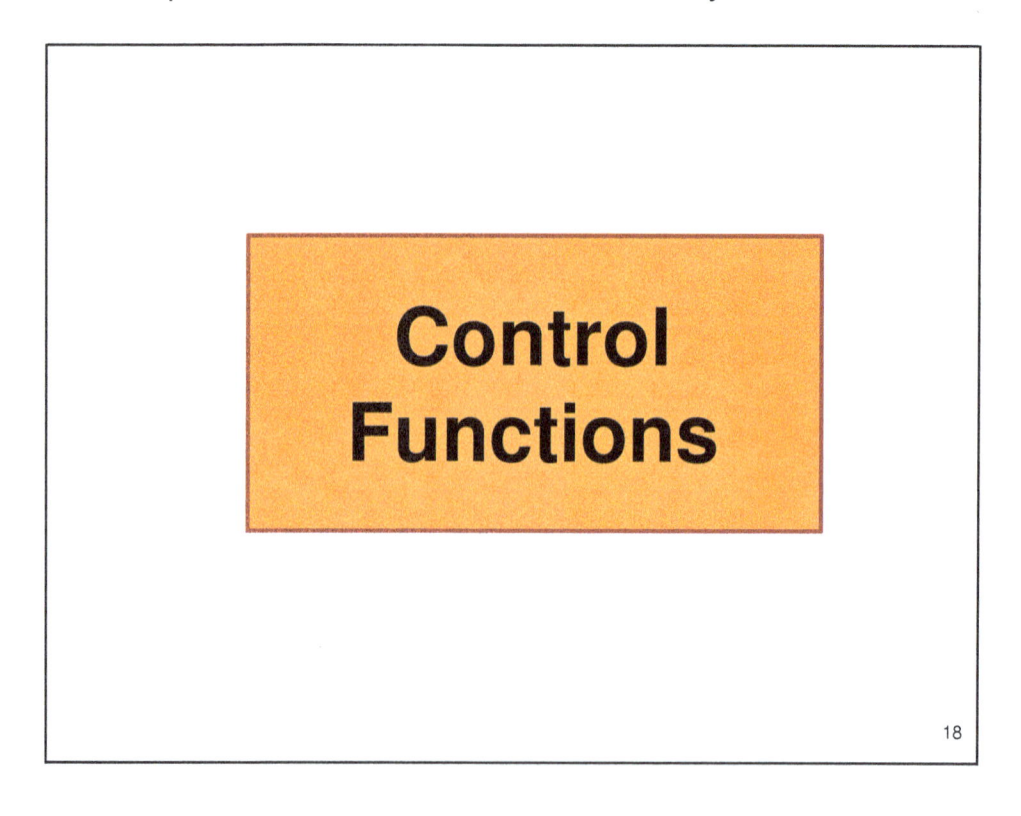

9.6 - Input Signal Generator

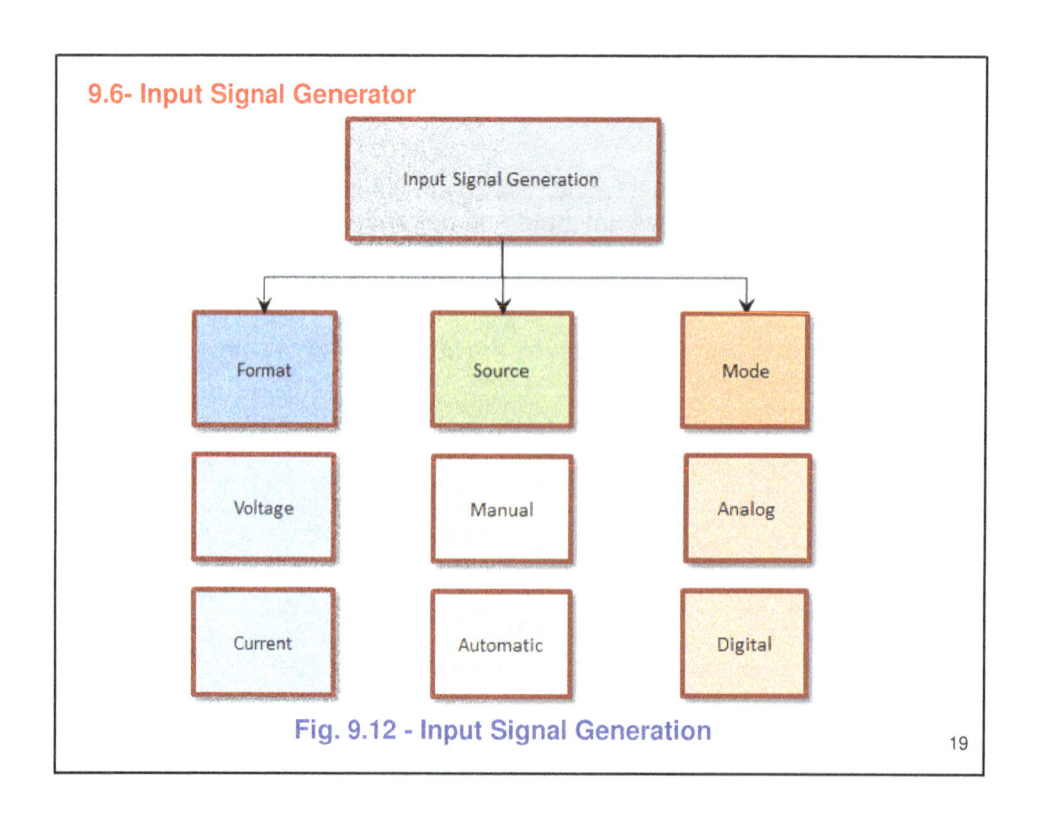

```
                    Input Signal Generation
                   /         |          \
              Format       Source        Mode

              Voltage      Manual        Analog

              Current      Automatic     Digital
```

Fig. 9.12 - Input Signal Generation

9.6.1- Format of Input Signal

- **Voltage Input Signal:**

  - Common for all industrial controllers.
  - Easy to produce and to troubleshoot.
  - Subject to voltage drop in long transmission cables.
  - Affected by noise.
  - (0-10) V for unidirectional parameter control, e.g. cylinder position.
  - ±10 V for bidirectional parameter control, e.g. bidirectional hydraulic motor speed.

- **Current Input Signal:**

  - Resists noise and losses.
  - Preferred for transmission through long cables
  - (4-20) mA or (0-20) mA.

9.6.2- Source and Mode of Input Signal

❑ **Manual:**

- **Manual Analog:** e.g. potentiometers and joysticks.

Fig. 9.13 - Examples of Manual Input Signal Devices

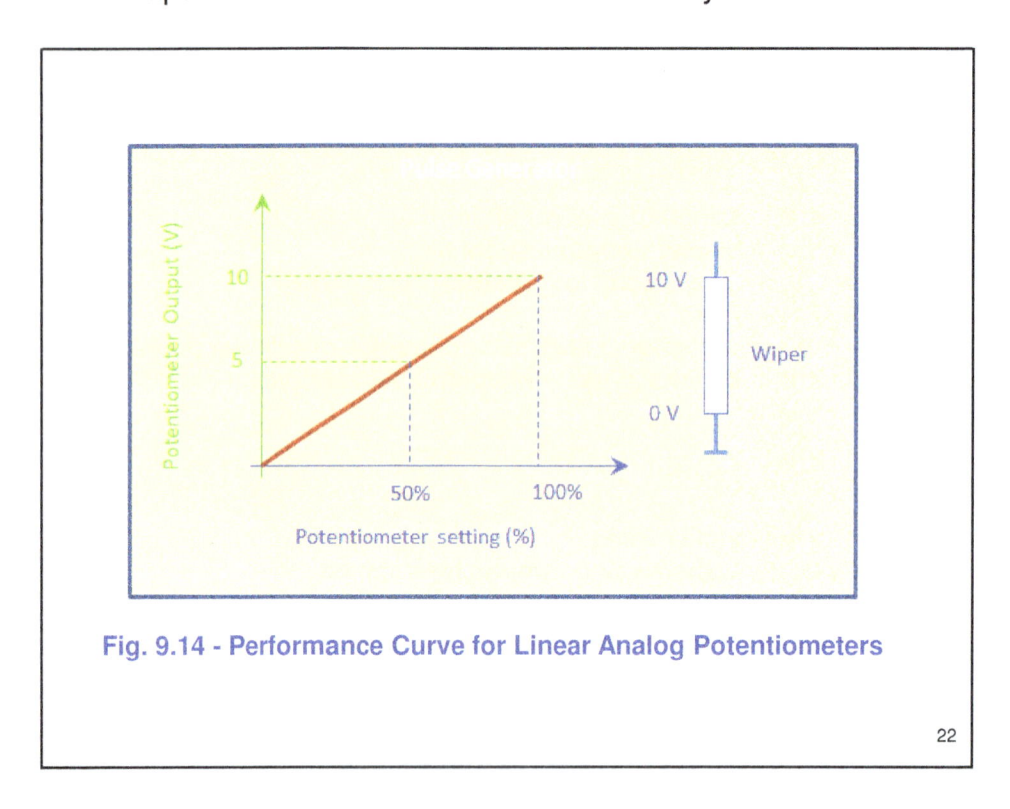

Workbook: Electro-Hydraulic Components and Systems
Chapter 9: Control Electronics for Electro-Hydraulic Valves

Fig. 9.14 - Performance Curve for Linear Analog Potentiometers

- **Manual Digital:** e.g. Human Machine Interface (HMI)

Fig. 9.15 - Touch Screen Operator Interface

❏ **Automatic:**

- **Automatic Analog:** e.g. potentiometers and joysticks.

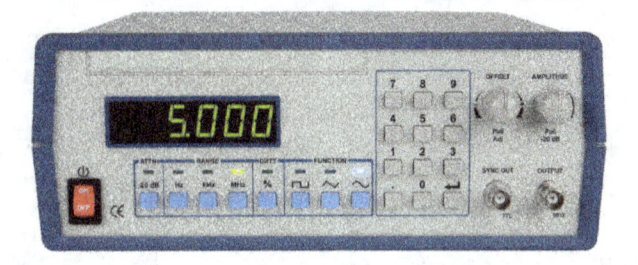

Fig. 9.16 - Example of an Analog Function Generator

❏ **Automatic:**

- **Automatic Digital:** e.g. PLCs and Software-based.

PLC-Based

PC-Based

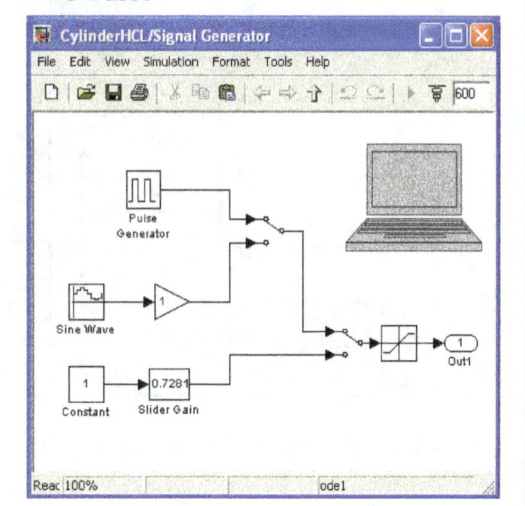

Fig. 9.17 - Automatic and Digital Input Signal Generation by PLC or PC

9.7- Feedback Sensors

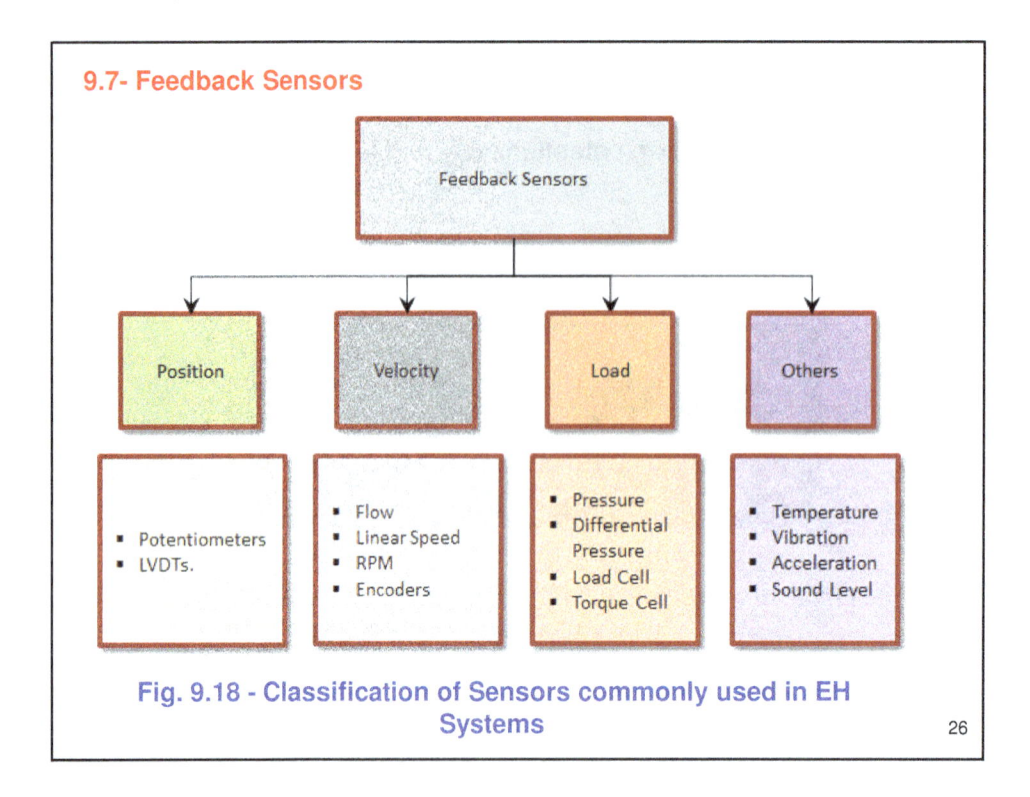

```
                    Feedback Sensors
         ┌──────────────┬──────────────┬──────────────┐
      Position       Velocity         Load          Others
```

- Potentiometers
- LVDTs.

- Flow
- Linear Speed
- RPM
- Encoders

- Pressure
- Differential Pressure
- Load Cell
- Torque Cell

- Temperature
- Vibration
- Acceleration
- Sound Level

Fig. 9.18 - Classification of Sensors commonly used in EH Systems

9.8- PID Controller

Sign Invertor

U_{in} → [▷] → $-U_{in}$

Example:
+ 5 V results in − 5 V

Summing Junction and Sign Invertor

U_R, U_F → [Σ] → U_E

Example:
$U_E = U_R - U_F$

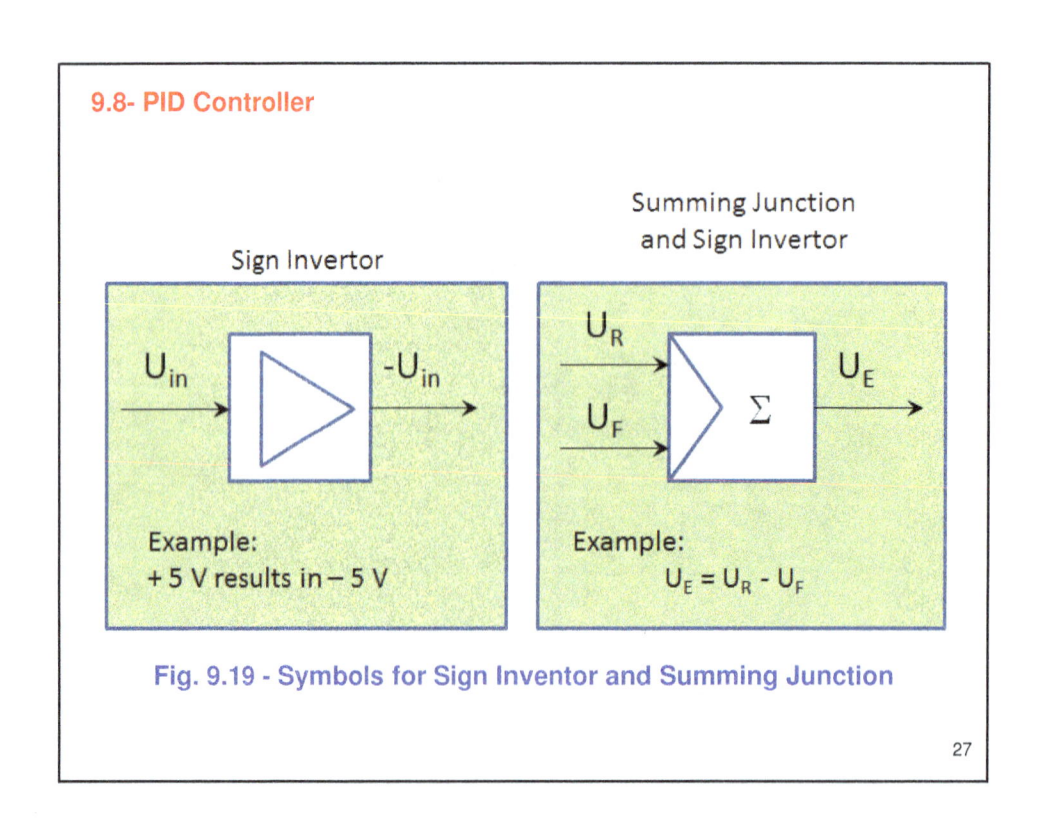

Fig. 9.19 - Symbols for Sign Inventor and Summing Junction

- U_R: Reference "Input" Signal.
- U_F: Feedback Signal.
- U_E: Error Signal.
- U_C: Control Signal.

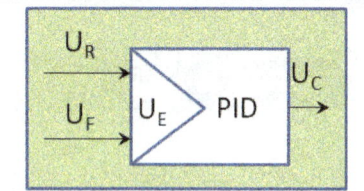

Fig. 9.20 - Symbol for PID Controller

$$U_c = k_P U_E + k_I \int U_E + k_D \frac{dU_E}{dt} \qquad 9.1$$

- **PID Controller has 3 different gains: P, I and D.**

- k_P = Proportional Gain:
- Responsible for the <u>valve shifting displacement</u> and consequently the <u>speed of the actuator</u>.

- k_I = Integral Gain:
- Responsible for eliminating the <u>steady state error</u>.

- k_D = Derivative Gain:
- Responsible for <u>removing oscillation and instability</u> of the controlled parameters.

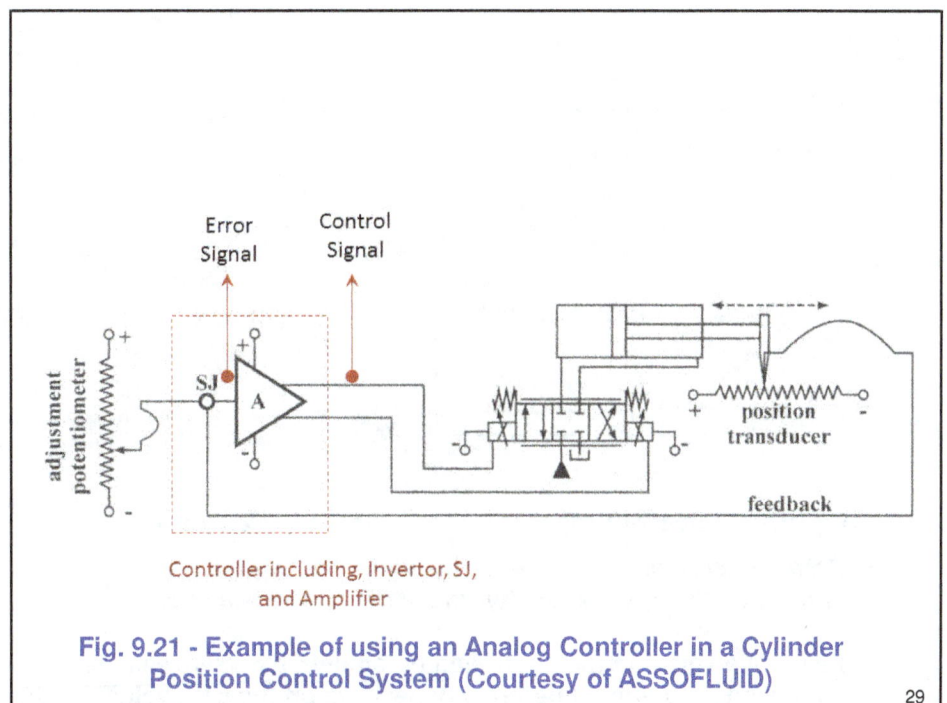

Fig. 9.21 - Example of using an Analog Controller in a Cylinder Position Control System (Courtesy of ASSOFLUID)

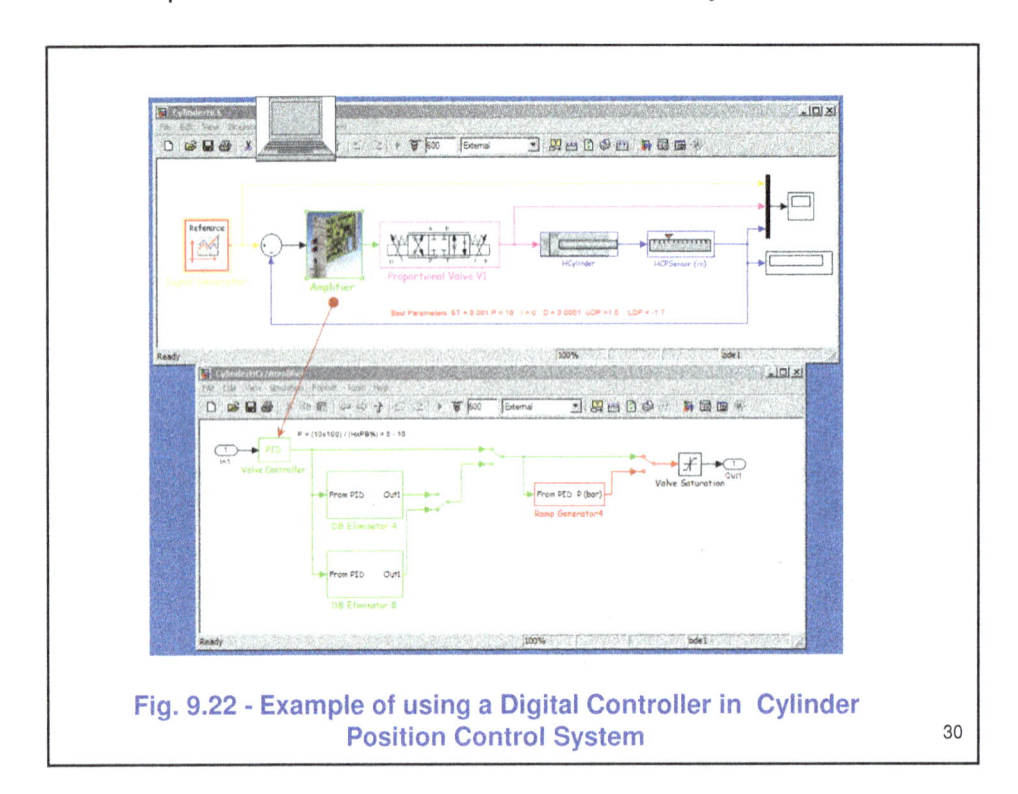

Fig. 9.22 - Example of using a Digital Controller in Cylinder Position Control System

9.9- Gain Adjustor

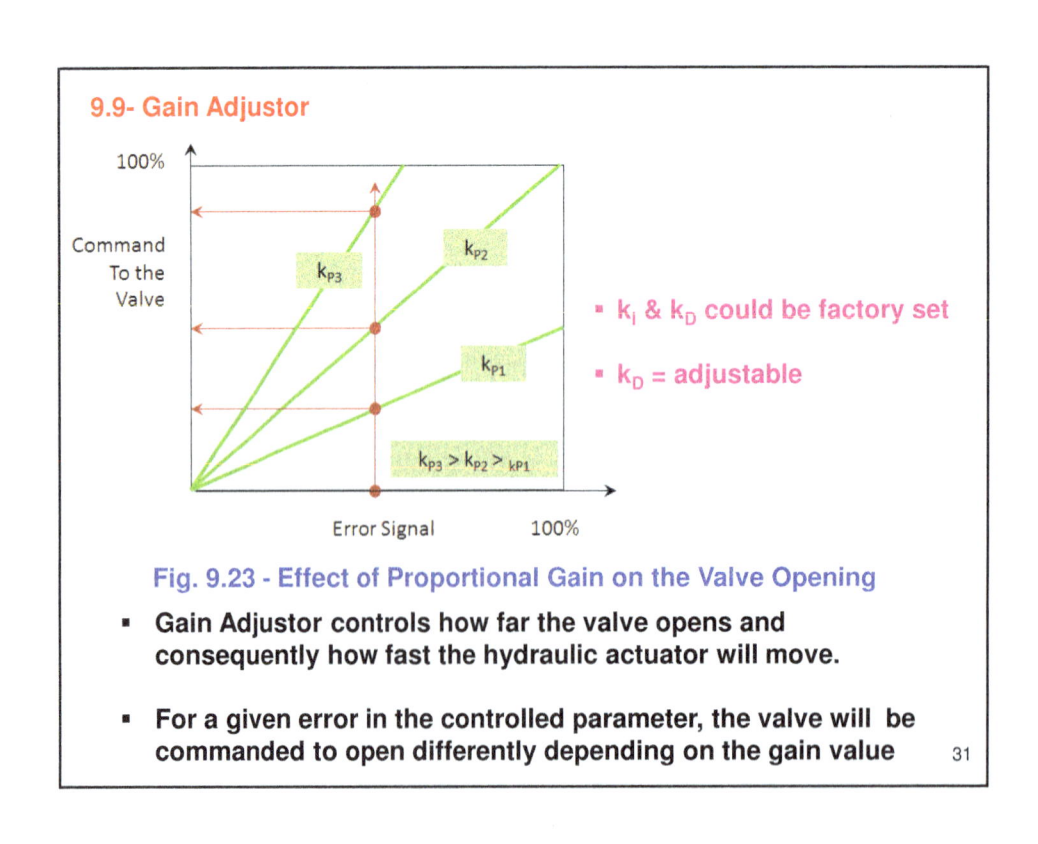

Fig. 9.23 - Effect of Proportional Gain on the Valve Opening

- k_I & k_D could be factory set
- k_D = adjustable

$k_{P3} > k_{P2} > k_{P1}$

- Gain Adjustor controls how far the valve opens and consequently how fast the hydraulic actuator will move.

- For a given error in the controlled parameter, the valve will be commanded to open differently depending on the gain value

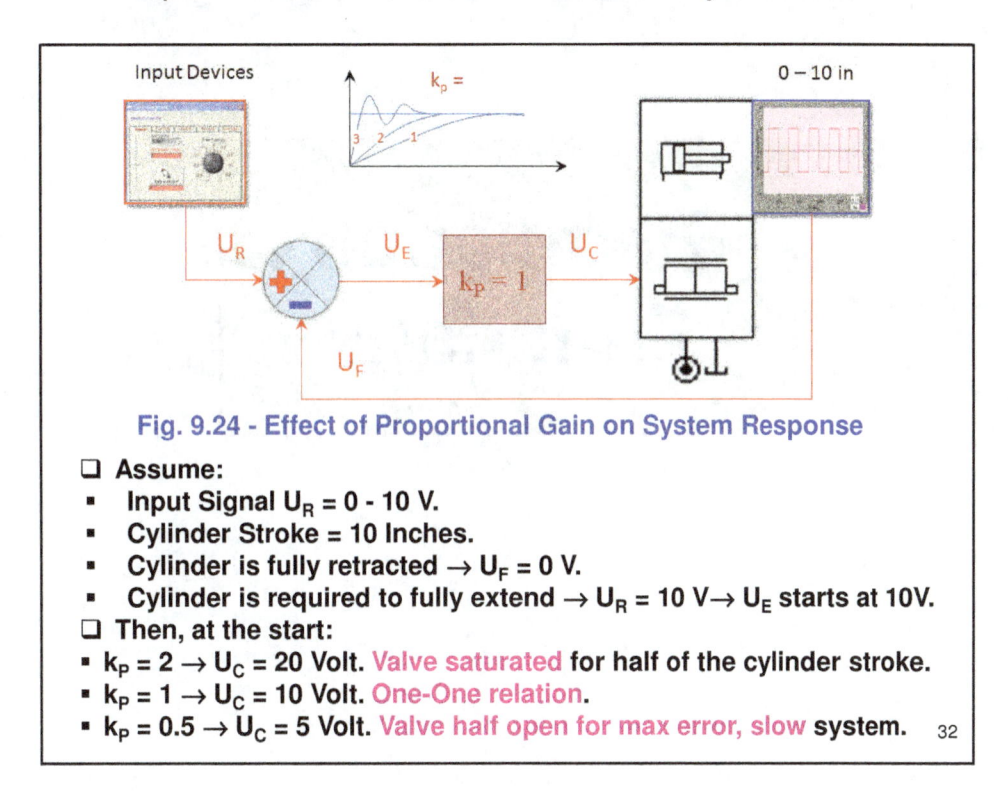

Workbook: Electro-Hydraulic Components and Systems
Chapter 9: Control Electronics for Electro-Hydraulic Valves

Fig. 9.24 - Effect of Proportional Gain on System Response

❏ Assume:
- Input Signal U_R = 0 - 10 V.
- Cylinder Stroke = 10 Inches.
- Cylinder is fully retracted → U_F = 0 V.
- Cylinder is required to fully extend → U_R = 10 V → U_E starts at 10V.

❏ Then, at the start:
- k_P = 2 → U_C = 20 Volt. **Valve saturated** for half of the cylinder stroke.
- k_P = 1 → U_C = 10 Volt. **One-One relation**.
- k_P = 0.5 → U_C = 5 Volt. **Valve half open for max error, slow** system.

Device view

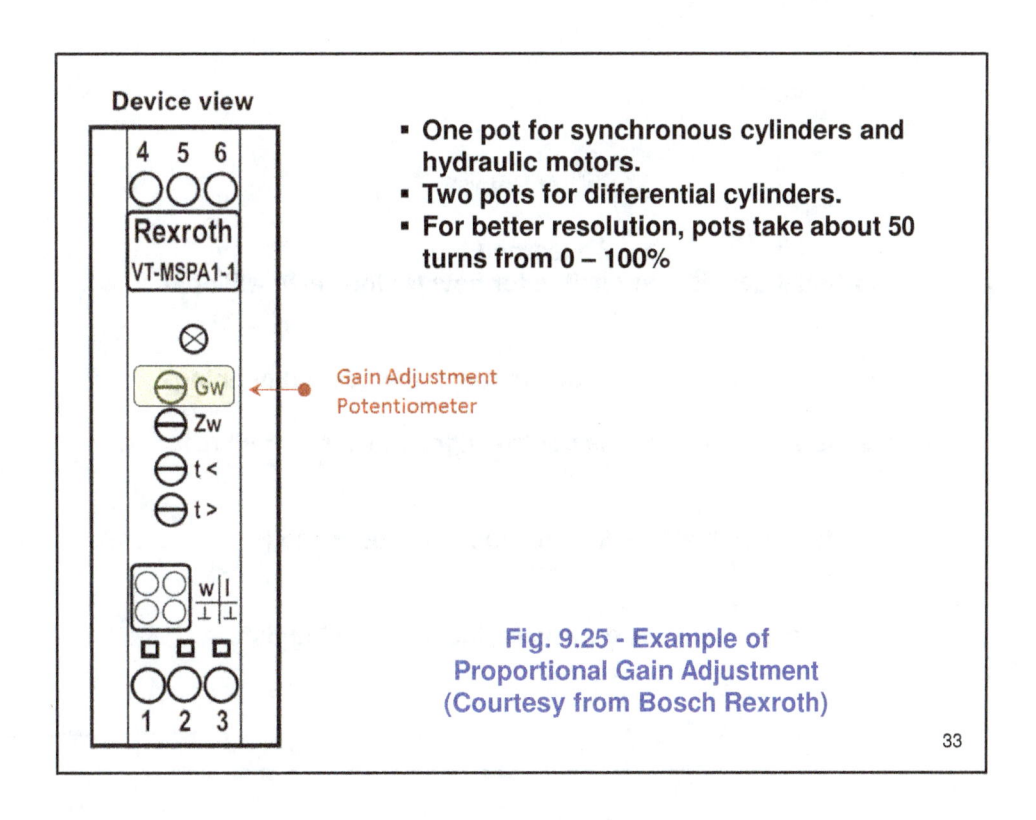

- One pot for synchronous cylinders and hydraulic motors.
- Two pots for differential cylinders.
- For better resolution, pots take about 50 turns from 0 – 100%

Gain Adjustment Potentiometer

Fig. 9.25 - Example of Proportional Gain Adjustment (Courtesy from Bosch Rexroth)

Regulation Functions

9.10 - Ramp Adjustor

Why Ramp?

We learned that
- Proportional gain is responsible for how far the valve will open.

BUT
how long the valve will take to reach the desired opening?

- Once the valve receives the control signal, the spool will respond very fast acting like an On/Off valve.

- The hydraulic actuator will consequently experience harsh start and jerky start.

- Reducing the P gain will only make the system sluggish.

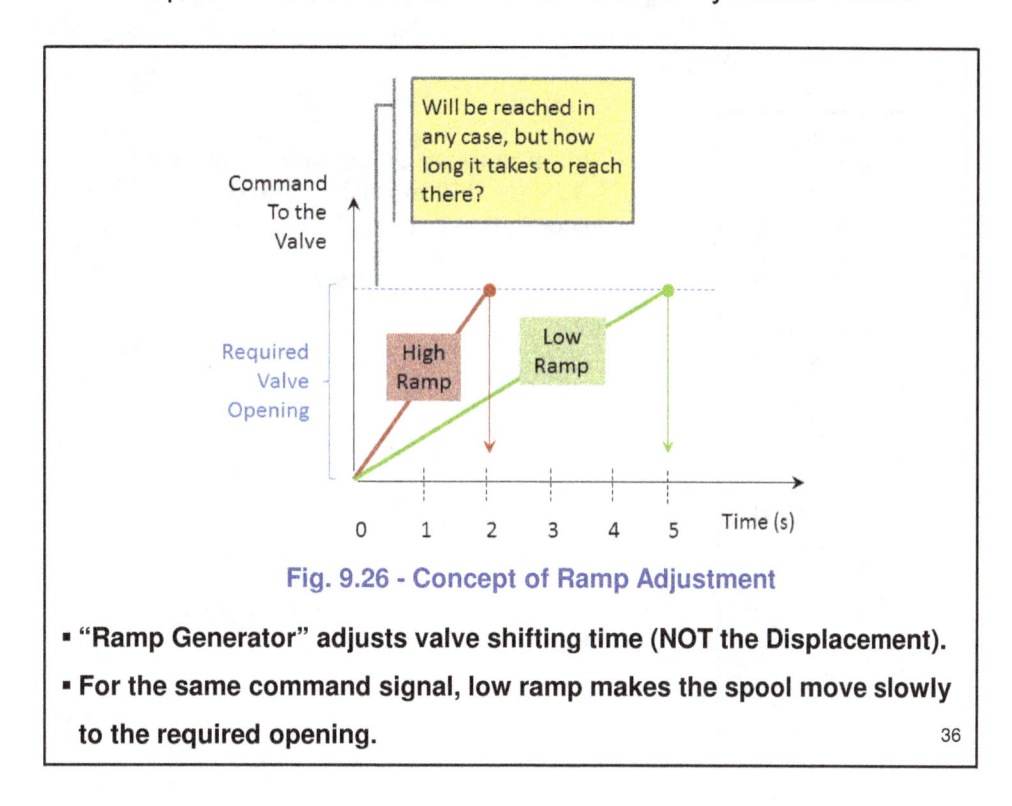

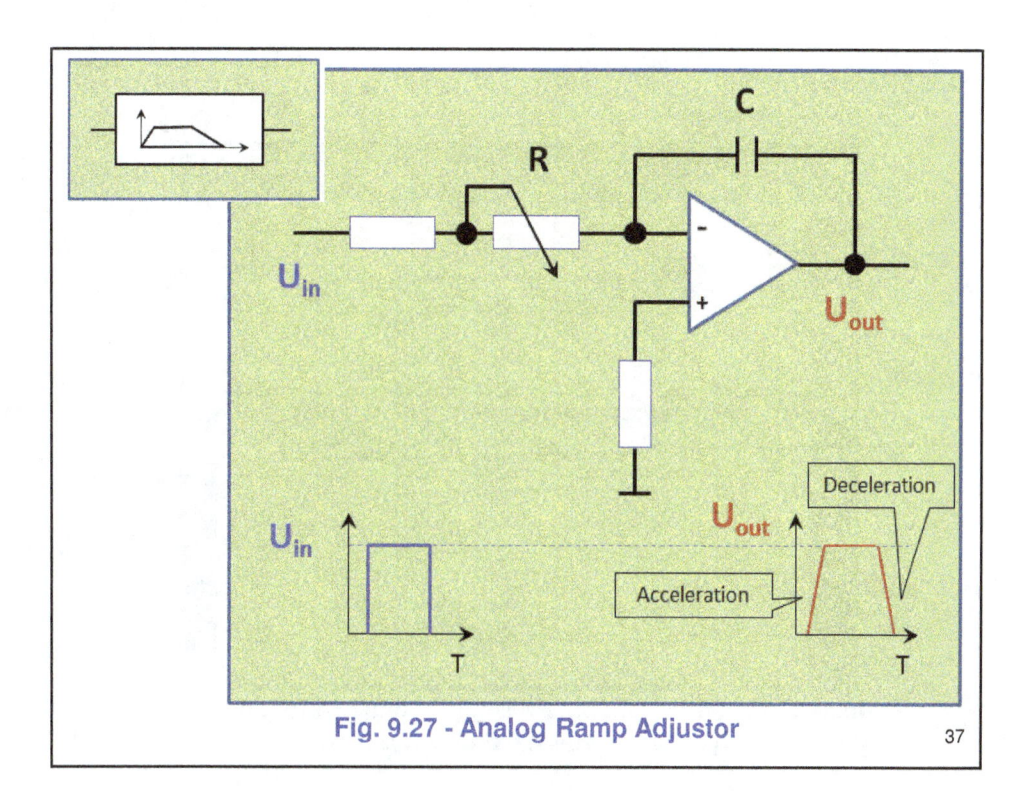

Workbook: Electro-Hydraulic Components and Systems
Chapter 9: Control Electronics for Electro-Hydraulic Valves

Fig. 9.26 - Concept of Ramp Adjustment

- "Ramp Generator" adjusts valve shifting time (NOT the Displacement).
- For the same command signal, low ramp makes the spool move slowly to the required opening.

Fig. 9.27 - Analog Ramp Adjustor

Workbook: Electro-Hydraulic Components and Systems
Chapter 9: Control Electronics for Electro-Hydraulic Valves

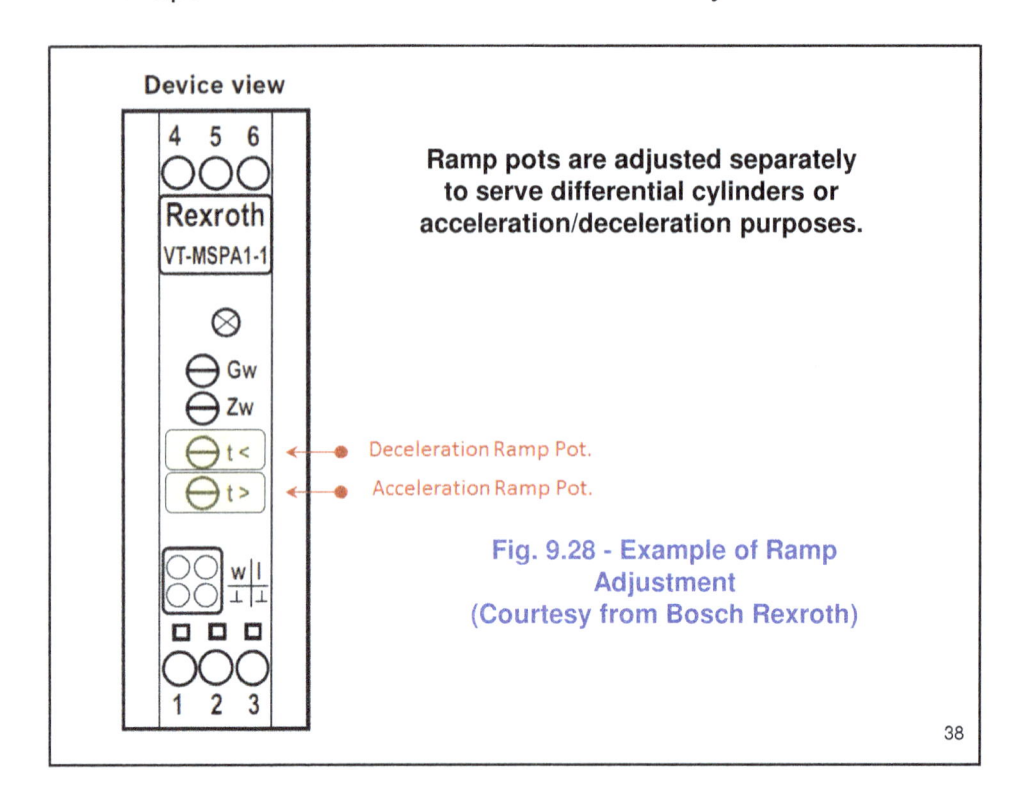

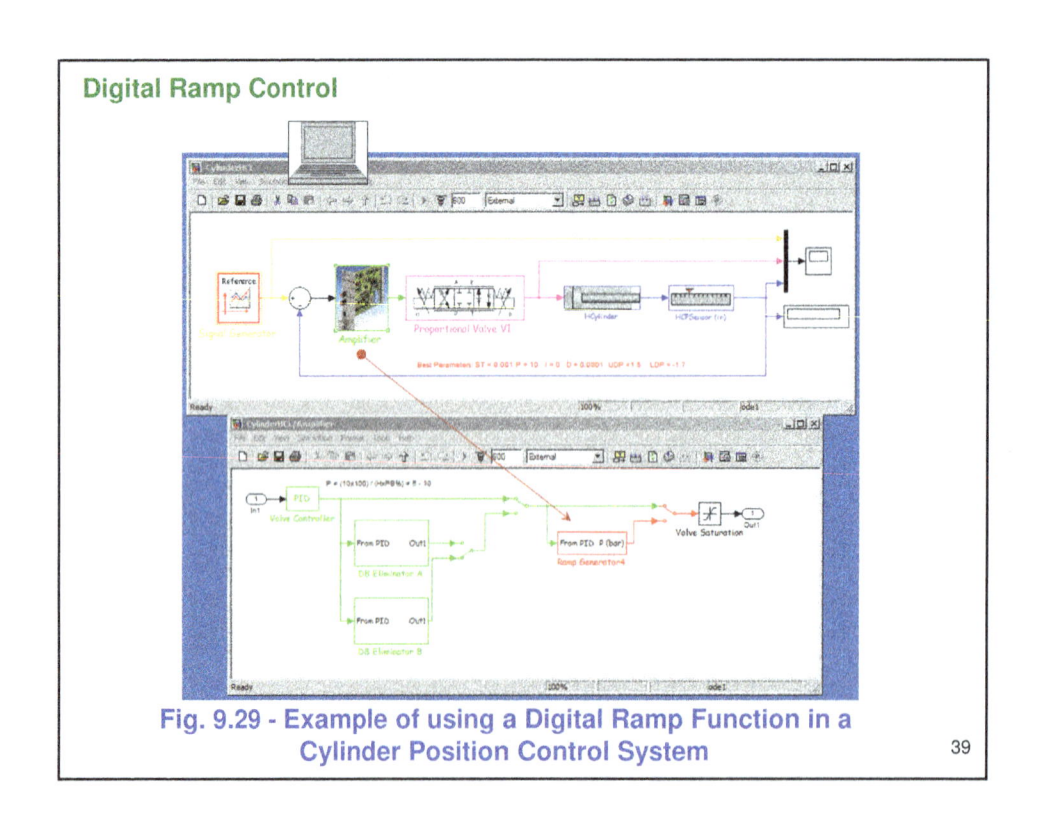

Device view

Ramp pots are adjusted separately to serve differential cylinders or acceleration/deceleration purposes.

Deceleration Ramp Pot.
Acceleration Ramp Pot.

Fig. 9.28 - Example of Ramp Adjustment
(Courtesy from Bosch Rexroth)

Digital Ramp Control

Fig. 9.29 - Example of using a Digital Ramp Function in a Cylinder Position Control System

9.11- Null Adjustor (I_{Bias})

- **Without Null Adjustment:** load creeps.
- Null Adjustment also referred to as " "Offset Setting".
- **Mechanical:**
- Flapper-Nozzle null adjustment (7) is a factory set process.
- Main spool null adjustment (8) is in field tuning process.

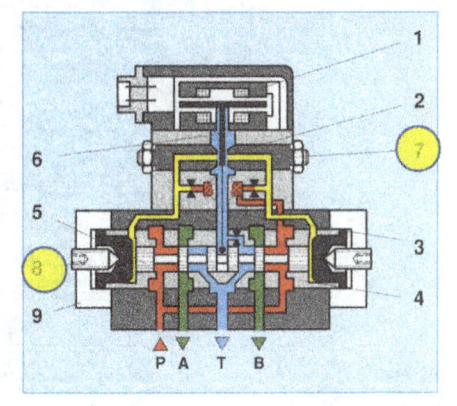

Fig. 9.30 - Mechanical Null Adjustment for a Servo Valve
(Courtesy of Bosch Rexroth)

- **Electrical:**
- By adding a "Bias" current. Also referred as "Basic Current".
- Typically ±5% input Signal.

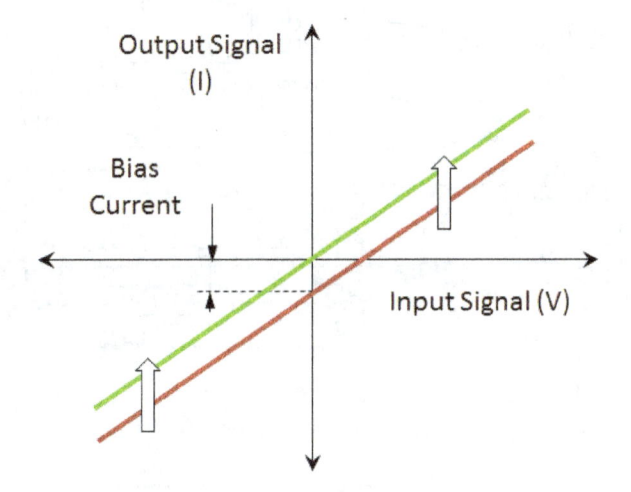

Fig. 9.31 - Electrical Null Adjustment by Bias Current

- **Example 1: Null Adjustment for a Two-Stage servo Valve with an OBE.**

Null Adjustment Screw

Servo Valves Series D765

Fig. 9.32 - Electrical Null Adjustment, Example 1 (Courtesy of Moog)

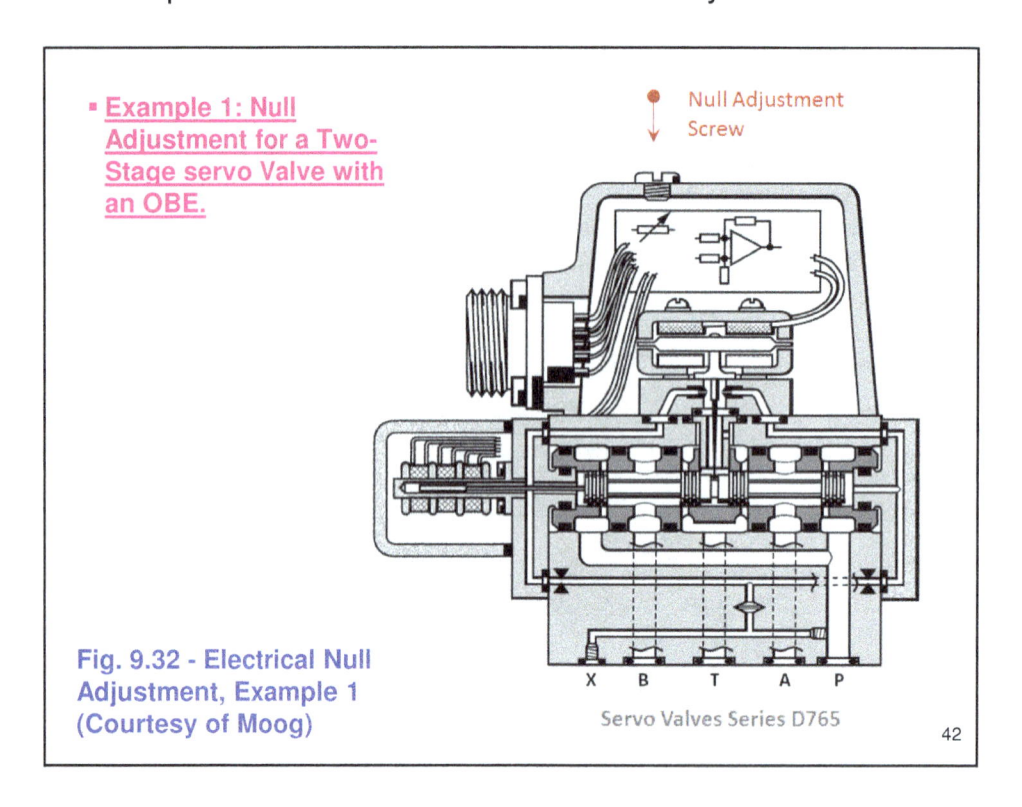

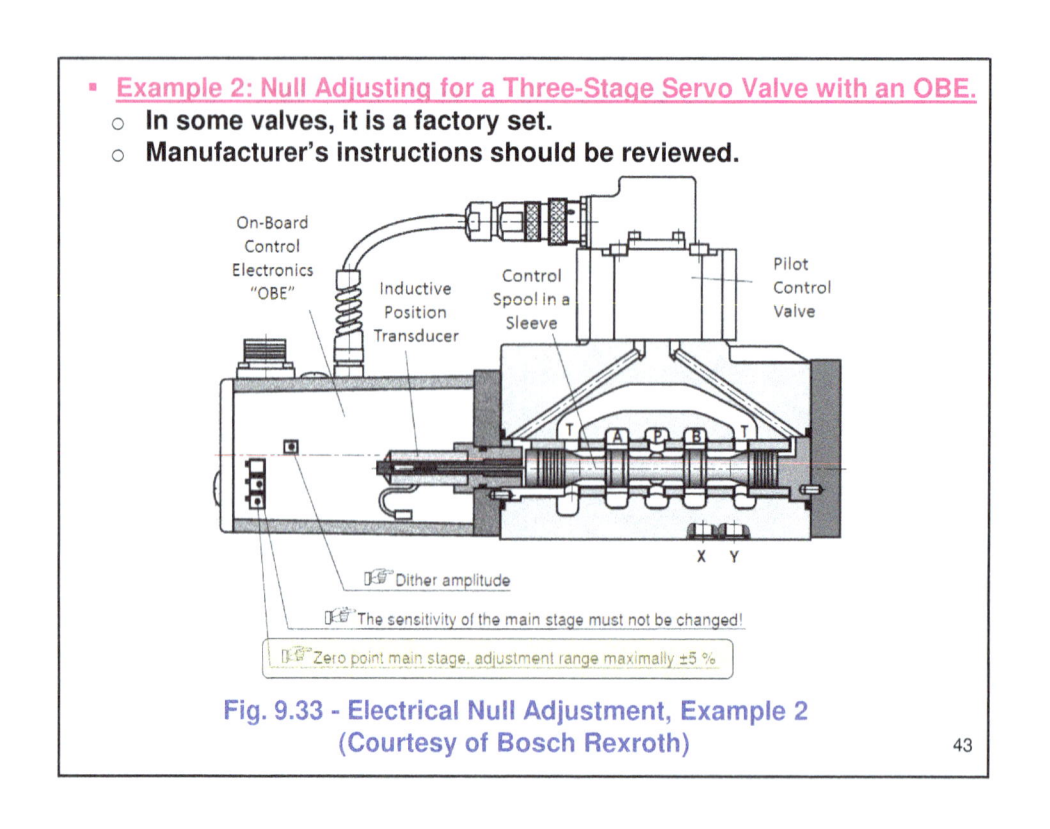

- **Example 2: Null Adjusting for a Three-Stage Servo Valve with an OBE.**
  - **In some valves, it is a factory set.**
  - **Manufacturer's instructions should be reviewed.**

On-Board Control Electronics "OBE"
Inductive Position Transducer
Control Spool in a Sleeve
Pilot Control Valve

- Dither amplitude
- The sensitivity of the main stage must not be changed!
- Zero point main stage, adjustment range maximally ±5 %

Fig. 9.33 - Electrical Null Adjustment, Example 2 (Courtesy of Bosch Rexroth)

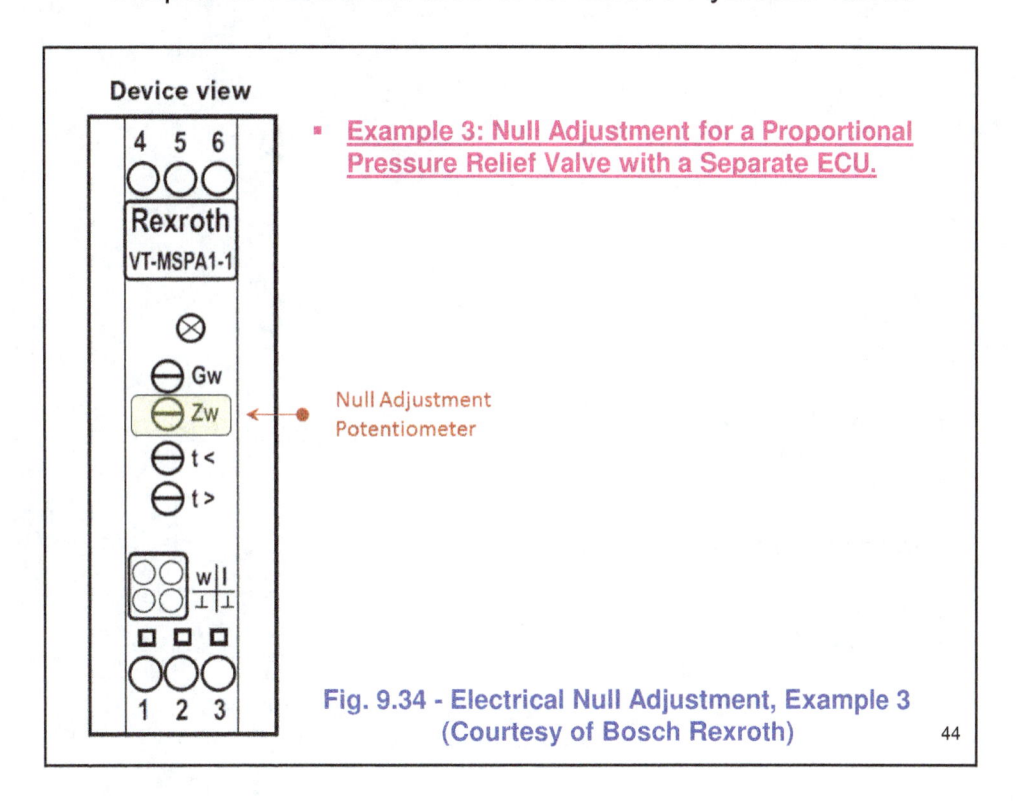

Example 3: Null Adjustment for a Proportional Pressure Relief Valve with a Separate ECU.

Device view

Null Adjustment Potentiometer

Fig. 9.34 - Electrical Null Adjustment, Example 3 (Courtesy of Bosch Rexroth)

9.12 - Dead Band Eliminator (I_{min})

Fig. 9.35 - Effect of Dead Band on Steady State Error

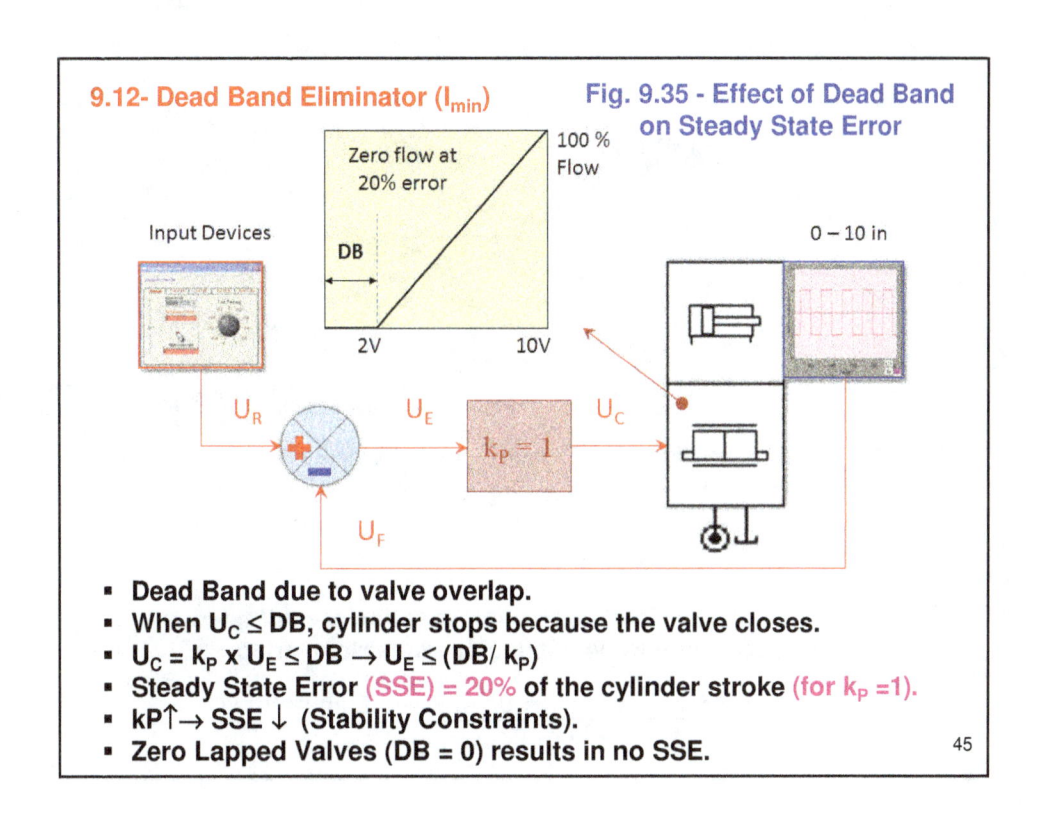

- Dead Band due to valve overlap.
- When $U_C \leq DB$, cylinder stops because the valve closes.
- $U_C = k_P \times U_E \leq DB \rightarrow U_E \leq (DB/k_P)$
- Steady State Error (SSE) = 20% of the cylinder stroke (for $k_P = 1$).
- $k_P \uparrow \rightarrow SSE \downarrow$ (Stability Constraints).
- Zero Lapped Valves (DB = 0) results in no SSE.

Apply Jump Current to Eliminate Dead Band Effect

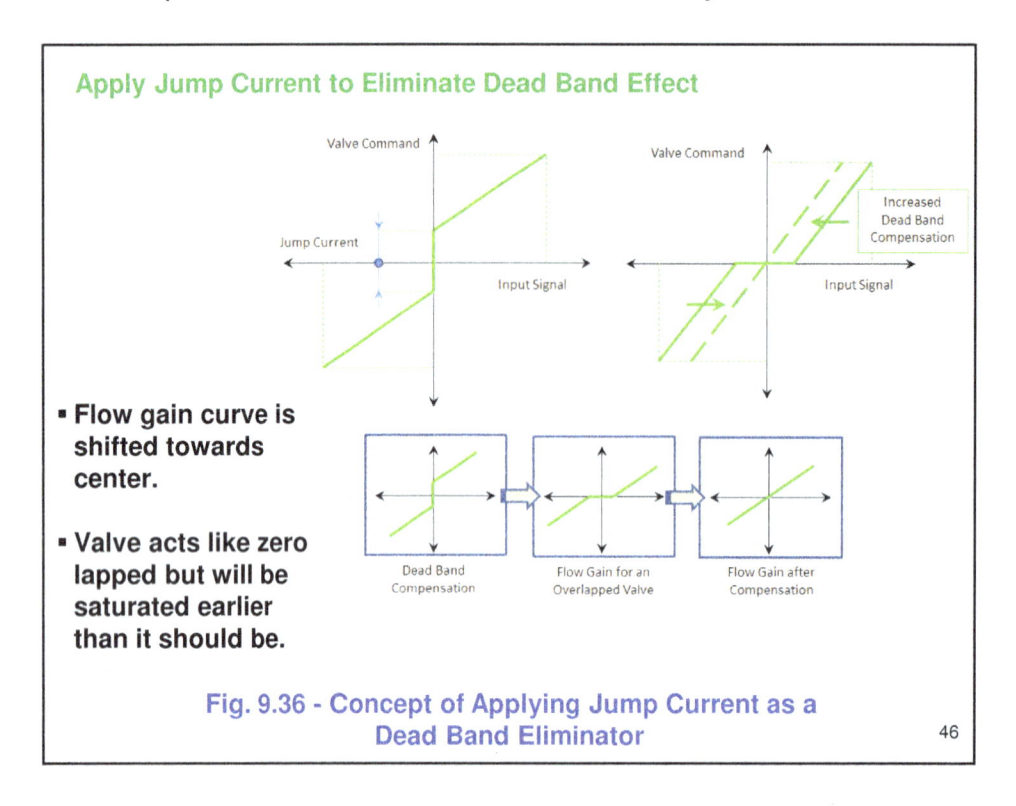

- Flow gain curve is shifted towards center.

- Valve acts like zero lapped but will be saturated earlier than it should be.

Fig. 9.36 - Concept of Applying Jump Current as a Dead Band Eliminator

Analog Jump Current

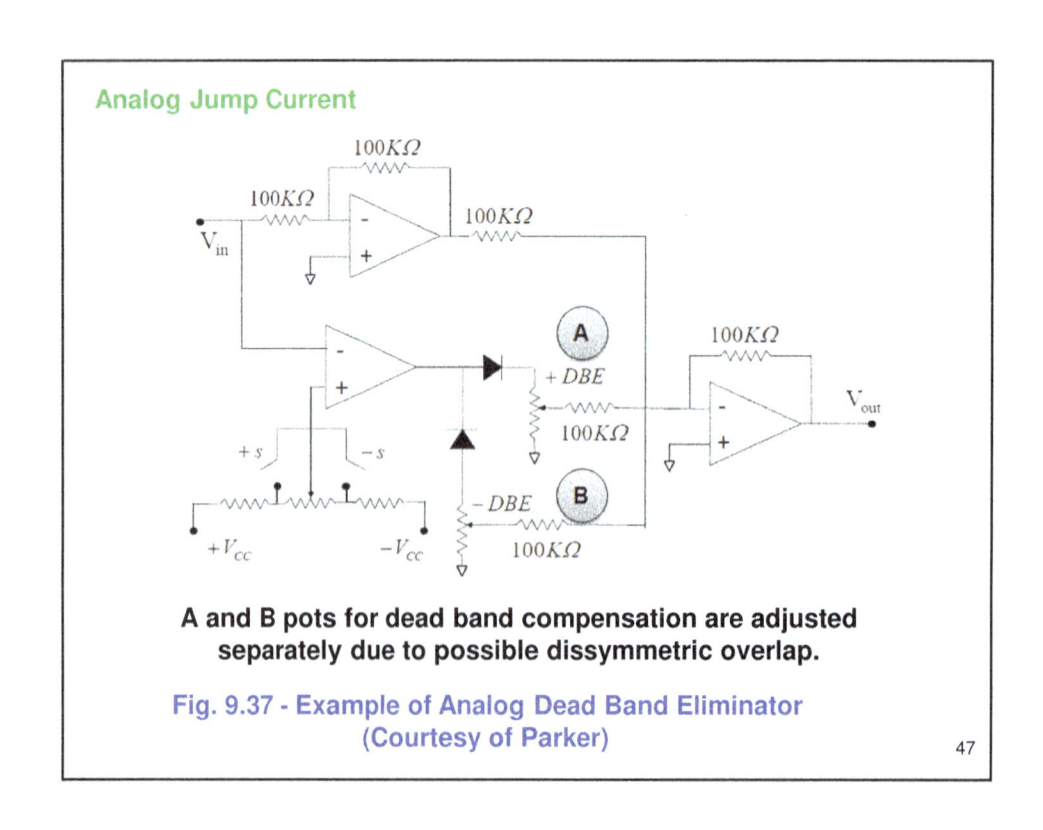

A and B pots for dead band compensation are adjusted separately due to possible dissymmetric overlap.

Fig. 9.37 - Example of Analog Dead Band Eliminator (Courtesy of Parker)

Digital: Method A (Jump Current)

If $U_E < DB$, then $U_C = U_E + I_{min}$

Digital: Method B (Magnifying the Error)

If $U_E < DB$, then $U_C = U_E \times K$ where $(K > 1)$

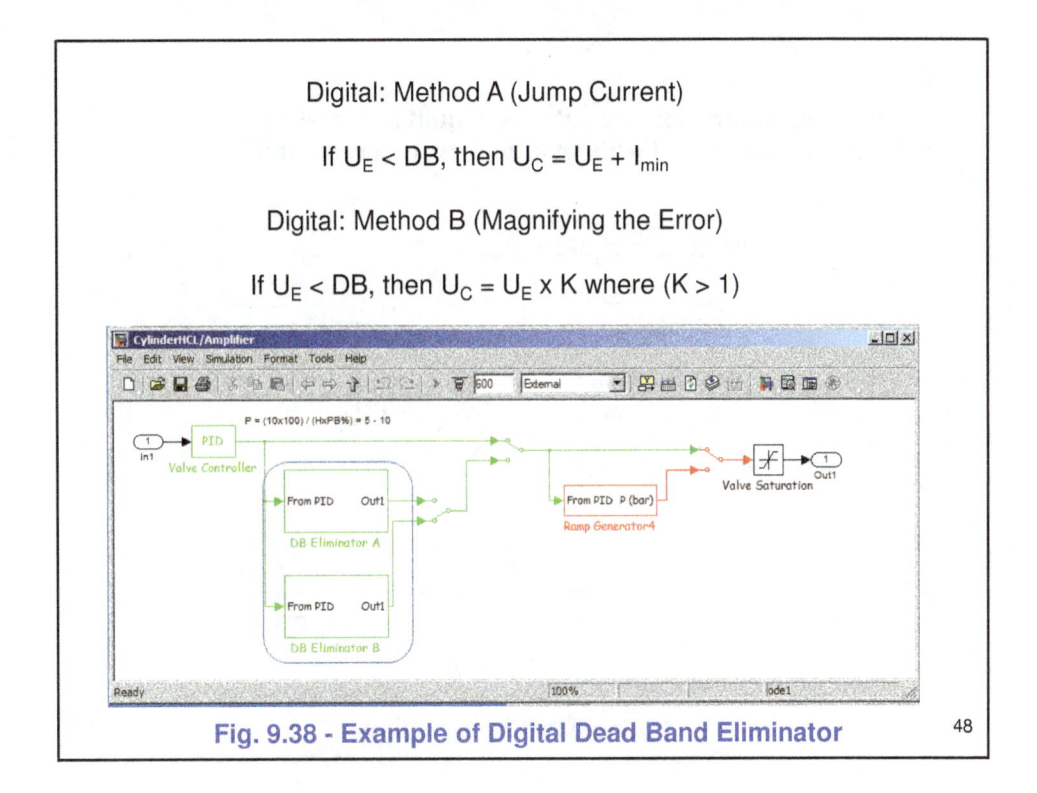

Fig. 9.38 - Example of Digital Dead Band Eliminator

9.13- Saturation Adjustor (I_{max})

Overload Protection – Saturation Adjustor – Maximum Current Limiter.

- Based on different regulations and control actions, an input signal may increase beyond the maximum allowable for a proportional solenoid or a torque motor.

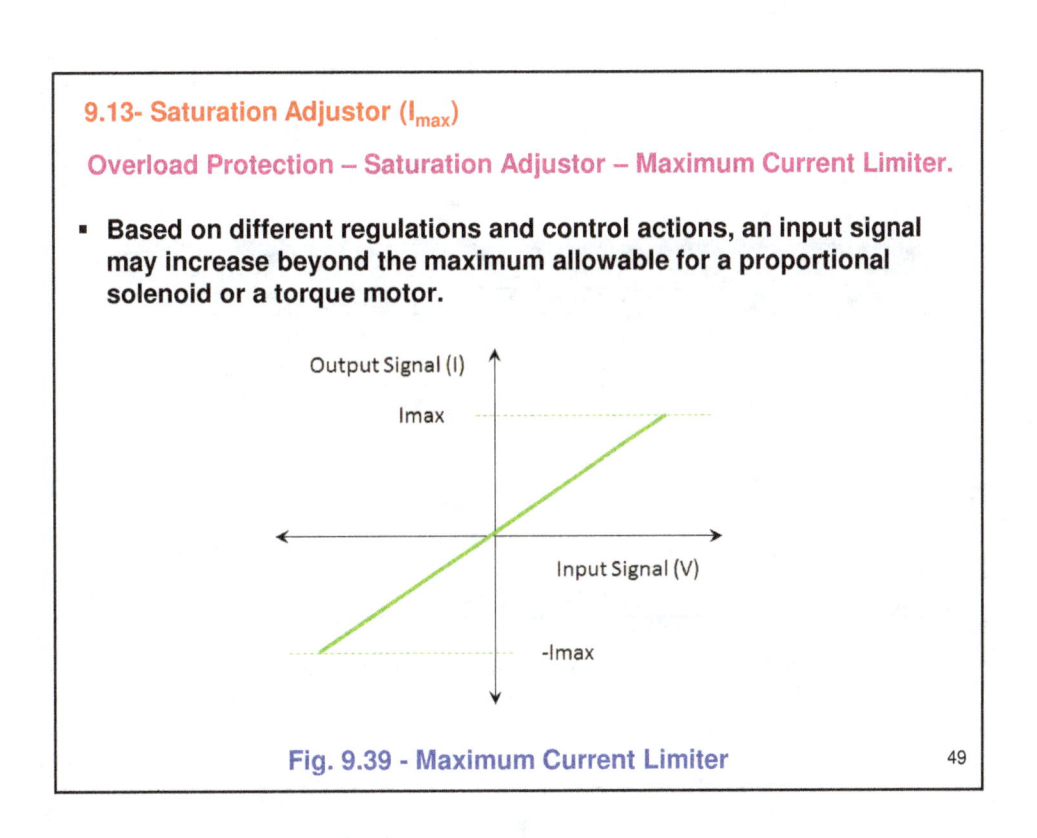

Fig. 9.39 - Maximum Current Limiter

- **Analog Overload Protection**
  - Could be factory set and automatic built in function.
  - Could be associated with overload protection monitor.

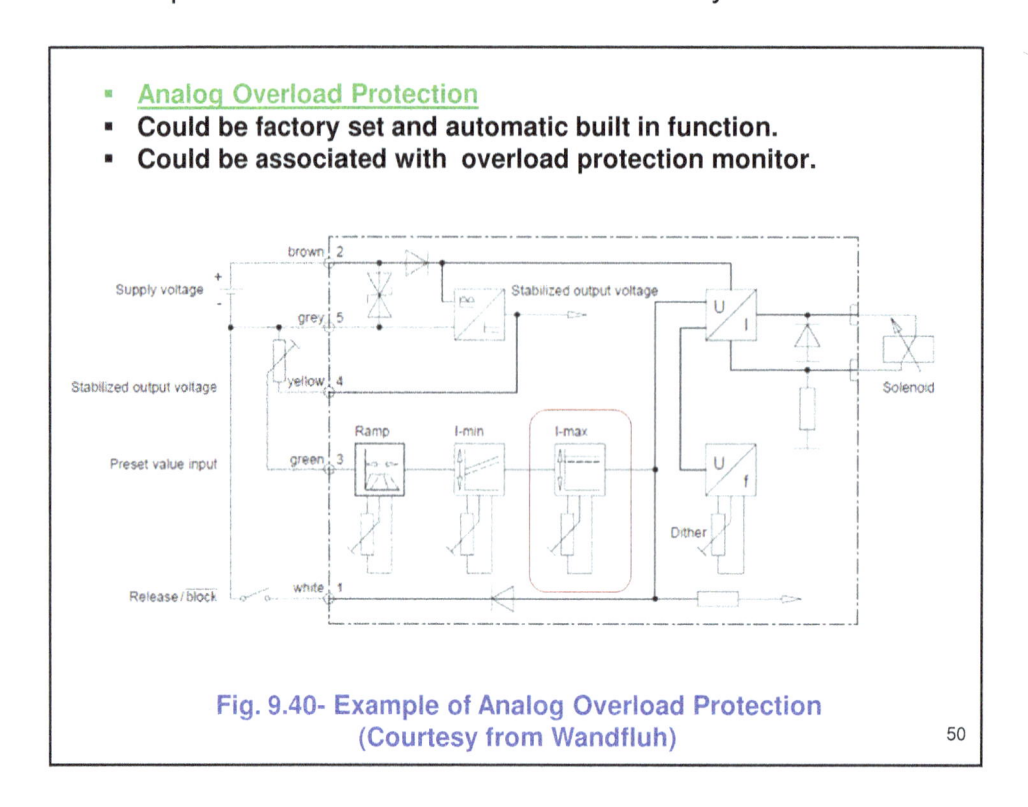

Fig. 9.40- Example of Analog Overload Protection
(Courtesy from Wandfluh)

- **Digital Overload Protection**

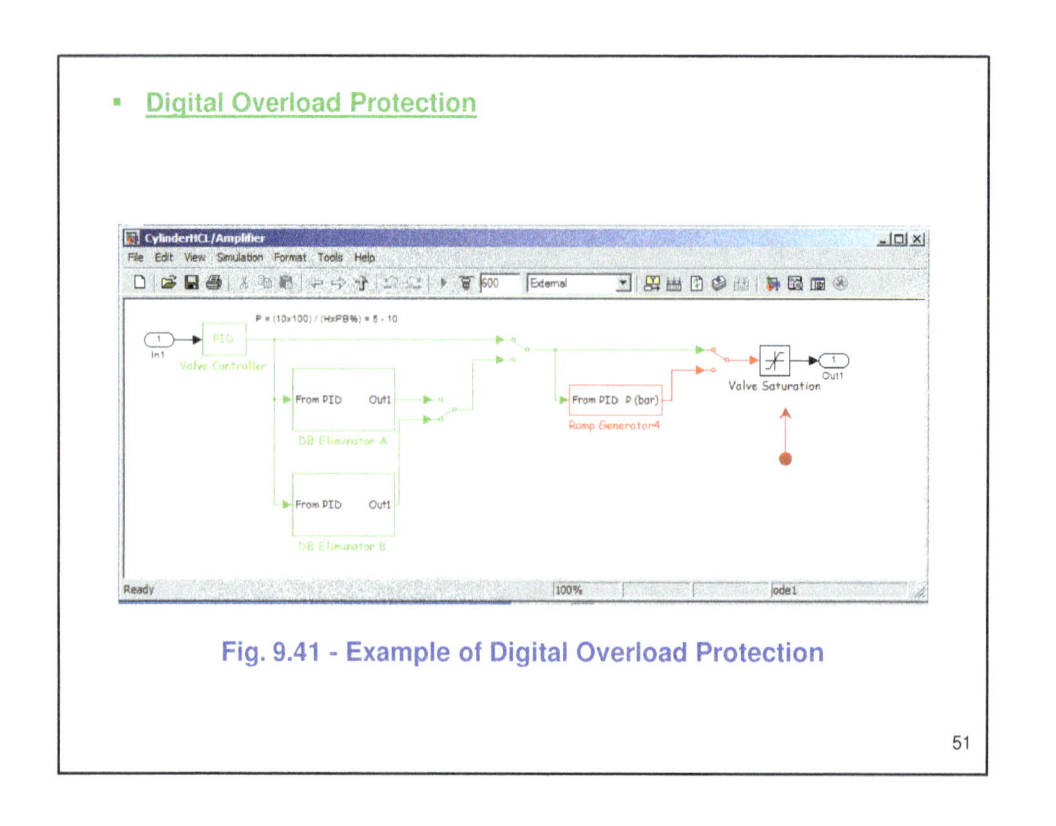

Fig. 9.41 - Example of Digital Overload Protection

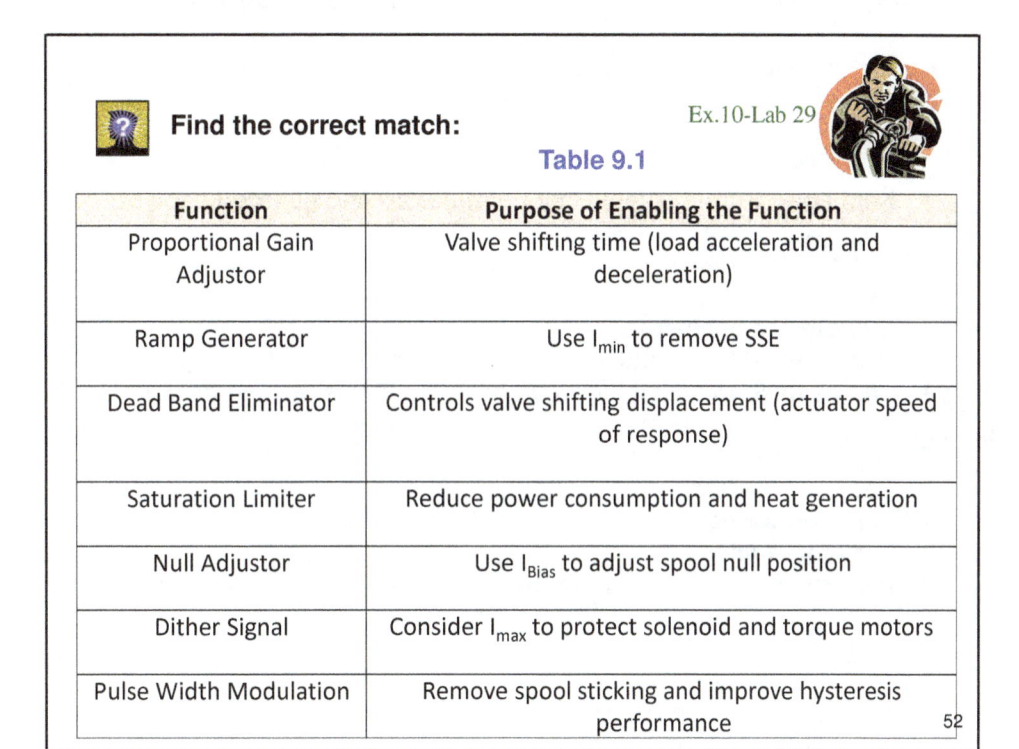

Workbook: Electro-Hydraulic Components and Systems
Chapter 9: Control Electronics for Electro-Hydraulic Valves

Find the correct match: Ex.10-Lab 29

Table 9.1

Function	Purpose of Enabling the Function
Proportional Gain Adjustor	Valve shifting time (load acceleration and deceleration)
Ramp Generator	Use I_{min} to remove SSE
Dead Band Eliminator	Controls valve shifting displacement (actuator speed of response)
Saturation Limiter	Reduce power consumption and heat generation
Null Adjustor	Use I_{Bias} to adjust spool null position
Dither Signal	Consider I_{max} to protect solenoid and torque motors
Pulse Width Modulation	Remove spool sticking and improve hysteresis performance

9.14- Control Electronics Basic Configurations

Control Electronics "Basic Configurations"

- Hardware Configuration
  - Panel Mounted
  - Modular
  - Onboard "OBE"
- Spool Control
  - Force Control
  - Stroke Control
- System Control
  - Valve Only
  - Single Axis Drive
  - Multiple Axis Drive
- Control Mode
  - Analog
  - Digital
  - Both
- Interfacing
  - No Interface
  - Network Interface

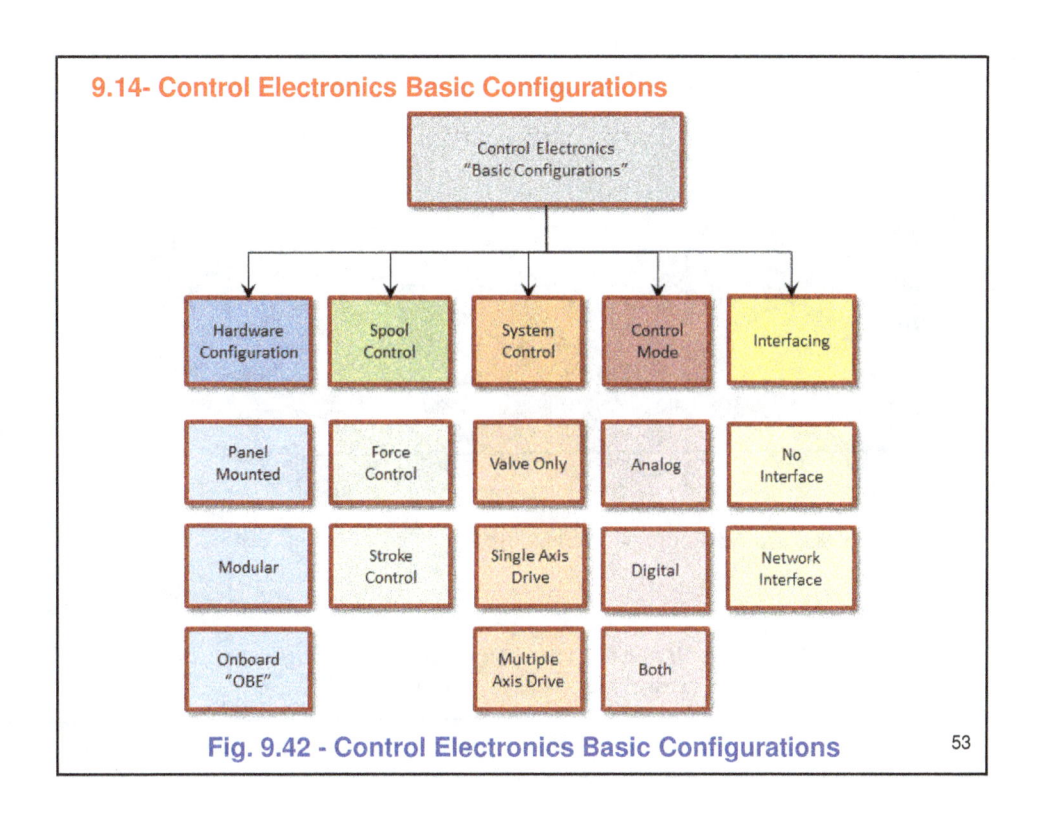

Fig. 9.42 - Control Electronics Basic Configurations

Workbook: Electro-Hydraulic Components and Systems
Chapter 9: Control Electronics for Electro-Hydraulic Valves

9.15- Hardware Configuration

- **Panel Mounted:**
  - Kept apart from the valve in a conditioned and guarded cabinet.

- **Modular:**
  - Customized enclosure.
  - Can be used for industrial/mobile applications.
  - Can be used to control valve only or multi-axis system control.

- **Integrated - Onboard "OBE":**
  - Compact system.
  - Reduced wiring.
  - Easy interfacing.
  - Performance limitation at high temperature (good for T<60°C).

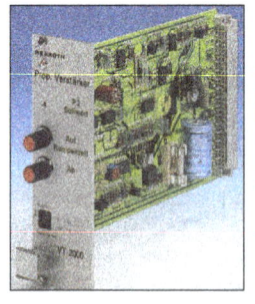

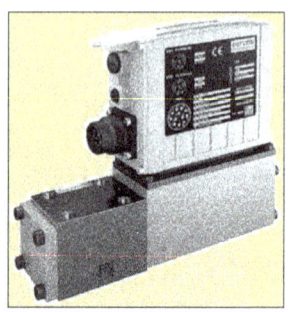

Panel Mounted Type   Modular Type   OBE

Fig. 9.43 - Control Electronics Hardware Configurations (Courtesy of Bosch Rexroth)

Workbook: Electro-Hydraulic Components and Systems
Chapter 9: Control Electronics for Electro-Hydraulic Valves

9.16- Typical Electronic Schematics for Continuous Valves

Example 1: ECU for controlling force-Controlled proportional pressure valves

Hardware Configuration	
Panel Mounted	
Modular	✔
OBE	
Spool Control	
Force Controlled	✔
Stroke Controlled	
System Control	
Valve Only	✔
Single Axis	
Multiple Axis	
Control Mode	
Analog	✔
Digital	
Interfacing	
No Interface	✔
Network Interfacing	

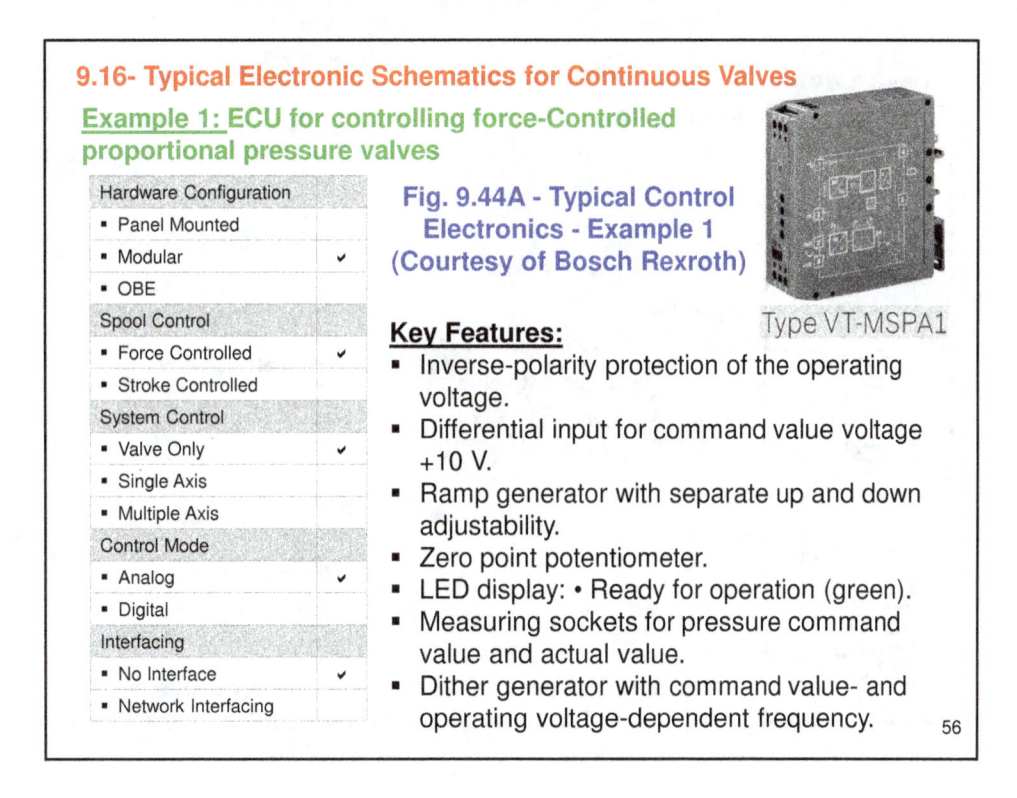

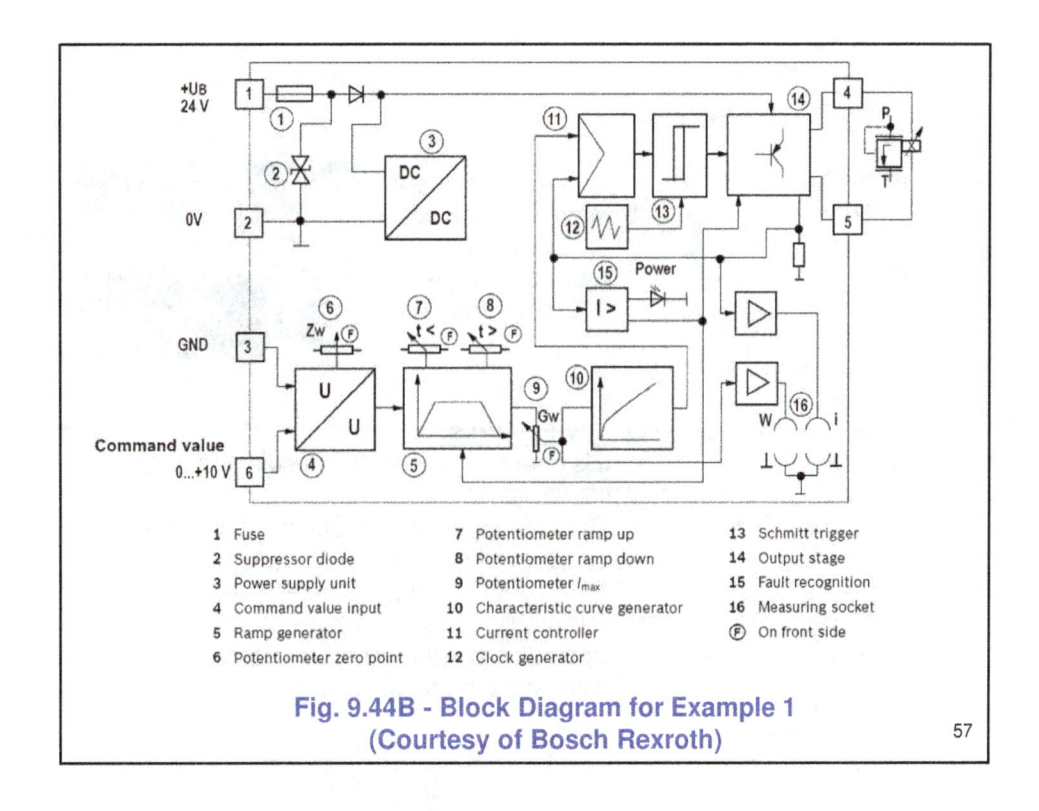

Fig. 9.44A - Typical Control Electronics - Example 1 (Courtesy of Bosch Rexroth)

Type VT-MSPA1

Key Features:
- Inverse-polarity protection of the operating voltage.
- Differential input for command value voltage +10 V.
- Ramp generator with separate up and down adjustability.
- Zero point potentiometer.
- LED display: • Ready for operation (green).
- Measuring sockets for pressure command value and actual value.
- Dither generator with command value- and operating voltage-dependent frequency.

1 Fuse	7 Potentiometer ramp up
2 Suppressor diode	8 Potentiometer ramp down
3 Power supply unit	9 Potentiometer I_{max}
4 Command value input	10 Characteristic curve generator
5 Ramp generator	11 Current controller
6 Potentiometer zero point	12 Clock generator
13 Schmitt trigger	
14 Output stage	
15 Fault recognition	
16 Measuring socket	
(F) On front side	

Fig. 9.44B - Block Diagram for Example 1 (Courtesy of Bosch Rexroth)

Workbook: Electro-Hydraulic Components and Systems
Chapter 9: Control Electronics for Electro-Hydraulic Valves

Device view

Potentiometer:
- "Gw" Pressure command value
- "Zw" Zero point
- "t <" Ramp time up
- "t >" Ramp time down

Sockets:
- "w" Pressure command value
- "I" Actual current value
- "⊥" Measurement zero

Terminal assignment

Terminal	
1	$+U_B$
2	Ground
3	$-U_{command}$
4	Solenoid +
5	Solenoid −
6	$+U_{command}$

Fig. 9.44C - Terminal Assignment for Example 1 (Courtesy of Bosch Rexroth)

Example 2: OBE for controlling single-solenoid force-controlled proportional valves

Hardware Configuration	
• Panel Mounted	
• Modular	
• OBE	✓
Spool Control	
• Force Controlled	✓
• Stroke Controlled	
System Control	
• Valve Only	✓
• Single Axis	
• Multiple Axis	
Control Mode	
• Analog	✓
• Digital	
Interfacing	
• No Interface	✓
• Network Interfacing	

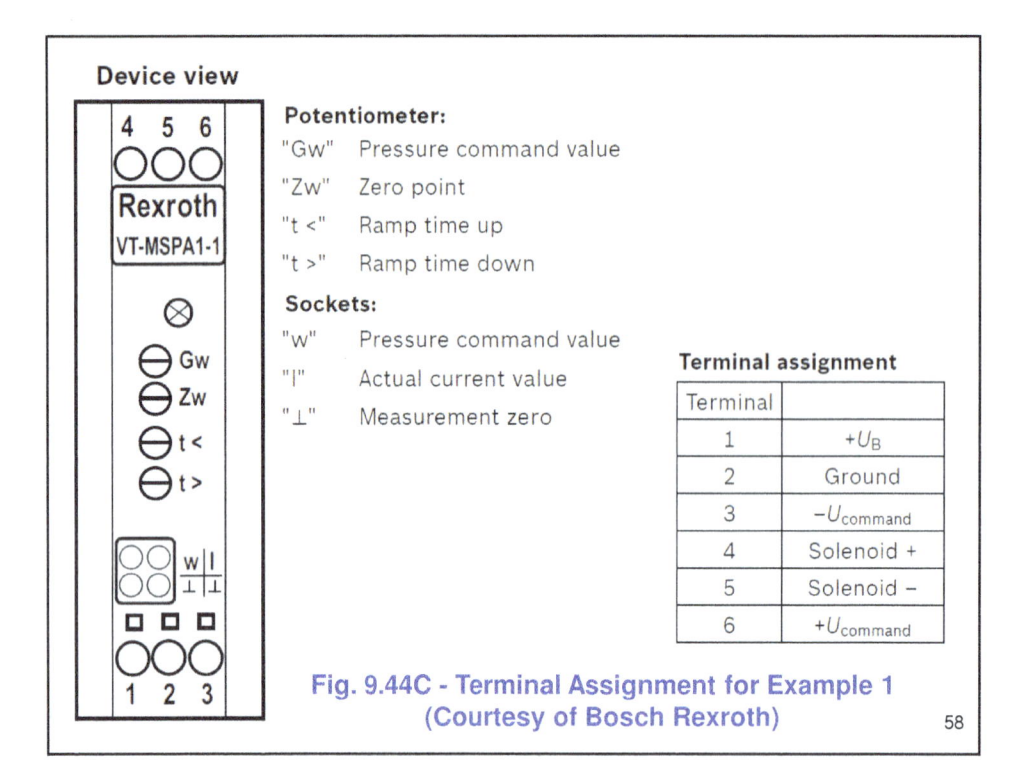

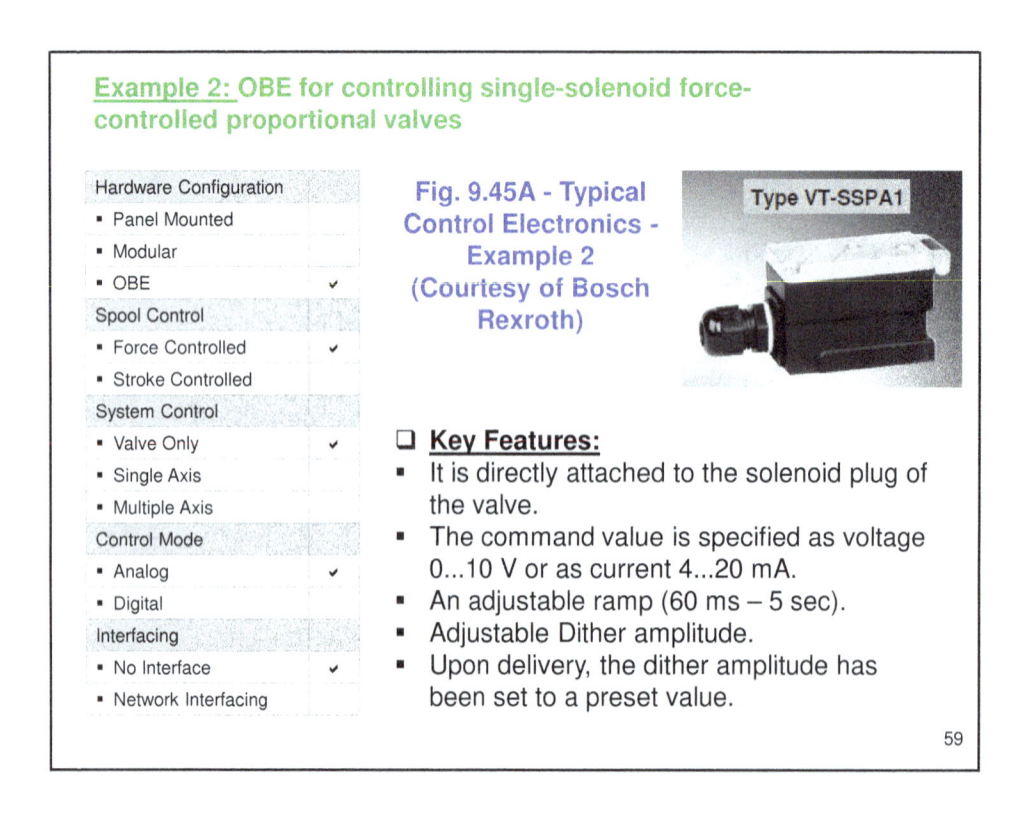

Fig. 9.45A - Typical Control Electronics - Example 2 (Courtesy of Bosch Rexroth)

Type VT-SSPA1

❑ **Key Features:**
- It is directly attached to the solenoid plug of the valve.
- The command value is specified as voltage 0...10 V or as current 4...20 mA.
- An adjustable ramp (60 ms – 5 sec).
- Adjustable Dither amplitude.
- Upon delivery, the dither amplitude has been set to a preset value.

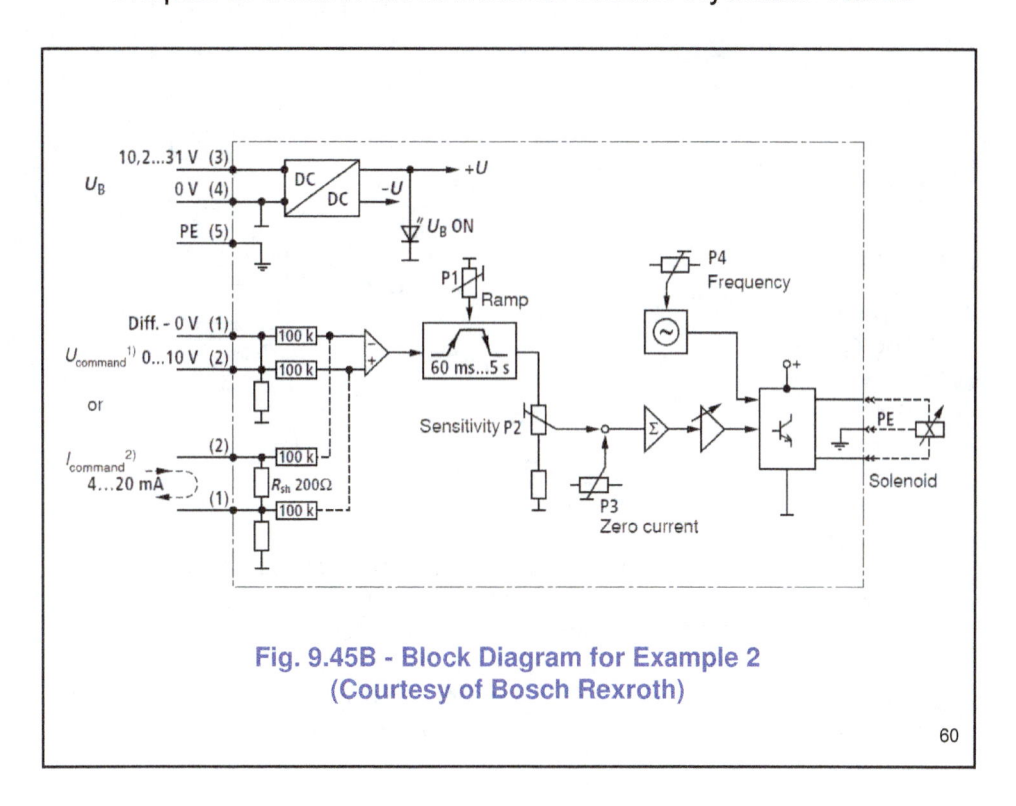

Workbook: Electro-Hydraulic Components and Systems
Chapter 9: Control Electronics for Electro-Hydraulic Valves

**Fig. 9.45B - Block Diagram for Example 2
(Courtesy of Bosch Rexroth)**

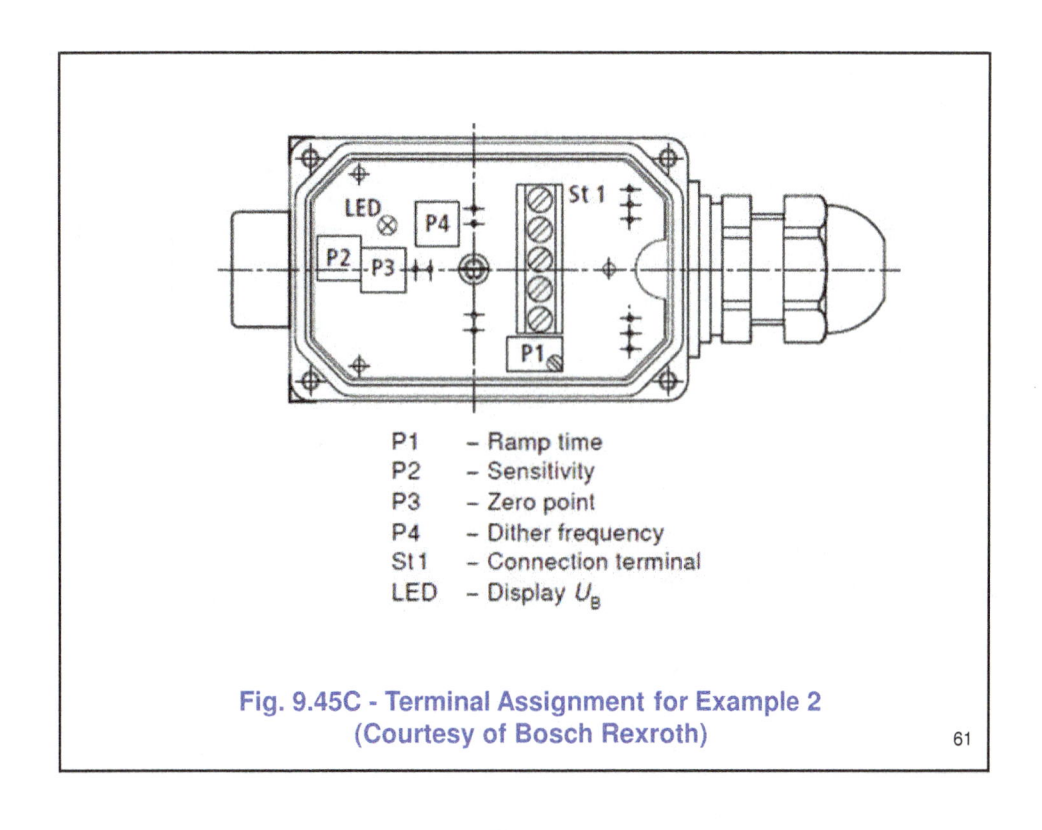

P1 — Ramp time
P2 — Sensitivity
P3 — Zero point
P4 — Dither frequency
St 1 — Connection terminal
LED — Display U_B

**Fig. 9.45C - Terminal Assignment for Example 2
(Courtesy of Bosch Rexroth)**

Workbook: Electro-Hydraulic Components and Systems
Chapter 9: Control Electronics for Electro-Hydraulic Valves

Example 3: OBE for controlling single-solenoid force-controlled proportional valves

Hardware Configuration	
• Panel Mounted	
• Modular	
• OBE	✓
Spool Control	
• Force Controlled	✓
• Stroke Controlled	
System Control	
• Valve Only	✓
• Single Axis	
• Multiple Axis	
Control Mode	
• Analog	✓
• Digital	✓
Interfacing	
• No Interface	✓
• Network Interfacing	

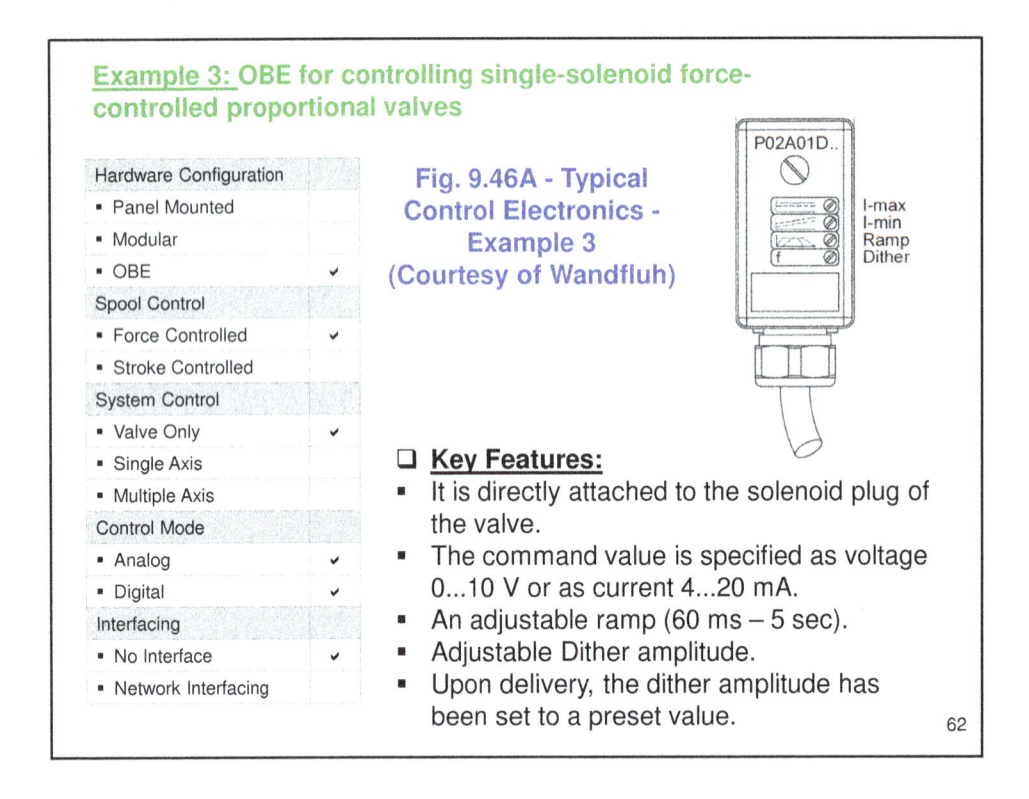

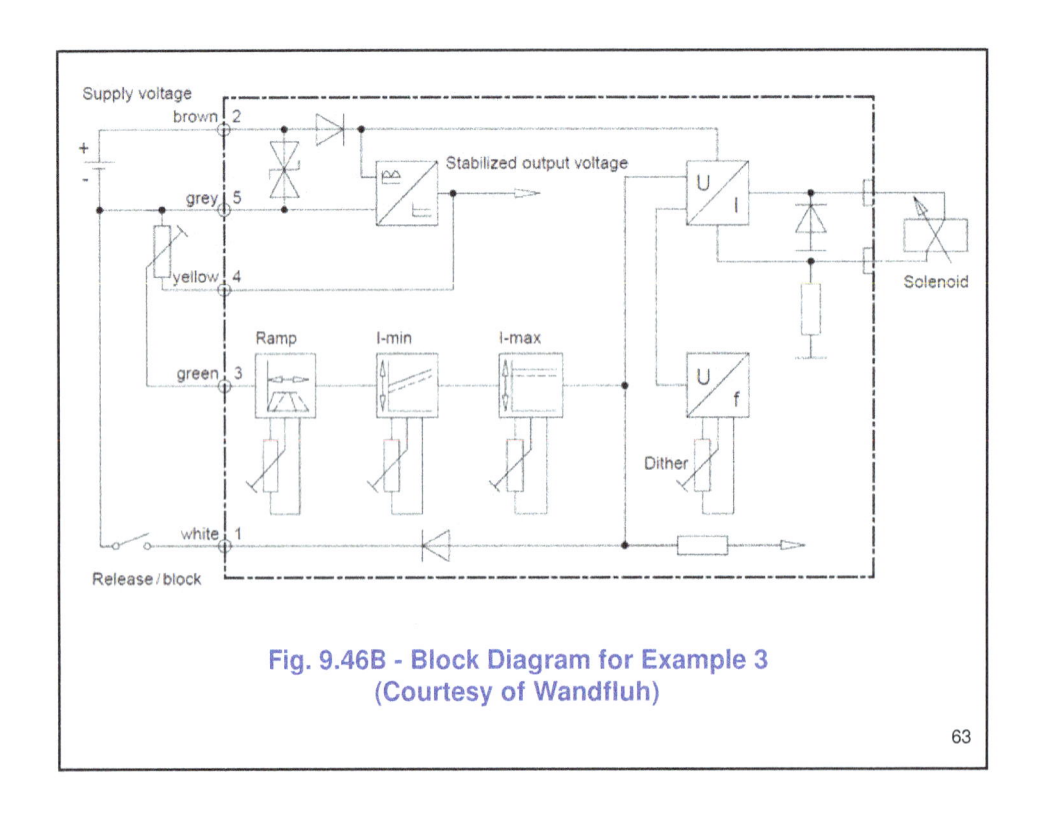

Fig. 9.46A - Typical Control Electronics - Example 3 (Courtesy of Wandfluh)

❑ **Key Features:**
- It is directly attached to the solenoid plug of the valve.
- The command value is specified as voltage 0...10 V or as current 4...20 mA.
- An adjustable ramp (60 ms – 5 sec).
- Adjustable Dither amplitude.
- Upon delivery, the dither amplitude has been set to a preset value.

Fig. 9.46B - Block Diagram for Example 3 (Courtesy of Wandfluh)

Workbook: Electro-Hydraulic Components and Systems
Chapter 9: Control Electronics for Electro-Hydraulic Valves

Connection with external power source release/$\overline{block}$ with PLC, PC or NC

Analog output 0...10 V PLC

Digital output PLC

Fig. 9.46C - Terminal Assignment for Example 3 (Courtesy of Wandfluh)

Example 4: Valve amplifier for controlling double-solenoid force-controlled proportional valves

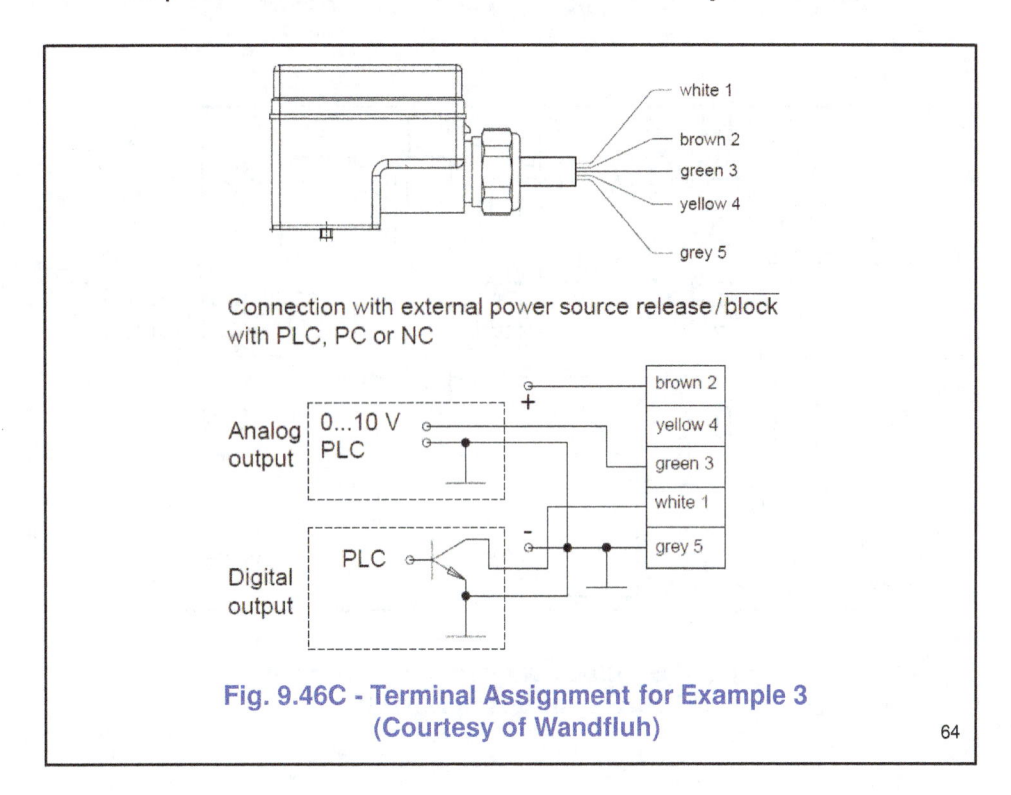

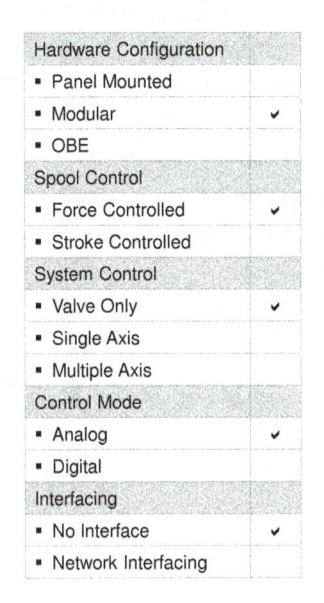

Fig. 9.47A - Typical Control Electronics - Example 4 (Courtesy of Bosch Rexroth)

Type VT 11118

Hardware Configuration	
• Panel Mounted	
• Modular	✓
• OBE	
Spool Control	
• Force Controlled	✓
• Stroke Controlled	
System Control	
• Valve Only	✓
• Single Axis	
• Multiple Axis	
Control Mode	
• Analog	✓
• Digital	
Interfacing	
• No Interface	✓
• Network Interfacing	

❑ **Key Features:**
- Selection of the valve type by means of change-over switch at the front.
- Differential command value voltage ±10 V.
- Enable inputs.
- Adjustable ramp generator.
- 2 output stages with fixed-frequency clocking.
- DC/DC converter (L0 = M0).
- Reverse polarity protection for operating voltage.
- Short-circuit-proof outputs.

Workbook: Electro-Hydraulic Components and Systems
Chapter 9: Control Electronics for Electro-Hydraulic Valves

**Fig. 9.47B - Block Diagram for Example 4
(Courtesy of Bosch Rexroth)**

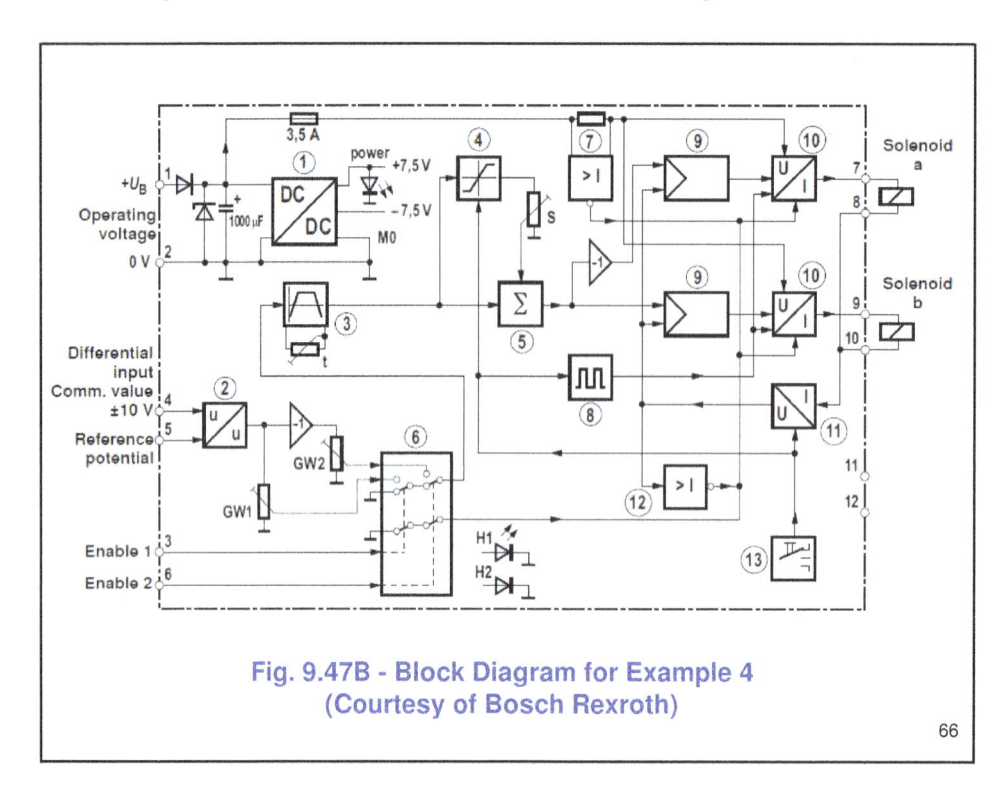

1 Power supply	11 Solenoid current measurement
2 Differential amplifier	12 Overcurrent detector
3 Ramp generator	13 Valve type selector switch
4 Step function generator	GW1 Command value attenuator 1
5 Summator	GW2 Command value attenuator 2
6 Command value changeover and output stage enable	t Ramp time setting
	S Jump height at $U_{Comm} = \pm10$ V
7 Short-circuit detector	H1 Enable logic mode 1
8 Clock-pulse generator	H2 Enable logic mode 2
9 Current regulator	
10 Output stage	

Workbook: Electro-Hydraulic Components and Systems
Chapter 9: Control Electronics for Electro-Hydraulic Valves

Potentiometer:

Gw1 I_{max} at Mode 1
Gw2 I_{max} at Mode 2
S   jump height
t   ramp time

LED-lamps:

power   operating voltage ON
H1   Enable logic mode 1
H2   Enable logic mode 2

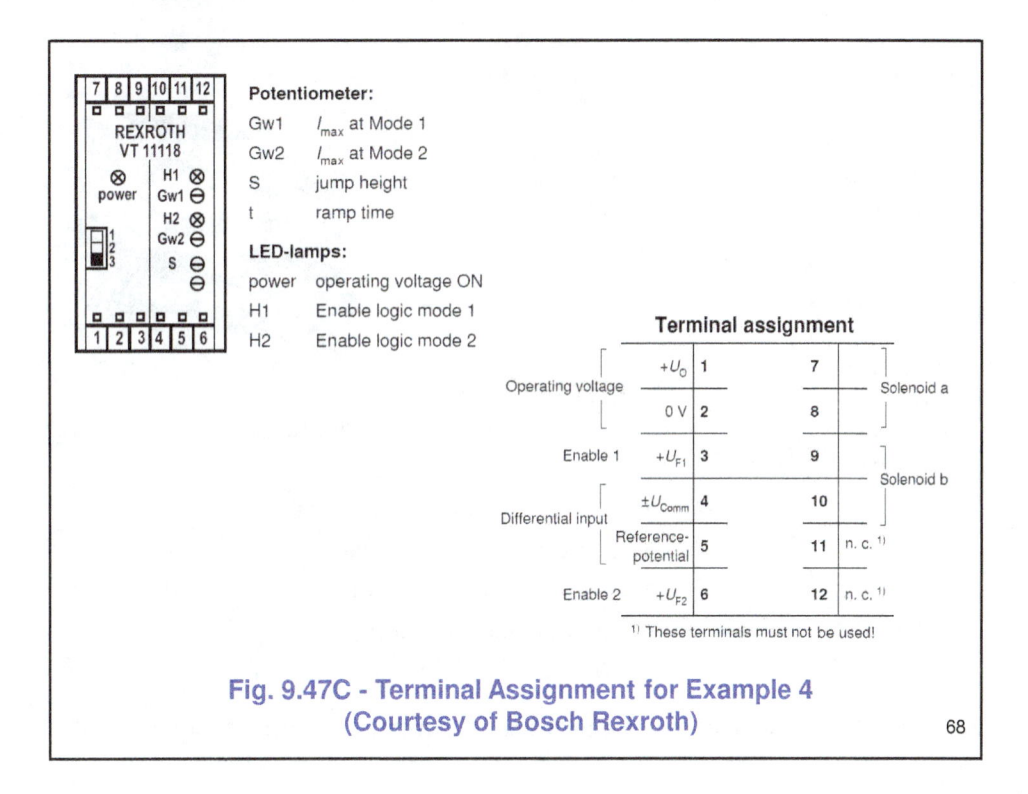

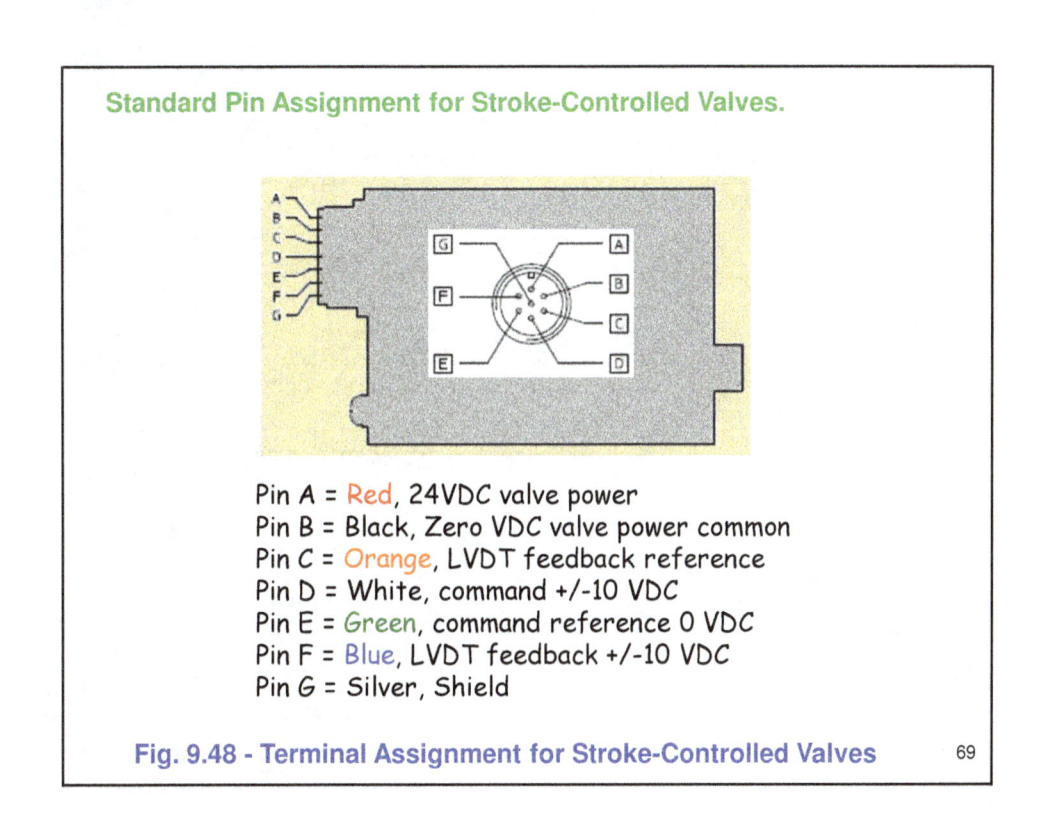

Terminal assignment

Operating voltage	$+U_0$	1	7	Solenoid a
	0 V	2	8	
Enable 1	$+U_{F1}$	3	9	Solenoid b
Differential input	$\pm U_{Comm}$	4	10	
Reference-potential		5	11	n. c. [1]
Enable 2	$+U_{F2}$	6	12	n. c. [1]

[1] These terminals must not be used!

**Fig. 9.47C - Terminal Assignment for Example 4
(Courtesy of Bosch Rexroth)**

68

Standard Pin Assignment for Stroke-Controlled Valves.

Pin A = Red, 24VDC valve power
Pin B = Black, Zero VDC valve power common
Pin C = Orange, LVDT feedback reference
Pin D = White, command +/-10 VDC
Pin E = Green, command reference 0 VDC
Pin F = Blue, LVDT feedback +/-10 VDC
Pin G = Silver, Shield

Fig. 9.48 - Terminal Assignment for Stroke-Controlled Valves

69

335

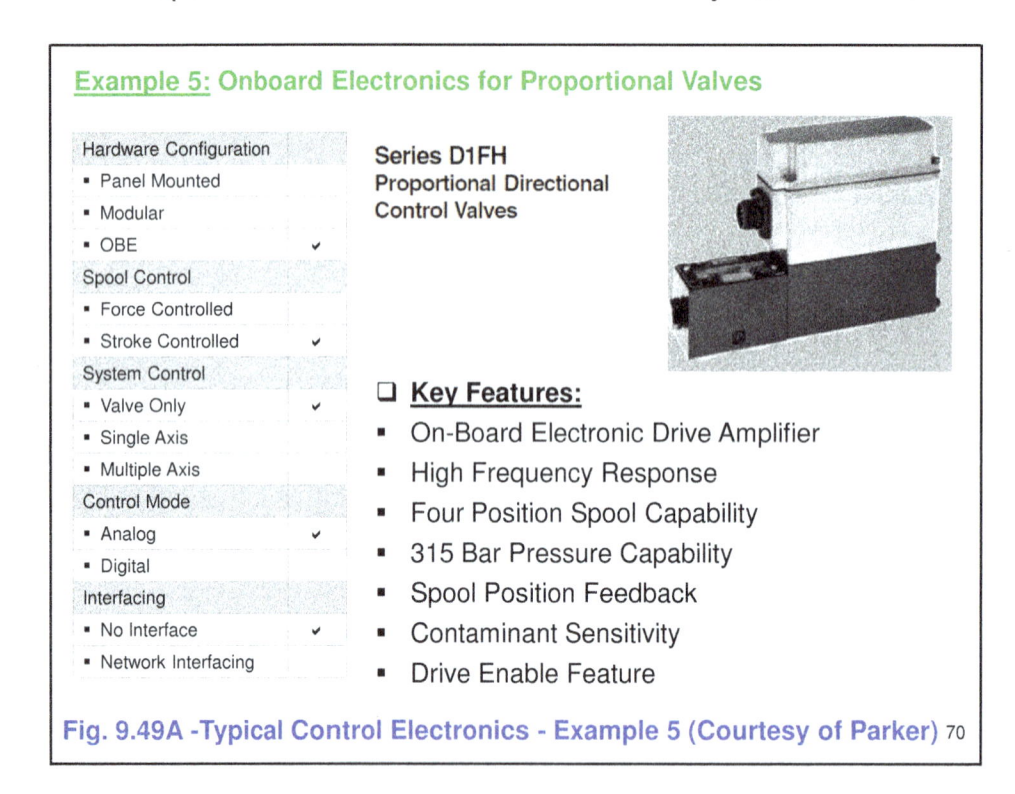

Workbook: Electro-Hydraulic Components and Systems
Chapter 9: Control Electronics for Electro-Hydraulic Valves

Example 5: Onboard Electronics for Proportional Valves

Hardware Configuration
- Panel Mounted
- Modular
- OBE ✓

Spool Control
- Force Controlled
- Stroke Controlled ✓

System Control
- Valve Only ✓
- Single Axis
- Multiple Axis

Control Mode
- Analog ✓
- Digital

Interfacing
- No Interface ✓
- Network Interfacing

Series D1FH
Proportional Directional Control Valves

❑ **Key Features:**
- On-Board Electronic Drive Amplifier
- High Frequency Response
- Four Position Spool Capability
- 315 Bar Pressure Capability
- Spool Position Feedback
- Contaminant Sensitivity
- Drive Enable Feature

Fig. 9.49A - Typical Control Electronics - Example 5 (Courtesy of Parker)

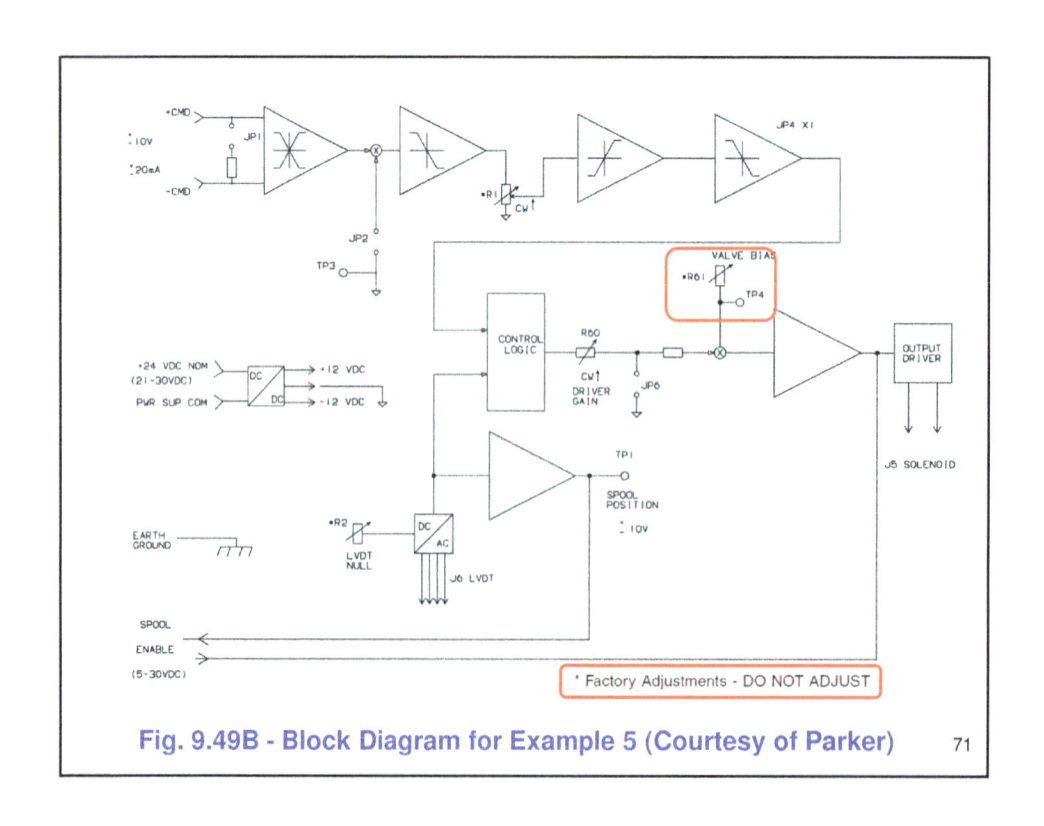

Fig. 9.49B - Block Diagram for Example 5 (Courtesy of Parker)

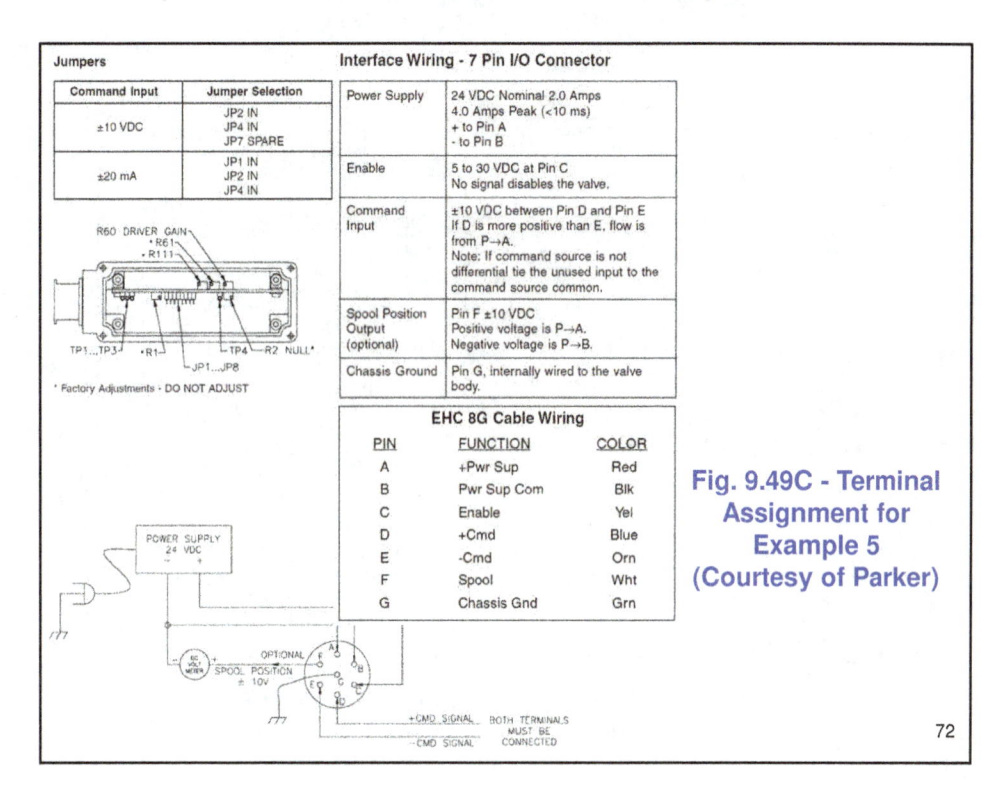

Workbook: Electro-Hydraulic Components and Systems
Chapter 9: Control Electronics for Electro-Hydraulic Valves

Jumpers

Command Input	Jumper Selection
±10 VDC	JP2 IN / JP4 IN / JP7 SPARE
±20 mA	JP1 IN / JP2 IN / JP4 IN

Interface Wiring - 7 Pin I/O Connector

Power Supply	24 VDC Nominal 2.0 Amps 4.0 Amps Peak (<10 ms) + to Pin A - to Pin B
Enable	5 to 30 VDC at Pin C No signal disables the valve.
Command Input	±10 VDC between Pin D and Pin E If D is more positive than E, flow is from P→A. Note: If command source is not differential tie the unused input to the command source common.
Spool Position Output (optional)	Pin F ±10 VDC Positive voltage is P→A. Negative voltage is P→B.
Chassis Ground	Pin G, internally wired to the valve body.

EHC 8G Cable Wiring

PIN	FUNCTION	COLOR
A	+Pwr Sup	Red
B	Pwr Sup Com	Blk
C	Enable	Yel
D	+Cmd	Blue
E	-Cmd	Orn
F	Spool	Wht
G	Chassis Gnd	Grn

*Factory Adjustments - DO NOT ADJUST

Fig. 9.49C - Terminal Assignment for Example 5 (Courtesy of Parker)

Example 6: Digital valve amplifier for Stroke-Controlled Proportional Valves

Hardware Configuration	
• Panel Mounted	✓
• Modular	
• OBE	
Spool Control	
• Force Controlled	
• Stroke Controlled	✓
System Control	
• Valve Only	✓
• Single Axis	
• Multiple Axis	
Control Mode	
• Analog	
• Digital	✓
Interfacing	
• No Interface	
• Network Interfacing	✓

Fig. 9.50A - Typical Control Electronics - Example 6 (Courtesy of Bosch Rexroth)

Type VT-VRPD-2

❏ **Key Features:**
- Use of a powerful microcontroller.
- Valve selection using operating software.
- Command value input as voltage or current.
- Ramp generator
- Digital inputs for calling pre-set command values.
- Enable input and fault output.
- Freely configurable measuring sockets
- Serial interface using PC software

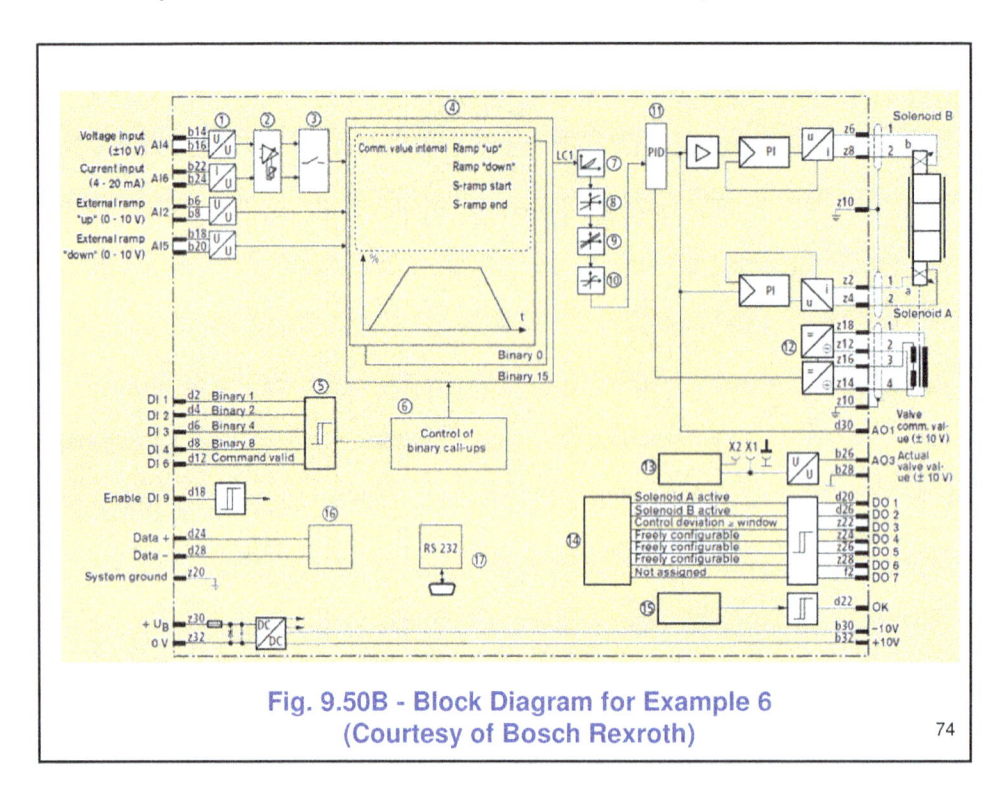

Workbook: Electro-Hydraulic Components and Systems
Chapter 9: Control Electronics for Electro-Hydraulic Valves

**Fig. 9.50B - Block Diagram for Example 6
(Courtesy of Bosch Rexroth)**

1. U/U or I/U converter
2. Analog input adjustment
3. Switching matrix
4. Switched-mode power supply unit
5. Binary command value call-ups
6. Control of binary call-ups
7. Characteristic curve adaptation
8. Step function generator
9. Offset
10. Limitation
11. PID controller
12. Oscillator / demodulator
13. Measuring sockets
14. Digital outputs
15. Fault logic
16. Local bus
17. Serial interface

Measuring sockets:
X1   Actual valve value
X2   Valve command value (default)
⊥   Reference potential (0 V)

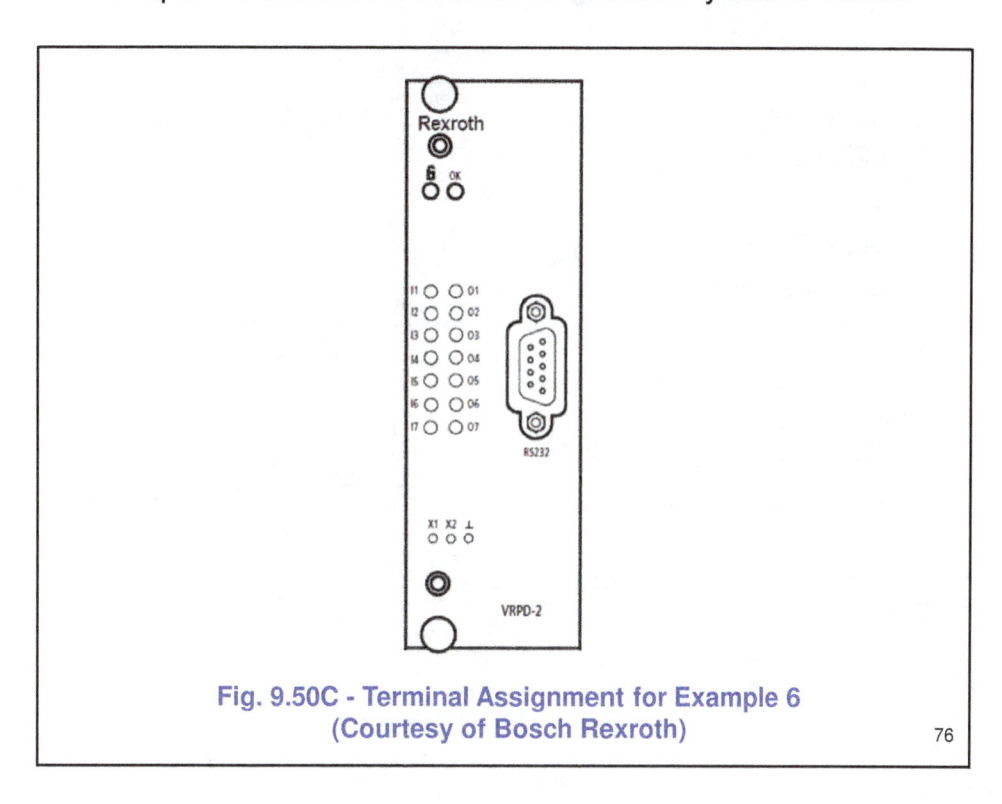

Workbook: Electro-Hydraulic Components and Systems
Chapter 9: Control Electronics for Electro-Hydraulic Valves

Fig. 9.50C - Terminal Assignment for Example 6
(Courtesy of Bosch Rexroth)

Example 7: Electric amplifier for displacement control of Piston Pumps.

Hardware Configuration	
• Panel Mounted	✓
• Modular	
• OBE	
Spool Control	
• Force Controlled	✓
• Stroke Controlled	
System Control	
• Valve Only	
• Single Axis	✓
• Multiple Axis	
Control Mode	
• Analog	✓
• Digital	
Interfacing	
• No Interface	
• Network Interfacing	✓

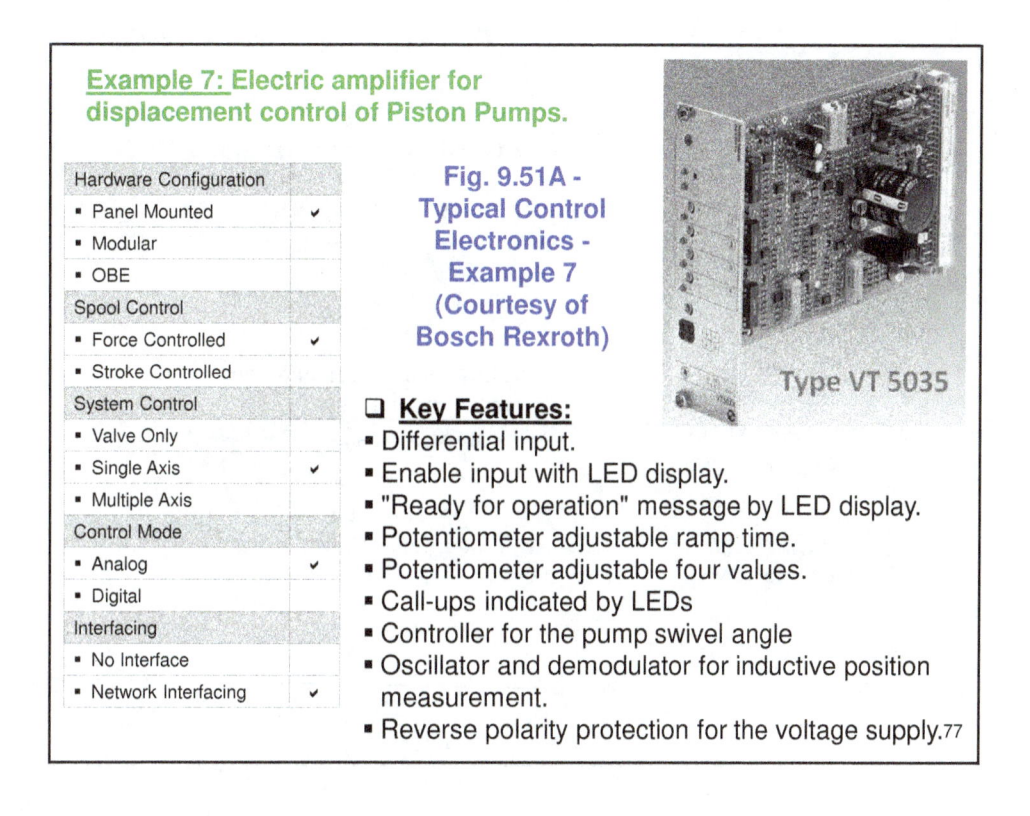

Fig. 9.51A - Typical Control Electronics - Example 7 (Courtesy of Bosch Rexroth)

Type VT 5035

❑ **Key Features:**
- Differential input.
- Enable input with LED display.
- "Ready for operation" message by LED display.
- Potentiometer adjustable ramp time.
- Potentiometer adjustable four values.
- Call-ups indicated by LEDs
- Controller for the pump swivel angle
- Oscillator and demodulator for inductive position measurement.
- Reverse polarity protection for the voltage supply.

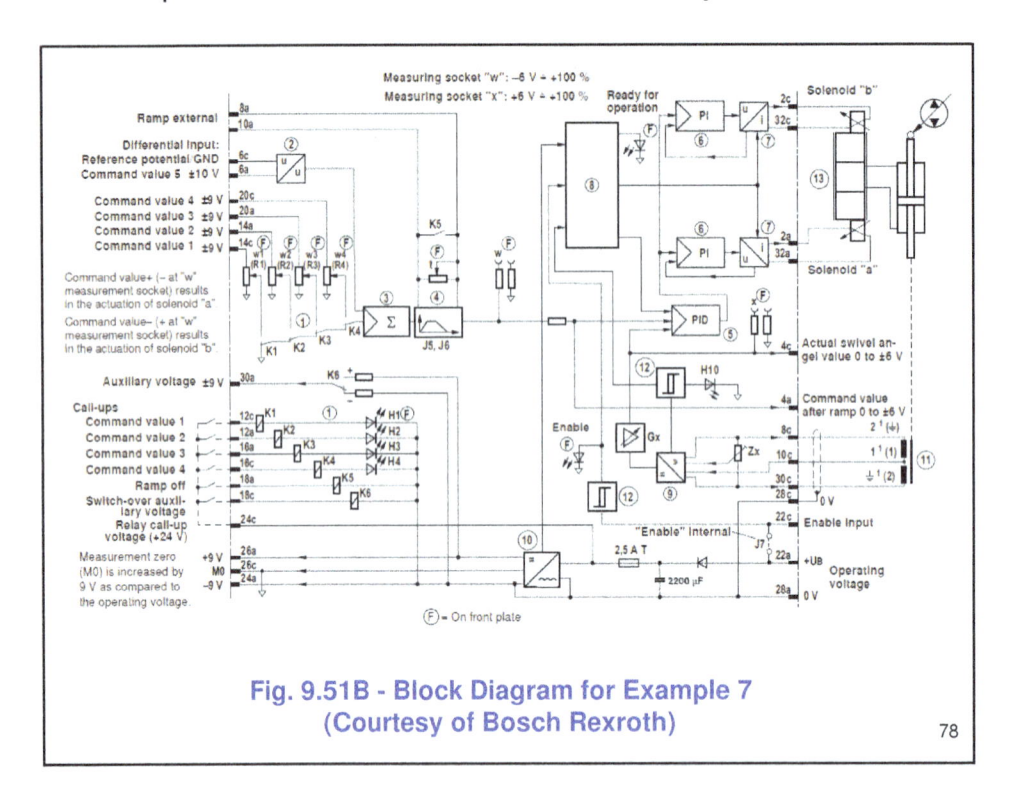

Workbook: Electro-Hydraulic Components and Systems
Chapter 9: Control Electronics for Electro-Hydraulic Valves

**Fig. 9.51B - Block Diagram for Example 7
(Courtesy of Bosch Rexroth)**

1 Command values
2 Differential amplifier
3 Summing device
4 Ramp generator
5 Swivel angle controller
6 Flow controller
7 Output stage
8 Monitoring
9 Oscillator/demodulator
10 Power supply unit
11 Position transducer
12 Actual value identification
13 Proportional valve

H1 to H4 = LED displays for command value call-ups
K1 to K6 = Call-up relay
R1 to R4 = Command value potentiometer
t = Ramp time

Notice for connection of the position transducer:

1 Applies to pump with clockwise rotation

( ) Applies to pump with counterclockwise rotation

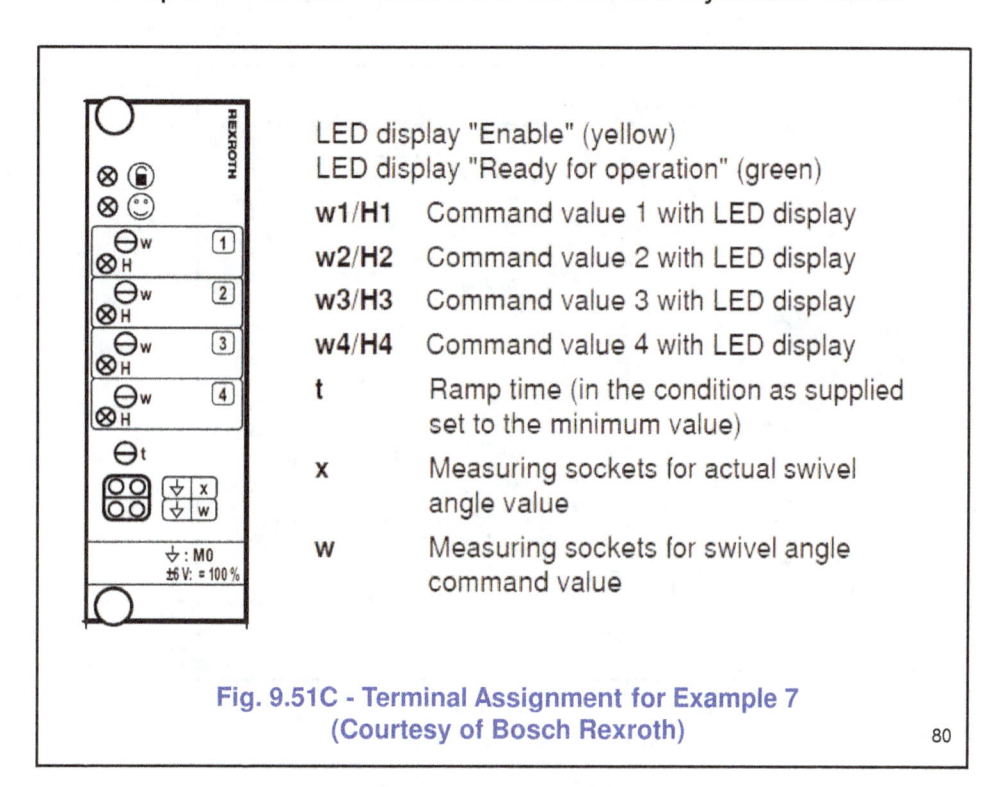

Workbook: Electro-Hydraulic Components and Systems
Chapter 9: Control Electronics for Electro-Hydraulic Valves

LED display "Enable" (yellow)
LED display "Ready for operation" (green)

w1/H1	Command value 1 with LED display
w2/H2	Command value 2 with LED display
w3/H3	Command value 3 with LED display
w4/H4	Command value 4 with LED display
t	Ramp time (in the condition as supplied set to the minimum value)
x	Measuring sockets for actual swivel angle value
w	Measuring sockets for swivel angle command value

**Fig. 9.51C - Terminal Assignment for Example 7
(Courtesy of Bosch Rexroth)**

Example 8: Onboard electronics for High Performance Proportional Valve.

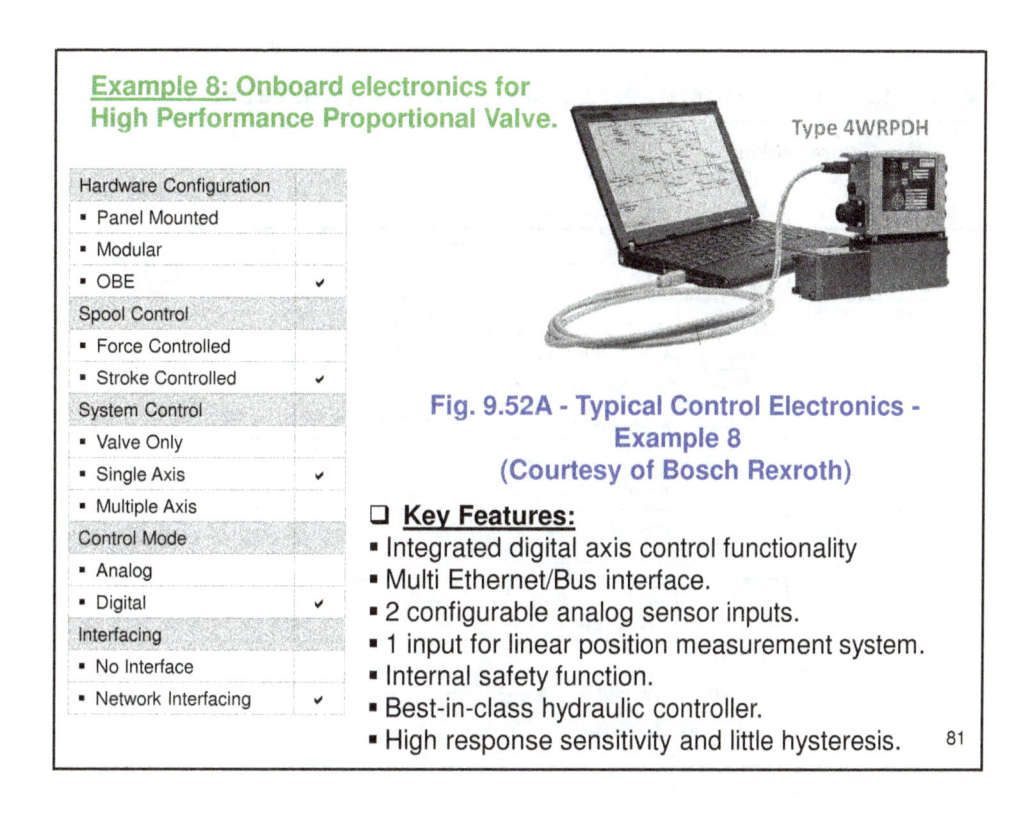

Type 4WRPDH

Hardware Configuration	
• Panel Mounted	
• Modular	
• OBE	✓
Spool Control	
• Force Controlled	
• Stroke Controlled	✓
System Control	
• Valve Only	
• Single Axis	✓
• Multiple Axis	
Control Mode	
• Analog	
• Digital	✓
Interfacing	
• No Interface	
• Network Interfacing	✓

**Fig. 9.52A - Typical Control Electronics -
Example 8
(Courtesy of Bosch Rexroth)**

❑ **Key Features:**
- Integrated digital axis control functionality
- Multi Ethernet/Bus interface.
- 2 configurable analog sensor inputs.
- 1 input for linear position measurement system.
- Internal safety function.
- Best-in-class hydraulic controller.
- High response sensitivity and little hysteresis.

Workbook: Electro-Hydraulic Components and Systems
Chapter 9: Control Electronics for Electro-Hydraulic Valves

Fig. 9.52B - Block Diagram for Example 8
(Courtesy of Bosch Rexroth)

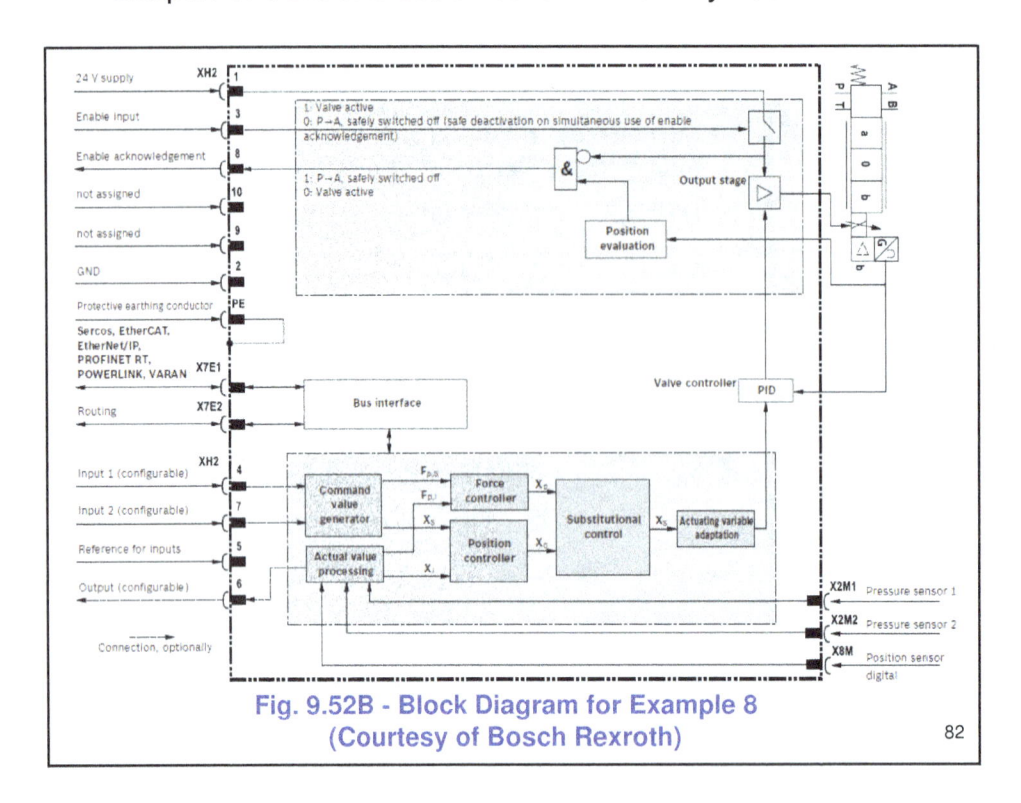

Connector pin assignment XH2, 11-pole + PE according to EN 175201-804

Pin	Core marking Cable, one-part [1]	Core marking Cable, split [2]	Interface D6 assignment
1	1	1	24 V DC supply voltage
2	2	2	GND
3	3	White	Enable input 24 V DC (high ≥ 15 V; low < 2 V)
4	4	Yellow	Command values 1 (4 ... 20 mA/±10 V) [3]
5	5	Green	Reference for command values
6	6	Violet	Actual value/4 ....20 mA/±10 V) [3; 4]
7	7	Pink	Command value 2 (4 ... 20 mA/±10 V) [3]
8	8	Red	Enable acknowledgement 24 V DC (I_{max} 50 mA) [5]
9	9	Brown	not assigned
10	10	Black	not assigned
11	11	Blue	Switching output 24 V, configurable (fault-free operation (24 V)/error (0 V) or power circuit signal), maximum 1.5 A [3; 5]

Fig. 9.52C - Terminal Assignment for Example 8
(Courtesy of Bosch Rexroth)

Workbook: Electro-Hydraulic Components and Systems
Chapter 9: Control Electronics for Electro-Hydraulic Valves

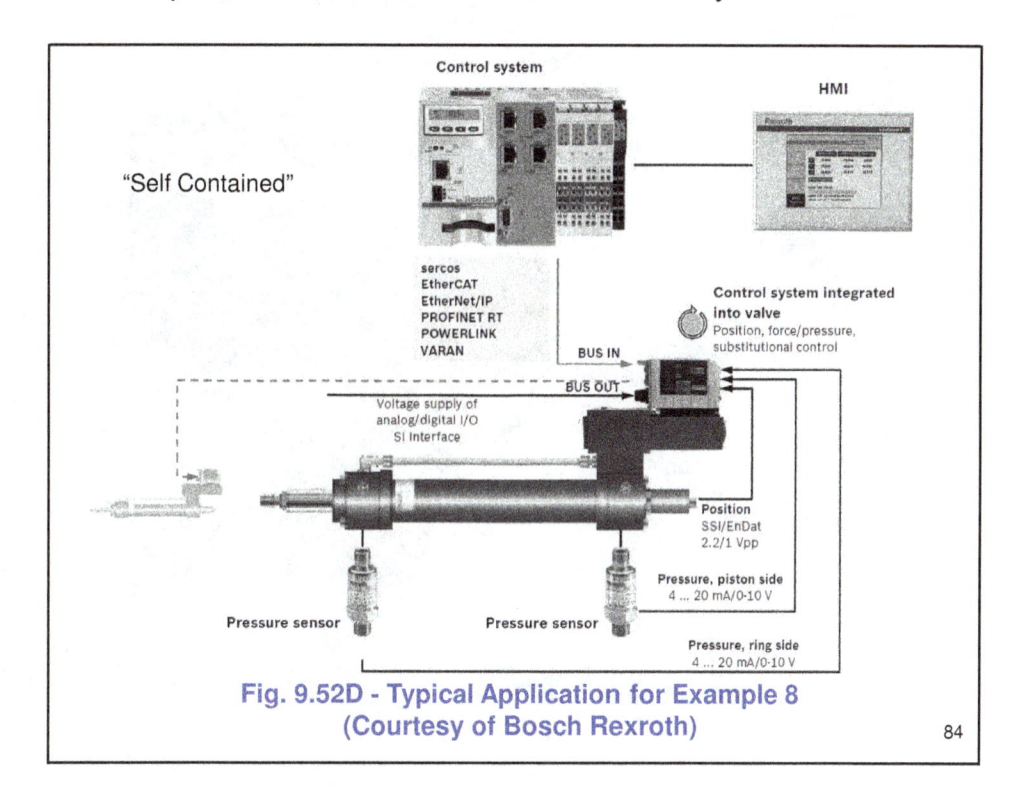

**Fig. 9.52D - Typical Application for Example 8
(Courtesy of Bosch Rexroth)**

Example 9: Onboard electronics for High Performance Proportional Valve.

Hardware Configuration	
• Panel Mounted	
• Modular	
• OBE	✓
Spool Control	
• Force Controlled	
• Stroke Controlled	✓
System Control	
• Valve Only	
• Single Axis	
• Multiple Axis	✓
Control Mode	
• Analog	✓
• Digital	✓
Interfacing	
• No Interface	
• Network Interfacing	✓

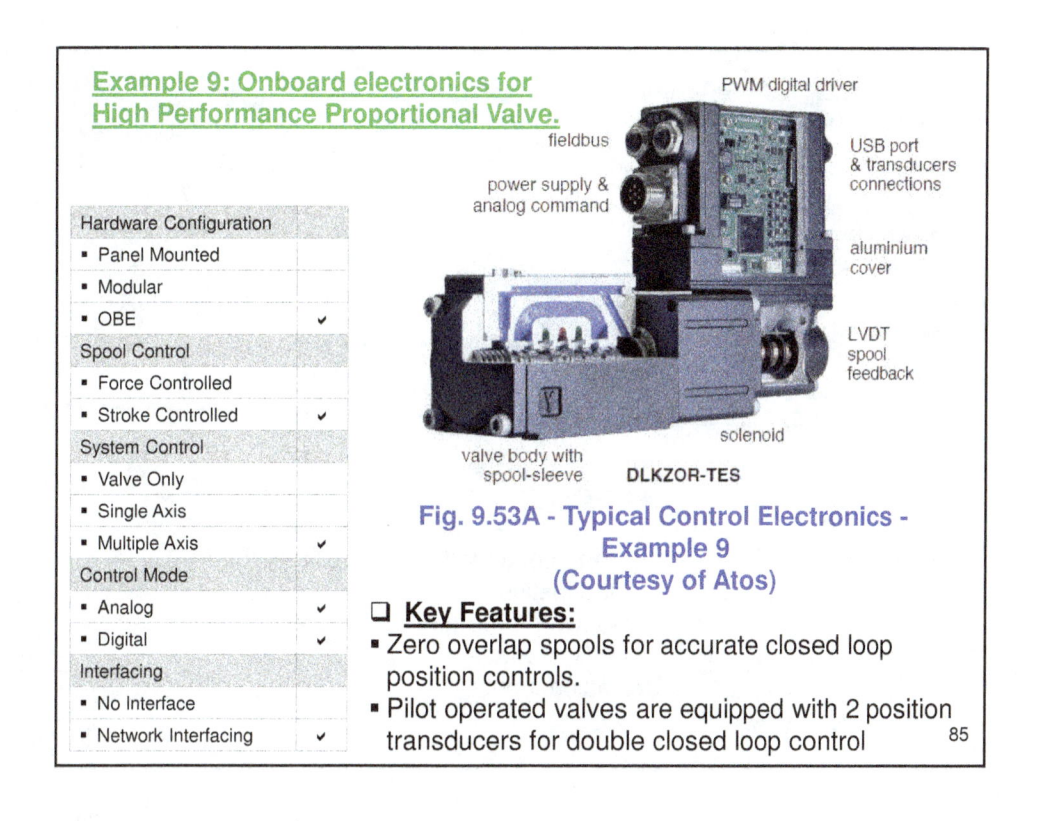

**Fig. 9.53A - Typical Control Electronics - Example 9
(Courtesy of Atos)**

❑ **Key Features:**
- Zero overlap spools for accurate closed loop position controls.
- Pilot operated valves are equipped with 2 position transducers for double closed loop control

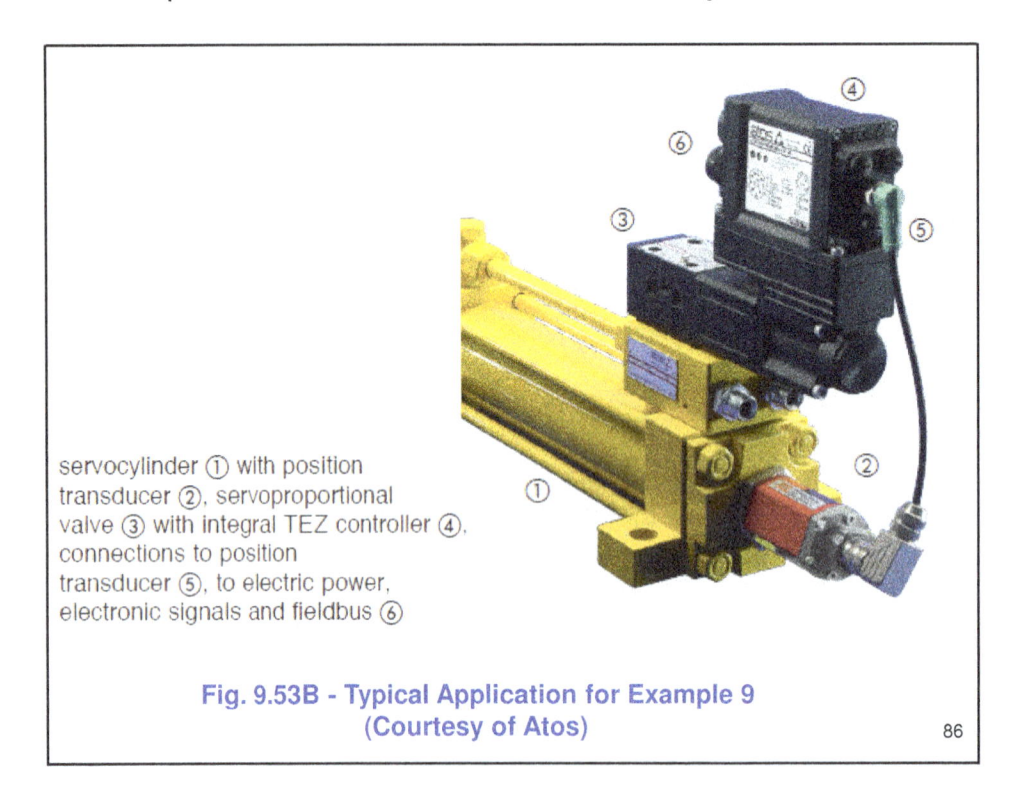

servocylinder ① with position transducer ②, servoproportional valve ③ with integral TEZ controller ④, connections to position transducer ⑤, to electric power, electronic signals and fieldbus ⑥

**Fig. 9.53B - Typical Application for Example 9
(Courtesy of Atos)**

Example 10: Digital Single-Axis Controller for driving proportional valves with two solenoids.

Hardware Configuration	
• Panel Mounted	
• Modular	✓
• OBE	
Spool Control	
• Force Controlled	✓
• Stroke Controlled	
System Control	
• Valve Only	
• Single Axis	✓
• Multiple Axis	
Control Mode	
• Analog	✓
• Digital	✓
Interfacing	
• No Interface	
• Network Interfacing	✓

Type SD6

**Fig. 9.54A - Typical Control Electronics - Example 10
(Courtesy of Wandfluh)**

❏ **Key Features:**
- Controls pressure, flow or position.
- Menu-controlled parameterization and diagnostics.
- Software programming.
- USB-interface.
- Analog inputs.

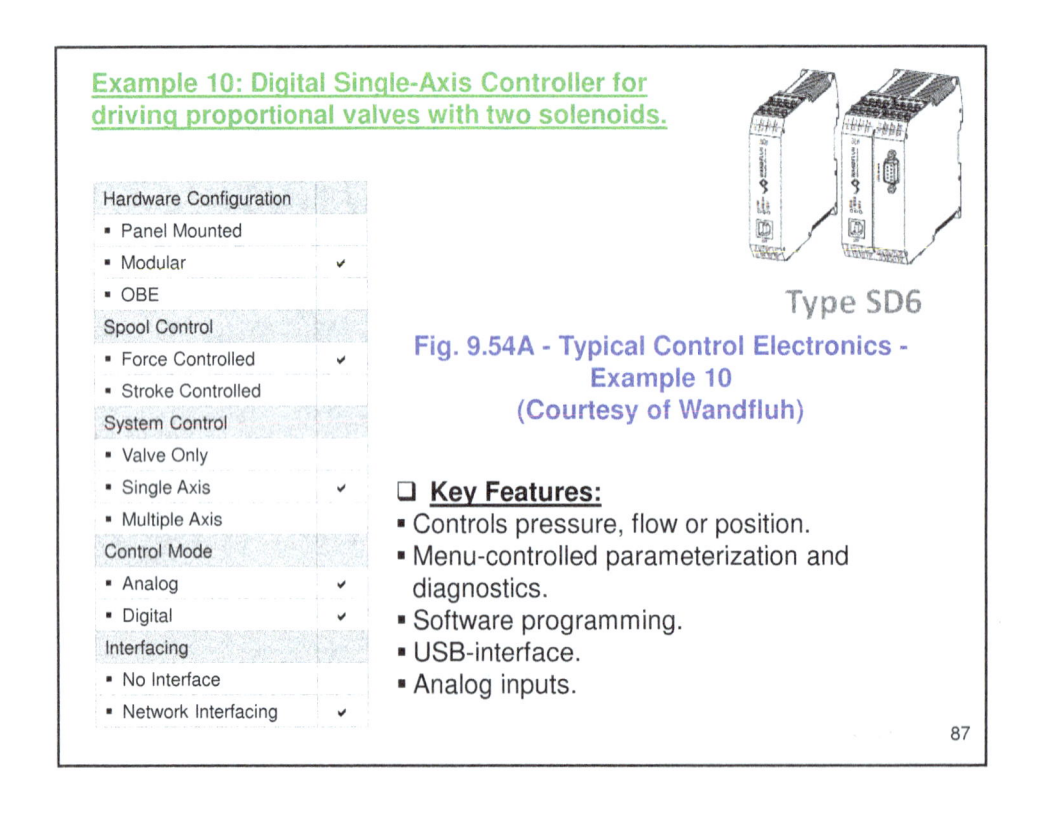

Workbook: Electro-Hydraulic Components and Systems
Chapter 9: Control Electronics for Electro-Hydraulic Valves

BLOCK DIAGRAM

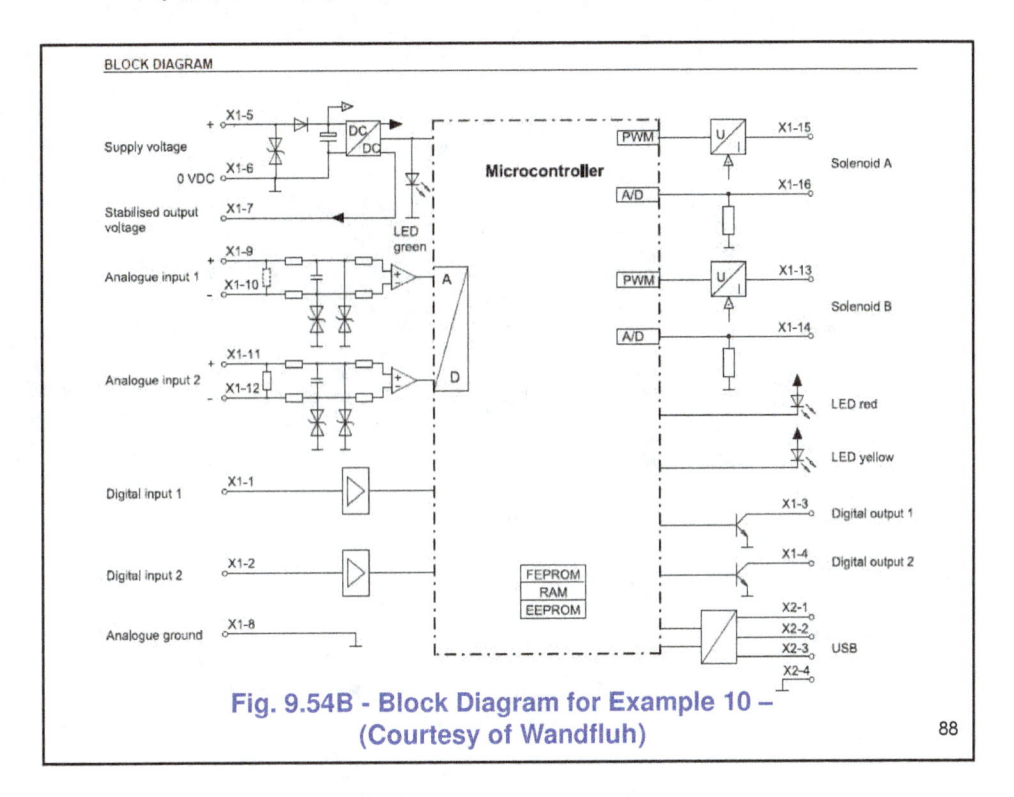

Fig. 9.54B - Block Diagram for Example 10 – (Courtesy of Wandfluh)

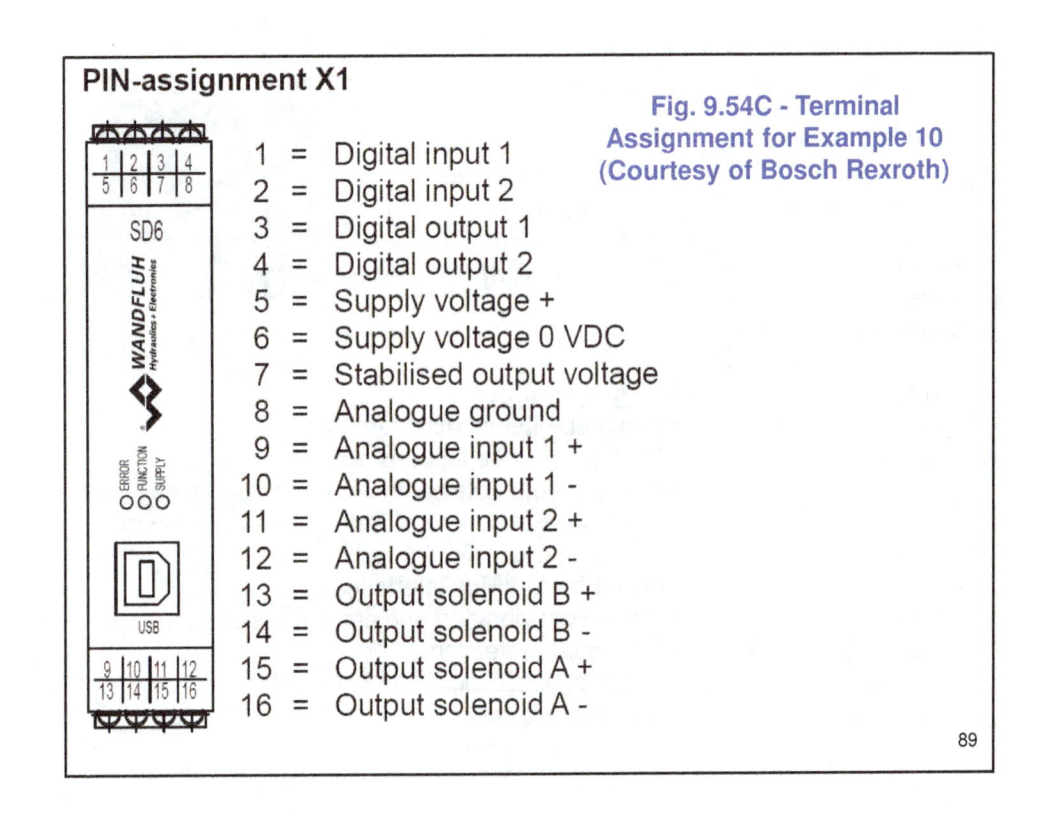

PIN-assignment X1

1 = Digital input 1
2 = Digital input 2
3 = Digital output 1
4 = Digital output 2
5 = Supply voltage +
6 = Supply voltage 0 VDC
7 = Stabilised output voltage
8 = Analogue ground
9 = Analogue input 1 +
10 = Analogue input 1 -
11 = Analogue input 2 +
12 = Analogue input 2 -
13 = Output solenoid B +
14 = Output solenoid B -
15 = Output solenoid A +
16 = Output solenoid A -

Fig. 9.54C - Terminal Assignment for Example 10 (Courtesy of Bosch Rexroth)

Workbook: Electro-Hydraulic Components and Systems
Chapter 9: Control Electronics for Electro-Hydraulic Valves

Fig. 9.54D - Typical Application for Example 10 (Courtesy of Wandfluh)

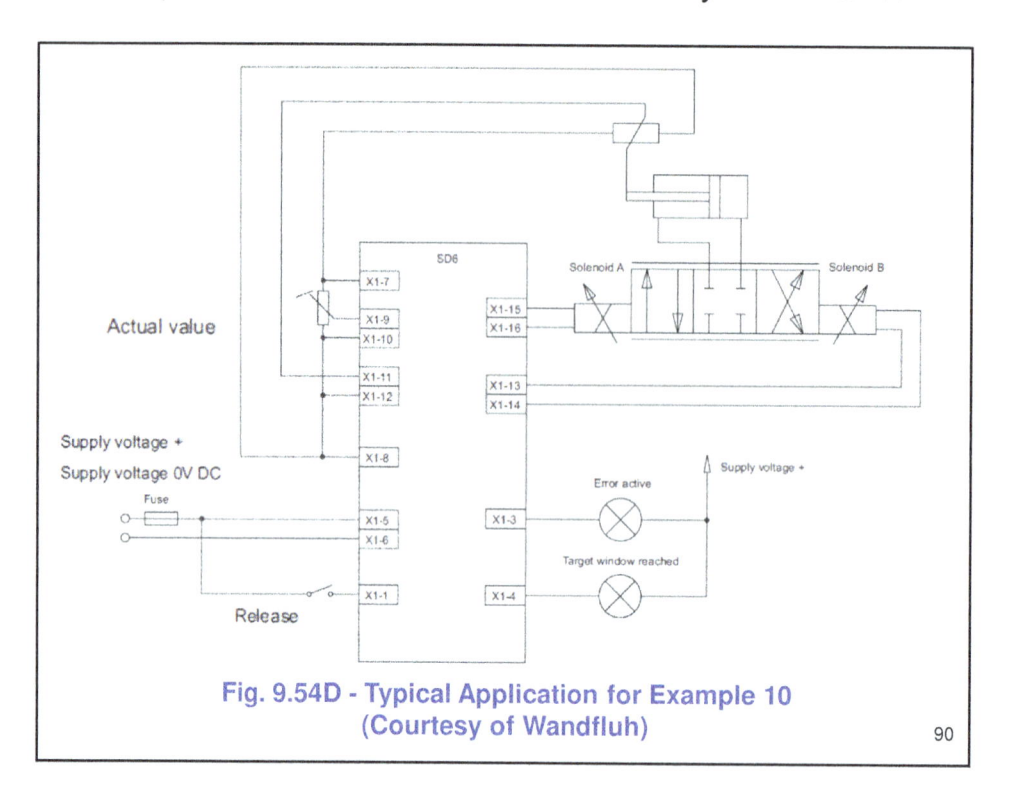

Example 11: Digital Multiple-Axis Controller.

Fig. 9.55A - Typical Control Electronics - Example 11 (Courtesy of Bosch Rexroth)

Type VT-HNC100

Hardware Configuration	
• Panel Mounted	
• Modular	✓
• OBE	
Spool Control	
• Force Controlled	✓
• Stroke Controlled	✓
System Control	
• Valve Only	
• Single Axis	✓
• Multiple Axis	✓
Control Mode	
• Analog	✓
• Digital	✓
Interfacing	
• No Interface	
• Network Interfacing	✓

❑ **Key Features:**
- Analog/Digital Inputs/Outputs.
- Command value input as voltage or current.
- PC programmable controller for up to 4 axes.
- Complies with the requirements for closed-loop control of hydraulic drives.
- Multi Ethernet/Bus interface
- Protection: vibration and climate resistance.
- Password protection.
- Fields of application: Machine tools, injection molding, presses.

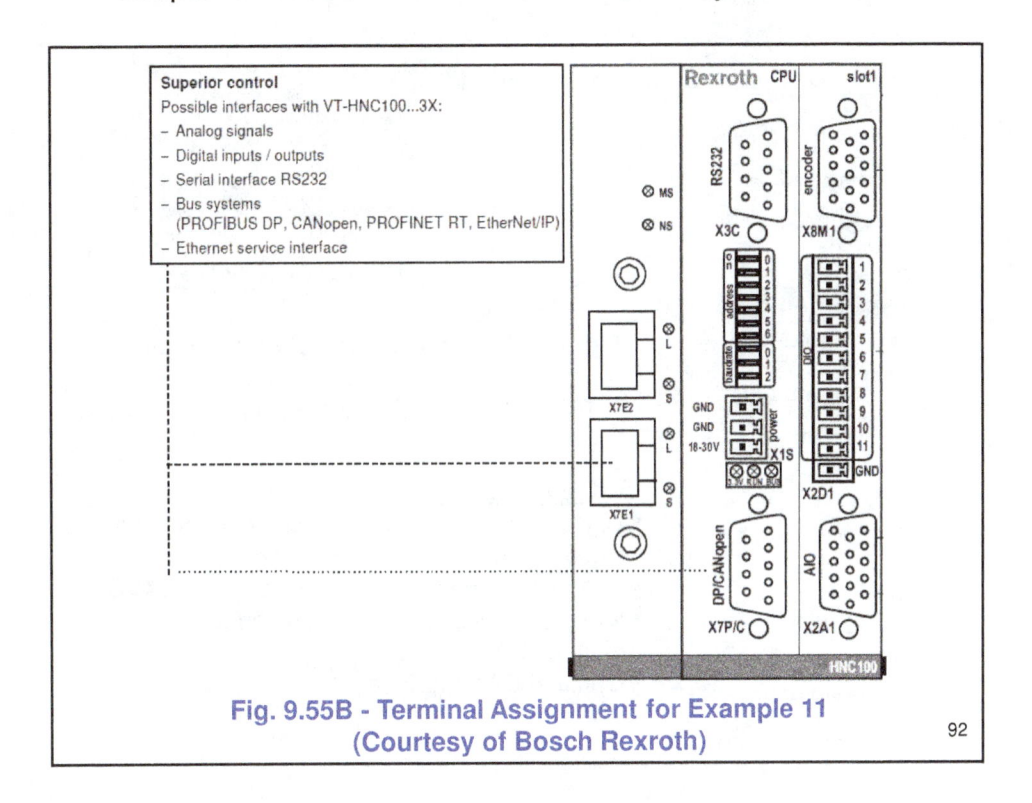

Workbook: Electro-Hydraulic Components and Systems
Chapter 9: Control Electronics for Electro-Hydraulic Valves

Fig. 9.55B - Terminal Assignment for Example 11
(Courtesy of Bosch Rexroth)

Fig. 9.55C - Typical Application for Example 11 (Courtesy of Bosch Rexroth)

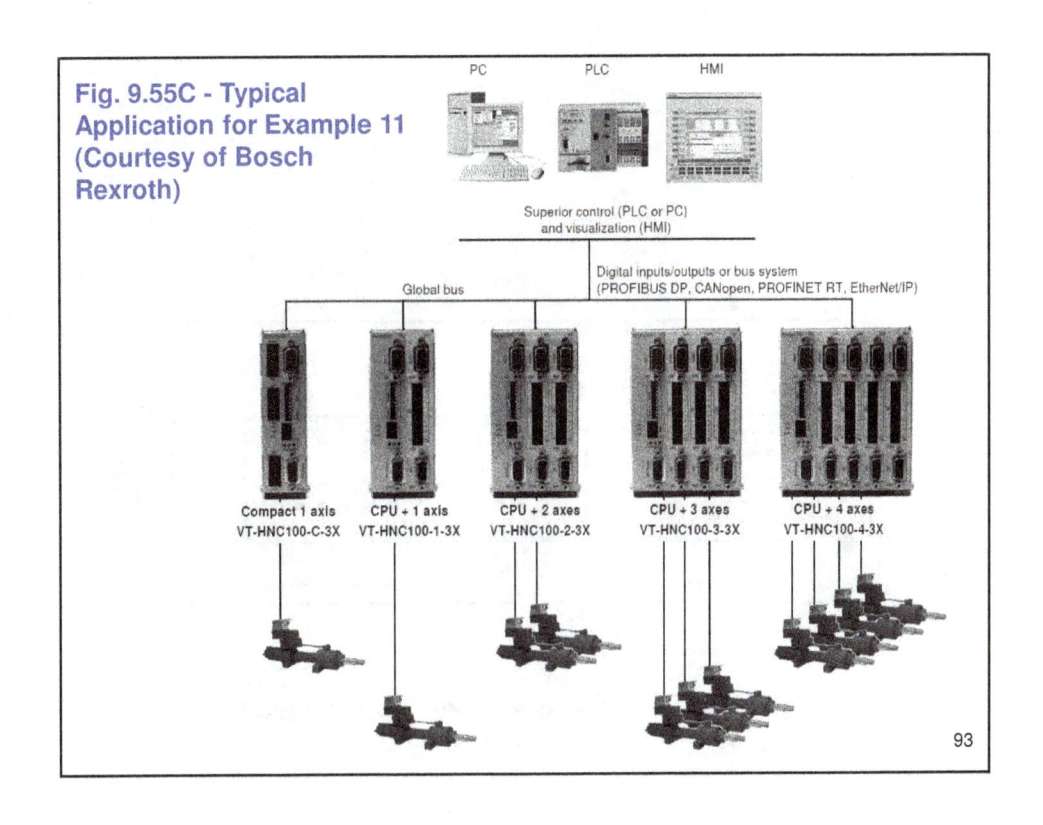

Example 12: Digital Controller for mobile applications.

Hardware Configuration	
• Panel Mounted	
• Modular	✓
• OBE	
Spool Control	
• Force Controlled	
• Stroke Controlled	✓
System Control	
• Valve Only	
• Single Axis	
• Multiple Axis	✓
Control Mode	
• Analog	✓
• Digital	✓
Interfacing	
• No Interface	
• Network Interfacing	✓

Type MD2

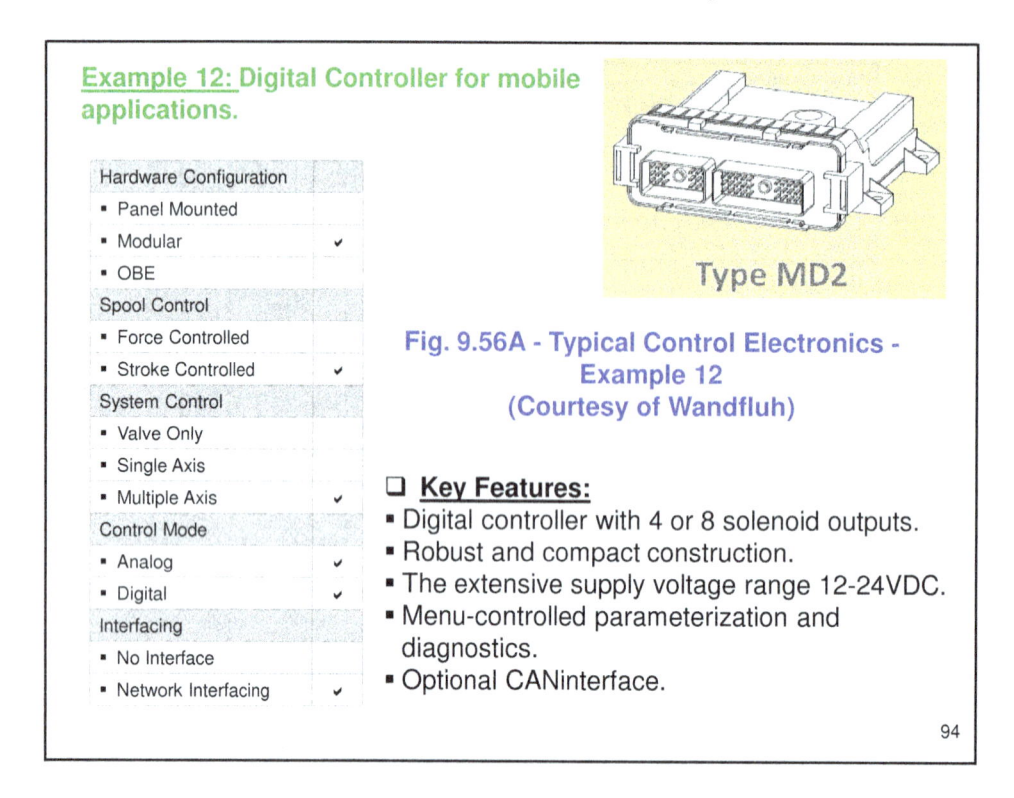

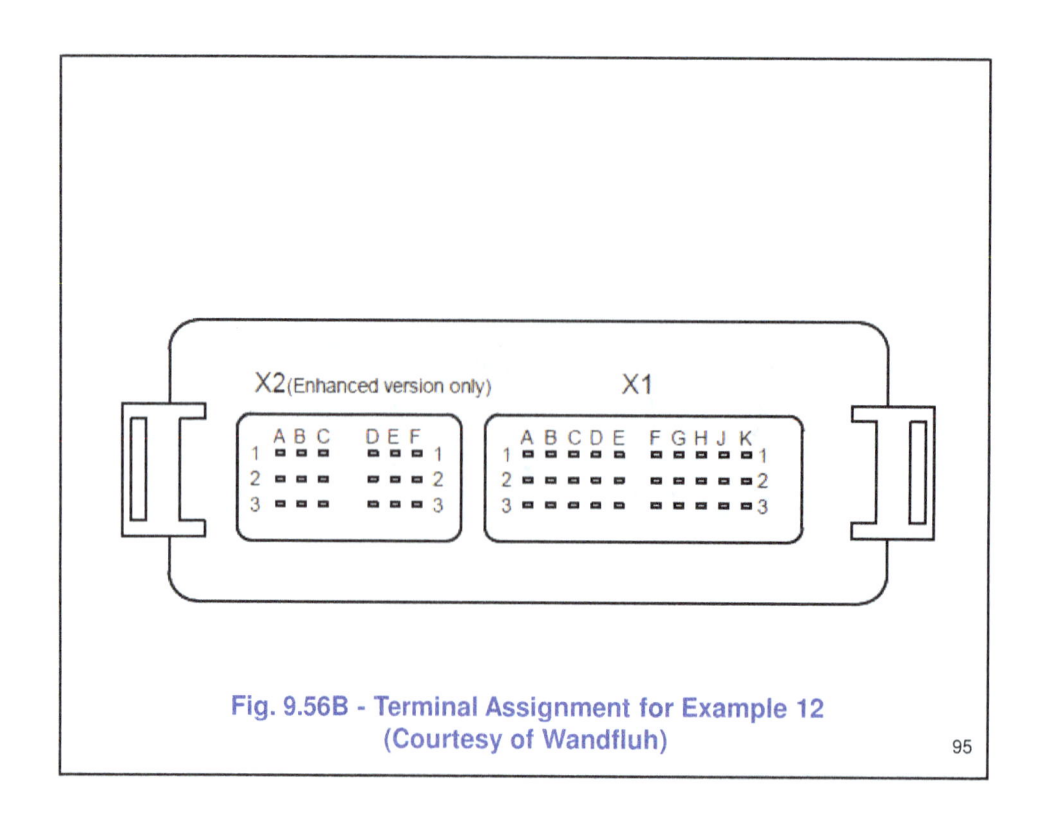

Fig. 9.56A - Typical Control Electronics - Example 12
(Courtesy of Wandfluh)

❑ **Key Features:**
- Digital controller with 4 or 8 solenoid outputs.
- Robust and compact construction.
- The extensive supply voltage range 12-24VDC.
- Menu-controlled parameterization and diagnostics.
- Optional CANinterface.

Fig. 9.56B - Terminal Assignment for Example 12
(Courtesy of Wandfluh)

DEVICE PLUG (X1)
A1 = Stabilised output voltage
A2 = Supply voltage + (Solenoid outputs)
A3 = Supply voltage 0 VDC (Solenoid outputs)
B1 = Stabilised output voltage
B2 = Supply voltage + (Logic part)
B3 = Supply voltage 0 VDC (Logic part)
C1 = Analogue ground
C2 = Digital input 1
C3 = Digital input 2
D1 = Analogue ground
D2 = VBUS (USB)
D3 = GND (USB)
E1 = Analogue input 1
E2 = D- (USB)
E3 = D+ (USB)
F1 = Analogue input 2
F2 = Digital output 1
F3 = Digital output 2
G1 = Output solenoid A +
G2 = Output solenoid A -
G3 = Digital ground
H1 = Output solenoid B +
H2 = Output solenoid B -
H3 = Reserved
J1 = Output solenoid C +
J2 = Output solenoid C -
J3 = Reserved
K1 = Output solenoid D +
K2 = Output solenoid D -
K3 = Reserved

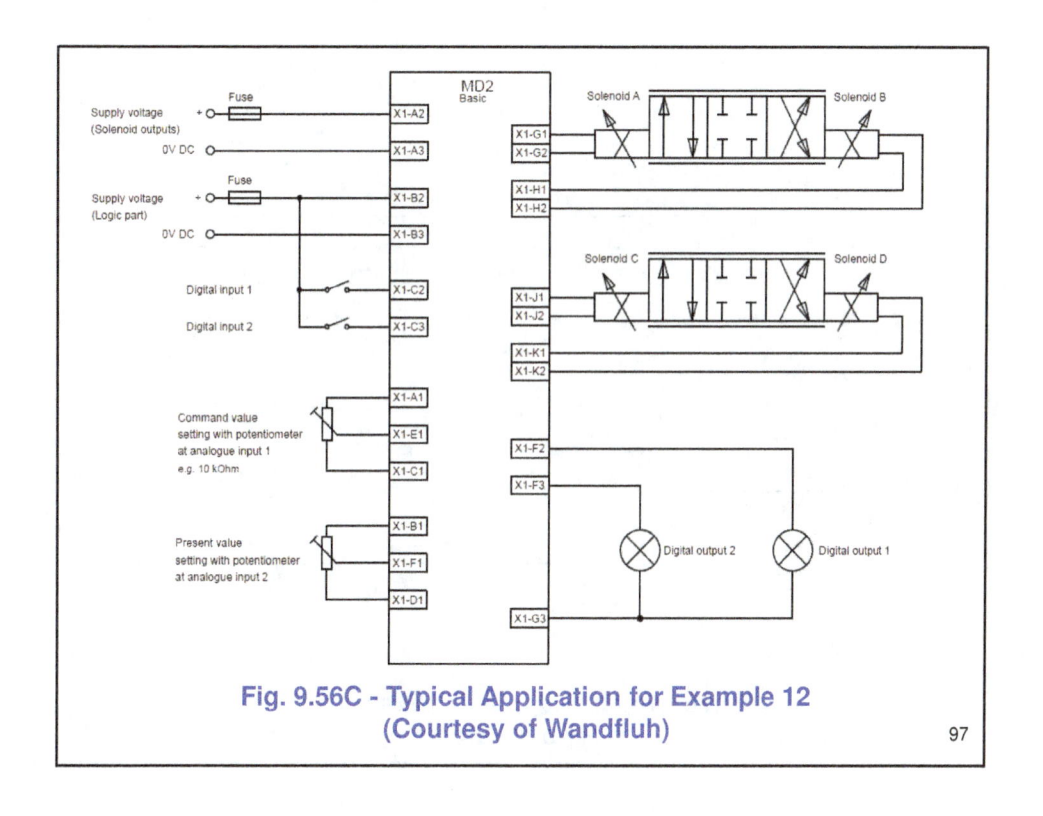

DEVICE PLUG (X2)
A1 = Output solenoid E +
A2 = Output solenoid E -
A3 = CAN High
B1 = Output solenoid F +
B2 = Output solenoid F -
B3 = CAN Low
C1 = Output solenoid G +
C2 = Output solenoid G -
C3 = CAN Gnd
D1 = Output solenoid H +
D2 = Output solenoid H -
D3 = CAN High
E1 = Digital input 3
E2 = Digital input 4
E3 = CAN Low
F1 = Analogue input 3
F2 = Analogue input 4
F3 = CAN Gnd

Fig. 9.56C - Typical Application for Example 12
(Courtesy of Wandfluh)

Workbook: Electro-Hydraulic Components and Systems
Chapter 9: Control Electronics for Electro-Hydraulic Valves

Example 13: Digital Controller for mobile applications.

CoreTek™ Benefits

- Controllers are optimized for Electro-Hydraulic System Integration
- Programmed with the powerful CoDeSys™ software tool, Backbone Configuration Tool and Impulse Service Tool.
- Fully-sealed, compact cast-aluminum housing is rated to IP67
- Advanced electronic design for reliability and accuracy.
- Reliable operation in real-world temperature conditions from -40° to +70°C (-40° to 158° F).
- Inputs and Outputs are protected against shorts.
- Outputs have diagnostic capability.
- No external cooling or heat dissipation required.

Fig. 9.57A - Typical Control Electronics - Example 13
(Courtesy of Hydraforce)

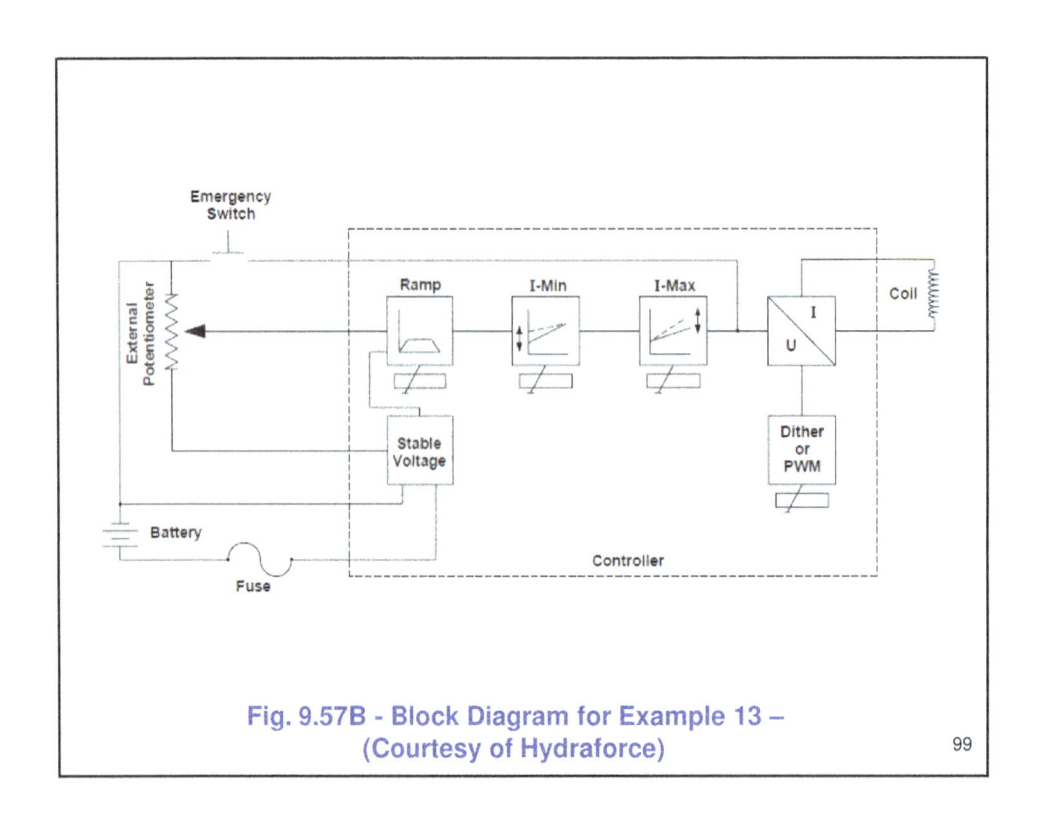

Fig. 9.57B - Block Diagram for Example 13 –
(Courtesy of Hydraforce)

Workbook: Electro-Hydraulic Components and Systems
Chapter 9: Control Electronics for Electro-Hydraulic Valves

Example 14: Digital Display Proportional Valve Driver for mobile applications.

Features & Benefits

- Compatible with any proportional valves.
- Multi inputs & outputs; per model.
- Independent adjustments.
- Field adjustable without specialty tools.
- Clear readings in any light.
- 3 to 4 digit extra bright displays.
- Displays actual values.
- Suitable for extreme -40°C conditions.
- Dual and single valve drivers available.
- Wide ramp time range (0 to 99.5 Sec).
- Wide dither range (40 to 450Hz).
- Short circuit reverse polarity protection.
- Command input overcurrent protection.

Fig. 9.58 - Block Diagram for Example 14 – (Excerption from Lynch)

Chapter 9 Reviews

1. Which of the flowing statements represents the correct match between the element's names and the element's basic functions shown below?

    A. (I=1, II=2, III=3, IV=4)
    B. (I=3, II=2, III=4, IV=1)
    C. (I=1, II=3, III=2, IV=4)
    D. (I=2, II=3, III=4, IV=1)

    I- Valve shifting displacement and consequently the actuator's speed.
    II- Valve shifting time and consequently the actuator's rate of acceleration.
    III- Jump current.
    IV- Maximum command value.

    1- Saturation Limiter.
    2- Proportional Gain.
    3- Ramp Generator.
    4- Dead Band Eliminator.

2. In the figure shown below, a closed loop system is used to control a hydraulic cylinder position. If the cylinder is fully retracted and it is commanded to fully extend, which of the following statement is True?

    A. Control signal starts with value of 20 and reduced with the cylinder extension, until it becomes 0 when the cylinder fully extend.
    B. Valve will be saturated until the cylinder reaches mid of its stroke then gradually closes.
    C. Cylinder will extend with maximum speed until it reaches 5 inches of its stroke then gradually decelerates.
    D. All of the above it true.

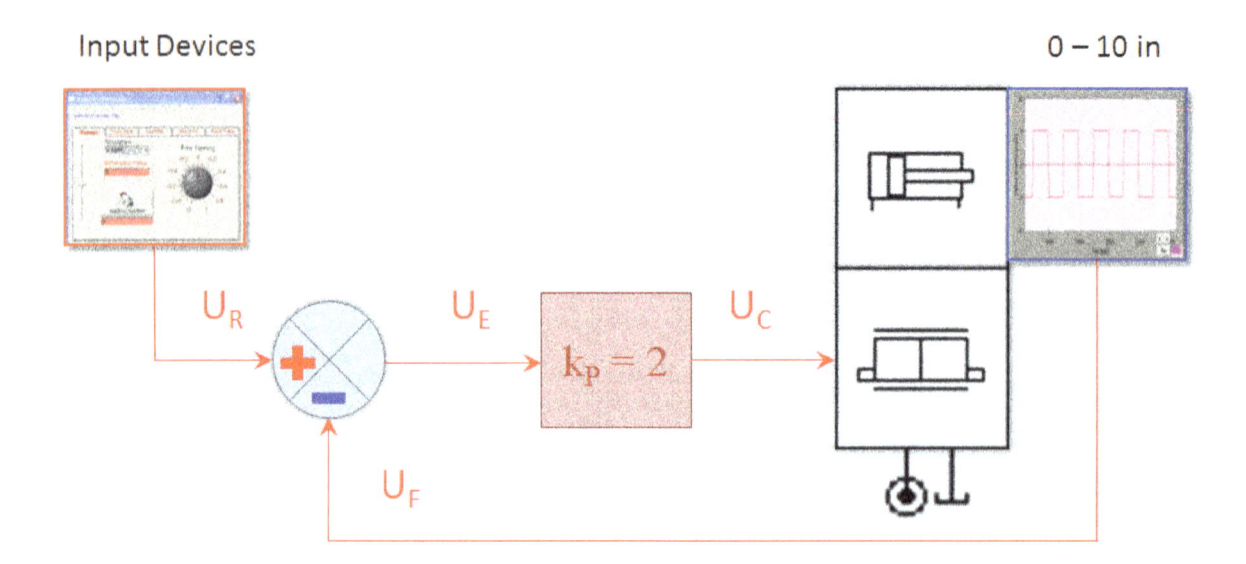

Input Devices   0 – 10 in

U_R, U_E, $k_p = 2$, U_C, U_F

Workbook: Electro-Hydraulic Components and Systems
Chapter 9: Reviews and Assignments

3. Increasing the proportional gain results in?
    A. Limiting the maximum current given to the valve.
    B. Reducing the acceleration and deceleration of the actuator.
    C. Possible system instability.
    D. Making the valve acts like a zero lapped valve.

4. Enabling the ramp generator results in?
    A. Limiting the maximum current given to the valve.
    B. Reducing the acceleration and deceleration of the actuator.
    C. Possible system instability.
    D. Making the valve acts like a zero lapped valve.

5. Enabling the dead band eliminator results in?
    A. Limiting the maximum current given to the valve.
    B. Reducing the acceleration and deceleration of the actuator.
    C. Possible system instability.
    D. Making the valve acts like a zero lapped valve.

6. Enabling the overload protector?
    A. Limiting the maximum current given to the valve.
    B. Reducing the acceleration and deceleration of the actuator.
    C. Possible system instability.
    D. Making the valve acts like a zero lapped valve.

7. Dither signal is used to?
    A. Assure the input-output linearity and eliminate the steady state error in positioning a cylinder when overlapped spool is used.
    B. Prevents spool sticking and reduce the hysteresis.
    C. Protect the valve solenoid by limiting the current to a maximum value.
    D. Assure maximum acceleration of the hydraulic actuator.

8. Pulse width modulation is used to?
    A. Assure the input-output linearity and eliminate the steady state error in positioning a cylinder when overlapped spool is used.
    B. Reduce the power consumption at the solenoid and reduce the hysteresis.
    C. Protect the valve solenoid by limiting the current to a maximum value.
    D. Assure maximum acceleration of the hydraulic actuator.

9. Steady state error in positioning a hydraulic cylinder can be reduced by?
    A. Using a high resolution feedback sensor.
    B. Increasing the proportional gain within the stability range of the system.
    C. Apply dead band eliminator if an overlapped spool is used.
    D. All of the above.

10. In a PWM signal of 50% duty cycle and 24V power supply, average voltage is?
    A. 0 Volt.
    B. 6Volt.
    C. 12 Volt.
    D. 18 Volt.

Workbook: Electro-Hydraulic Components and Systems
Chapter 9: Reviews and Assignments

Chapter 9 Assignment

Student Name: -- Student ID: -------------------

Date: -- Score: -----------------------

In the figure shown below, define the elements with the circled numbers 1 through 16

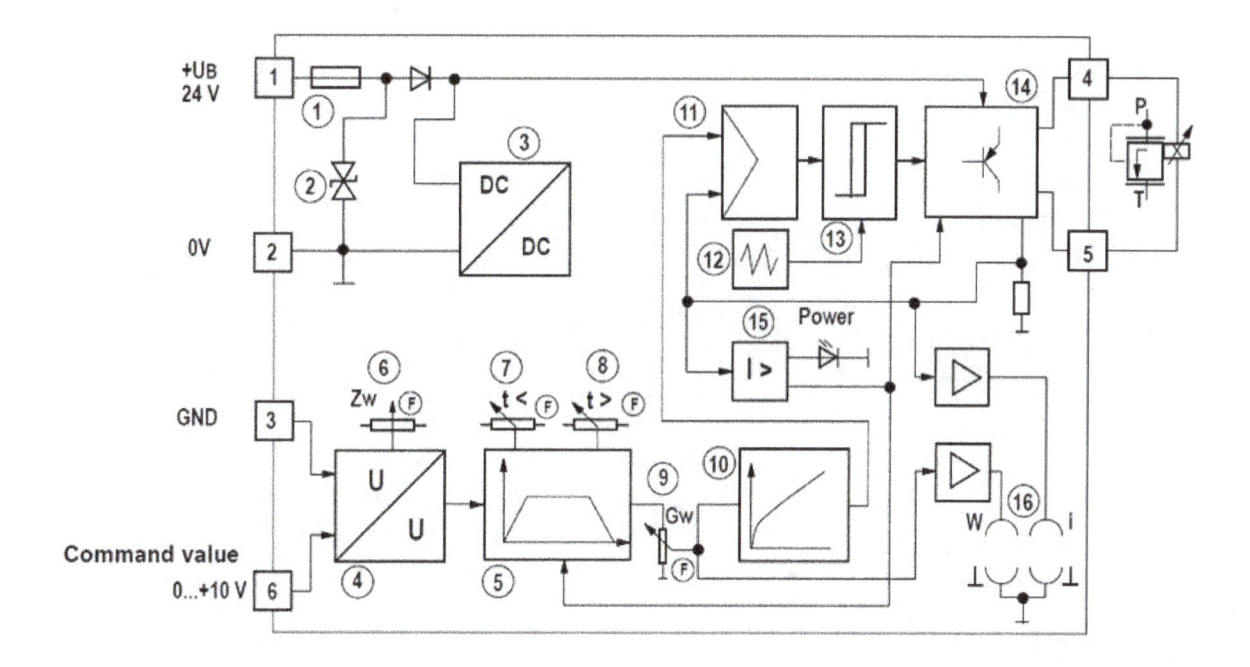

Chapter 10
Electro-Hydraulic Valves Commissioning and Maintenance

Objectives:

This chapter introduces guidelines for commissioning and maintenance of EH valves.

Brief Contents:

10.1- Installation and Commissioning Instructions

10.2- Filtration Requirements

10.3- Electro-Hydraulic Valve Tests and Maintenance

10.1- Installation and Commissioning Instructions

10.1.1- Cleanliness

- **Surroundings:** Keep surrounding area clean and organized.
- **Cleaning Towels:** only use lint-free material or special material (e.g. 100% cotton) for cleaning purposes.
- **Protective Plates and Plastic Caps:** should only be removed immediately prior to installation.

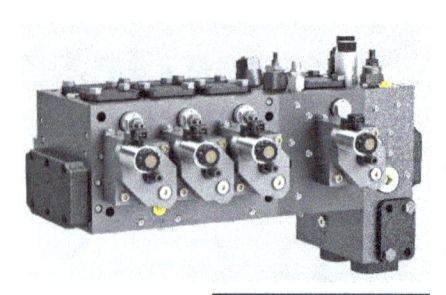

Fig.10.1- Instructions to Keep the Valves Clean

10.1.2- Installation Position
- Preferably horizontal.
- **Valve is mounted on an actuator**, particularly cylinders, the valve should be positioned perpendicular to the acceleration direction of the actuator.
- That reduces the effect of induced vibration by the actuator on the valve performance.

10.1.3- Mounting Screws
- Mounting screws must be tightened to the torque specified in the data sheets from the manufacturer.

10.1.4- Seals
- Make sure that the valve is equipped with proper sealing elements are based on the type of fluid and working temperature. It is highly advisable to review the manufacturer on this subject.

10.1.5- Electrical Connection
- Enable overload protection.

- Avoid running electrical cables within hydraulic plumbing or near sources of electrical noise such as electrical motors and VFDs.

- DO NOT test valve with only electrical signal without hydraulic fluid through the valve. Otherwise, the wire spring in a flapper nozzle stage may be severely stretched.

- Same is applicable for proportional valve in order to prevent spool friction without lubrication.

10.1.6- Storage Space
- Dry & dust-free room with low humidity.

- For storage longer than 3 months, fill housing with preservative oil and seal the valve with protective covers.

- The valves must be checked from time to time to ensure that they are stored correctly.

- Always store the valves with protective covers in place.

10.2- Filtration Requirements

10.2.1- Reservoir Breather Filter:

- Air breather filter is recommended with water vapor absorption ability.
- Air breather filter should have the same β ratio as the oil filter.

10.2.2- Filtration

- **Non-bypass Pressure filters**: assures long service life for proportional and servo valves.
- For non-bypass pressure filter, element collapse rating must be greater than system operating pressure.
- **Last Chance Filters.**

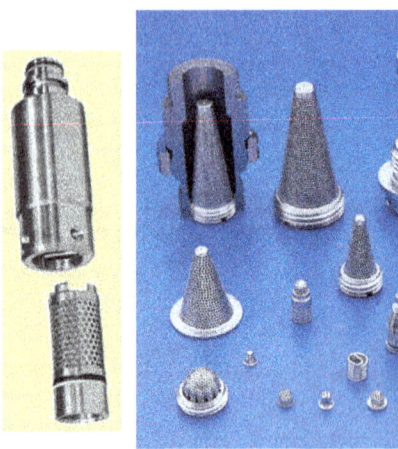

Fig.10.2- Examples of Last Chance Filters

❏ **Clogging indicator.** Possibly with trip signal

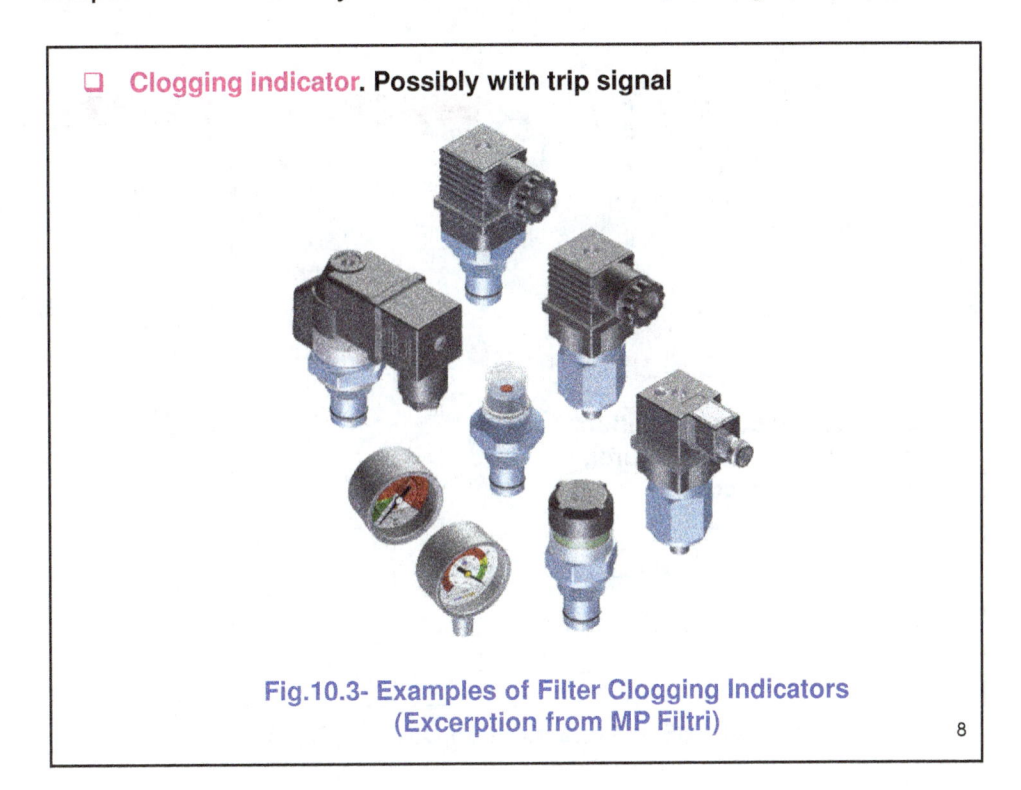

Fig.10.3- Examples of Filter Clogging Indicators
(Excerption from MP Filtri)

❏ **Filter Rating:**

❏ It is always advisable to refer to manufacturers maximum recommended ISO4406 contamination level.

❏ If no information were found, 5 micron absolute or better for servo valves

$$(\beta_5 = 75, \text{ i.e. Efficiency} = 98.7\%).$$

❏ If the work environment is so harsh, additional off-line filtration may be necessary.

Workbook: Electro-Hydraulic Components and Systems
Chapter 10: Electro-Hydraulic Valves Commissioning and Maintenance

```
  * / * / *
  ↓   ↓   ↓
 4μm  6μm 14μm
```

ISO 4406 Chart		
Range Code	Particles per milliliter	
	More than	Up to/including
24	80000	160000
23	40000	80000
22	20000	40000
21	10000	20000
20	5000	10000
19	2500	5000
18	1300	2500
17	640	1300
16	320	640
15	160	320
14	80	160
13	40	80
12	20	40
11	10	20
10	5	10
9	2.5	5
8	1.3	2.5
7	0.64	1.3
6	0.32	0.64

18/16/13 — Proportional Valves

15/13/10 — Servo Valves

If no specific recommendation provided by the manufacturer, follow the shown code

Fig. 10.4- ISO Contamination Code* 4406-1999

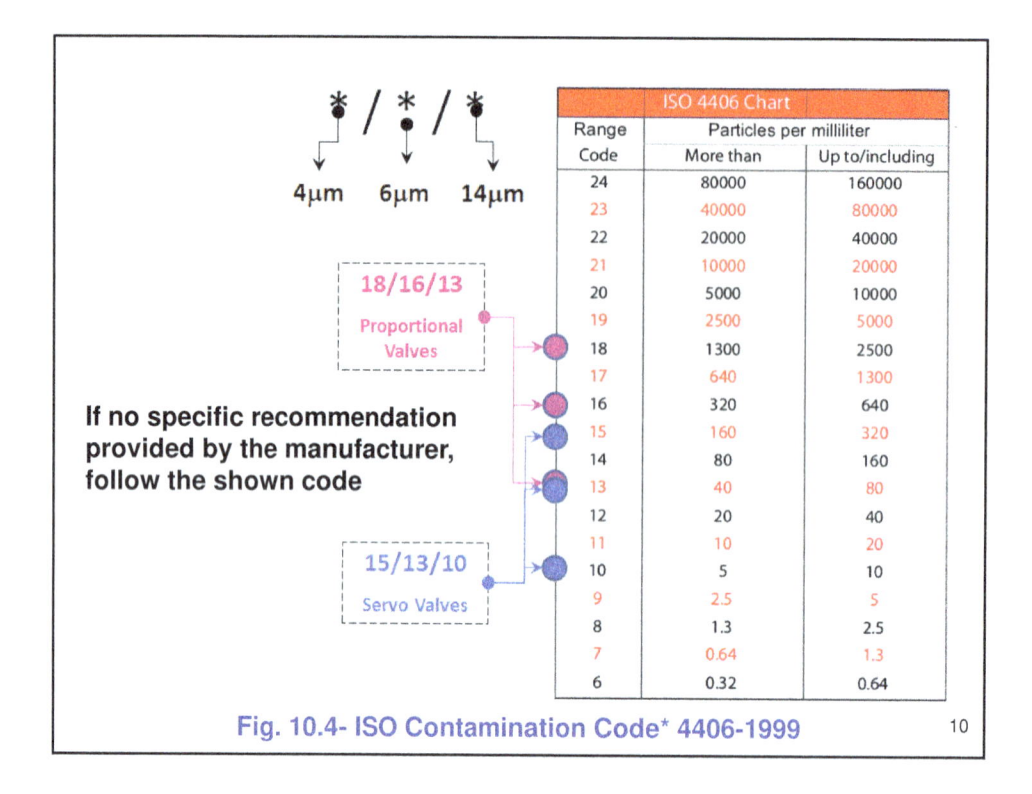

Contamination class	Particle size in μm				
	5–15	15–25	25–50	50–100	> 100
00	125	22	4	1	0
0	250	44	8	2	0
1	500	89	16	3	1
2	1000	178	32	6	1
3	2000	356	63	11	2
4	4000	712	126	22	4
5	8000	1425	253	45	8
6	16000	2850	506	90	16
7	32000	5700	1012	180	32
8	64000	11400	2025	360	64
9	128000	22800	4050	720	128
10	256000	45600	8100	1440	256
11	512000	91200	16200	2880	512
12	1024000	182400	32400	5760	1024

Contamination classes to NAS 1638

- Based on NAS class 6 is recommended.
- Max number of dirt particles found in 100 ml of fluid.
- Being phased out in favor of ISO 4406.
- Some limited use in Aerospace. Not used in current mobile and Industrial applications.

Fig. 10.5- NAS Contamination Code 1638

Workbook: Electro-Hydraulic Components and Systems
Chapter 10: Electro-Hydraulic Valves Commissioning and Maintenance

10.2.3- System Flushing

- Less viscous fluid + high temperature + Reynolds # > 5000
- Before installing one of these close-clearance valves on a new or a refurbished system, it is essential to carry out an effective flushing procedure on the entire system to remove "built-in" contamination particles (for example: weld scale, rust, dust, lint, metal shavings, blast-sand, paint chips, pipe dope, etc.)
- Before flushing, valves and other critical components are removed and replaced with an inexpensive flushing module (sub-plate).
- During flushing process, oil samples must be taken and cleanliness class is to be checked as per the standard test protocol.
- ❑ Conductors: must be flushed and dried prior to installation.
- ❑ Tank: must be sealed after cleaning.

- The oil in the system should be flushed through the filter at least 150 to 300 times.

- Flushing time can be calculated as follows

$$t = \frac{V}{Q} \times X \qquad 10.1$$

t = Time in hours
V = Tank volume in liters
Q = Pump delivery rate in L/min
X = [(150-300)/60] = 2.5 to 5

Fig. 10.6- Example of Hydraulic Filtration Unit

- This is just a guide only and may not be sufficient depending on filter efficiency and contamination level prior testing.

- Flushing must continue until the ISO 4406 level is below the manufacturer recommendations.

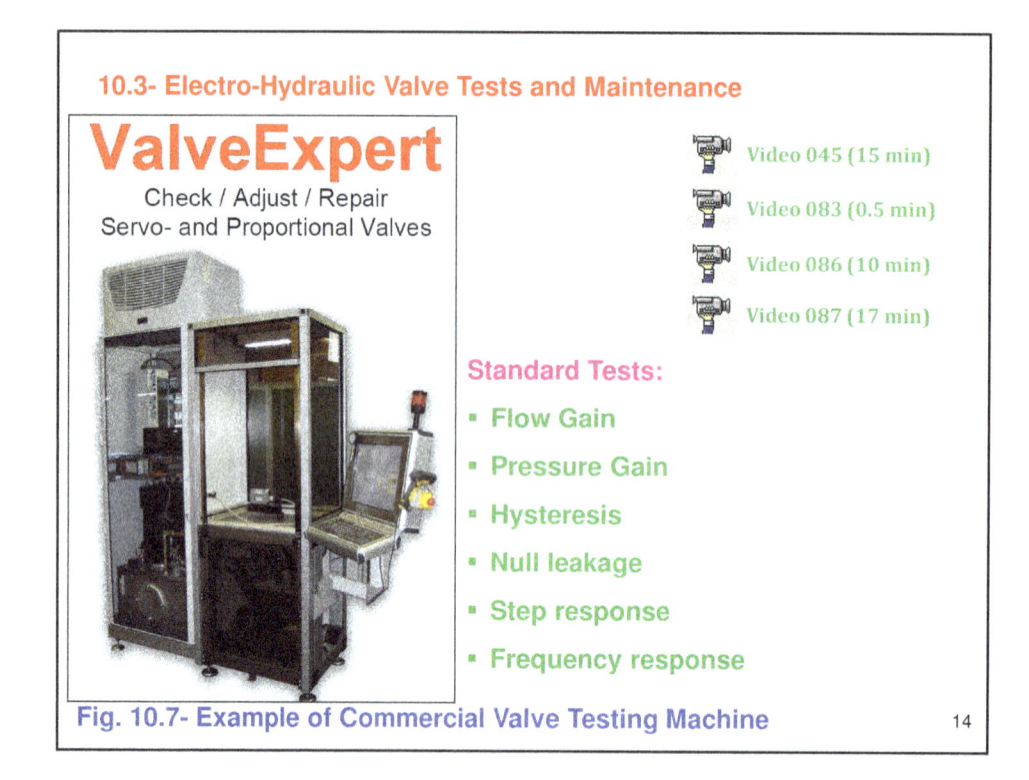

10.3 - Electro-Hydraulic Valve Tests and Maintenance

ValveExpert
Check / Adjust / Repair
Servo- and Proportional Valves

Video 045 (15 min)
Video 083 (0.5 min)
Video 086 (10 min)
Video 087 (17 min)

Standard Tests:
- Flow Gain
- Pressure Gain
- Hysteresis
- Null leakage
- Step response
- Frequency response

Fig. 10.7- Example of Commercial Valve Testing Machine

Chapter 10 Reviews

1. Flow gain valve test is performed to show the?
    A. Response of the valve to a sudden input signal.
    B. Valve band width.
    C. Valve flow versus input command.
    D. Valve flow gain deviation from original characteristics.

2. Frequency test is performed to show the?
    A. Response of the valve to a sudden input signal.
    B. Valve band width.
    C. Valve flow versus input command.
    D. Valve flow gain deviation from original characteristics.

3. Hysteresis valve test is performed to show the?
    A. Response of the valve to a sudden input signal.
    B. Valve band width.
    C. Valve flow versus input command.
    D. Valve flow gain deviation from original characteristics.

4. Step response test is performed to show the?
    A. Response of the valve to a sudden input signal.
    B. Valve band width.
    C. Valve flow versus input command.
    D. Valve flow gain deviation from original characteristics.

5. Null leakage test is performed to show the?
    A. Response of the valve to a sudden input signal.
    B. Leakage around the central position of the valve..
    C. Valve flow versus input command.
    D. Valve flow gain deviation from original characteristics.

Workbook: Electro-Hydraulic Components and Systems
Chapter 10: Reviews and Assignments

Workbook: Electro-Hydraulic Components and Systems
Chapter 10: Reviews and Assignments

Chapter 10 Assignment

Student Name: -- Student ID: -------------------

Date: -- Score: ------------------------

Given the fact that oil is required to pass 200 times through a system to be cleaned, find out the time needed to flush a system that has a flushing pump of 20 liter/min and a tank of 200 liters.

Workbook: Electro-Hydraulic Components and Systems
Answers to Chapter Reviews

Answers to Chapter Reviews

Chapter 1:

1	2	3	4	5	6	7	8	9	10
B	D	A	B	D	D	B	A	A	C

Chapter 2:

1	2	3	4	5	6	7	8	9	10
D	C	B	D	C	B	D	C	A	C

Chapter 3:

1	2	3	4	5	6	7	8	9	10
D	D	D	C	A	B	A	D	B	D

11	12	13	14	15	16	17	18	19	20
D	A	C	A	B	C	D	A	D	C

Chapter 4:

1	2	3	4	5	6	7	8	9	10
B	C	D	A	C	A	D	B	B	A

Chapter 5:

1	2	3	4	5	6	7	8	9	10
A	B	C	B	A	C	B	D	A	B

11	12	13	14	15	16	17	18	19	20
A	D	C	D	C	A	D	B	A	C

Chapter 6:

1	2	3	4	5	6	7	8	9	10
A	D	C	B	C	D	C	B	C	D

Chapter 7:

1	2	3	4	5	6	7	8	9	10
D	C	A	C	C	B	D	D	C	A

Chapter 8:

1	2	3	4	5	6	7	8	9	10
B	D	C	A	A	B	C	A	C	B

Chapter 9:

1	2	3	4	5	6	7	8	9	10
D	D	C	B	D	A	B	B	D	C

Chapter 10:

1	2	3	4	5
C	B	D	A	B

CPSIA information can be obtained
at www.ICGtesting.com
Printed in the USA
FSHW02n2141170818
51310FS